Resilience Assessment and Evaluation of Computing Systems

T0181246

Katinka Wolter · Alberto Avritzer
Marco Vieira · Aad van Moorsel
Editors

Resilience Assessment and Evaluation of Computing Systems

 Springer

Editors
Katinka Wolter
Institute of Computer Science
Free University Berlin
Berlin, Germany

Alberto Avritzer
Siemens Corporate Research and
  Technology
Princeton, NJ, USA

Marco Vieira
Faculdade de Ciências e Tecnologia
Departamento de Engenharia Informática
Universidade de Coimbra
Coimbra, Portugal

Aad van Moorsel
Newcastle University
Newcastle upon Tyne, UK

ISBN 978-3-642-43674-1   ISBN 978-3-642-29032-9 (eBook)
DOI 10.1007/978-3-642-29032-9
Springer Heidelberg New York Dordrecht London

ACM Computing Classification (1998): C.4, G.3, D.2

© Springer-Verlag Berlin Heidelberg 2012
Softcover reprint of the hardcover 1st edition 2012
This work is subject to copyright. All rights are reserved by the Publisher, whether the whole or part of the material is concerned, specifically the rights of translation, reprinting, reuse of illustrations, recitation, broadcasting, reproduction on microfilms or in any other physical way, and transmission or information storage and retrieval, electronic adaptation, computer software, or by similar or dissimilar methodology now known or hereafter developed. Exempted from this legal reservation are brief excerpts in connection with reviews or scholarly analysis or material supplied specifically for the purpose of being entered and executed on a computer system, for exclusive use by the purchaser of the work. Duplication of this publication or parts thereof is permitted only under the provisions of the Copyright Law of the Publisher's location, in its current version, and permission for use must always be obtained from Springer. Permissions for use may be obtained through RightsLink at the Copyright Clearance Center. Violations are liable to prosecution under the respective Copyright Law.
The use of general descriptive names, registered names, trademarks, service marks, etc. in this publication does not imply, even in the absence of a specific statement, that such names are exempt from the relevant protective laws and regulations and therefore free for general use.
While the advice and information in this book are believed to be true and accurate at the date of publication, neither the authors nor the editors nor the publisher can accept any legal responsibility for any errors or omissions that may be made. The publisher makes no warranty, express or implied, with respect to the material contained herein.

Printed on acid-free paper

Springer is part of Springer Science+Business Media (www.springer.com)

Preface

This Springer Verlag book is a natural consequence of the workshop on Resilience Assessment and Evaluation organized in the seminar series of Schloss Dagstuhl in July 2010. As such, the book got its inspiration from the high quality and serene professional facilities at the Leibniz Centre for Informatics and the relaxing and inspiring Saarland country side near the Dagstuhl castle. For one week, about 25 scientists and engineers in resilience assessment and evaluation came together at Dagstuhl and discussed the latest trends in the field, in highly informal manner. You can find information about the original seminar by googling for 'Dagstuhl seminar10292'.

The aim of the book is to provide an extensive overview of past, current, and future trends in resilience assessment and evaluation. Most participants at the seminar have contributed to this book, discussing case studies, general concepts, and their latest research. This book also leverages another effort that ran for the 2 years preceding the seminar, namely the EU FP7 sponsored coordination action AMBER: Assessing, Measuring and Benchmarking Resilience. AMBER aimed at providing a research agenda for the EU in resilience assessment and evaluation. Several chapters in this book are extensions of earlier versions that were made available publicly in the form of deliverables in the AMBER project.

We perceived the need for a book that targets engineers in dependable computer systems as well as academics and their PhD and MSc students. Modern-day computer systems integrate increasingly many components and systems, with growing demands from users and increasingly diverse failure and attack modes from which the system requires to be protected. This holds in varying degrees for our home and entertainment networks, for increasingly integrated enterprise systems, for safety critical computers in planes and plants, and for the critical infrastructure that serves us water, energy, communication, and other basic elements of our daily lives.

Our society would want us to be able to assess the resilience of these computer systems: what types of accidental failures and malicious attacks are these systems subject to, and how do they deal with the failures and attacks? Can we quantify and measure the resulting resilience for a system in a meaningful way, and if we can

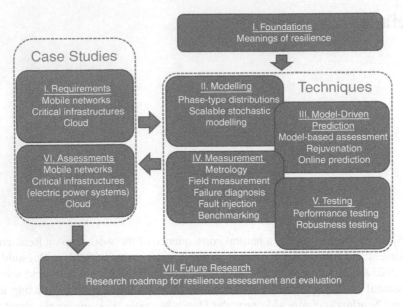

Fig. 1 How to use this book

quantify it, can we already predict it at design time, before we deploy and run the systems? These are the main questions researchers and engineers in resilience assessment try to answer, trying to invent and improve methods, techniques, and tools to answer these questions.

The process in creating this book. At the Dagstuhl workshop the participants created an outline table of contents for the book, subdivided into main parts and individual chapters. An open call for contributions was then launched and sent to the resilience community through widespread e-mails. The editors of this book selected the chapters that they considered the best fit with the purpose of this book and invited the authors to submit their proposed paper for peer review. Two rounds of peer review were used to assure the quality of the contributions. The result, we believe, is a set of high-quality papers that cover the most important aspects of resilience assessment and evaluation.

How to use this book? The accompanying diagram in Fig. 1 sketches the structure of this book, and should help the student and general reader in making use of this book. The book is divided in seven parts, I–VII.

Each part provides a natural grouping of related topics, such as challenges in Part I and Testing in Part V. Moreover, the parts can be grouped further as given in Fig. 1. The core of the book is a set of 12 chapters on Techniques, as depicted in the box on the upper right-hand side. These have been organized into four blocks, which we will comment on in some more detail.

The techniques are both motivated and demonstrated two times by three chapters on Case Studies for a number of areas highly relevant in modern times. The case studies are grouped together on the left-hand side of Fig 1. For each case study (mobile, critical infrastructure, and cloud) there is a chapter on resilience assessment challenges in Part I as well as a chapter on assessment results in Part VI.

Each of the chapters is stand-alone, that is, a reader will be able to learn from each individual chapter without having to consult any of the other chapters. As a consequence, the reader can consult chapters in somewhat arbitrary order. This makes the book particularly suitable for seminar series or group reading discussions. With respect to the case studies, it would be natural to read the challenges and results in tandem, although to appreciate how results have been derived, a deeper study of the techniques discussed in the other part is recommended for those wanting to penetrate deeper into the material. The arrows back and forth between Case Studies and Techniques illustrate that chapters from both parts can be read in various orders depending on the specific needs and background of the reader.

The set of techniques and case studies then culminates in the chapter on Future Research, which presents a version of the research roadmap delivered by the AMBER EU coordination action. The material discussed in the research roadmap chapter is immediately accessible to the experienced resilience engineer. However, for a general audience, the diagram shows a directed arrow pointing from Case Studies to the Future Research box. We believe that for more novice readers the discussion on future research is especially useful once the reader has gained some appreciation for the challenges and advances discussed in the chapters on techniques and case studies.

Acknowledgments

Many people have contributed to the process of creating this book and deserve our gratitude. First of all, we thank all contributors for their effort and endurance, the technical contributions, and also the organizational effort needed to assemble the chapters and the time invested for cross-reviewing chapters for quality control.

We owe many thanks to Springer Verlag for publishing this manuscript as a book and we especially thank Ralf Gerstner for his steady support and endless patience. The anonymous reviewers of the book proposal and Boudewijn Haverkort have made many helpful comments that improved the book and its organization.

Similarly, we thank Marc Herbstritt and his team at Dagstuhl castle for hosting us for one week in July 2010. It was a stimulating time that started our work on this book.

It is clear that we as editors could not have succeeded in producing this book and assuring the quality of this book without the help of the reviewers of the chapters. We sincerely thank the following colleagues who were willing to spend some of their valuable time on creating this book and help achieving the necessary quality through their reviewing efforts:

Ermeson Andrade; Alberto Avritzer; Vlastimil Babka; Andre Bondi; Andrea Ceccarelli; Lucia Cloth; Ricardo Czekster; Alexandra Danilkina; Felicita Di Giandomenico; Nicholas Dingle; Salvatore Distefano; João Durães; José Fonseca; Rahul Ghosh; Stephen Gilmore; Gabor Horvath; Nikolaus Huber; Kaustubh Joshi; Leila Kloul; Samuel Kounev; Istvan Majzik; Zoltan Micskei; Philipp Reinecke; Anne Remke; Martin Riedl; Kai Sachs; Francesca Saglietti; Johann Schuster; Markus Siegle; Kishor Trivedi; Jing Zhao;

We are much indebted to Philipp Reinecke from Freie Universität Berlin for technical and organization support, setting up the SVN version control site and helping in many stages of putting together this book. In addition to that, in what probably was his remaining spare time, Philipp managed also to contribute to several technical chapters. Philipp: we have greatly appreciated your enthusiastic support!

We also sincerely thank Johari Abdullah and Rob Cain from Newcastle University for their help in the unification of a very heterogenously created set of chapters into proper latex-type setting.

We have received solid technical support from the IT staff at Freie Universität Berlin that allowed us to maintain a version control to aid collaboration of a group of 52 authors who have been working on this book.

Finally, we like to thank you, the reader, for your interest in this book. We hope that you will find the collection of articles inspiring and useful for your work or study and will use the book as a lasting source of information on resilience engineering.

Berlin, Germany   Katinka Wolter
Princeton, NJ, US   Alberto Avritzer
Coimbra, Portugal   Marco Vieira
Newcastle upon Tyne, UK   Aad van Moorsel

December 2011

Contents

Contributors

Nuno Antunes CISUC, Department of Informatics Engineering, University of Coimbra, Coimbra, Portugal, e-mail: nmsa@dei.uc.pt

Alberto Avritzer Siemens Corporate Research and Technology, Princeton, NJ, USA, e-mail: alberto.avritzer@siemens.com

Vlastimil Babka Faculty of Mathematics and Physics, Charles University in Prague, Prague, Czech Republic, e-mail: babka@d3s.mff.cuni.cz

Raul Barbosa Faculty of Sciences and Technology, University of Coimbra, Coimbra, Portugal, e-mail: rbarbosa@dei.uc.pt

Samir Bellahsene PRiSM, Université de Versailles, Versailles, France, e-mail: sabe@prism.uvsq.fr

Luca Berardinelli Dipartimento di Informatica, Università dell'Aquila, Italy, e-mail: luca.berardinelli@univaq.it

Levente Bodrog Department of Telecommunications, Budapest University of Technology and Economics, Budapest, Hungary, e-mail: bodrog@webspn.hit.bme.hu

Andrea Bondavalli University of Firenze, Italy, e-mail: bondavalli@unifi.it

Andre B. Bondi Siemens Corporate Research and Technology, Princeton, NJ, USA, e-mail: andre.bondi@siemens.com

Jeremy T. Bradley Imperial College London, London, UK, e-mail: jb@doc.ic.ac.uk

Fabian Brosig Karlsruhe Institute of Technology, Karlsruhe, Germany, e-mail: fabian.brosig@kit.edu

Andrea Ceccarelli University of Firenze, Firenze, Italy, e-mail: andrea.ceccarelli@unifi.it

Silvano Chiaradonna ISTI Department, Italian National Research Council, Pisa, Italy, e-mail: chiaradonna@isti.cnr.it

Lucia Cloth Department of Applied Information Technology, GU Tech, Oman, e-mail: lucia.cloth@gutech.edu.om

Vittorio Cortellessa Dipartimento di Informatica, Università dell'Aquila, Italy, e-mail: vittorio.cortellessa@univaq.it

Ricardo M. Czekster PUCRS/Faculdade de Informatica, Avenida Ipiranga, 6681, Predio 32, Sala 505, CEP 90619-900 Porto Alegre, Brazil, e-mail: ricardo.czekster@pucrs.br

Alexandra Danilkina Institute of Computer Science, Free University Berlin, Germany, e-mail: danilkin@zedat.fu-berlin.de

Nicholas Dingle School of Mathematics, University of Manchester, Manchester, UK, e-mail: nicholas.dingle@manchester.ac.uk

Salvatore Distefano University of Messina Engineering Faculty, Messina, Italy, e-mail: sdistefano@unime.it

Joao Duraes DEI/CISUC, Polytechnic Institute of Coimbra, Coimbra, Portugal, e-mail: jduraes@isec.pt

Lorenzo Falai Resiltech S.R.L, Pontedera (Pisa), Italy, e-mail: lorenzo.falai@resiltech.com

José Fonseca DEI/CISUC, University of Coimbra & UDI, Polytechnic Institute of Guarda, Coimbra, Portugal, e-mail: josefonseca@ipg.pt

Felicita Di Giandomenico ISTI Department, Italian National Research Council, Italy, e-mail: digiandomenico@isti.cnr.it

Katja Gilly Universidad Miguel Hernandez, Elche, Spain, e-mail: katya@umh.es

Stephen Gilmore University of Edinburgh, Edinburgh, UK, e-mail: Stephen.Gilmore@ed.ac.uk

Richard Hayden Imperial College London, London, UK, e-mail: rh@doc.ic.ac.uk

Nikolaus Huber Karlsruhe Institute of Technology, Germany, e-mail: nikolaus.huber@kit.edu

Kaustubh Joshi AT&T Labs Research, Florham Park, NJ, USA, e-mail: kaustubh@research.att.com

Carlos Juiz Universitat de les Illes Balears, Palma, Spain, e-mail: cjuiz@uib.es

Johan Karlsson Department of Computer Science and Engineering, Chalmers University of Technology, Gothenburg, Sweden, e-mail: johan@chalmers.se

Soila P. Kavulya Carnegie Mellon University, Pittsburgh, PA, USA, e-mail: spertet@ece.cmu.edu

Leïla Kloul PRiSM, Université de Versailles, Versailles, France, e-mail: kle@prism.uvsq.fr

Samuel Kounev Karlsruhe Institute of Technology, Karlsruhe, Germany, e-mail: kounev@kit.edu

Paolo Lollini University of Firenze, Italy, e-mail: lollini@unifi.it

Henrique Madeira DEI/CISUC, Polytechnic Institute of Coimbra, Coimbra, Portugal, e-mail: henrique@dei.uc.pt

István Majzik Budapest University of Technology and Economics, Budapest, Hungary, e-mail: majzik@mit.bme.hu

Zoltán Micskei Budapest University of Technology and Economics, Budapest, Hungary, e-mail: micskeiz@mit.bme.hu

Leonardo Montecchi University of Firenze, Firenze, Italy, e-mail: lmontecchi@unifi.it

Priya Narasimhan Carnegie Mellon University, Pittsburgh, PA, USA, e-mail: priya@cs.cmu.edu

Ramon Nou Barcelona Supercomputing Center, Barcelona, Spain, e-mail: ramon.nou@bsc.es

Philipp Reinecke Institute of Computer Science, Free University Berlin, Germany, e-mail: philipp.reinecke@fu-berlin.de

Anne Remke University of Twente, Enschede, The Netherlands, e-mail: anne@cs.utwente.nl

Martin Riedl Department of Computer Science, Universität der Bundeswehr München, Neubiberg, Germany, e-mail: martin.riedl@unibw.de

Carlo Rosa Dipartimento di Informatica, Università dell'Aquila, Italy, e-mail: carlo.rosa@univaq.it

Kai Sachs SAP AG, Walldorf, Germany, e-mail: kai.sachs@sap.com

Markus Siegle Department of Computer Science, Universität der Bundeswehr München, Neubiberg, Germany, e-mail: markus.siegle@unibw.de

Anton Stefanek Imperial College London, London, UK, e-mail: as1005@doc.ic.ac.uk

Lorenzo Strigini Centre for Software Reliability, City University London, London, UK, e-mail: Strigini@csr.city.ac.uk

Nigel Thomas School of Computing Science, Newcastle University, Newcastle, UK, e-mail: nigel.thomas@ncl.ac.uk

Kishor S. Trivedi Department of Electrical and Computer Engineering, Duke University, Durham, NC, USA, e-mail: kst@ee.duke.edu

Marco Vieira DEI/CISUC, Polytechnic Institute of Coimbra, Coimbra, Portugal, e-mail: mvieira@dei.uc.pt

Michele Vadursi University of Naples "Parthenope", Centro Direz. Is. C4, Naples, Italy, e-mail: michele.vadursi@uniparthenope.it

Katinka Wolter Institute of Computer Science, Free University Berlin, Germany, e-mail: katinka.wolter@fu-berlin.de

Part I
Introduction and Motivating Examples

Chapter 1
Fault Tolerance and Resilience: Meanings, Measures and Assessment

Lorenzo Strigini

Abstract To assess in quantitative terms the "resilience" of systems, it is necessary to ask first what is meant by "resilience", whether it is a single attribute or several, which measure or measures appropriately characterise it. This chapter covers: the technical meanings that the word "resilience" has assumed, and its role in the debates about how best to achieve reliability, safety, etc.; the different possible measures for the attributes that the word designates, with their different pros and cons in terms of ease of empirical assessment and suitability for supporting prediction and decision making; the similarity between these concepts, measures and attached problems in various fields of engineering, and how lessons can be propagated between them.

1.1 Introduction

Measuring or assessing a quality for any object, e.g., "resilience" for a system, requires clarity about what this quality is.

The word "resilience" has become popular in recent years in the area of information and communication technology (ICT) and policy related to ICT, as part of a more general trend (for instance, the word "resilience" is in favour in the area of critical infrastructure protection). The increasing use of this word creates the doubt whether it is just a new linguistic fashion, for referring to what is commonly studied, pursued and assessed under names like "fault tolerance", "dependability" (a term mostly restricted to ICT usage), "security", "reliability, availability, maintainability and safety" (RAMS), "human reliability", and so on, or it actually denotes new concepts. While there may be a component of fashion, the increased use of the word

L. Strigini (✉)
Centre for Software Reliability,
City University London,
Northampton Square, London EC1V 0HB, UK
e-mail: strigini@csr.city.ac.uk

K. Wolter et al. (eds.), *Resilience Assessment and Evaluation of Computing Systems*,
DOI: 10.1007/978-3-642-29032-9_1, © Springer-Verlag Berlin Heidelberg 2012

"resilience" is often meant to highlight either novel attention to these problems or a plea for a shift of focus in addressing them. It is useful to consider what these new foci may be and whether they require new concepts and new measures. Technical, and especially quantitative, reasoning about "resilience" requires clear definitions of these concepts, whether old or new.

Without reviewing in detail the multiple uses of "resilience", it is useful to recognise how the technical problems and debates in which it appears in different areas of application are related, highlighting similarities and differences in the problems they pose for quantitative reasoning, including measurement and benchmarking, and retrospective assessment as well as prediction.

The word "resilience", from the Latin verb *resilire* (*re-salire*: to jump back), means literally the tendency or ability to spring back, and thus the ability of a body to recover its normal size and shape after being pushed or pulled out of shape, and therefore figuratively any ability to recover to normality after a disturbance. Thus the word is used technically with reference to materials recovering elastically after being compressed, and also in a variety of disciplines to designate properties related to being able to withstand shocks and deviations from the intended state and go back to a pre-existing, or a desirable or acceptable, state. Other engineering concepts that are related to resilience therefore include for instance fault tolerance, redundancy, stability, feedback control.

A review of scientific uses of the word "resilience" for the European project *ReSIST* ("Resilience for Survivability in IST") [770] identified uses in child psychology and psychiatry, ecology, business and industrial safety. In many cases, this word is used with its general, everyday meaning. Some users, however, adopt specialised meanings, to use "resilience" as a technical term.

The premise for calling for an everyday word to be used with a new specialised meaning is that there is a concept that needs to have its own name, for convenience of communication, and lacks one. The concept is sometimes a new one ("entropy", for instance), or a new refinement of old concepts ("energy", for instance), or just a concept that needs to be referred to more often than previously (because the problems to be discussed have evolved) and thus requires a specialised word. Sometimes, the motivation is that words previously used for the same concept have been commandeered to denote, in a certain technical community, a more restricted meaning: for instance, after the word "reliability" acquired a technical meaning that was much more restrictive than its everyday meaning, the word "dependability" came to be used, by parts of the ICT technical community, to denote the everyday meaning of "reliability" [63].

For the word "resilience", a tendency has been to use it, in each specific community, to indicate a more flexible, more dynamic and/or less prescriptive approach to achieving dependability, compared to common practices in that community. Thus the above-cited document [770], for instance, concluded that a useful meaning to apply to "resilience" for current and future ICT is "ability to deliver, maintain, improve service when facing threats and evolutionary changes": that is, the important extension to emphasise in comparison with words like "fault tolerance" was the fact that the perturbations that current and future systems have to tolerate include change. While

existing practices of dependable design deal reasonably well with achieving and predicting dependability in ICT systems that are relatively closed and unchanging, the
tendency to making all kinds of ICT systems more interconnected, open, and able to
change without new intervention by designers, is making existing techniques inadequate to deliver the same levels of dependability. For instance, evolution itself of the
system and its uses impairs dependability: new components "create" system design
faults or vulnerabilities by feature interaction or by triggering pre-existing bugs in
existing components; likewise, new patterns of use arise, new interconnections open
the system to attack by new potential adversaries, and so on [769].

For a comparison with another field of engineering, a document on "infrastructure resilience" [632] identifies "resilience" as an extension of "protection". As an
example of the direction for this extension, this paper questions whether burying the
cables of a power distribution grid to prevent hurricane damage is "resilience", but
suggests that installing redundant cabling is.

An important specialised use of the word "resilience" has emerged with "resilience
engineering", a movement, or a new sub-discipline, in the area of safety (or, more
generally, performance under extreme conditions) of complex socio-technical systems. Here, the word "resilience" is used to identify enhanced ability to deal with the
unexpected, or a more flexible approach to achieving safety than the current mainstream approaches. The meaning is somewhat different between authors, which need
not cause confusion if we consider "resilience engineering", rather than "resilience",
as the focal concept for these researchers, and actually a neologism, designating an
area of studies and the ongoing debate about it. This area will be further discussed
below.

From the viewpoint of the problems of quantitative assessment, measurement and
benchmarking, the goals of these activities and the difficulties they present, there
is no sharp boundary between the socio-technical systems that are of concern to
ICT specialists and those addressed by "resilience engineering". There are undoubt-
edly differences in the typical scales of the systems considered, but the progress
in ICT towards the "future Internet" and greater interconnection of ICT with other
infrastructures and activities are cancelling these differences [769]. Most dependability problems in ICT have always involved some social and human factors influencing
dependability, for instance through design methods and constraints, or through the
maintenance or use of technical systems. In this sense, ICT dependability is about
socio-technical systems. As ICT becomes more pervasive and interlaced with human
activities, the dependability of the technical components in isolation may become a
minor part of the necessary study of dependability and thus of resilience. For example, this occurs in a hospital or air traffic control system, where automated and human
tasks interact, and contribute redundancy for each other, on a fine-grain scale. It also
occurs where large scale systems involve networks of responsibilities across multiple organisations, as in the provision of services (possibly through open, dynamic
collaboration) on the present or future Internet.

In view of these similarities and disappearing boundaries between different categories of systems, this short survey, written from the vantage point of practices in
the technical side of ICT dependability assessment, tries to emphasise the possible

new problems, or desirable new viewpoints, that may come from the progressive extension of the domain that ICT specialists have to study towards systems with a more important and more complex social component.

1.2 The "Resilience Engineering" Movement

The title "resilience engineering" has been adopted recently by a movement, or emerging discipline or community, started around a set of safety experts dealing mostly with complex socio-technical systems, like for instance industrial plant, railways, hospitals. A few symposia have taken place focusing on this topic and books have been published. This movement uses the term "resilience engineering" to designate "a new way of thinking about safety" [767]. The focus of these researchers is on moving beyond limitations they see in the now-established forms of the pursuit of safety: too much focus on identifying all possible mechanisms leading to accidents and providing pre-planned defences against them; too little attention to the potential of people for responding to deviations from desirable states and behaviours of the system. Thus the resilience engineering authors underscore the needs for reactivity and flexibility, e.g., "The traits of resilience include experience, intuition, improvisation, expecting the unexpected, examining preconceptions, thinking outside the box, and taking advantage of fortuitous events. Each trait is complementary, and each has the character of a double-edged sword" [681].

In using the term "resilience", there is a range between authors focusing on the resilient *behaviour* of the socio-technical system—its visibly rebounding from deviations and returning to (or continuing in) a desirable way of functioning—and those who focus on the characteristics they believe the system must have in order to exhibit such behaviour, like for instance the cultural characteristics and attitudes in the above quote. This degree of ambiguity need not cause confusion if we simply use the "resilience engineering" phrase to designate a set of related concerns, rather than "resilience" as a specific technical term. It points, however, at the variety of attributes—whatever we may call them—that are inevitably of interest to measure or predict.

Importantly, authors in "resilience engineering" underscore the difference between "resilience" and "safety", the former being just one of the possible means to achieve the latter. Their concern is often one of balance, as they see excessive emphasis on (and perhaps complacency about the effectiveness of) static means for achieving safety, designed in response to accidents, while they see a need for a culture of self-awareness, learning how things really work in the organisation (real processes may be very different from the designed, "official" procedures), taking advantage of the workers' resourcefulness and experience in dealing with anomalies, paying attention to the potential for unforeseen risks, fostering fresh views and criticism of an organisation's own model of risk, and so on. On the other hand, safety can be achieved in organisations that do not depend on "resilience" in this sense of the word, but on rigid, pre-designed and hierarchical approaches [405].

1.3 The Appeal of Resilience and Fault Tolerance

Before discussing issues of measurement and quantitative assessment, it is useful to identify some concepts and historical changes that are common to the various technical fields we consider.

When something is required to operate dependably (in a general sense, including "being secure against intentional harm"), the means available for ensuring this dependability include mixes of what in the ICT world are often called "fault avoidance" and "fault tolerance" [63]. The former means making components (including, by a stretch of the word "component", the design of the system, with its potential defects that may cause failures of the system) less likely to contain or develop faults, the latter means making the system able to tolerate the effects of these faults.

1.3.1 Historical Shifts Between "Fault Avoidance" and "Fault Tolerance"

Historically, the balance between the two approaches is subject to shifts, as is the level of system aggregation at which fault tolerance is applied. For instance, to protect the services delivered by a computer, a designer may add inside the computer redundant component(s) to form a fault-tolerant computer. Alternatively, the designer of a system using the computer (say, an automated assembly line) might provide a rapid repair service, or stand-by computers to be switched in by manual intervention, or manual controls for operators to take control if the computer fails: all these latter provisions make the control function of the assembly line fault-tolerant (to different degrees). This is a case of shift from fault tolerance in the architecture of a system component (the computer) to fault tolerance in the architecture of the system (the assembly line).

Fault tolerance (for various purposes, e.g., masking permanently disabled components, preventing especially severe effects of failures,[1] recovering from undesired transients) is a normal feature of much engineering design as well as organisation design. Fault tolerance against some computer-caused problems is nowadays a normal feature within computer architecture, but over time, as computers in an organisation or engineered plant become more numerous, the space for forms of fault tolerance "outside the computer" increased. Much of the computer hardware and software is obtained off-the-shelf, meaning that for the organisation achieving great confidence in their dependability may be unfeasible or expensive, but on the other hand there is a choice of alternatives for error confinement and degraded or reconfigured operation (relying on mixes of people and technology) if only some of these

---

[1] Including "system design failures": all components function as specified, but it turns out that in the specific circumstances the combination of these specified behaviours ends in system failure: the system's *design* was "faulty".

components fail, and for selectively deploying redundant automation (or redundant people) where appropriate.

Shifts of balance between fault tolerance and fault avoidance, and across levels of application of fault tolerance, occur over time with changes in technology, system size and requirements. Shifts away from fault tolerance are naturally motivated by components becoming more dependable, or their failure behaviour better known (so that fault tolerance is revealed to be overkill), or the system dependability requirements becoming (or being recognised to be) less stringent. Shifts towards more fault tolerance are often due to the observation that fault avoidance does not seem to deliver sufficient dependability, or has reached a point of diminishing returns, and in particular that good fault tolerance will tolerate a variety of different anomalous situation and faults, including unexpected ones. Thus, fault tolerance for instance often proves to be an effective defence against faults that the designers of components do not know to be possible and thus would not have attempted to avoid.

Examples of these factors recur in the history of computing, and can be traced to some extent through the arguments presented at the time to argue that the state of technology and application demanded a shift of emphasis: for instance in the papers by Avizienis in the 1970s [61] arguing for a return to more fault tolerance in computers; those of the "Recovery Oriented Computing" project in the early years of the twenty-first century [109] arguing for attention to more dynamic fault tolerance, in systems comprising multiple computers and operators. In the area of security, similar reasons motivated arguments for more of a "fault tolerance" oriented approach [302], later reinforced by concerns about the inevitable use of off-the-shelf computers and operating systems [63]. Similar considerations have applied to the proposals for fault tolerance against software faults [191, 740]. More recently, a call for papers on "Resiliency in High Performance Computing" [768] points at how the scaling up of massively parallel computations implies that the likelihood of at least one component failing during the computation has become too high if the computation is not able to tolerate such failures; similar considerations have arisen for the number of components in chips, or networks, etc, repeatedly over the years. For an example in larger systems that go beyond ICT, we may consider titles like "Moving from Infrastructure Protection to Infrastructure Resilience" [383], advocating a shift from a perceived over-emphasis on blocking threats before they affect critical infrastructure (e.g., electrical distribution grids) to making the latter better able to react to disruption. All these arguments must rely implicitly on some quantification of the risk involved by each alternative defensive solution—a sound argument about which solution entails the least risk, even without giving explicit numerical risk estimates for the individual solutions—although this quantification is not very visible in the literature.

1.3.2 Evolving versus Unchanging Redundancy

A related, recurrent line of debate is that advocating more flexible and powerful fault tolerance, in which fault tolerance mechanisms, rather than following narrowly pre-defined strategies, can react autonomously and even evolve in response to new situations, like the human mind or perhaps the human immune system [15, 62]. Some of the recent "autonomic computing" literature echoes these themes [454]. The trade-off here is that one may have to accept a risk that the fault-tolerant mechanisms themselves will exhibit, due to their flexibility and complexity, unforeseen and sometime harmful behaviour, in return for an expectation of better ability to deal with variable, imperfectly known and evolving threats. The challenge is to assess this balance of risks, and to what extent a sound quantitative approach is feasible.

In the social sciences' approach to these problems, observations about the importance of redundancy and flexibility underpin the literature about "high reliability organisations" [783] and to some extent about "safety cultures". In this picture, the "resilience engineering" movement could be seen as just another shift in which dynamic reaction (fault tolerance) to anomalies is seen as preferable to prior provisions against them, as a precaution against unexpected anomalies. Its claim to novelty with respect to the community where it originated is in part a focus on the importance of the unexpected. This summary of course does not do justice to the wealth of specific competence about safety in organisations in the "resilience engineering" literature, or about computer failure, human error, distribution networks etc to be found in the other specialised literature mentioned above. Our goal here is to identify broad similarities and differences and their implications on assessment, measuring and benchmarking.

Much current emphasis in "resilience engineering" is about flexibility of people and organisations, not just in reacting to individual incidents and anomalous situations, but also in learning from them and thus developing an ability to react to the set of problems concretely occurring in operation, even if not anticipated by designers of the machinery or of the organisation. There is for instance an emphasis, marking recent evolution in the "human factors" literature, on the importance of understanding work practices as they are, as opposed as to how they have been designed to be via procedures and automation of tasks. The real practices include for instance "workarounds" for problems of the official procedures, and may contribute to resilience and/or damage it, by creating gaps in the defences planned by designers and managers. It is appropriate to consider differences identified by "resilience engineering" authors between the "resilience engineering" and the older "high reliability organisation" movement. Perhaps the most cited paper [783] from the latter discussed how flight operations on U.S. Navy aircraft carriers achieved high success rates with remarkably good safety. This paper focused on four factors: "self-design and self-replication" (processes are created by the people involved, in a continuous and flexible learning process), the "paradox of high turnover" (turnover of staff requires continuous training and conservatism in procedures—both seen as generally positive influences—but also supports diffusion of useful innovation), "authority overlays" (distributed authority allowing local decisions by low-ranking people as well as producing higher level decisions through co-operation and negotiation), "redundancy" (in the machinery and supplies but also in overlapping responsibilities for monitoring and in built-in extra staffing with adaptability of people to take on different jobs as required). In contrast, a paper about how "resilience engineering"

[680] differs from this approach refers to healthcare organisations and how their culture and lack of budgetary margins severely limit the applicability of the four factors claimed to be so important on aircraft carriers; it points at the potential for improving resilience by, for instance, IT systems that improve communication within the organisation and thus distributed situational awareness and ability to react to disturbances. Another valuable discussion paper [586] emphasises the steps that lead from the general sociological appreciation of common issues—exemplified by the "high reliability organisation" literature—to an engineering approach with considerations of cost and effectiveness in detail.

1.4 Resilience and Fault Tolerance Against the Unexpected

We see that a frequently used argument for both fault tolerance (or "resilience", seen as going beyond standard practices of fault tolerance in a given community) in technical systems and more general "resilience" in socio-technical systems is based on these being broad-spectrum defences. Given uncertainty about what faults a system may contain or what external shocks and attacks it has to deal with, it seems better to invest in flexible, broad-spectrum defensive mechanisms to react to undesired situations during operation, rather than in pre-operation measures (stronger components, more design verification) that are necessarily limited by the designers' incomplete view of possible future scenarios.

1.4.1 Competing Risks; the Risk of Complex Defences

This argument can, however, be misleading. It is true that general-purpose redundancy and/or increased resources (or attention) dedicated to coping with disturbances as they arise, or to predicting them, can often deal with threats that designers had not included in their scenarios. But there will also be threats that bypass these more flexible defences, or that are created by them. An example can be found in the evolution of modular redundancy at the level of whole computers. The "software implemented fault tolerance" (SIFT) concept in the 1970s [384], the precursor of many current fault-tolerant solutions, responded to the fact that one could affordably replicate entire computations running on separate computers, so that the resulting system would tolerate any failure of any hardware or software component within a single computer (or communication channel). This was certainly a more general approach than either more expenditure on fault avoidance without redundancy, or ad-hoc fault tolerance for foreseen failures of each component in a single computer. It was a more powerful approach in that it may well tolerate the effects of more faults, e.g., some design faults in the assembly of the computer or in its software (thanks to loose synchronisation between the redundant computers [390]). But the SIFT approach also ran into the surprise of "inconsistent failures": the same loose, redundant organisation that gives the system some of its added resilience makes it vulnerable to a specific failure mode. A faulty unit, by transmitting inconsistent messages to other units, could prevent the healthy majority of the system from enforcing correct system behaviour. To tolerate a single computer failure might require four-fold redundancy (and a design that took into account this newly discovered problem) rather than three-fold as previously believed. This was an unexpected possibility, although now, with experience grown from its discovery, it is easy to demonstrate it, using a simple model of how such a system could operate.

Other events that may surprise designers may be unexpected hardware failure modes; operators performing specific sequences of actions that trigger subtle design faults; new modes of attack that "create" new categories of security vulnerabilities; threats that bypass the elaborate defences created by design (ultra-high availability systems go down because maintenance staff leave them running on backup batteries until they run out, testing at a nuclear power plant involves overriding safety systems until it is too late for avoiding an accident (the Chernobyl disaster), attackers circumvent technical security mechanisms in ICT via social engineering); in short, anything that comes from outside the necessarily limiting model of the world that the designers use. Some such surprises arise from incomplete analysis of the possible behaviours of a complex system and its environment (cf the Ariane V first-flight accident [592]). Perhaps the incompleteness of analysis is inevitable given complexity, and indeed there is a now common claim that accidents—at least in "mature" organisations and engineered systems—tend to originate from subtle combinations of circumstances rather than direct propagation from a single component failure [722]).[2] On the other hand, designers also *choose* "surprises" to which their systems will be vulnerable: they explicitly design fault tolerance that will not cope with those events that they consider unlikely, trading off savings in cost or complexity against increases in risk that are (to their knowledge) acceptable.

In the ICT area, it is tempting to see "surprises" as manifestations of designer incompetence, and indeed, in a rapidly evolving field with rapidly increasing markets, many will be ignorant about what for others is basic competence. But there is also a component of inevitable surprises. In other areas of engineering it has been observed that the limits of accepted models and practices are found via failure [725, 921], usually of modest importance (prototype or component tests showing deviations from model predictions, unexpected maintenance requirements in operation, etc), but sometimes spectacular and catastrophic (the popular textbook examples—the Tacoma Narrows bridge, the De Havilland Comet).

---

[2] Although many authors point out that accidents caused by single component failures are still common. A component failure occurs in a system design that happens to omit those defences that would prevent that specific failure from causing an accident.

1.4.2 Quantifying Surprises?

Thus, the argument that a more "resilient" design—more open-ended forms of redundancy—offers extra protection is correct, but when it comes to estimating *how much* extra protection, or *which form* of redundancy will be more effective—when we need measurement and quantitative assessment—there is a difference between threats. There is a range of degrees to which quantitative reasoning is useful, perhaps best illustrated via examples. For a well known and frequent hardware failure mode, we may be able to trust predictions of its frequency, and thus predict the system reliability gain afforded by a specific redundant design, if some other modelling assumptions are correct. For other forms of failure, we may have very imprecise ideas about their frequency—for instance, this usually applies, at the current state of practice, to software failures in highly reliable systems—and yet, we can decide which designs will tolerate specific failure patterns, and via probabilistic modelling even decide whether a design is more resilient than another one given certain plausible assumptions. Last, there are surprises that violate our modelling assumptions. Designers can try to reduce them by keeping an open mind, and making the system itself "keep an open mind", but have no indication of how successful they are going to be. In the case of organisations, it may well be, for instance, that organisational choices that improve resilience against certain disturbances will be ineffective or counterproductive against others [933].

Insofar as resilience is obtained by making available extra resources, limits on resources demand that designers choose against which threats they will deploy more redundant resources. Limits on resources also recommend more flexible designs, in which these resources can deal with more different challenges. Again, these qualitative considerations demand, to be applicable to concrete decisions, quantification (at least adequate to support rough comparisons) of the risk and costs of different solutions.

This set of considerations has highlighted many areas where measurement and assessment of resilience or fault tolerance are desirable, and started to evoke a picture of measures that may be useful and of the difficulties they may involve. The discussion that follows looks at choices of attributes to measure, and difficulties of measurement and prediction, in some more detail, taking a viewpoint inspired by "hard" quantification approaches in engineering and considering some of the issues created by extension towards more complex socio-technical systems.

1.5 Quantifying Resilience: Its Attributes, and Their Possible Measures

In quantitative assessment there are always two kinds of potential difficulties: defining measures that usefully characterise the phenomena of interest; and assessing the values (past or future) of these measures.

About the first difficulty, dependability and resilience are broad concepts encompassing multiple attributes, so that there are multiple possible measures. The discussion that follows will take for granted that there are many dependability attributes of potential interest, which are different and may well be in conflict under the specific constraints of a certain system: for instance, pursuing safety—ability to avoid specific categories of mishaps—may conflict with the pursuit of availability—the ability to deliver service for a high fraction of the time (see for instance [63] for a high-level set of definitions). Irrespective of the specific dependability attribute of interest, we will summarily characterise categories of measures related to fault tolerance and resilience, with some discussion of their uses and difficulties in measurement and prediction.

The categories will be introduced in terms of "systems" (meaning anything from a small gadget to a complex organisation) that have to behave properly despite "disturbances" (an intentionally generic term, to cover component faults inside the system, shocks from outside, overloads, anomalous states, no matter how reached).

The sections that follow

- first discuss categories of measures in common use in quantitative reasoning about ICT, both as measures at whole-system level and as parameters, describing components and their roles, in mathematical models for deriving such whole-systems measures:

  - measures of dependability in the presence of disturbances, which may be estimated empirically in operation or in a laboratory, or through probabilistic models (as functions of measures at component level), as discussed in other chapters of this book
  - measures of the amount of disturbances that a system can tolerate, typically obtained from analysing a system's design
  - measures of probability of correct service given that a disturbance occurred ("coverage factors"), typically estimated empirically, often in a laboratory

- and then proceed to examine more speculative areas:

  - proposed predictors of resilience in socio-technical systems
  - more detailed measures that discriminate between different forms of "resilient" behaviours.

While pointing out differences between categories of systems and types of "resilience", the discussion will identify problems that they share and that may recommend importing insights from some areas of study to others.

1.5.1 Measures of Dependable Service Despite Disturbances

The first category of measures that give information about resilience are simply measures of dependability of the service delivered by a system that is subject to disturbances. The better the system worked despite them, the more resilient it was.

Indeed, a question is why we would want to measure "resilience" or "fault toler-ance" attributes, rather than "dependability" attributes. The former are just means for achieving the latter.

For instance, an availability measure for a function of a system, obtained over a long enough period of use in a certain environment (pattern of usage, physical stresses, misuse, attacks etc), will be a realistic assessment of how well that function tolerates, or "is resilient" to, that set of stresses and shocks.[3]

This kind of measure is certainly useful when applied to documenting past depend-ability. It will be useful, for instance, in invoking a penalty clause in a contract, if the achieved availability falls short of the level promised. It will also have some uses in prediction. Suppose that the system is a computer workstation used for well-defined tasks in a relatively unchanging environment. A robust measure of past availability ("robust" may imply for instance repeating the measure over multiple workstations of the same type, to avoid bias from variation between individual instances) will be trusted to be a reasonable prediction of future availability (if the environment does not change). Measures on two types of workstations will be trusted to indicate whether one will offer substantially better availability than the other.

The Difficulty of Extrapolation

If we wish to compare systems (workstations, in this example), that have not been operated in the same environment, we will sometimes define a reference load (of usage as well as stresses etc)—a "benchmark" workload and stress (or *fault*) load, in the current IT parlance (see Chap. 14). Here, the broader "resilience" literature about engineering and socio-technical systems has to confront difficulties that are also evident for strict computer dependability evaluation [616], but with differences of degree. These difficulties can be generally characterised as *limits to the extrapolation of measures* to environments that are different from those where the measures were obtained. If a system copes well in the presence of one type of disturbances but

---

[3] A conceptual problem arises here, which will recur in different guises throughout this discussion. To use an example, suppose that two computers are made to operate in an environment with high levels of electromagnetic noise. Of the two, computer A is heavily shielded and mostly immune to the noise. The other one, computer B, is not, and suffers frequent transient failures, but always recovers from them so that correct service is maintained. The two thus prove equally dependable under this amount of stress, but many would say that only B is so dependable *thanks to* its "resilience": A just avoids disturbances; only B "bounces back" from them. Should we prefer B over A? Suppose that over repeated tests, B sometimes fails unrecoverably, but A does not. Clearly, A's lack of "resilience" is then not a handicap. Why then should we focus on assessing "resilience", rather than dependability? Or at least, should we not define the quality of interest (whether we call it "resilience" or not) in terms of "correct behaviour despite pressure to behave incorrectly"? An answer might be that the resilience mechanisms that B has demonstrated to have will probably help it in situations in which A's single-minded defence (heavy shielding) will not help. But then the choice between A and B becomes an issue of analysing how much better than A B would fare in various situations, and how likely each situation is. Measures of "resilience" in terms of recovery after faltering are just useful information towards estimating measures of such "dependability in a range of different situations".

less well with another type, changing the relative weights of these two types of disturbances will change the degree of dependability that will be observed. There will not even be a single indicator of "stressfulness" of an environment, so that we can say that if a system exhibited—say—99% availability under the benchmark stress, it will exhibit *at least* 99% availability in any "less stressful" environment [739]. Likewise, we won't be able to trust that if system A is more dependable (from the viewpoint of interest: e.g., more reliable) than system B in the benchmark environment, it will still be more dependable in another environment. An extreme, but not unusual case of the extrapolation problem is the difficulty of predictions about systems that are "one of a kind" (from a specific configuration of a computer system, to a specific ship manned by a specific crew, to a specific spontaneous, temporary alliance of computers collaborating on a specific task in the "future internet") or will be exposed to "one of a kind" situations: that is (to give a pragmatic definition), systems or situations for which we have no confidence that the measures taken elsewhere, or at a previous time, will still prove accurate. Again, extreme examples are easily found for the human component of systems: an organisation that appears unchanged, after some time from a previous observation, in terms of staff roles, machinery, procedures, may in reality have changed heavily due to staff turnover, or ageing, or even just the experience accumulated in the meantime (for instance, a period without accidents might reduce alertness). Here arises the first reason for going beyond whole-system dependability measures: they do not produce an understanding of *why* a system exhibits a certain level of dependability in a given environment—how each part of the system succumbed or survived the disturbances, which behaviours of which parts accomplished recovery, why they were effective—which could turn into a model for predicting dependability as a function of the demands and stresses in other environments.

Another problem with extrapolation is often created intentionally, as a necessary compromise. If we want a benchmark to exercise the whole set of defences a system has, we need the environment to "attack" these defences. This may require the benchmark load to condense in a short time many more stress events than are to be expected in real use; but some aspects of resilience are affected by the frequency of stresses. If the system being "benchmarked" includes people, their alertness and fatigue levels are affected. If it involves slow recovery processes (say, background processes that check and correct large bodies of data), an unrealistically high frequency of disturbances may defeat these mechanisms, although they would work without problems in most realistic environments.

Last, there is the problem of resilience against *endogenous stresses*. These exist in all kinds of systems: a computer may enter an erroneous state due to a software design fault being activated or an operator entering inappropriate commands; a factory may suffer from a worker fainting, or from a fire in a certain piece of machinery; and so on. If we wish a common benchmark to measure resilience against these kinds of disturbance, it will need to include some simulation of such events. But this may produce unfair, misleading measures. Perhaps a computer that has very little tolerance to errors caused by internal design faults has been designed this way for the right reasons, since it has no design faults of the types that it cannot tolerate; the less a computer interface tends to *cause* operator errors, the less the computer needs to tolerate them; the less a factory tends to cause workers to become ill on the job, the less it needs to operate smoothly through such events; etc.

This unfairness also has a beneficial aspect, though: it allows a benchmark to give at least some information about resilience against the unexpected or unplanned-for disturbances. The benchmark deals with hypothetical situations. What if in a factory where nobody ever becomes ill, one day somebody does? What if the computer does have unsuspected design flaws? Likewise, modern regulations require many safety measures for all systems of a certain kind, irrespective of the probability, for a specific system, of the situations in which they would be useful. In these circumstances, a dependability or safety "benchmark" (from a fault injection experiment in a computer to an emergency drill in a factory) verifies that certain precautions are in place, and thus certain stresses are likely be tolerated if they were ever to happen. However, engineering for better dependability under a benchmark situation does not necessarily improve dependability in any operational situation different from the benchmark.

1.5.2 Measures of Tolerable Disturbances

A type of attributes that often allow simple and intuitive measures, and thus are heavily used, is the extent of deviation (or damage or disturbance) that a system can tolerate while still later returning to the desired behaviour or state (or still preserving some invariant property about its behaviour, e.g., some safety property: choosing different invariants will define different measures).

Thus, in ICT it is common to characterise a certain fault-tolerant computer design as able to mask[4] (without repair) up to k faulty components; or a communication code as able to detect (or to reconstruct the original message despite) up to t single-bit errors; or that a user interface will tolerate up to m erroneous inputs in one transaction; etc. Likewise, in the world of larger systems, we can rate a ship as being able to self-right from a tilt of so many degrees from the upright position; or a factory's staffing level as being calculated to allow for so many absences without loss of productivity. In ecology, a proposed measure of "resilience" of an environment is the size of a basin of attraction, in its state space: the distance by which the environment's state may be moved from a stable point without becoming unstable and moving into another basin of attraction (this distance measure is proposed to be used with a complementary measure of "resistance": the "force" needed for a given shift in the state space) [174].

To generalise, this set of attributes, and their measures, are about how far the object of interest can be pushed without losing its ability to rebound or recover; or how quickly it will rebound, or how closely its state after rebounding will resemble the state before the disturbance. To reason properly about these attributes of a system, it is important to recognise them as separate: system A may be "more resilient" than

---

[4] "Masking" usually meaning that the externally observed behaviour of the system shows no effect of the fault.

system B from one of these viewpoints, and "less resilient" from another one; for instance, A may be slower than B in recovering from a disturbance of a certain size, but able to recover from a more extreme disturbance than B can.

A great advantage of this type of measures is that for many ICT systems they are easy to obtain directly from their designs: so long as the implementation matches the design in some essential characteristics, we know that certain patterns of faults or disturbances are tolerated. These measures are also typically robust to the extrapolation problem.

If "measuring" on the design is unsatisfactory (for instance we expect the implementation to have flaws; or the required measure is too complex to calculate), we would rely on observations of the system in operation. There may be difficulties in obtaining enough observations of "disturbances" close to the limit, in knowing where the limit is (for systems that should not be tested to destruction), and in deciding whether the system's resilient reaction is deterministic, that is, whether observing successful recovery from a certain extent of disturbances allows us to infer 100 % probability of recovery. Again, socio-technical systems offer the most striking examples of the doubts that can affect estimates of these measures.

A limitation of these "maximum tolerable disturbance" measures, even for systems where they are easy to obtain, is that we may well be interested in characterising how well a system rebounds from *smaller* disturbances. For instance, given a form of fault tolerance that allows for some degradation of service, we may then want to measure not just how far the system can be pushed before failing altogether, but the relationship between the size of disturbances and the degradation of performance. For instance, for a network (of any kind) one might measure the residual throughput (or other measure of performance) as a function of the amount of network components lost (or other measure of faults or disturbances); this kind of function has been proposed [365] for resilience of critical infrastructures, leaving open the question of which single-number characterisation (if any) of these curves would be useful in practice. We will return later to characterisations of resilience as a function rather than a single, synthetic measure.

1.5.3 Measures of "Coverage Factors"

If we recognise that for most systems of interest the resilient behaviour is non-deterministic in practice,[5] we are no longer interested in *whether* the system will rebound from a disturbance but in the *probability* of it successfully rebounding; or perhaps the distribution of the time needed for it to return to a desired state; or other probabilistic measures. Thus in fault-tolerant computing we talk about the

---

[5] That is, even for many deterministic systems, their behaviour is complex enough that the knowledge we can build about them is only statistical or probabilistic. For instance, many software systems (deterministic in intention) have a large enough state space that many failures observed in operation appear non-deterministic—they cannot be reproduced by replicating the parts of the failure-triggering state and inputs that are observable [390].

"coverage" factor of a fault-tolerant mechanism, defined as the probability of the mechanism successfully performing its function in response to a disturbance (e.g., detecting a data error, or recovering from it), conditional on the disturbance (e.g., the data error) occurring; or we talk about the probability distribution of the latency of a component fault (i.e. of the time needed to detect it) rather than of a single numerical estimate.

Coverage factors are especially attractive as true measures of resilience. For instance, if we estimate a probability of a disturbance being tolerated so as not to cause system failure (a coverage factor, c), and know the frequency f of disturbances, then, in a simple scenario with rare disturbances, $(1 - c) * f$ would give us the frequency of system failures. The frequency of system failures is a measure of dependability (reliability), and one can see, for instance, that to improve reliability in this scenario one needs either to reduce the frequency of disturbances or to increase the coverage factor: the latter does indeed represent resilience, how well the system responds to adversity. And, as a concrete advantage, this relationship between a coverage factor and failure frequency seems to support extrapolation of dependability assessment to a different environment: I could estimate the former in the laboratory, usually with artificially frequent disturbances, to make measurement easier, and then I could extrapolate to any environment where the same disturbances occur more or less frequently. This is the basis for the predictive use of fault injection as described in Chap. 13 or dependability benchmarking as described in Chap. 14.

Even with complex systems in which multiple components and mechanisms co-operate to achieve resilience, probabilistic models, as described elsewhere in this volume, allow predictions of the probability of successfully resilient behaviour and hence of dependability measures as functions of coverage factors and of frequency of disturbances (internal faults or externally generated shocks).

However, this possibility of extrapolation is actually severely limited. Importantly, the probability of tolerating a disturbance will be a function of the type disturbance that occurred. So, all "coverage" measures have to be defined with respect to some stated type, or mix, of faults or disturbances; and the difficulties of extrapolation that characterised measures of dependability under stress also affect, in principle, measures of coverage. In particular, the desirability but also the limits of "benchmark" scenarios apply when estimating coverage factors just as when measuring a dependability measure [739].

1.5.4 Measures of Socio-Technical Resilience

Since we are comparing the understanding of resilience with respect to different categories of systems, and the categorisation above is derived from examples at the simple end of the spectrum, it is useful to compare with proposed measures in the areas of complex socio-technical systems. We take as an example the list of attributes of resilience in socio-technical systems proposed by Woods [941]; we can relate them to the categories given above, as well as consider how amenable they are to precisely defined measures. These attributes are:

- "buffering capacity", which is essentially an "extent of tolerable disturbances" as discussed above. The potential difficulties only concern how easily this can be captured in practically usable measures;
- "flexibility versus stiffness: the system's ability to restructure itself in response to external changes or pressures". It is not clear how this could be measured. For instance, to measure flexibility in the observed operation of a system, we would need to decide which forms of "restructuring" were actually useful, without the benefit of checking how the crisis would develop if the restructuring had not taken place. So, the literature tends to describe this form of "flexibility" through scenarios or anecdotes;
- "margin: how closely or how precarious the system is currently operating relative to one or another kind of performance boundary", again related to "extent of tolerable disturbances". This has often useful definitions in technical systems, for instance we can define an acceptable maximum load on a network before it goes into congestion, or the minimum required set of functioning components necessary for basic services, while in socio-technical systems it is often difficult to identify what terms like "stretched to breaking point" may mean, and what measures of "distance" from this point may be appropriate;
- "tolerance: how a system behaves near a boundary—whether the system gracefully degrades as stress/pressure increase or collapses quickly when pressure exceeds adaptive capacity". This has parallels in many technical areas, and certainly in ICT, where "graceful degradation" is a frequent requirement, but for which no textbook, standardised measure exists.

1.5.5 Measuring the Supposed Determinant Factors of Resilience

When trying to assess dependability (and resilience) in the face of threats that cannot be predicted in detail, a proposed approach relies on identifying factors that are believed to enhance resilience. When dealing with well-understood risks, this exercise may take the form of simple design analysis. In many cases, assessment can rely on the combination—via a probabilistic model—of analysing which defensive mechanisms are in place, estimates of their coverage factors, and estimates of the probability distributions of disturbances to which they will need to react. There are of course difficulties with all these estimates, which qualify the confidence one can have in predictions obtained this way. But when dealing with the human and social determinants of system behaviour, the conjectured determinant factors of resilience often have a "softer" or at least more complex character. The coverage and component reliability parameters of a model for a complex socio-technical system, and even the model itself, would be often too difficult to establish with any confidence. Only empirical observations of system resilience would then be trusted.

A concern in the "resilience engineering" literature is that one tends to judge organisations on their past performance, but these "measures of outcomes" may lack predictive power: success in the past is no guarantee of success in the future, due to the extreme extrapolation problems mentioned above. Hence a search for "leading indicators" that can be used to assess future resilience. Many of these are cited in the literature. For instance, a review [432] lists measures of "Management commitment, Just culture, Learning culture, Opacity, Awareness, Preparedness and Flexibility", of "Empowerment, Individual responsibility, Anonymous reporting, Individual feedback, [...]" for individual workers and of "Organizational structure, Prioritizing for safety, Effective communication" for organisations (citing [388]); and others.

Such factors are commonly believed to be important in determining how well an organisation will perform from the safety and resilience viewpoints. So, informed judgements about how "resiliently" organisations will react to stresses will benefit from considering these "indicators"; if the indicators were reliable, an organisation might want to identify reasonable target values and levels of trade-offs among them. But objective measures of such attributes are difficult to define. Different systems can be ranked on ordinal scales with regard to attributes of interest, or specific numerical, objective measures can be used as proxy measures if shown empirically to correlate with desired behaviours. For instance, [389], studying the safety of ship operation, reports a massive effort in which factors believed to indicate "safety culture" were estimated by anonymous surveys of individuals; the research goal is to check how well these proxy measures correlate to observed safety performance (such as records of accidents, near misses, negative reports by competent authorities). If good correlation were found, some function of the "leading indicators" could be used for early warnings of accidents being too likely on a certain ship. On the other hand, predictive models akin to those described in this volume (e.g., Chap. 7), based on such measures and suitable for informing design of these systems, for instance answering questions like "To what extent should power be devolved to workers in this system so that the positive effects outweigh the negative effects?", appear unfeasible.

Precedents for emphasis on "determinant factors" of desired characteristics exist in all areas of engineering, as sets of mandated or recommended practices. A pertinent example is in standards for safety-critical computing, e.g., [461], where sets of good practices are recommended or required to be applied in developing and verifying software, as a function of the criticality of the software's functions. This is a reasonable approach, in principle, and yet checking that these practices were applied is a poor substitute for directly checking that the product has acceptably low probability of behaving unsafely: the former (good practice) does not imply the latter (safe enough behaviour). The difficulties are twofold: there is no clear knowledge of how much these practices, and their possible combinations, tend to help; and we should expect that (comparable) systems that are equal in the extent of application of these practices may still differ in the achieved results (the levels of dependability).

Indeed, many authors in the "resilience engineering" literature are wary of attempts at quantification, applied to complex systems, as liable to oversimplify the issues and divert management effort towards achieving required values of measures that have the "advantage" of concrete measurement procedures but no guaranteed relationship to outcomes. Others have used quantitative modelling for illustration and general insight, borrowing physics-inspired formalisms for modelling complex systems at a macroscopic level [311].

1.5.6 More Detailed Characterisations of Resilience

Two important topics that have emerged in the discussion so far are: the difference between tolerance/resilience for "design base", expected disturbances and for unexpected or extraordinary (excluded by design assumption) ones; and the possible need to characterise not just the size of the tolerable stresses, but more detail about the resilient behaviour in response to different levels or patterns of stresses.

In this latter area, one can look for measures like *performability*, defined [646] as the set of probabilities of the "levels of accomplishment" of a system's function,[6] or functions like network throughput as a function of loss of components. These options are no sharp departure from dependability modelling approaches that are well established in ICT.

While these measures are meaningful, authors have been looking, as exemplified in the previous section, for ways to characterise "resilient" behaviour in a more detailed fashion, although accepting that the result may be qualitative insight rather than models suitable for prediction.

To discuss the various parameters that may characterise resilience in an organisation, Woods and Wreathall [942] use the "stress-strain" diagram used in material science, as in Fig. 1.1. With materials, the y axis represents the "stress" applied to a sample of the material (e.g., tensile force stretching a bar of metal), and the x axis represents the degree of stretch in the material ("strain"). When tested, the typical building material will exhibit a first region of linear response (the stretch is proportional to the force applied), followed by a less-than-linear region, and finally by quick yielding that leads to breaking. As it moves from the linear to the sub-linear region, the material also moves from elastic behaviour, where the original size will be regained when the stress is removed, to permanent deformation. A qualitative analogy with organisations is made, in terms of "a uniform region where the organization stretches smoothly and uniformly in response to an increase in demands; and an extra region (x-region) where sources of resilience are drawn on to compensate for non-uniform stretching (risks of gaps in the work) in response to increases in demands". Thus in the "extra region" it is assumed that an organisation that successfully self-modifies shifts onto a new curve, that departs from the now-decreasing main curve and gives some extra amount of increase in tolerated "stress" for extra "strain", so as to be able to tolerate stresses beyond its "normal" maximum.

So, these authors identify a region of "orderly" adaptation to increasing stress (in some cases one might identify measures of both stress and strain with an

---

[6] If "accomplishment" has a numerical measure, e.g., throughput of a system, the system's performability is defined by the probability distribution function of this measure.

Fig. 1.1 Stress-strain
diagram

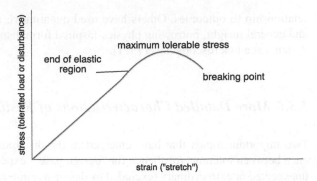

approximately linear relationship, e.g., increased inflow of patients to a hospital
being covered by increasing work hours within established procedures). Beyond this
maximum, the cost-effectiveness of use of resources decreases and a maximum exists, beyond which extra stress can only be tolerated by some kind of reconfiguration
of the organisation, e.g., mustering extra resources or freeing them by changes of
operation mode.

This view suggests sets of attributes that can be measured to characterise the
response of the system, like the size of the "uniform" range, and the extra stress that
can be tolerated before the degeneration into failure. The above authors identify as
especially important the ability of an organisation to manage smoothly transitions
between regions, and its "calibration", defined as its ability to recognise in which
region it is operating, so that reconfiguration is invoked when necessary (and presum-
ably not too often: we note that in many real situations, the ability to assess how well
calibrated they were for past decisions is limited. One cannot always tell whether a
decision to restructure to avoid catastrophic failure was really necessary—especially
in view of the uncertainty that the decision maker normally faces in predicting the
future). They rightly claim that the stress-strain analogy for organisation behaviour
is a first step in clarifying some of the attributes that characterise resilient behaviour
(hence also a first step towards quantitative modelling) and importantly highlight the
difference between "first-order" and "second-order" adaptive behaviour—the "normal stretching" of the organisation's design in the uniform region, versus the more
radical restructuring to work beyond the "normal" limit—but note the limitation of
representing "stress" as a unidimensional attribute, and the need for further work.
A limitation that seems important is that this kind of graph implicitly assumes that
the stress-strain relationship can be plotted as independent of time. This matches
well those measurement processes for the strength of materials in which stress is increased slowly, moving between states of equilibrium at least up to the maximum of
the curve. If the timing of the applied stimulus (as e.g., with sharp impact or repetitive
stress) makes a difference in how the material reacts, additional properties can be
studied, possibly requiring additional measures. In organisations (or for that matter
in computers), many of the stresses may need to be characterised in terms of dynamic

characteristics, or need to be defined in practice in terms of timing characteristics of events.

Considering the time factor may also bring into play other aspects of self-stabilisation, and other necessary design trade-offs, which can be illustrated by analogy with other engineering examples, outside the science of materials. For instance, making a ship more "stable" (increasing its metacentric height, so that it will self-right more promptly after heeling to one side) makes it also more liable to roll at higher frequency following the tilt of the waves, a characteristic that can reduce the effectiveness of the crew, make a warship unable to use its weapons, etc. Likewise, all "resilience" that relies on detecting (or predicting) component failures or shocks must strike a compromise between the risk of being too "optimistic"—allowing the situation to deteriorate too far before reacting—or too "pessimistic"—reacting too promptly, so that false alarms, or reactions to disturbances that would resolve themselves without harm, become too much of a drain on performance or even damage resilience itself.

1.6 Conclusions

A theme running through this survey has been that as fault tolerance (or resilience), that is, dynamic defences, exist in all kinds of systems, the measures that may be appropriate for studying them also belong to similar categories and the difficulties in defining measures, in performing measurements, and in predicting the values of measures also belong to common categories. Interest in studying and/or in extending the use of fault tolerance or resilience has expanded of late in many areas,[7] and we can all benefit from looking at problems and solutions from different technical areas. I gave special attention to the "resilience engineering" area of study, since its choice of topic problems highlights extreme versions of measurement and prediction problems about the effectiveness of "resilience" that exist in the ICT area. In all these areas there are spectra of prediction problems from the probably easy to the intractable. The "resilience engineering" movement has raised important issues related to the measurement and prediction of "resilience" attributes. One is simply the recognition of the multi-dimensionality of "resilience". For instance, Westrum [933] writes: "Resilience is a family of related ideas, not a single thing. [...] A resilient organization under Situation I will not necessarily be resilient under Situation III [these situations are defined as having different degrees of predictability]. Similarly, because an organization is good at recovery, this does not mean that the organization is good at foresight".

The boundaries between strict technical ICT systems and socio-technical systems are fuzzy, and for many applications the recognition of social components in deter-

---

[7] U.S. Navy aircraft carriers exploited redundancy for safety long before Rochlin and his co-authors studied it. On the other hand, their study prompted more organisations to recognise forms of redundancy in their operation, and protect them during organisational changes, and/or to consider applying redundancy.

mining meaningful assessment of dependability is important [769]. Concerns about improving measurement and quantitative prediction are often driven by the concrete difficulties in applying existing methods in new systems: just as increasing levels of circuit integration and miniaturisation made it unfeasible to monitor circuit operation at a very detailed level via simple probes and oscilloscopes, so the deployment of services over large open networks and through dynamic composition may create new difficulties in measuring their dependability. More general problems may arise, however: do we need to choose appropriate new measures for characterising the qualities of real interest? If they are amenable to measurement in practice, to what extent will they support trustworthy predictions? To what extent may the benefit of "reasonably good" measures (perhaps acceptable proxies for the "truly important" ones) be offset by natural but undesirable reactions to their adoption: designers and organisations focusing on the false target of achieving "good" values of these measures, perhaps to the detriment of the actual goal of dependability and resilience?

These questions underlie all assessment of resilience and dependability, but more markedly so as the socio-technical systems studied become less "technical" and more "social". Authors in "resilience engineering" have identified research problems in better characterising, even at a qualitative, descriptive level, the mechanisms that affect resilience. Quantitative measurement may follow. Quantitative predictive models may or may not be feasible, using results from the abundant research in modelling—at various levels of detail—the dependability of complex infrastructure and ICT; quantitative approaches from mathematical physics [311] may also yield insight even without predictive power. Research challenges include both pushing the boundary of the decision problems that can be addressed by sound quantitative techniques, and finding clearer indicators for these boundaries. There are enough historical examples of quantitative predictions proving misleading, and perhaps misguided, but we often see these with the benefit of hindsight. Perhaps most important would be to define sound guidance for "graceful degradation" of quantitatively driven decision making when approaching these limits: more explicit guidance for exploiting the advantages of measurement and quantitative prediction "as far as they go" but avoiding potential collapse into unrealistic, "pure theory"—driven decision making.

Acknowledgments This work was supported in part by the "Assessing, Measuring, and Benchmarking Resilience" (AMBER) Co-ordination Action, funded by the European Framework Programme 7, FP7-216295. This article is adapted from Chap. 15 of the "State of the Art" report produced by AMBER, June 2009.

Chapter 2
Resilience in Mobile Networks: A Need and a Challenge

Alberto Avritzer, Luca Berardinelli, Vittorio Cortellessa, Leïla Kloul, Carlo Rosa and Katinka Wolter

Abstract In this chapter, we describe the most important network protocols supporting modern applications in mobile cellular networks, wireless sensor networks (WSN) and mobile ad hoc networks (MANETs). We first focus on the handover procedure in mobile cellular networks and the network failures due this procedure. The current solutions to enable seamless handover in the existing networks technologies are presented. We then present, in the context of WSN, a framework for model-based design and performance analysis of WSN software applications. WSN are often used for applications with strict non-functional requirements. They, thus, require solid modelling approaches for non-functional attributes of WSN software applications. We finally discuss approaches and cross-layer protocols addressing the problems of security, wireless communications, and network topology changes in

A. Avritzer
Siemens Corporate Research and Technology, 755 College Road East,
Princeton, NJ 08540, USA
e-mail: alberto.avritzer@siemens.com

L. Berardinelli · V. Cortellessa · C. Rosa
Dipartimento di Informatica, Università dell'Aquila,
Via Vetoio 1, Coppito, AQ, Italy
e-mail: luca.berardinelli@univaq.it

V. Cortellessa
e-mail: vittorio.cortellessa@univaq.it

C. Rosa
e-mail: carlo.rosa@univaq.it

L. Kloul (✉)
PRiSM, Université de Versailles, 45 Avenue des Etats Unis,
78000 Versailles, France
e-mail: kle@prism.uvsq.fr

K. Wolter
Institute of Computer Science, Free University Berlin,
Takustr. 9, 14195 Berlin, Germany
e-mail: katinka.wolter@fu-berlin.de

K. Wolter et al. (eds.), *Resilience Assessment and Evaluation of Computing Systems*,
DOI: 10.1007/978-3-642-29032-9_2, © Springer-Verlag Berlin Heidelberg 2012

MANETs before presenting a framework which aims to study both the benefits of increased awareness between software layers on the same node (inter-layer awareness) and between the same layer over different nodes (intra-layer awareness), and the combination of new/existing protocols designed with a cross-layer criterion.

2.1 Introduction

Modern mobile networks are being engineered to support applications with very different quality of service requirements for performance, reliability, and security. For example, some of these applications require support for different quality of service types: voice over IP, video streams generated from cameras and/or from video servers, real-time short messages for real-time communication, reliable message delivery mechanisms to support distributed systems middleware, and best effort message delivery supporting notification, or e-mail service.

The frame sizes, traffic shape, and quality of service/resilience requirements of these applications are very different. Mobility and wireless transmission protocols introduce additional challenges for supporting these quality of service/resilience requirements, such as, impact of power transmission on mobile network performance, interference from hidden terminals, shadows, cost of wireless licenses, and interference when using unlicensed bands. In addition, wireless roaming from access point to access point require resources (e.g., channel allocation) and time (e.g., allocation of new IP address) that can impact the quality of service offered by the network infrastructure to the application.

Therefore, understanding the domain of applicability of different wireless network protocols and developing new wireless protocols supporting broadband have been some of the major areas of activity of the IEEE 802 LAN/MAN standards committee (http://www.ieee802.org). The detailed specifications of IEEE 802.11 and IEEE 802.16 can be obtained from IEEE at http://www.ieee802.org/11 and at http://www.ieee802.org/16.

The ubiquitity of the low cost 802.11g, WiFi, at 54 mbps rates has motivated many users to consider deploying multi-media applications supporting mission-critical systems, without consideration to the ability of WiFi of supporting traffic shaping, priority, and quality of service of shared voice and video applications. However, 801.11g has no provisions to enforce priorities among data streams, which makes it very challenging to support the required quality of service requirement for voice and video over a mobile 802.11g network.

In contrast, 802.11P defines eight priorities classes for data streams originating from a given station. The priority hierarchy is to give top priority to voice data streams, followed by video stream, voice control streams, best effort streams, transaction data, and bulk data. These eight priorities are aggregated into four major priority classes: voice, video, best effort, and background. 802.11P priority schemes cannot guarantee quality of service requirements of the different data streams, because of the lack of central bandwidth control, as 802.11P is based on multiple access-collision avoidance (CSMA-CA) algorithm.

802.16, WiMAX, *Worldwide Interoperability for Microwave Access*, is an IEEE standard that was designed to provide broadband access, while guaranteeing the contracted quality of service requirements. The 802.16 standard is a family supporting several design options for the physical layer, media access layer, frequency bands, etc. 802.16 was initially designed to support broadband using fixed wireless access. However, it was extended to support mobility. WiMax is able to enforce quality of service guarantees because the WiMax base station has as one of its architecture features the task of managing and controlling all connections under its supervision. The performance characteristics of each connection can be managed through bandwidth allocation per user, variable size frames, and frame aggregation/deaggregation from different connections to improve overall system performance. WiMax defines five quality of service classes that can support applications that are sensitive to delay and jitter, like voice over IP. However, WiMax based mobile networks can also suffer from roaming delays and lack of resources if the network is not carefully engineered to support the quality of service and scale of the application to be deployed. For example, some of the handover mechanisms supported by WiMax may introduce handover delays of up to 1 s, which is unacceptably high for mission-critical applications.

Therefore, requirements specifications, architecture, modelling, design, and testing of the application ecosystem that will be sharing the wireless mobile environment is the recommended engineering approach to ensure that the required quality of service, resilience, and scale is provided to customers.

In this chapter we illustrate some of the needs and challenges for resilience assessment in mobile networks in three different domains: mobile cellular networks, sensor networks and mobile ad-hoc networks.

In Sect. 2.2 we describe mobility management and time-synchronisation in cellular networks. In Sect. 2.3 we describe the software design and deployment support provided by sensor networks. In Sect. 2.4 we describe cross-layer protocols support for mobile ad-hoc networks. Section 2.5 contains our conclusions.

2.2 Mobile Cellular Networks

Following the development of electronic and computer technologies, new multimedia applications such as real-time conversational video conferences and interactive video games, have been developed. These applications which are, by nature, bandwidth consumers are made available over mobile networks that have finite resources and may experience some end-to-end transmission delays.

The advent of new mobile technologies, like long term evolution (LTE) in the side of 3rd generation partnership project (3GPP) and IEEE 802.16m in WiMAX forum's side introduce new access technologies that offer a high bandwidth rate and reduce both transmission delays and error ratio. However, the call dropping rate, which is an important quality of service (QoS) index in mobile networks, mostly related to the users mobility, remains a challenging issue. The changes of radio channel during movement of mobile users between network cells, namely handovers, impose short sessions disconnection. Unfortunately, for some classes of traffic such as multimedia applications, handovers are not transparent to mobile subscribers and the continuity of users session are not guaranteed. Consequently, offering these services is not an easy task to achieve and remains the current challenge in mobile networking.

In the following we focus on the network failures due to the handover procedure. We first give more details on the notion of handover in mobile cellular networks and present the current solutions to enable seamless handover in the existing networks technologies.

2.2.1 Handover and the Quality of Service

The handover is defined as the change of radio channel used by a wireless terminal. The new radio channel can be either with the same base station (intracell handover) or with a new base station (intercell handover). In the case of intercell handover, for example, where the subscriber crosses cell boundaries and moves to an adjacent cell while the call is in progress, the call must be handed off to the new cell in order to provide uninterrupted service to the mobile subscriber. If the new cell does not have enough channels to support the handover, or if the session disconnection between the mobile user and the old base station lasts longer than a critical time while the connection with the new base station is not established yet, the call is dropped.

Globally, the handover procedure is the same in all the existing mobile technologies but the network signalling and the involved network entities differ [97]. The handover procedure consists of three main phases: the measurement, the decision and the execution phases. During the measurement phase, the mobile terminal scans the received signals from all the neighbouring cells and a measurement report is established. This report is then used to make the handover decision and to select the next cell to which the mobile terminal will be connected. In the case of LTE architecture [317], the handover decision is made directly by the base station, that is the evolved node B (eNB). However, in UMTS and WiMAX technologies, higher entities are responsible of this task. In UMTS [899], the decision is made by the radio network controller (RNC), whereas the access serving network gateway (ASN-GW) is responsible for this decision in WiMAX [4].

Once the handover decision is made and the handover request is accepted by the target cell, the execution phase is initiated. During this phase, the admission control and required resources configuration are performed in the target cell. The user terminal detaches from the source cell and synchronises with the new serving cell.

The handover procedure is considered as a *break-before-make* approach, that is the connection of a mobile terminal with its serving cell is interrupted before being connected to a new cell. Failures due to the handover procedure are more noticeable by a mobile subscriber in urban areas because of the high coverage density;

the number of cells in the neighbourhood of each cell is very important. Thus, for a given cell, measuring the signal quality of all its neighbouring cells, during the measurement phase of the handover procedure, may be time-consuming. In these conditions, respecting the QoS requirements of real-time services, such as real-time conversational audio services, may become impossible. The result is either the degradation of the offered service quality or simply the communication dropping. Thus, the handover procedure has an important effect on the performances of the system and the probability of forced communication termination must be limited because from the point of view of a mobile user, forced termination of an ongoing communication is less acceptable than blocking a new one.

2.2.2 Handover Failure Resilience

In order to reduce the failure rate of the handover procedure and guarantee a certain service continuity, several solutions have been proposed [97]. One solution to enable soft handover consists in maintaining a simultaneous connection between the mobile terminal and a set of selected serving base stations called the *diversity set*. The selection of the base stations to be included in the set is based on the signal quality criteria. This solution, which is known as the macro diversity handover (MDHO) [4], has been proposed in the context of WiMAX. A similar idea is used in the fast base station switching (FBSS) [4], another solution proposed in the context of WiMAX architecture. However, in this solution, the mobile terminal is connected to a single base station from the diversity set, during a frame period. If the signal quality is not good enough to maintain an acceptable connection, the mobile terminal may change its serving base station from the diversity set.

In both MDHO and FBSS, continuous measurements should be made in order to maintain the diversity set. These measurements result in an important number of control messages exchanged between the user terminal, the base stations in the diversity set, and ASN-GW for the authentication and registration procedures. Moreover, maintaining simultaneous connections between one user terminal and several base stations increases the blocking probability of the new calls.

In order to speed up the handover procedure and thus reduce the network failure rate in the context of LTE architecture, a new interface, called $X2$, between the base stations (eNBs) has been introduced [317]. This interface is dedicated to signalisation message exchanges and data forwarding during the handover procedure. Such an interface allows improving the execution phase of the handover and the communications between eNBs. However, there is still a need of regular signal quality measurements and this introduces delays in the mobile communications.

Mobility prediction, if well performed, may constitute the solution to enable seamless handovers and reduce the call dropping rate. It anticipates the preparation of the handover in the next cell to be visited by a mobile user, avoiding any disconnection of the mobile terminal from its current serving cell before it gets connected to the new cell (*make-before-break* approach). The efficiency of a mobility prediction model relies on both the model itself, and the network and the mobile user data used in this model. Among the data used in the different mobility prediction approaches developed, we have the mobility history of the user, and the signal strength. When a prediction model fails to predict the next cell to be likely visited by the mobile the handover multicast is the ultimate solution to enable seamless handovers. In this case, instead of one cell, a group of cells is predicted. The network prepares, in advance, the handover of user to all the cells belonging to this group. A mobility prediction approach is presented in Chap. 17.

2.2.3 Resilient Network Services

Computer networks provide a large and diverse set of services. When considering mobile cellular networks, as done in the previous sections, issues such as handover of calls can be resolved reliably only if the backhaul network is dependable and highly resilient. Resilience of the network is a challenge since mobile backhaul networks carry not only traffic necessary to operate the cellular network but also different types of user initiated load. From the perspective of network operation, the network payload can be considered background load that causes impairments on the network management packets travelling through the backhaul network.

Time and frequency synchronisation can be considered a service operated in the backhaul network that is especially relevant for seamless handover of calls. In modern packet-switched networks time and frequency synchronisation is no longer achieved using dedicated wires but rather through efficient and reliable protocols. Such protocols rely on time stamps transmitted with dedicated small data packets and hence they are sensitive to poor network conditions. High load causes delays, and, even worse, increases the variance in transmission delays. To counteract such problems, the resilient network must either avoid situations of high load or offer mechanisms that allow the proper operation of the time and frequency synchronisation service in the presence of disturbances.

High accuracy requirements therefore seem to translate into the demand for a highly resilient network. The precision time protocol (PTP) according to the standard IEEE 1588 has been designed to achieve extremely precise time and frequency synchronisation even in an unresilient network. The main features of the protocol are simple and straightforward. Accuracy is implemented using high priority for small timing packets versus low priority for all background traffic, i.e., network payload. Timing packets are usually sent at a rate of 32 packets per second, assuming that they will not notably load the network. An interesting and important question is whether these priorities can assure resilience of the time and frequency synchronisation service when the underlying network is not resilient. According to statements from industry, the experimental analysis of the protocol is difficult and expensive, since emulation of realistic conditions requires considerable effort and the necessary oscillators and measurement equipment are very expensive and exist only in a few experimental laboratories world-wide. But the question can be addressed using model-based techniques. However, since a high level of detail in the model is required, analytical models are not suitable and even the simulation models exhibit extremely long runtimes. This demands for suitable approximation methods as illustrated in Chap. 6. As will be shown in the case study in Chap. 17, detailed simulation studies reveal that the desired accuracy of frequency synchronisation in an unresilient network can only be achieved with suitable network or protocol design.

2.3 Sensor Networks

The potential of model-driven engineering (MDE) techniques in software development has been largely demonstrated in many application domains [537], whereas model-driven approaches have started to appear only recently in the wireless sensor networks (WSN) domain [359, 382]. This is probably related to the fact that software (and model-driven) engineering experts have addressed their interest to this domain since short time, although it is evident the key role that software plays in WSN.

For example, the battery consumption of sensor nodes is one of the most critical problems, so the nodes have to be managed by "energy saving" algorithms and protocols. A badly designed algorithm or protocol could inhibit the use of the whole WSN. As another example, the network stack is not immutable and independent from the applications, as in traditional distributed systems. It is partly integrated in the WSN software applications, and strictly linked to the network topology.

In addition, "code-and-fix" approaches, that are mostly followed for the WSNs software development, limit the maintainability and reusability of software and represent a barrier towards the applicability of (model-based) non-functional analysis approaches. In [730] the need of adopting a standard process to drive the development of the WSNs software is strongly highlighted, as well as the fact that this goal can be naturally achieved with the application of software engineering and model-driven engineering techniques.

Sensors are devices with limited resource capabilities, although in many cases quite complex computational tasks are assigned to them. Also, communication and coordination are notoriously critical tasks in WSN, so they have to be efficiently managed. In addition, WSN are often utilised for applications with strict non-functional requirements (e.g., eHealth, alarm systems, etc.). All these motivations claim for solid approaches to the modelling and analysis of non-functional attributes of WSN software applications.

We retain that model-driven engineering techniques can be very helpful for this goal in the WSN context, as well as they do in other contexts [83]. Nevertheless, still very few approaches have appeared to tackle non-functional issues in WSN with model-based techniques [876]. Whereas in the functional world model-based approaches are typically aimed at automated code generation, model checking, model-based testing, etc., in the non-functional world a typical benefit of model-based approaches is to enable model transformations that can generate non-functional models (e.g., for reliability, performance, etc.) conforming to the original design model.

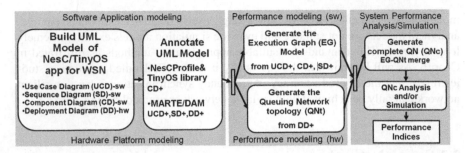

Fig. 2.1 The proposed approach

The automated generation of non-functional models undertakes the most common obstacles to the adoption of non-functional validation of software, that are: (i) the extra-time needed to build by scratch a non-functional model, and (ii) the special skills needed to accurately do this task (i.e., to build "by scratch" a model that conforms to the design model).

In this section we briefly describe a framework for model-based design and performance analysis of software applications for WSN. It makes use of UML design models extended with a UML profile strongly inspired to the NesC programming language. Besides, thanks to MARTE [1] (i.e., the UML standard profile for modelling performance and other non-functional elements) the UML models are further annotated with performance parameters. Following the experience and the results achieved for general purpose [239] and context-aware software applications [108], transformations can be applied to this augmented UML model to generate performance models. Finally, the performance model solution provides the indices of interest.

Our approach is depicted in Fig. 2.1. The UML model comprises two views:

- a **structural view** includes the software architecture in terms of components and connectors, and the hardware platform where such components are deployed;
- a **dynamic view** illustrates the system functionalities as a set of component-based interactions.

We use a reduced set of UML modelling elements or *language packages*: *components* and *deployments* for the structural view, *use cases* and *interactions* for the dynamic view. Such elements are then depicted on their appropriate UML diagrams, as follows.

The component diagram includes components, ports and connectors. A port specify a distinct interaction point that contains all the interfaces a component offers to/requires from its environment. Then a connector wires ports by establishing a correspondence between provided and required interfaces. The component diagram in our approach comprises both NesC/TinyOS-based components and generic components.

---

[1] http://www.omgmarte.org/

The deployment diagram describes the hardware platform where the executables are deployed. The DD of a wireless sensor network should represent *motes* including their specific hardware components (e.g., sensors, radio module, CPU, battery and timer). Similarly to CD, the DD also comprises other generic execution hosts and communication networks.

Besides a structural description, the system functionalities have been represented as component-based interactions in the dynamic view. Each sequence diagram (SD) specifies the behaviour of a system functionality.

Beside this, UML provides an extension mechanism, named *profiling*, to tailor it to specific domains. Domain specific elements are introduced through the definition of *stereotypes*, *tagged-values* and *constraints*. Our framework exploits this mechanism to combine three different domains: (i) the software modelling of NecS/TinyOS applications, (ii) the hardware modelling of WSNs and (iii) the performance modelling.

The *UML profile for modelling and analysis of real-time and embedded systems* (MARTE) is used to represent the latter two domains. It is an OMG standard profile that specialises UML by adding domain-specific concepts to support the modelling and analysis of real-time and embedded systems like WSNs are. It is organised in several related sub-profiles that can be individually tailored for specific purposes. In particular, a working collection of MARTE sub-profiles that fits our modelling and analysis needs includes:

- the *general resource model* (*GRM*) suitably specialised for static and dynamic modelling of *software* (*SRM*) and *hardware* (*HRM*) resources;
- the *generic quantitative analysis model* (*GQAM*), that supports the representation of hw/sw resource usages by the system, and different sub-profiles devoted to specific non-functional analyses (e.g., *PAM* for *performance*);
- the textual *value specification language* for the annotation of quantitative and qualitative tagged values of non-functional properties (*NFP*).

Then we devised a new UML profile, named NesC-WSN, to suitably specialise UML for the modelling of (i) NesC/TinyOS-based applications through the *software architecture model* sub-profile and (ii) hardware through the *hardware platform model* sub-profile (see Fig. 2.2).

The former sub-profile is used for annotating both the structural (CD, DD) and dynamic views (SD) of NesC [371] applications. The latter introduces few WSN-specific hardware stereotypes, like «mote» and «nodeGroup», for distinguishing them from generic nodes.

In particular the NesC-WSN profile includes two model libraries, *TinyOS* and *motes*. They contain a set of ready-to-use stereotyped model elements that are intended to be imported in any NesC/TinyOS annotated UML models (similarly to classes of a software library in some programming languages) to speed up the modelling activity. In particular *TinyOS* includes «configuration» and «module» components whereas *motes* contains a ready-to-use set of «mote» nodes.

Table 2.1 summarises the envisaged mappings between the NesC keywords and WSN concepts on the one side, and UML metaclasses on the other side: whenever

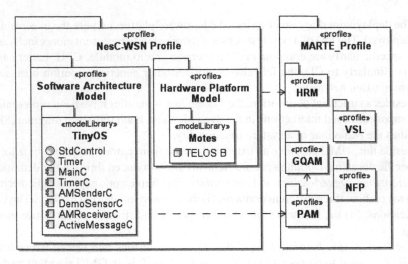

Fig. 2.2 NesC-WSN and MARTE profiles

domain specific elements cannot be clearly identified in UML "as is", a new stereo-type has been devised. For example «command» and «event» are used for distinguishing NesC-specific from generic operations as well as «mote» and «sensor» are used for characterising WSN specific hardware elements.

The peculiarity of our NesC-WSN profile is that it depends on MARTE (see Fig. 2.2): the latter is always required and applied to a UML Model whenever our new profile is used. Indeed the rightmost column of Table 2.1 contains the MARTE stereotypes that should be applied together with the corresponding NesC-WSN ones. Such a combination results in a UML Model that is ready to be transformed in a queueing network [239].

The performance analysis we base on requires the following additional information to be annotated on sequence and deployment diagrams:

- The workload («GaWorkloadEvent») for each functionality.
- The resource demand vector («GaAcqStep») that represents the amount of *resources* (e.g., the number of CPU *instr*uctions executed or the amount of data transferred) that a NesC call requires to be completed.
- The multiplicity (*resMult* tag) and performance capabilities of hardware resources such as «motes» and «hwProcessors».

Moreover the stereotypes of the *hardware platform model* sub-profile are applied on the DD: (i) to logically group nodes performing similar tasks («nodeGroup»), (ii) to identify and quantify the number of «mote»nodes and their inner hardware components (e.g., «sensor»).

Finally, NesC/TinyOS components are annotated on the CD. Their operations are «command»that directly implement the provided interfaces («module») or delegate the implementation to wired components («configuration»).

Table 2.1 NesC-WSN, UML and MARTE mappings

Keywords	UML	NesC-WSN profile	Marte profile
	Base class	Stereotype	Stereotype
Software architecture model			
Call	Message	«call»	«GaAcqStep»
Command	Operation	«command»	«GaStep»
Configuration	Component	«configuration»	
Event	Operation	«event»	«GaStep»
Module	Component	«module»	
Hardware platform model			
Node group	Node	«nodeGroup»	«resource»
Mote	Node	«mote»	«HwReosurce»
Sensor	Node	«sensor»	«HwI/O»

The same operations are invoked in the SD. Since a «call» is also a «GaAcqStep»in UML MARTE, the amount of resources that need to be acquired by the receiving component to execute the «call» can be annotated on each of them.

Once the UML model has been annotated, execution graphs can be obtained from the annotated component and sequence diagrams following PRIMAUML [239]. An execution graph (EG) is a model that represents the software dynamics along with their resource demands.

The generation of the QN model consists instead of two steps (Fig. 2.1):

1. A QN topology (*QNt*) is firstly obtained from the annotated DD: indeed the service centers of the QN represent CPUs and disks whereas network connections are represented by delay centers.
2. A complete QN (*QNc*) is obtained by adding the workload and a set of service requests for each class of jobs. In [837] an algorithm is provided to transform the resource requests of an EG into service time requests to QN service centers.

Performance indices of interest can be obtained from analytical solution or simulation of a complete QN.

In conclusion, the WSN domain presents very suitable aspects for a successful application of such techniques. The WSN software applications are quite complex because they run on heterogeneous (and sometime open) platforms, but from a modelling viewpoint only the components running on sensor nodes and the surrounding devices are actually new. When the elaboration reaches canonical network nodes (such as gateways, servers, etc.), performance modelling can be based on existing well-assessed techniques. Hence, the introduction of modelling instruments to represent, let say, the NesC part of WSN applications can be sufficient to enable the realization of models of the whole system architecture. Being limited the world to be modelled, it seems not too ambitious to envision a library of WSN-related modelling bricks, each representing a specific element type of a WSN, that can be instantiated and composed in larger models of WSN applications. The same applies to intermediate layers of the network stack, such as WSN middleware.

2.4 Mobile Ad Hoc Networks

A mobile ad hoc network (MANET) is a network made of wireless devices, also referred as nodes. A device can join a MANET and be part of it until it falls within the radio range of one o more devices belonging to the MANET. Each device can move at variable speed depending on the transportation mean that carries it. No infrastructure is assumed in MANETs, both physical (e.g., firewalls, routers, etc.) and/or logical (e.g., Certification Authorities, addressing/name assignment authorities, etc.).

Due to absence of physical infrastructure, each node can communicate directly with its neighbours (i.e., devices that are in the node radio range) and needs the collaboration of other nodes to carries its messages to non-neighbour destinations. Furthermore, the lack of a logical infrastructure devoted to node identity, both in assignment and control, makes the MANET nodes essentially anonymous. Finally the nodes can be resource-limited in terms of CPU power, wireless transmission range, power unit life, etc.

Hence MANETs suffer of several problems mainly related to security weakness, wireless communications, and network topology change. Typical approaches to these issues either are based on relaxing some specific constraints of MANETs (e.g., they assume the existence of a trusted certification authority), or they show successful results only for limited ranges of their parameters (such as density, speed and number of nodes). From a software viewpoint, most approaches aim at reusing in a wireless environment the software interfaces of the TCP/IP stack layers used in a wired environment. They basically tackle the above issues by modifying the internal mechanisms of certain layers.

Cross-layer protocols have been introduced to address the above problems in MANET, and they lay on a common design idea: the exchange of additional information between layers enables better strategies due to an increase of awareness among layers.

This idea is not new by itself, because several existing approaches exploit awareness between network and transport layers [861, 862, 953, 961]. However in these approaches the TCP protocols always play a passive role in that they are explicitly notified about the network events from the routing protocols, like path losses, traffic congestion, packet salvaged, etc. More recently, cross-layer design has been applied between network and MAC layers, where more evident performance improvements can be achieved (see, for example, [140]).

In this section we briefly introduce a framework aimed to study not only the benefits of increased awareness between software layers on the same node (inter-layer awareness) and between the same layer over different nodes (intra-layer awareness), but also the combination of new/existing protocols designed with a cross-layer criterion.

Although intended to address performance issues, this framework is also based on considerations among the MANET availability. For availability we intend that each node, regardless its resource capabilities, would have some probability of success in joining any existent MANET. MANETs composed by nodes running the same routing and transport protocol, that agree on some security mechanism are out of our design due to two reasons: (i) at each software level it does not exist the optimal solution but only solutions that work well in a fixed range of nodes mobility, nodes resource capability, network size, etc.; all parameters that can appreciably vary on different MANETs, on different areas of the same MANET and on the same one area during the MANET life; (ii) nodes, perhaps constrained by its resource capabilities, that cannot adopt an unique, and not necessarily optimal, MANET strategy could not join to it. Furthermore we remove the assumption of the existence of an out of line authority that assigns identities to nodes, because this can be a particular instance of a MANET[2] but this assumption is unrealistic in general. So in our view the nodes are essentially anonymous and the distributed solutions (IDS and/or *reputation-based control*) loose most of their effectiveness.

Apart the above assumptions, cross layer projects are natural approaches in the MANET environment, also if the stack layer interfaces have to be modified with respect to the well-assessed standards used in wired networks. Such augmented intra-layer awareness brings relevant advantages in MANETs, where very often an event can be detected at some layer and profitably consumed in a different layer, as illustrated in the following examples.

First, in order to acquire more channel capacity, a node can break near communication paths by forging and sending a message with a source MAC address of a victim node. Only the MAC layer of the victim node can detect this event, whereas it may be useful at the network layer to be aware of this event in order to support finer security policies (e.g., the victim node could start a key exchange with neighbours for sake of authentication).

Second, a node, through a violation of the MAC protocol, can carry a rough jam attack and destroy all communication around it for a radius depending on the power of its antenna. Even worse it can jam only packets containing transport data and pass through packets devoted to search or update paths. Similarly to the previous example, only the MAC layers of victims have the capability to detect this event, but the only solution would be to move the victims out of the radio influence of attacker. In order to implement such solution this event awareness should reach the user of the victim node.

Third, the paths established between two peers in a MANET are (by definition) time limited. One case of path removal is due to node mobility that can break the link connection between two adjacent nodes, thus causing all paths containing this link to become useless. If the MAC layer informs the network layer about the failure of packets transmission toward its neighbours (i.e., a RTS/CTS failure in the MAC protocol) the latter can decide if the link, and all paths containing it, have to be considered useless. Then the network layer can try to repair the paths and/or notify the event to the interested peers. Furthermore, the data link will keep trying to send packets in its queue across the broken link, so wasting the channel capacity and delaying other packets that have a higher probability to be successfully transmitted.

---

[2] Such as a MANET conference deployed in absence of infrastructure, where each newcomer gets its certificate in the conference secretariat before joining it.

An enhanced awareness between data link and network layers would allow to drop these packets or suspend their transmission while waiting for a new next hop in the path.

A diverse type of augmented awareness regards information exchanged by the same layer deployed on different nodes (inter-layer approach). An inter-layer approach can be "local" when the information is directly exchanged between couples of nodes, as examples in [818], where queue information is exchanged by the MAC layers of neighbour nodes in order to decouple the single channel constraint in multiradio option. A deeper inter-layer approach regards information generated at some point of network and carried toward other nodes in distributed IDS (es. [752]) or in routings protocols generally during the maintenance and repair phase. This latter information is subject to modification message attack and needs strong security mechanisms.

Figure 2.3 schematically shows our framework. Transport and routing layers are equipped with a certain number of modules and a manager. A module can be devoted to implement: (i) a specific protocol (e.g., the transport protocol X), (ii) a stub protocol, that is a procedure that decides how to process messages sent with a protocol that is not currently deployed on the node, (iii) a certain type of decisional policy (e.g., a policy to decide when discarding a certain path), or (iv) a simple software connector between different modules.[3] Each manager coordinates the operations within its layer.

Other frameworks have been developed in order to allow cross-layer interactions in MANET. In Mobileman framework [237] cross-layer interaction is implemented through a shared area. At this purpose a vertical component is added to the classical stack, where all protocols layer (i.e., MAC, network, transport, middleware, application) can accede. This approach allows to add or remove protocols belonging to different layer without modifying other layers. In MANETKit [760] the authors claim that "no single protocol is well suited to a subset of operating conditions to be found in any given MANET environment at any given time", thus this framework permits to hold more routing protocols and to switch from each others at run time. Other frameworks have the simple goal of simplifying the protocol development and testing on real systems (e.g., [169, 525]). At this purpose they offer the set of APIs necessary to running the routing protocol in user space. This framework entails all stack layers and each module is allocated in its appropriate layer, instead of being constructed on the top of the application layer.

2.5 Conclusion

Following the development of electronic and computer technologies, new applications such as real-time conversational video conferences and interactive video games,

---

[3] In the figure only the first two types of modules have been shown, because they represent the main focus of our work.

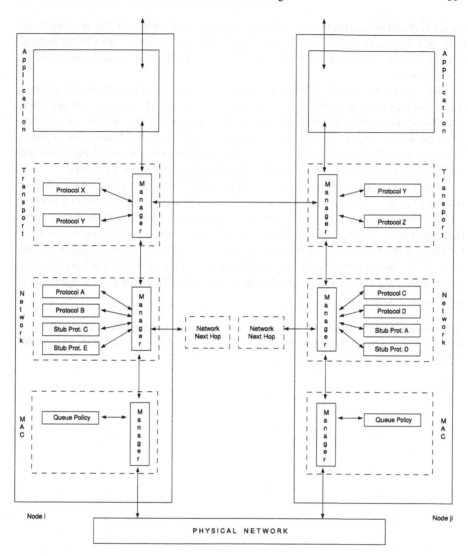

Fig. 2.3 The structure of our framework

have been developed. These applications which are, by nature, bandwidth consumers are made available over mobile networks that have finite resources and may experience some end-to-end transmission delays.

In this chapter, we have described the most important network protocols supporting modern applications in mobile cellular networks, wireless sensor networks (WSN) and mobile ad hoc networks (MANETs). We first focussed on the handover procedure in mobile cellular networks and the network failures due to this break-before-make approach as during a handover the connection of a mobile terminal with its serving cell is interrupted before being connected to a new cell. We presented the current solutions to enable seamless handover in the existing networks technologies, pointing out the importance of a dependable and highly resilient backhaul network to this issue.

Sensors are devices with limited resource capabilities, although in many cases quite complex computational tasks are assigned to them. Moreover, WSN are often used for applications with strict non-functional requirements. These networks require solid modelling approaches for non-functional attributes of WSN software applications. Thus, in the second part of this chapter, a framework for model-based design and performance analysis of software applications for WSN was presented.

Due to absence of physical infrastructure, MANETs suffer of several problems mainly related to security weakness, wireless communications, and network topology change. Typical approaches and cross-layer protocols to address these issues were discussed in the last part of this chapter before presenting a framework which aims to study both the benefits of increased awareness between software layers on the same node (inter-layer awareness) and between the same layer over different nodes (intra-layer awareness), and the combination of new/existing protocols designed with a cross-layer criterion.

Acknowledgments Leïla Kloul is supported by the European Celtic project HOMESNET [8]. Katinka Wolter is partly supported by the German Research Council under grant number Wo 898/3-1.

Chapter 3
Assessing Dependability and Resilience in Critical Infrastructures: Challenges and Opportunities

Alberto Avritzer, Felicita Di Giandomenico, Anne Remke and Martin Riedl

Abstract Critical infrastructures (CI) are very complex and highly interdependent systems, networks and assets that provide essential services in our daily life. Most CI are either built upon or monitored and controlled by vulnerable information and communication technology (ICT) systems. Critical infrastructures are highly interconnected systems and often use common ICT components and networks. Therefore, cascading faults and failures are likely events in critical infrastructures. Moreover, such failures can easily spread to other infrastructures and can possibly span to other countries or even continents. Assessing resilience is thus a cornerstone for improving the dependability in critical infrastructures. Due to the complexity and interdependency of such systems many different challenges and opportunities surface when developing methods and tools for resilience assessment. During the last decade both academia and industry developed an increased interest in this research area and a variety of projects with different focus started to emerge. This chapter gives an overview about the main requirements for resilience assessment and discusses the state of the art and emerging research directions. To exemplify the diversity of this

A. Avritzer (✉)
Siemens Corporate Research and Technology,
755 College Road East, Princeton, NJ 08540, USA
e-mail: alberto.avritzer@siemens.com

F. Di Giandomenico
ISTI Department, Italian National Research Council,
via Moruzzi 1, 56124 Pisa, Italy
e-mail: digiandomenico@isti.cnr.it

A. Remke
University of Twente,
Enschede, The Netherlands
e-mail: anne@cs.utwente.nl

M. Riedl
Universität der Bundeswehr München,
Neubiberg, Germany
e-mail: martin.riedl@unibw.de

K. Wolter et al. (eds.), *Resilience Assessment and Evaluation of Computing Systems*,
DOI: 10.1007/978-3-642-29032-9_3, © Springer-Verlag Berlin Heidelberg 2012

research area a special focus is put on different sub-fields with increasing granularity from the fairly general interdependency modeling to the reliability modeling of a Smart-Grid distributed automation network.

3.1 Introduction

More and more, our society and economy rely on the well-operation of a number of infrastructures, which regulate critical processes and provide vital services. Critical infrastructures (CI) span a number of critical sectors, including: public health and safety, energy, water, information and telecommunications, emergency services, agriculture and food, transportation, banking and finance, government, and many others. Over the last 10–15 years, the role of information and communication technology (ICT) in society has dramatically changed. Where some 15 years ago, ICT supported some business and stand-alone office processes, ICT now forms the heart of these processes. Moreover, ICT now plays an important role in most processes and services in our economy, however, often hidden behind a non-ICT-like user interface. A key example of the latter is the role of ICT in critical infrastructures. The non-functioning of critical infrastructures has a vast impact on economic and social welfare. Hence, for such infrastructures it is essential to be resilient against faults and failures and to survive catastrophic events.

This chapter focuses on ICT-based critical infrastructures from the point of view of their dependability and resilience assessment. As reported in the "critical infrastructure resilience final report and recommendations" produced by the US NIAC [677], resilience has become an important dimension of the critical infrastructure protection mission, and a key element of the value proposition for partnership with the government because it recognizes both the need for security and the reliability of business operations. Although each critical infrastructure sector operates differently, a common definition of infrastructure resilience is needed for public policies and governance to be effective. Toward this end, the NIAC has developed the following definition based on discussions with executives and security experts across many sectors: *infrastructure resilience is the ability to reduce the magnitude and/or duration of disruptive events. The effectiveness of a resilient infrastructure or enterprise depends upon its ability to anticipate, absorb, adapt to, and/or rapidly recover from a potentially disruptive event.*

Research has shown that many critical infrastructures are interrelated. Especially ICT infrastructures play an important role in the vitality of other infrastructures, e.g., the transport infrastructure does not operate properly without a reliable ICT infrastructure. These dependencies can cause cascades of failures that start with simple defects in one type of system, and may finally lead to disasters in other infrastructures. Therefore, there is a growing need to analyze the chains of influence that cross multiple sectors and that can induce potentially unforeseen secondary effects. This reinforces the importance to consider dependability and resilience as a component of critical infrastructure protection strategy and to devise appropriate methodologies and techniques to promote its analysis and assessment.

This chapter on critical infrastructures points out the requirements for resilience assessment of this challenging and crucial sector, pointing out relevant studies already performed and indicating promising directions for future work. A more concrete approach to resilience analysis in the CI field is provided in the Chap. 18 in this book.

The outline of the remainder of the chapter is as follows. Section 3.2 discusses the requirements for resilience assessment of critical infrastructures. Section 3.3 presents an overview of the literature on critical infrastructures assessment approaches. A significant effort is nowadays devoted to interdependencies modeling and analysis. Section 3.4 reviews some relevant studies in this field. Section 3.5 focuses on how stochastic hybrid models can be used for the dependability evaluation of fluid critical infrastructures. This section provides an interesting example of the similar feature characteristics between critical infrastructures in different domains. Increasing the level of detail, Sect. 3.6 presents a concrete case study to illustrate a reliability assessment approach for a distributed automation smart-grid distribution network. Emerging directions for research in critical infrastructures are presented in Sect. 3.7.

3.2 Requirements for Resilience Assessment of Critical Infrastructures

Critical infrastructures are often controlled by a supervisory control and data analysis (SCADA) [857] system, which is potentially vulnerable to attacks and misuse. SCADA systems consist of sensors, actuators, controllers and a human-machine interface (HMI) through which human operators control the physical process. It is important to correctly capture interdependencies that arise between the SCADA network and the physical network, but also interdependencies between different critical infrastructures. Interdependency assessment is discussed in Sect. 3.2.1. Modeling formalisms have to be able to capture the complex nature of critical infrastructures; requirements with respect to, e.g., scalability, heterogeneity and compositionality are presented in Sect. 3.2.2. Measures that can be used to evaluate the resilience of such systems have to be defined in a sound and unambiguous way. Different types of evaluation are highlighted in Sect. 3.2.3. Possible faults range from the malfunctioning of SCADA components to cyber attacks and are summarized in Sect. 3.2.4.

3.2.1 Interdependencies

There is a consensus in the literature on critical infrastructures that interdependency analysis is of paramount importance to improve the resilience, survivability and security of these vital systems. An interdependency is a bidirectional relationship between two infrastructures through which the state of each infrastructure influences or is correlated to the state of the other [778]. Infrastructure interdependencies can be categorized according to various dimensions in order to facilitate their identification, understanding and analysis. Among the most important dimensions identified in [778] are: (a) the couplings among the infrastructures and their effects on their response behaviour (loose or tight, inflexible or adaptive), (b) the state of operation (normal, stressed, emergency, repair), and (c) the type of failure affecting the infrastructures (common-cause, cascading, escalating).

Interdependencies increase the vulnerability of the corresponding infrastructures as they give rise to multiple error propagation channels from one infrastructure to another that increase their exposure to accidental as well as to malicious threats. Consequently, the impact of component failures in critical infrastructures can be exacerbated due to interdependencies and the overall severity of a failure is generally much larger and more difficult to foresee, compared to failures confined to single infrastructures. As reported in [742], past major power grid blackouts have been initiated by a single event (or multiple related events such as an equipment failure of the power grid that is not properly handled by the SCADA system) that gradually led to cascading outages and eventually to the collapse of the entire system.

Analyzing interdependencies allows a greater understanding of the effects of failures. Three types of failures are of particular interest when analyzing interdependent infrastructures: (1) cascading failures, (2) escalating failures, and (3) common cause failures. Cascading failures occur when a failure in one infrastructure causes the failure of one or more component(s) in a second infrastructure. Escalating failures occur when an existing failure in one infrastructure exacerbates an independent failure in another infrastructure, increasing its severity or the time for recovery and restoration from this failure. Finally, common cause failures occur when two or more infrastructures are affected simultaneously because of some common cause.

3.2.2 Modeling Formalism

The large and complex nature of critical infrastructures with a multiplicity of interactions and types of interdependencies involved requires a very flexible compositional modeling framework that is able to accommodate different levels of abstraction. To analyze their safety and survivability in the presence of disasters advanced structuring, monitoring and assessment methods are necessary. From the modelling point of view, abstraction layers and modular, hierarchical and compositional approaches are viable directions to cope with these aspects. New model classes and languages are necessary to accurately describe the structure and behavioral dependencies in critical infrastructures. Which modeling methods are suitable for which infrastructure? Which are the crucial system issues to accurately model per infrastructure? Expert knowledge will be necessary to establish critical subsystems and sensible parameters settings; sensitivity analyses can be used to distinguish the crucial parameters, thereby

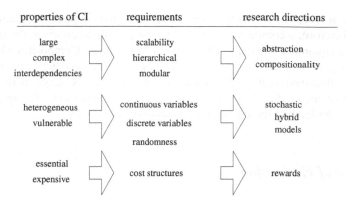

Fig. 3.1 Properties of critical infrastructures, requirements for modeling and emerging research directions

keeping the state space of the models as small as possible. As critical ICT infrastructures are very big systems, scalability is an important issue in modeling and analysis.

Figure 3.1 illustrates how the most prominent properties of critical infrastructures lead to certain modeling requirements and hence, to research directions that are expected to improve the state of the art in the field. Details are given in the following.

When modeling these complex systems [300] not all parameters and not all usage patterns are known exactly. Moreover specific details of vulnerabilities and failures will be unknown at design time, such as the mean time to failure and the impact of a given vulnerability. In such cases it is appropriate to make stochastic assumptions about the system and the disaster behavior. Hence, modeling formalisms that have been shown useful to model large-scale computer and communication systems, e.g., stochastic process algebra and stochastic Petri net, can be used to formally describe critical infrastructures, their inter-dependencies, and their cost structure.

The heterogeneity of typical critical infrastructures may require a combination of different formalisms/techniques to describe the various components of a system and their dependencies. For example, the combination of continuous and discrete phenomena may need to be captured in the modeling framework. Examples of discrete quantities are the number of spare parts and the state of sensors, actuators and ICT-components, whereas the continuous variables represent quantities, like the amount of produced energy, or the quality of treated water in terms of temperature and pressure. Hence, a modeling formalism is needed that allows describing both discrete and continuous quantities. Due to the flexible combination of discrete and continuous state components, Stochastic Hybrid Models (SHMs) can be a natural choice to accurately model both the process automation and the production process which is the essential part of several critical infrastructures.

The cooperation among subsystems of different nature inside the same Critical Infrastructure or among cooperating critical infrastructures requires advanced methods to reconcile different aspects under a common development and assessment framework. In this context, the studied infrastructures are assembled from many heterogeneous subsystems with different specifications, operation phases, and regimes. Therefore, a common framework has to be able to combine the different structural and quantitative aspects of critical infrastructures. Compositional modeling [141, 859] can simplify the modeling process and can lead to intuitive formalisms. Compositional analysis reduces the complexity of verification. Changes in the system then only affect the modified component and not the complete model. Compositional analysis is a challenging topic that requires additional research.

3.2.3 Type of Evaluation

The Evaluation of critical infrastructures has to address functional properties like inter-dependencies, deadlocks, etc., as well as extra-functional (quantitative) properties. As an example of the latter, what is the probability that 10 min after the occurrence of a given disaster, a basic service level is again available for 80 % of the user population? Not all infrastructures are critical and not all critical infrastructures have the same level of criticality. An evaluation process is required to identify vulnerabilities, interdependencies and interoperabilities between systems, to understand what specific assets of the addressed CI are utmost critical and need to be protected the most. Following this evaluation, steps can be taken to mitigate the identified vulnerabilities. For example, if an electric substation is damaged leading to a blackout, complications are experienced by a number of other systems/infrastructures and by the services they provide, like railroad operations causing a decreased movement of commodities and potential complications for emergency services. Thus, that electric substation must be protected not only for the Energy Sector, but also for the safeguarding of other sectors' infrastructure. Clearly, properties to be considered as indicators of the resilience of the Critical Infrastructure under study may vary significantly depending on the specificity of the targeted sector. The evaluation method should therefore be able to specify and assess resilience indicators according to the sector's needs, addressing both the interest of system designers and operators as well as users requesting services to the infrastructure.

Safety is defined as the absence of catastrophic consequences on the user(s) in their environment [63]. Safety analysis of vital infrastructures encompasses the classification of different types of disasters, according to their probability of occurrence and their effect on the controlled process; one can think of the degree of damage regarding time, space and monetary costs, in addition to the probability of cascading failures. Generally speaking, disasters can also be the result of malicious attacks or even terrorism. Thus, security of critical infrastructures is an important issue to deal with. Hence, combining the necessary expertise on network security (especially intrusion detection) and on system modeling and analysis is necessary to forecast the consequences of security attacks. This will help in finding the right counter measures and to develop recovery strategies.

Dependability has been defined as the ability to deliver service that can justifiably be trusted [63]. Given detailed models of the critical infrastructures, the dependabil-

Fig. 3.2 Building a more
resilient infrastructure comes
at a certain cost

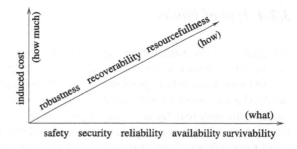

ity of the system can be evaluated with analytical techniques or using simulations. It is of utmost importance to use clear and formally defined notions of dependability. For example, survivability [300] is defined as a system's ability to recover predefined levels of service in a timely manner. Survivability evaluation then encompasses the evaluation of the probability (distribution) that predefined service levels are reached within a certain time, given the recent occurrence of a disaster of some form. Survivability evaluation deliberately only addresses the system recovery process; the process toward the disaster is explicitly not modeled, but taken as given. After the classification of disasters through safety analysis, a so-called Given Occurrence Of Disaster (GOOD) model of the system under study can be built and used for survivability analysis. Clearly, the recovery following a disaster highly depends on the type and extent of the disaster as well as on the affected infrastructure and built-in recovery mechanisms.

Other approaches from the field of dependable system design to achieve high-dependability (reliability and availability) can be useful in the context of critical infrastructures, such as implementing smart recovery strategies or introducing redundancy of some form [543]. As introduced in [677], it can be useful to divide resilience into *robustness*, *recoverability* and *resourcefulness*. Clearly, building a more resilient system comes at a higher price, so what is the relation between increased costs and increased resilience? Where does the point of diminishing returns lie? What is technically possible at which costs?

Figure 3.2 illustrates the relationship between a system's extra-functional requirements/properties, i.e., (**what**), the architecture features that are constructed to make the system more resilient, i.e., (**how**), and the associated costs to build a resilient system, i.e., (**how much**). As an example of this tradeoff, one might consider intrusion tolerance techniques and compare them to the objective of completely avoiding intrusions. Intrusion tolerance techniques are more likely to be successful in practice and may be less costly and more practical to implement, e.g. through redundancy. The implementation of a more practical approach increases the system's robustness and can be measured, e.g., in safety and security. Minimizing the time to recovery through smart repair schedules increases the recoverability of the system and will lead to higher availability and survivability. Comparing infrastructure designs alternatives with respect to their survivability and dependability will lead to more informed design decisions and hence, to more resourceful infrastructures.

3.2.4 Type of Faults

For each vital infrastructure a variety of possible disasters has to be considered. In case of a network infrastructure, a disaster can be a power outage, or it can be an explosion demolishing parts of the system. Both accidental and malicious faults need to be accounted for in the analysis. In previous decades, accidental threats were basically the only real threats facing infrastructure, especially natural disasters, which tend to be localized to one region and have a fixed and, at times, predictable duration. Until the bombing of the Murrah Federal Building in Oklahoma City in 1994, low attention was devoted to malicious acts targeting these critical components. In more recent years, preparation for Y2K (2000), fall-out from post-9/11 events, and a series of blackouts of the power systems experienced both in the US and in Europe have all reinforced the evidence of how vulnerable these systems are or can become to human attacks. Cyber attacks to the ICT systems that are controlling critical infrastructures are becoming more and more prominent. As an example, consider what happened in Australia [754] in 2001: a hacker broke into the network of a water treatment plant, opened an emergency valve and started to pump one million liters of raw sewage into the city parks. We could provide many other similar examples, such as the recent Stuxnet-worm [331, 520] which poses a serious thread to computers controlling industrial processes in the energy sector, or the Aurora attack on power generators [645] where the system could hurt itself via unauthorized SCADA commands.

A so-called threat or failure model [495] can be built, encompassing information on the type of failures that can be expected, their frequency, their duration and their intensity (e.g., computational strength). Because failures may be dependent on the system state, such a dependence has to be formulated as part of the model as well. Similarly, countermeasure models can be created, taking into account the incurred costs (monetary, or otherwise) of taking the countermeasure and its effect on the productivity of the infrastructure.

Heterogeneity also needs to be addressed at the level of vulnerability exposed by the different subsystems composing a critical infrastructure, e.g. the use of subsystems, such as Wireless SCADA, which are known to be typically vulnerable to error and misuse. In fact, advances in technology and SCADA systems have enhanced critical sector operations but created additional vulnerabilities, which must be addressed to adequately protect the critical infrastructure.

3.3 State of the Art in Resilience Assessment of CI: General Overview

The last decade saw significant research opportunities in resilience assessment of critical infrastructures. One of the important characteristics of critical infrastructure that contributes to its complexity is heterogeneity. Therefore, related work focuses on different aspects, such as the spatial distribution, interdependencies, uncertainty, non-linearity or hybrid systems. The evaluation is mainly focused on vulnerability, risk, and recoverability. It is performed using qualitative and quantitative assessment methodologies.

In the following we overview existing methods and techniques for CI assessment, providing a rough classification into qualitative and quantitative approaches. Due to the sheer amount of related work, completeness cannot be achieved - for more overview papers, please refer to [339, 376, 962].

Qualitative assessment approaches are discussed in Sect. 3.3.1 and quantitative approaches in Sect. 3.3.2. Section 3.3.3 gives an overview of ongoing projects in critical infrastructures.

3.3.1 Qualitative Assessment Approaches

Qualitative assessment approaches are descriptive in nature and aim at an in-depth understanding of the critical infrastructure under study. In the following, we discuss several approaches for qualitative vulnerability and risk assessment.

Qualitative vulnerability assessment can help to better understand the nature of vulnerabilities and to identify common causes of major failures. In the following we present two approaches to illustrate the application of qualitative vulnerability assessment to water cleaning systems and to electric power systems.

The Infrastructure Vulnerability Assessment Model (I-VAM) [328] proposes a qualitative treatment of the vulnerability of water cleaning systems. System experts have to establish value functions and weights and they have to assess protection measures of the system. Simulation (Monte-Carlo and Latin Hypercube) is used to analyze the sensitivity of the model and to obtain a vulnerability density function.

In [45] it is claimed that blackouts in electrical power systems are seldom caused by the failure of a single equipment but instead caused by cascading effects that cannot be predicted. There, structural vulnerability is classified into *node-* (e.g., substation) and *plain vulnerability* (e.g., transmission line). Moreover, different types of vulnerability indices are proposed for different kind of operating parameters based on over-limit information, regulating information, loss of load, and sensitivity analysis. The authors provide methods to calculate different kinds of indices to assess an electrical power system concerning its vulnerability after an out of service condition for a substation or a group of tie lines.

Qualitative risk assessment aims to identify the qualitative value of risk regarding a certain situation and threat under certain assumptions and uncertainties. Qualitative risk assessment approaches are often used to identify targeted threats, e.g., cyber-attacks. In the following we present a review of related work on inductive and deductive risk assessment, fault and attack trees, tableau-based and ontology-based approaches.

Inductive risk assessment methods, such as event tree analysis (ETA), start from certain observations (e.g., a particular hazard) within the system and try to inductively find its consequences. In [16] event tree analysis has been applied as a systematic

approach to investigate scenarios within a 3-step methodology. There, the consequences of the event of hurricane Hazel have been modeled. Each branch of the
tree leads from a prior event to more specific outcome. The elements rainfall, flood
and wind have been depicted in the first level, resulting in consequences concerning certain rivers or infrastructural damage. Based on the insights a flooding model
for the Humber river has been developed based on GIS information (i.e., digital
elevation maps). Moreover a knowledge base with interdependency relations has
been developed.

Deductive risk assessment methods on tree-based structures (e.g., fault-tree,
attack-tree effect, tree analysis, cause-consequence analysis) work in the opposite
way, i.e., from general to more specific. In [874] a new algorithm based on attack-
trees as simplified methodologies for impact analysis has been developed for the
evaluation of cyber-security incorporating password policies and port auditing. The
root node represents the ultimate goal (e.g., getting access) of an attack. Different
subgoals can be formulated as intermediate nodes that are combined over "AND" and
"OR" nodes. A leaf node can be an element of an intrusion scenario including defense
nodes as successors. First, cyber-security conditions are measured corresponding to
a number of assumed values reflecting the severity of vulnerability. Then, the attack
tree is formulated concerning the identified attack objectives, cyber-security conditions and countermeasures. Finally, the system vulnerability is derived from the
computed leaf vulnerabilities of the attack tree and the specific scenario associated
with the attack.

To incorporate both random failures and deliberate acts, [672] combines *fault-
trees and attack trees*. Fault-trees are traditionally used to calculate the reliability
of systems, whereas attack trees enrich the fault tree by malicious attack patterns.
Attacks can cause several types of events that are classified as basic, intermediate,
or top events, depending on the event position in the attack tree. Top events result in
an overall critical infrastructure failure. The authors argue that an attack tree goal is
equivalent to a fault tree event, where the difference lies in its origin (i.e., originated
by a malicious agent or random events). To use extended fault-trees, the measures
resulting from the attack trees must be quantified in terms of event probabilities. This
is done by deriving the probability of basic events from assumptions concerning the
motivation and resources of threat agents, environmental conditions and subjective
probabilities associated to the elements of the attack tree.

Often, simple *tableau-based* approaches are used for qualitative risk assessment.
For example, in [13] the FEMA defines a tableau based approach to mitigate terrorist
attacks against buildings. Global vulnerability is quantified according to a reference
scale. Asset values are assessed based on a parameters scaling between very low and
very high. Event profiles for terrorism and technological hazards are defined and a
matrix representation of the asset vulnerability is presented. The tools provided can
help decision makers decide which types of threats they want to counter after risk
assessment is made for each threat.

There is also work available that uses *ontology-based* decision support tools for
critical infrastructures, i.e., with underlying description logics. Ontologies provide a
way to represent domain knowledge in the form of classifications and relationships.

Automated deduction algorithms allow for reasoning about potential vulnerabilities and threats. In [211] such an ontology-based approach is taken as part of the INSPIRE project, which aims to increase security and protection through infrastructure resilience. Vulnerabilities due to the integration of IT and SCADA systems are explored together with connections, dependencies, and security aspects of SCADA systems (e.g., inherent vulnerabilities and effects). An OWL-DL ontology representing physical and logical assets, safeguards, threats, sources of attack, and vulnerabilities is described. These elements are connected through different types of relations (e.g., `Vulnerability` "is exploited by" `Threats`, `Asset` "has vulnerability" `Vulnerability`). Instances of certain CI infrastructures can be developed and queried using a query language by applying the underlying deduction algorithms. One example of a query language is SWRL. Questions such as "If I have resources A,B and C what kinds of attacks should I expect?" can be posed and answered by the implemented query mechanism.

3.3.2 Quantitative Assessment Approaches

Quantitative assessment uses measurable data to analyze and improve resilience or aims at computing quantitative performance measures, such as survivability, reliability, or efficiency. In the following, we review some quantitative assessment approaches using statistical analysis, stochastic models, and testbeds.

Statistical analysis. Critical infrastructure disruption events that cascaded across CI boundaries are examined in [612]. A *disruption event database* has been built as an empirical database, using publicly available data, which can be used for the understanding of cascading effects. It has been established that such effects mainly originate from energy and telecommunication. The dependencies are very focused and directional. The authors therefore question the domino theory since they only found very rare events which result in deep level cascades.

An extension to *data farming* is used in [280] to generate different types of network topologies. Simulation experiments with random and targeted terrorist attacks are performed. The results show that most networks start to fail when the number of attacks is larger than an empirically defined measure of node connectivity. Random network topologies seem not to be as robust as scale-free networks.

Stochastic models. An important formalism used in stochastic modeling is Stochastic Activity Networks [805]. This formalism has been applied in [115] as an approach for quantitative interdependency analysis in the context of large and complex CIs. The papers focus is on how the occurrence and size of cascades changes, when the strength of interdependencies is varied. The Möbius tool [225] is used to simulate the Stochastic Activity Network models using event-driven Monte-Carlo Simulation. The modeling process consists of determining the distribution of the cascade size, followed by an assessment of the impact of model abstraction and refinement on the quality of the results obtained in the analysis.

In [873] the risk of cyber attacks on the power system is calculated as the product of two factors: the probability of a successful intrusion and the impact of the intrusion as loss of power due to an unexpected loss of electric load. The two risk factors are evaluated by two separate techniques. The cyber layer underlying the substation control systems is analysed through stochastic firewall and password models, while the impact factor for the attack upon a SCADA system is measured by the ratio *loss of load/total load* through power flow simulation. Experiments are conducted on a case study via simulation of the power flow and dynamic analysis. The integration of the cyber and power models is based on the simplifying assumption that cyber attacks can provoke unexpected opening of circuit breakers and the associated loss of electric load.

The assessment of *survivability* of a network with virtual connections exposed to node failures is discussed in [424]. Survivability modeling assumes that an undesired event already has occurred and therefore the frequency of such an event is dispensable. Survivability models objective is to quantify the level of service degradation during system recovery periods. Both a time-space decomposed analysis based on continuous time Markov chains (CTMCs) and a simulative approach have been used to cross validate. A number of scenarios with different network sizes have then been analyzed with respect to the survivability of the network.

Testbeds are used to conduct empirical studies of resilience assessment. A security testbed [753] has been built to emulate a SCADA network that is going through a Denial of Service (DoS) attack. Several national and international collaborative approaches for testbed implementations exist [212]. Testbeds have also been developed and evaluated as part of ongoing research programs on critical infrastructures, such as CRUTIAL [112, 247] and IRRIIS [470].

Network theory and graph-based representations of the infrastructures' topology are often applied to study interdependencies and relevant properties of the structure of a system. Such representations can be used for both qualitative and quantitative approaches.

The focus of [28] is on data survivability in pipeline systems. Two weighted graphs are constructed, one representing the pipeline structure and the other representing the set of sensors and their interconnections. Different types of constraints have to be respected such as source/sink balance, flow conservation, maximum bandwidth, and the availability of energy. The optimal network topology problem is solved using known algorithms for the solution of the Maximum Concurrent Multicommodity Flow problem.

In [863] a *graph-based* approach is used in combination with *statistical analysis*. Directed multigraphs augmented by response functions represent the interactions between the network components, and are used to analyze the interdependent effects of random failures and targeted attacks. Graph elements exist for non-storable resources (e.g., in the electric grid network), storable resources (e.g., in gas or oil pipeline), reliability (e.g., in the telephony transport layer), types of failures, repair time, and logistic delay. Graph statistics and analytical approaches are used to identify critical components. The simulation experiments show that a failure in the gas distribution network leads to a total failure in the telecommunication network and to reduced functionality of the power distribution network.

The authors of [126] conduct a *structural analysis* of the power transmission grid by applying a topological approach that extends the traditional metrics derived from complex network theory (e.g., degrees of nodes and global efficiency) with two new metrics, the entropic degree and net-ability. The new metrics account for the physical and operational behavior of power grids in terms of real power-flow allocation over lines and line flow limits. This approach can be used to assess structural vulnerabilities in power systems in contrast to traditional, purely topological metrics. The impact analysis of control systems availability on managing power contingencies is not supported by this extended topological approach.

More techniques to cope with CI models naturally exist, such as agent-based [882] and Monte-Carlo simulation approaches [115, 788]. High Level Architecture (HLA) or spatial reasoning using GIS [557] can be applied to distributed simulation. Exhaustive methods such as model-checking or performance evaluation approaches can also be applied. Multiformalism modeling [340] incorporates different modeling formalisms and applies dedicated solvers to obtain results in heterogeneous environments.

3.3.3 Current Programs

There are a number of ongoing projects in the field of critical infrastructures that mainly focus on quantitative analysis and interdependency analysis of the power grid using simulation models. For example, Trustworthy Cyber-Infrastructure for Power (TCIP) [868, 869] models trust and security issues for power and SCADA systems. Placing SCADA data communication on the Internet creates an environment where providing a reliable computing base is a challenge. Therefore, TCIP connects simulation models and tools developed for the power grid with those developed for the internet. Quantitative and qualitative evaluation constitute major research efforts in TCIP [801], with focus on means to model, simulate, emulate, and experiment with the various subsystems in the power grid. A variety of evaluation tools are adopted to enable validation, including PowerWorld, RINSE, formal logic, PowerWeb and APT.

In the CRitical UTility InfrastructurAL resilience (CRUTIAL) project [112, 247], the emphasis lies on ICT infrastructures for electric power grids, the study of interdependencies and the analysis of critical scenarios. The Integrated Risk Reduction for Information Based Infrastructure Systems (IRRIIS) project [470] focuses on simulation approaches, with emphasis on interdependencies in information-based infrastructure systems. Both projects focus on the analysis of interdependencies and will be discussed in more detail in Sect. 3.4.

Vital Infrastructure Threats and Assurance (VITA) [747] aims to raise awareness to the vulnerabilities of critical infrastructures by creating simulations and using role-plays. The project developed methods, tools and techniques for infrastructure protection and a demonstrator experiment with a focus on energy that can be used to gain insight into protection mechanisms on an international level.

The focus of the reliable infrastructures sub-project of the Next Generation Infrastructures project [745] is on the design approach for damage prevention to infrastructures and on the avoidance of system instabilities in the presence of failures. For example, the research on distribution centers security aims at ensuring the survivability of vital nodes in a networked information infrastructure to prevent system-wide failures. Another goal of the research is the protection of integrated ICT departments.

The Power Systems Engineering Research Center (PSERC) [746] does research on power markets, power systems, transmission and distribution technologies. This research aims at increasing the efficiency and reliability of increasingly complex and dynamic power systems through modeling, evaluation, and control. One area of research is the development of estimation techniques that use past system-wide failure data to help in the prediction of future system-wide failure events.

3.4 Focus on Interdependencies Modeling and Analysis

As already discussed in Sect. 3.2.1, strong dependencies exist among infrastructures, which can easily become a vehicle through which faults, errors and attacks propagate. If not controlled, these dependencies can create a multiplicative effect, leading to cascading and escalating failures of one or more critical infrastructures. It is thus extremely important to understand the associated relationships, for the prevention and limitation of threats and vulnerability propagation, and for recovery and continuity in critical scenarios.

Among the most recent efforts in addressing the modeling and analysis of interdependencies in critical infrastructures, we briefly recall the activities developed in the context of the European initiatives IRRIIS [740] and CRUTIAL [245], and some other works from the literature.

3.4.1 The CRUTIAL Approach

The CRUTIAL project [245] has addressed new networked systems based on ICT for the management of the electric power grid, in which artefacts controlling the physical process of electricity transportation need to be connected with information infrastructures, through corporate networks (intra-nets) that are in turn connected to the Internet.

The project has developed new architectural patterns that are resilient to both accidental failures and malicious attacks, and comprehensive modelling approaches, supported by measurement based experiments, to analyse critical scenarios in which faults in the information infrastructure provoke a serious impact on the controlled electric power infrastructure.

In CRUTIAL the interdependencies between infrastructures have been investigated by means of models at different abstraction levels: (i) from a very abstract view expressing the essence of the typical phenomena due to the presence of interdependencies, (ii) to an intermediate level of detail representing in a rather abstract way the structure of the infrastructures, in some scenarios of interest, (iii) to a quite detailed level where the system components and their interactions are investigated at a finer grain, considering elementary events occurring at the components level and analyzing their impact at the system level. Accordingly, the proposed framework is based on a hierarchical modelling approach that accommodates the composition of different types of models and formalisms, including generalized stochastic Petri nets (GSPNs), fault trees (FT), Stochastic Well formed Nets (SWN), and Stochastic Activity Networks (SAN). Each of these formalisms brings particular benefits that motivated its selection into the CRUTIAL modelling approach. However, this choice is not exclusive, and other formalisms with equivalent characteristics could also be used. Significant contributions have been obtained by CRUTIAL considering the qualitative description of interdependencies related-failures (mainly, unified models considering accidental and malicious threats in a integrated way) and the quantitative assessment of their impacts on the dependability and security of electrical power systems services [247].

The approach has coped with the lack of data representing realistic probability estimates of the occurrence of cyber threats and consequent failure modes by creating two complementary testbeds. These have been set up to run controlled experiments and to collect otherwise unavailable data related to cyber misbehaviours on grid teleoperation and micro grid control scenarios. One platform, the telecontrol testbed, consisted of power station controllers on a real-time control network, interconnected to corporate and control centre networks. The other platform, the microgrid testbed, was based on power electronic converters controlled from PCs, interconnected over an open communication network. Both testbeds integrated elements from the electrical infrastructure as well as from the ICT infrastructure, in order to focus on their interdependencies, and specifically on the vulnerabilities that occur in the electric power system when a part of the information infrastructure breaks down [248].

3.4.2 The IRRIIS Approach

The IRRIIS project [470] aims at increasing dependability, survivability and resilience of EU ICT-based critical information infrastructures. The basis for this work is the knowledge elicitation focused on interdependencies between the two infrastructures "electricity" and "telecommunication including Internet". Several approaches have been pursued to model and analyze the interdependencies.

A theoretical framework has been developed in [689], where an approach equivalent to process modeling is adopted, which views a CI as a process and dependencies that are modeled as response functions. Quantitative interdependency analysis, in the context of large complex critical infrastructures, is presented in [115], where a discrete state-space continuous-time stochastic process is used to model the operation of the critical infrastructure, taking interdependencies into account. Of primary interest to the model are the implications of the level of abstraction and parameterization for the study of dependencies on the distribution of cascade-sizes within and across infrastructures. The Leontief input-output economical model representing market dynamics has been exploited and adapted to model critical infrastructures dependencies [471]. In addition, an empirical approach [612] has been applied to analyse a large set of critical infrastructures failure data to discover patterns across infrastructures failures.

The IRRIIS consortium has developed Simulation for Critical Infrastructure Protection (SimCIP), an agent-based simulation environment for controlled experimentation with a special focus on CIs interdependencies [536]. The simulator is intended to be used to deepen the understanding of critical infrastructures and their interdependencies and to identify possible problems. It is intended to be used to validate and test architectural solutions aiming at enhancing the dependability of large critical information infrastructures. The network model for SimCIP is based on a multi-layer simulation approach (technical, cyber, management).

3.4.3 Other Studies

In addition to the work reported in the previous sections, several other simulation models have been proposed to analyze interdependencies, in the context of Electric Power Systems [42, 190] and in connection with telecommunication networks [42, 282, 782, 787]. A study to identify the state-of-the-art in critical infrastructures interdependency modelling and analysis and the government/industry requirements for related tools and services has been described in [116], where a strategy aiming to bridge the gaps between existing capabilities and UK government/industry requirements is also presented. In the report [720], the field of infrastructure interdependency analysis is first presented, then a survey on modeling and simulation techniques used for the infrastructure and interdependencies is introduced together with the leading research efforts. Data was collected from open source material and when possible through direct contact with the individuals leading the research. The issue of identifying appropriate metrics for quantifying the strength of interdependencies has also been addressed in a few studies, such as [177, 791].

3.5 Focus on Fluid Infrastructures

Since different infrastructures have different characteristics, this section provides a survey of the modeling requirements for so-called *fluid critical infrastructures*, i.e., water, gas and oil treatment and distribution. In contrast to e.g., power transmission and distribution networks, fluid infrastructures have mainly linear characteristics and as opposed to power, the fluid can be easily stored. Taking these specifics into account, it is possible to come up with a suitable and scalable modeling formalism and analysis technique for fluid infrastructures, as presented in [393] and summarized below.

A recent report of TNO Defence and Security [610] analyzes the current situation in the water sector and found a large number of vulnerabilities. Based on this research a number of detailed measures have been proposed [611] to increase security in the water sector. Given the severe consequences of successful attacks on SCADA systems, it is very important to analyze the trade-off between the cost and the efficiency of such measures, already in the design process. The efficiency of these measures can be expressed in terms of survivability, i.e., the time it takes after a successful attack, before the system recovers to an acceptable level of service.

An example critical infrastructure that intensively uses SCADA systems for process control are wastewater-management systems. Water is cleaned in several chemical and physical cleaning steps, before it is distributed to the end users. A suitable modeling formalism for such systems needs to take into account continuous and discrete quantities, as well as random failure and repair times. SHMs combine discrete and continuous variables with stochastics, hence, allow to model random phenomena in a natural way. On the one hand, a very nice theory has been developed that takes into account the full expressiveness of Stochastic Hybrid Models [737]. However, the industrial application that we are considering is by far too large for state-of-the-art approaches; hence the focus of the presented approach is on scalability. On the other hand, several formalisms supporting SHMs have been defined [263, 395, 444], where each of them is suitable only in some very specific domain, and suffers from limitations that prevent it from being used in other applications.

Recall, that interdependencies between the physical process and the ICT control infrastructure are crucial in critical infrastructures. Therefore critical infrastructures are very big and complex systems and scalability is of utmost importance. State-of-the-art analysis methods from the area of SHM, however, do not scale. The systems under consideration are characterized by deterministic fluid transportation, however, with rates that change according to a stochastic process. Hence, Fluid Stochastic Petri Nets (FSPNs) [395, 444] and Piece-wise Deterministic Markov Processes (PDMPs) [263] appear to be suitable. However, the memory of continuous variables in PDMPs is lost upon stochastic transitions. Hence, they are not suitable to model the physical behaviour of fluid critical infrastructures. First and second order Fluid stochastic Petri nets (FSPNs) [395, 444] have a sound mathematical basis allowing for a completely formalized characterization of the state-evolution in terms of differential equations. However, such equations can be solved only when there are at most one or two continuous variables. Simulation is the only available alternative when considering larger models [221, 394]. Another limitation of current FSPN approaches is the lack of efficient compositional techniques.

The above clearly shows the need for a modeling and analysis framework that is specifically tailored towards fluid critical infrastructures. To tackle the issue of scalability, a new approach based on Hybrid Petri nets [261] was proposed, where the deterministic evolution is separated from the stochastic behaviour of the system [393], by exploiting the quasi-deterministic behaviour of the system under study, given that failure and repair events are stochastic. Therefore, there are relatively few stochastic transitions, which allows for separating the deterministic and the stochastic evolution of the system, using a conditioning/deconditioning argument. This will speed up the reachability analysis and will allow for a large number of continuous variables in the model, as opposed to previous approaches.

The Hybrid Petri Net formalism with General one-shot transitions (HPNG) as proposed in [393] is specifically tailored towards fluid critical infrastructures. It allows for an arbitrary number of continuous variables that can be connected via fluid transitions. These transitions can be controlled by discrete places that can be connected via deterministic and generally distributed transitions. Generally distributed transitions must respect the constraint that they can fire only once during the evolution of the model: for this reason we call them one-shot transitions.

Gribaudo and Remke [393] also introduces a new and efficient computation scheme for all reachable states of a model: parametric reachability analysis. This technique separates the deterministic and the stochastic components of a HPNG by conditioning the deterministic evolution on the samples drawn from the probability distributions associated to the general transitions. After all reachable parametric locations have been computed, important performance metrics (such as the distribution of fluid) can be derived by a deconditioning procedure. As opposed to similar SHM solution algorithms, the presented technique allows for an arbitrary number of fluid variables.

The algorithm as described in [393] presents a first step in the analysis of HPNGs and needs to be extended in several ways to realistically model and analyze fluid critical infrastructures. Currently, the algorithm only allows for one general one-shot transition, resulting in an underlying state-space of parametric locations that depend on the sample of the one general transition. The approach used can be made more scalable by extending the algorithm to allow for more generally distributed transitions resulting in parametric locations that depend on as many samples.

In [393] the effect of different failure and different repair time distributions is shown for a model of a water treatment facility. Possible results include the distribution of fluid during the recovery process and the probability to reach an unsafe state after a failure or attack. This helps system engineers to dimension storage tanks in a way that failures do not influence the continuity of water delivery. In industry, there is currently a trend towards combining the processing of drinking, surface and waste water into one integrated water network, which makes the system even more complex and hence, more vulnerable. Moreover, due to legal constraints in the Netherlands, by 2014 the operation of the water treatment and distribution has to be fully automated without direct human control. This requires the a priori development of optimal repair strategies. Hence, water companies are very interested in evaluating the dependability of their infrastructures and in comparing design alternatives based on a cost/benefit analysis.

3.6 Case Study: Reliability Modeling of a Smart-Grid Distributed Automation Network

This section illustrates a reliability assessment approach for a distributed automation Smart-Grid network. Specifically, we present the computation of the System Average Interruption Duration Index (SAIDI) metric for one specific power distribution circuit, which consists of 7200 feet of the main distribution line, encompassing 117 transformers and serving a total of 780 customers [931].

SAIDI is one of the most important performance metric for power utilities, as it evaluates the utilities' performance after a sustained power interruption. For example, SAIDI is used in the United States by public service commissions to monitor and control power utilities performance.

Some regions in the United States determine utility power rates based on the utility performance as measured by SAIDI and other metrics that are related to the time to restore power after a failure event and the number of customers affected by a power failure. Therefore, utilities are required to measure and report SAIDI to the controlling public service commissions. The public service commission has the power to investigate failure events and to order the power utility to improve performance [930].

The main circuit line is equipped with a distributed generator system at the end of the line that can be switched on when faults are detected on the main line or to provide an additional source of load. The peak load of the main line is 1692 KW, measured in August, i.e., during a hot midweek summer day. The distributed generator is designed to provide 100 % of the circuit load, i.e 1692 KW. The distribution network is composed of 34 feeder lines that are connected to the main circuit feeder. Figure 3.3 shows one feeder line divided into 40 sections. The back-up generator connects to the main feeder line through a Tie switch, not shown in Fig. 3.3. Each main feeder line can be divided into several sections, at a significant cost for construction and maintenance per section. The added benefit of increased number of sections is the increased level of granularity of power control.

Currently, the main feeder line is not divided into sections, so if a fault occurs on the main line, all the customers on the main feeder line would be impacted. The objective of implementing a Smart-Grid distributed automation approach is to decrease the customer impact of power failure events as assessed by the expected value to the SAIDI metric.

In a distributed automation approach, after a power failure event, the faulty section is switched off the main line, and the substation powers the upstream part of the feeder, while the distributed generator powers the downstream part of the feeder, reducing the outage impact only to the customers that are supported by the now isolated failed section. Therefore, while the customers in the faulty section still see an outage with an average repair time of 4 h, the other customers that are served by the main feeder line, experience a power interruption that will last only about 2 min. The tradeoff in this distributed automation design is the cost and complexity of having many

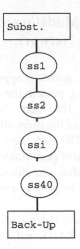

Fig. 3.3 Architecture model of one main feeder line divided into 40 sections

sections, against the large customer impact of the failure of a large section, if too few circuit sections are implemented.

In this example we use hours as the unit for the failures and repair rates. SAIDI is defined as $\sum (r_i \cdot N_i)/N_T$, where r_i is the actual service restoration time in hours, N_i is the total number of customer interrupted, and N_T is the total number of customers [930].

Figure 3.4 presents a Markov reward model of the states a given section can be in, due to failures, repairs, and section line switches, that are needed to isolate the failed section from the main circuit feeder line. Each state is described by a three tuple, where the first entry captures the power state (on/off), the second entry represents the state of the smart-grid communications network (on/off), and the third entry characterizes whether the section is currently in-line or out-of-line.

When the distributed automated network is operating correctly, the Markov chain is in state $s = (1, 1, IN)$, which is state 1 in Fig. 3.4. We have to consider two different types of failures:

- A power failure with impact on the section under study occurs with rate f1 and the Markov chain will transition to state $s = (0, 1, IN)$. The power failure has to be repaired in stages. If the smart-grid automated repair is functioning properly, the Markov chain moves to state $s = (0, 1, IN)$, after the power failure. Next, the Markov chain moves with rate sw1 to state $s = (0, 1, OUT)$, where the section is removed from the line. In this state all customers upwards and downwards from the failed section already have their service restored. Typically, the average time to switch a section off-line is between 1 and 2 min. The average time to repair a power failure manually is between 1 and 4 h, depending on several factors like, urban density, traffic congestion, cause of equipment failure, and the extent of the damage to the equipment [736]. In this example, we assume the average switching

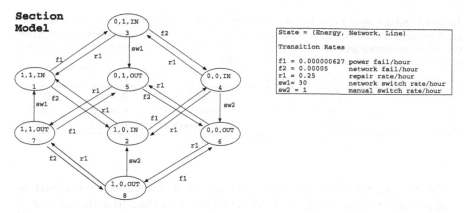

Fig. 3.4 Markov reward model, describing failures and repairs in a distributed automation power network

time, $1/sw1$, to be equal to 2 min, and the average manual power failure repair time, $1/r1$ to be equal to 4 h. When the power failure is corrected manually, the Markov chain moves to state $s = (1, 1, OUT)$, with rate r1 and from this state the section is switched back on with rate sw1. Then, the customers from the affected section have their power restored.

- A smart-grid failure occurs with rate f2 and which impacts the section under study, the Markov chain will transition to state $s = (1, 0, IN)$. In this state there is no impact on the power-grid service to the power customers. However, if a power failure occurs in this state, with rate f1, the Markov chain moves to state $s = (0, 0, IN)$. In this state all customers in the main feeder line suffer from the power failure and the line has to be manually switched off, with rate sw2, to isolate the section from the main feeder line. From state $s = (0, 0, OUT)$ a manual repair brings the Markov chain to state $s = (1, 0, OUT)$ and a manual smart-grid repair brings the Markov chain to state $s = (1, 0, IN)$. In this state the power is restored for all customers but the automated Smart-grid recovery is still not repaired. When the Smart-Grid is repaired the Markov chain moves to the initial state $s = (1, 1, IN)$. The other transitions follow a similar pattern and are shown in Fig. 3.4.

One of the challenges in Reliability modeling is the expression of the system reliability degradation as a function of time, which is often overlooked. Therefore, we solved the Markov chain model with rewards using the Tangram-II [272] transient analysis solver. A Markov chain with rewards analysis was used to represent one year of operation. The model captures only one section failure at a time. Therefore, an important assumption used in this Markov modeling example is that the smart-grid is designed to automatically restore one section failure at a time, and that the probability of occurrence of a second power failure while the first one is being repaired is very small.

Table 3.1 SAIDI after 1 year of operations for several section designs, for 0.1 power failures/km/year and 1/20, 000 failures per hour communications network equipment, manual repair time of 4 h, automated section switching time of 2 min, manual line switching of 1 h, total number of customers served by main feeder line equal to 780

Number of sections	Customers impacted	f1	f2	SAIDI
5	156	0.000005	0.00005	0.036
20	39	0.000001252	0.00005	0.0025
40	20	0.000000627	0.00005	0.00074

Table 3.1 presents the computed SAIDI metric for different alternatives of section designs after one year of operation. These results are derived from the solution of a transient Markov reward model for a section failure rate of 0.1 failures per km per year. This analysis is useful for the engineering of the topology of distribution automation networks, where the engineer needs to tradeoff between the investment in number of sections and the distribution automation reliability.

The empirical results shown in Table 3.1 illustrate the tradeoff between increased reliability and the additional cost of designing a larger number of sections into the main feeder line. Table 3.1 shows that to achieve increased reliability a larger number of sections has to be built, at additional cost for construction and maintenance. The benefit obtained from the construction of sections that control a smaller number of users that can be isolated quickly in the case of power failures is the improved power reliability, which is demonstrated by the smaller values of the SAIDI metrics for the main line feeder design when the feeder line is divided into 40 sections.

3.7 Conclusions and Emerging Research Directions

Following the more comprehensive approach of CI resilience as presented in [677], state of the art resilience assessment approaches for CIs should help improve the robustness of CIs. The increase in robustness of CIs can be achieved by increasing their absorptive capacity, their resourcefulness or their recoverability. The absorptive capacity, for example, can be improved by adding more redundancy. Resourcefulness is the ability of using the available resources efficiently in the presence of failures and disasters. It depends mainly on the adaptive capacity of the system. Optimized repair schedules can improve the resourcefulness of critical infrastructures. Finally, recoverability can be optimized by minimizing repair times. On one hand, improvements to the robustness of critical infrastructures come at a certain cost. On the other hand increased robustness will reduce costs due to systems down time. Hence, the resilience of critical infrastructures should be evaluated under a cost-minimization criteria.

The variety of initiatives in resilience assessment of critical infrastructures, some of which are briefly overviewed in this chapter, testify to the paramount role of resilience assessment in several critical sectors. However, when comparing requirements for resilience assessment, as identified in Sect. 3.2, with existing approaches, as discussed in Sect. 3.3, it becomes clear that further research is still required. Most of the modeling research in CIs uses simple handcrafted reliability block diagrams, fault-trees, or simplistic stochastic Petri nets. Recent research on fluid critical infrastructures suggests that stochastic hybrid systems can be tailored toward this new application field. The application of stochastic methods captures the continuous dynamics of the physical world and the discrete characteristics of the control infrastructure. However, further research is necessary to ensure the scalability of hybrid approaches.

Advances in industrial control systems and technology, such as SCADA systems, enhance sector operations but create additional vulnerabilities and increase interdependencies, whose effects are hard to detect and to mitigate. To make these systems resilient, research must integrate an understanding of resilience, security, human interaction, and complex network design to address the threats. The largeness and diversity of critical infrastructures and the different characteristics of their parts requires a compositional integrated formalism. The necessity of continuous assessment activities calls for a composite (i.e., holistic) evaluation framework, where the synergies and complementaries among several evaluation methods can be fruitfully exploited.

Further research is required to develop coherent resilience properties in different sectors of CIs. Sound and rigorous definitions of measures, such as recoverability and survivability, can be cast into temporal logics, e.g., continuous stochastic logic [228]. Research into recently developed methods for stochastic model checking needs to be adapted for the use within critical infrastructures. The use of abstraction, bi-simulation reduction, and symbolic state space representation techniques can help to tackle large state-spaces. In addition, the use of simulation-based statistical model checking can be a applied to the assessment of resilience of CIs.

Chapter 4
Providing Dependability and Resilience in the Cloud: Challenges and Opportunities

Samuel Kounev, Philipp Reinecke, Fabian Brosig, Jeremy T. Bradley,
Kaustubh Joshi, Vlastimil Babka, Anton Stefanek and Stephen Gilmore

Abstract Cloud Computing is a novel paradigm for providing data center resources as on-demand services in a pay-as-you-go manner. It promises significant cost savings by making it possible to consolidate workloads and share infrastructure resources among multiple applications resulting in higher cost- and energy-efficiency. However, these benefits come at the cost of increased system complexity and dynamicity posing new challenges in providing service dependability and resilience for applications running in a Cloud environment. At the same time, the virtualization of

S. Kounev (✉) · F. Brosig
Karlsruhe Institute of Technology, 76131 Karlsruhe, Germany
e-mail: kounev@kit.edu

F. Brosig
e-mail: fabian.brosig@kit.edu

P. Reinecke
Institute of Computer Science, Free University Berlin,
Takustr. 9, 14195 Berlin, Germany
e-mail: philipp.reinecke@fu-berlin.de

J. T. Bradley · A. Stefanek
Imperial College, London, UK
e-mail: jb@doc.ic.ac.uk

A. Stefanek
e-mail: as1005@doc.ic.ac.uk

K. Joshi
AT&T Labs Research, Florham, Park NJ, USA
e-mail: kaustubh@research.att.com

V. Babka
Faculty of Mathematics and Physics, Charles University in Prague, Prague, Czech Republic
e-mail: babka@d3s.mff.cuni.cz

S. Gilmore
University of Edinburgh, Edinburgh, UK
e-mail: Stephen.Gilmore@ed.ac.uk

K. Wolter et al. (eds.), *Resilience Assessment and Evaluation of Computing Systems*,   65
DOI: 10.1007/978-3-642-29032-9_4, © Springer-Verlag Berlin Heidelberg 2012

physical resources, inherent in Cloud Computing, provides new opportunities for novel dependability and quality-of-service management techniques that can potentially improve system resilience. In this chapter, we first discuss in detail the challenges and opportunities introduced by the Cloud Computing paradigm. We then provide a review of the state of the art in dependability and resilience management in Cloud environments, and conclude with an overview of emerging research directions.

4.1 Introduction

In today's data centers, IT services and applications are typically hosted on dedicated hardware in order to provide dependability guarantees. Server capacity is typically over-dimensioned to ensure adequate Quality-of-Service (QoS) under variable workloads and load fluctuations. The use of dedicated hardware with over-dimensioned capacity not only leads to poor resource efficiency, but also makes it hard to react to changes and conflicting demands in operating conditions, business processes or use practices. Moreover, the adoption of new applications and the increasing demand for IT services leads to an exponential growth in the number of servers and the required network infrastructure. Servers in data centers nowadays are estimated to have average utilization ranging from 5 to 20% [761, 826] which corresponds to their lowest energy-efficiency region [91]. The growing number of underutilized servers, often referred to as "server sprawl", translates into increasing data center operating costs including system management costs and power consumption costs of the server, network and cooling infrastructure. According to a study at Lawrence Berkley National Labs (2007), power consumption in data centers doubled from 2000 to 2005, and in 2006, the USA alone spent an estimated 61 TW-h in data centers. By 2025, power consumption in data centers is projected to grow by 1600% and energy will become the major factor in the Total-Cost-of-Ownership (TCO) for IT [922]. Already today, according to IDC, over 40% of data center customers report power demand outstripping supply, while cooling capacities at their threshold have become a limiting factor in deploying new systems [810]. In addition to driving costs up, the rising energy consumption of the ICT sector will have a significant impact on the global CO_2 emissions. While today, ICT accounts for 2–4% of the global CO_2 emissions, it is projected to reach 10% in 5–10 years [517]. Thus, reducing the costs of ICT and their environmental footprint while keeping a high growth rate of IT services is one of today's greatest challenges for society.

Driven by the pressure to improve energy efficiency and reduce data center operating costs, industry is looking towards Cloud Computing, which is a novel paradigm for providing data center resources (computing, network and storage) as on-demand services over a private or public network in a pay-as-you-go manner. Cloud Computing is normally considered at three different levels: i) Infrastructure-as-a-Service (IaaS) where raw compute, storage, and network resources are provided, ii) Platform-as-a-Service (PaaS) where an application environment on top of the bare bones infrastructure is provided, iii) and Software-as-a-Service (SaaS) where a working application is provided (e.g., NetSuite and SalesForce.com). Cloud Computing makes it possible for enterprises to consolidate their IT resources internally or to completely outsource their IT infrastructure taking advantage of the economies of scale of a shared infrastructure. In both cases, some substantial reductions in the TCO for IT can be achieved. Virtualization plays a key role in this process since it makes it possible to significantly reduce the number of servers in data centers by having each server host multiple independent virtual machines (VMs) managed by a Virtual Machine Monitor (VMM) often referred to as a Hypervisor. By enabling the consolidation of multiple applications on a smaller number of physical servers, virtualization promises significant cost savings resulting from higher energy efficiency and lower system management costs. Moreover, virtualization facilitates system evolution by enabling adaptability and scalability of service infrastructures.

With investments of billions of dollars, the fortunes of dozens of companies, and major research initiatives staked on its success, it is clear that Cloud Computing is here to stay. Cloud-based infrastructures are rapidly becoming a destination of choice to host a varify of applications ranging from high-availability enterprise services and online TV stations, to batch-oriented scientific computations. However, it is not yet clear whether Cloud services can be a dependable alternative to dedicated infrastructure. In the context of this chapter, we consider *dependability* to be the ability of a system to provide dependable services in terms of availability, responsiveness and reliability. As part of dependability, *resilience* is understood as the system's ability to continue providing available, responsive and reliable services under external perturbations such as security attacks, accidents, unexpected load spikes or fault-loads. The remainder of this chapter explores this question and is organized as follows. Section 4.2 describes the challenges and opportunities in providing dependability and resilience in the Cloud. In Sect. 4.3, we review the state of the art in dependability, performance and security management in Cloud infrastructures. An overview of the emerging research directions in Cloud Computing is provided in Sect. 4.4

4.2 Challenges and Opportunities

The increased complexity and dynamicity induced by Cloud Computing pose new challenges and opportunities in providing service dependability and resilience. On one hand, availability and privacy are serious challenges for applications hosted on Cloud infrastructure. On the other hand, a Cloud provider's economies of scale allow levels of investment in redundancy and dependability that are difficult to match for smaller operators. Furthermore, the ability to monitor large numbers of applications and users can enable "wisdom of crowds" approaches to provide enhanced security much in the same way that network providers have been able to do with worms and DDoS attacks.

4.2.1 Challenges

In spite of the many benefits Cloud Computing promises, today, the lack of trust in shared virtualized infrastructures is a major showstopper for its widespread adoption. According to [7], 74% of technological and financial decision makers in the UK would not put mission-critical applications in the Cloud. Service unavailability, performance unpredictability, and security risks are frequently cited as major reasons for the lack of trust [5, 58]. Some recent stress tests conducted by Sydney-based researchers revealed that the infrastructure-on-demand services offered by Amazon, Google and Microsoft suffer from regular performance and availability issues [936]. Response times of services varied by a factor of twenty depending on the time of day the services were accessed.Throttling, power failures and deliberate design constraints were among the reasons for the unpredictability. According to [58], concerns of organizations about service availability is the number one obstacle to the adoption of Cloud Computing. Service overload, hardware failures and software errors as well as operator errors are among the most common causes of service unavailability as experience with Google's AppEngine, GMail and Amazon's AWS services shows [12, 936].

The lack of trust in shared virtualized infrastructures is a major impediment which applies both to public and private Clouds. Indeed, virtualization comes at the cost of increased system complexity and dynamicity. The increased dynamicity is caused by the introduction of virtual resources and the lack of direct control over the underlying physical hardware. The increased complexity is caused by the complex interactions between the applications and workloads sharing the physical infrastructure. The inability to predict such interactions and adapt the system accordingly makes it hard to provide dependability guarantees in terms of availability and responsiveness as well as resilience to external perturbations such as security attacks. Thus, virtualization introduces new sources of failure and threats degrading the dependability and trustworthiness of Cloud Computing infrastructures. Service providers are faced with the following challenges:

- How much resources (e.g., CPUs, main memory, storage capacity, network bandwidth) should be allocated to a new application deployed in the Cloud infrastructure and how should the application be configured to satisfy its requirements for dependability (availability and reliability) and responsiveness avoiding the pitfalls of underprovisioning or overprovisioning resources?
- How much and at what rate and granularity (e.g., CPU cycles, cluster nodes) should resources be added or removed proactively to avoid Service Level Agreement(SLA) violations or inefficient resource usage due to varying customer workloads and load fluctuations?

Moreover, the consolidation of workloads translates into higher utilization of physical resources which makes the system much more vulnerable to threats resulting from unforeseen load fluctuations, hardware and software failures, and network attacks. The Cloud provider is faced with the challenge of how to efficiently share physical resources among hosted applications in the face of highly variable and unpredictable resource demands as well as operational failures.

An environment with a few large Cloud infrastructure providers not only increases the risk of common-mode outages affecting a large number of applications, but also provides highly visible targets for attackers. Community-driven sites such as [3] track outages in major Cloud providers and have documented a number of outages and security vulnerabilities over the last two years affecting hundreds of Internet sites.

Sharing of Cloud resources by entities that engage in a wide range of behaviors and employ best practices to varying degrees can expose Cloud applications to increased risk levels. For example, on April 26 2008, Amazon's Elastic Cloud (EC2) had an outage [1] across several instances due to a single customer applying a very large set of unusual firewall rules and instantiating a large number of instances at the same time, thereby triggering a performance degradation bug in Amazon's distributed firewall.

Multiple administrative domains between the application and infrastructure operators reduces end-to-end system visibility and error propagation information, thus making problem detection and diagnosis very difficult. Additionally, for competitive reasons, Cloud infrastructure providers may not provide full disclosure regarding the cause of outages or other detailed infrastructure design information, raising the question of the verifiability of claims regarding dependability.

The hosting of data on outsourced and shared infrastructure that may be in a different legal jurisdiction than the owner of the data has serious legal and privacy implications. Corporate accountability legislation such as the Sarbanes-Oxley Act (SOX) of 2002 and privacy clauses included in legislation such as the Health Insurance Portability and Accountability Act (HIPAA) of 1996 and the Telecommunications Act of 1996 create obstacles to the applicability of Cloud solutions in the financial, healthcare, and telecom industries. For example, BusinessWeek reported in Aug 2008 [535] that ITricity, a European provider of Cloud computing capacity, could not offer services to such companies until it began offering owner-hosted private Cloud services. The recently formed industrial consortium called the Cloud Security Alliance [2] includes in its charter several issues regarding the interplay of Cloud Computing and legal requirements.

4.2.2 Opportunities

Cloud computing enables economies of scale leading to large redundancy levels and wide geographical footprints. For example, Amazon's EC2 currently supports two regions in the US and Europe, each split into independent "availability zones", while AT&T's Synaptic Cloud computing offering provides five "super IDCs" located across the world. These can be leveraged through techniques such as virtual machine migration and cloning to provide better fault tolerance and disaster recovery, especially for operators of smaller applications that may not have been able to afford such capabilities.

New security and reliability services can be enabled or strengthened by virtue of being located in the Cloud. For example, popular cloud-based email services such as GMail amplify manual feedback from some users to provide automatic spam filtering for all users. Oberheide et al. describe in [696] a Cloud-based anti-virus solution that can not only utilize multiple vendors to provide better coverage, but also compares data blocks across users to improve efficiency and provides an archival service for forensic analysis.

Managed Cloud services that include OS level support can result in improved reliability and security due to consistent centralized administration and timely application of patches and upgrades.

4.3 State-of-the-Art Review

In this section, we provide an overview of the state of the art in dependability and resilience for Cloud Computing. We start with a discussion of dependability assessment techniques and then survey methods for managing dependability. The approaches in the first section cover the issue of how Cloud system dependability could be assessed, while the second section comprises methods that are used to manage the system with the goal of improving dependability.

4.3.1 Approaches for Dependability Assessment

Availability, performance and security can be evaluated using measurements on real deployments, measurements on test-beds, simulations, and analysis of models. While a lot of work exists in these areas, approaches specifically targeting Cloud systems are still rare.

Measurement studies on real Cloud systems are undertaken in order to understand the effect of the Cloud on the application. In [378], resilience of an Infrastructure-as-a-Service (IaaS) cloud is quantified as job rejection rate and response delay in situations where the cloud is subjected to changes in demand and available capacity. However, most existing studies focus on performance. The typical approach in such studies is to generate a workload and measure various performance indicators. The tools used and the indicators of interest depend on the application that the authors focus on. [322, 481, 673, 705, 924] serve as examples for evaluation of the performance of applications for scientific computing. These studies employ tools that generate a typical High Performance Computing (HPC) workload, and measure run-times. In [705], the authors also evaluate the time required for allocating and releasing virtual machines. Since flexible resource allocation is a major selling-point of Cloud systems, this aspect should not be ignored when considering overall performance. Furthermore, [705] studies the performance of disk I/O operations performed on virtualized disks. The findings in [279, 322, 481] show that applications running on Cloud systems have run-times that are longer and exhibit more variance than applications running on native systems. The authors of [705], however, state that extensive caching in a Cloud system may result in significantly faster disk I/O operations, compared to a native system. On the other hand, [705] also shows that a quick performance drop occurs once the cache size is exceeded.

Experimentation on test-beds is seldom performed for Cloud systems, since the complexity and costs of setting up a Cloud environment of realistic size become prohibitively large. Existing approaches thus tend to focus on special aspects of Cloud systems, in particular, specific programming models and virtualization technology. In [460], the performance of an application using MapReduce is evaluated on a small cluster of physical machines, each of which runs several virtual machines. The authors point out that performance may suffer from virtual machines competing for the physical I/O resources. As virtualization is a key component in Cloud Computing, its impact on dependability must be understood. Virtualization is rather amenable to experiments in test-beds. Existing work [452, 453, 637, 831] focuses on studying the performance impact of configuration options and workloads through experiments with benchmarks and standard performance measurement tools. Performance indicators are typically throughput and benchmark-specific aggregated metrics.

Benchmarks for virtualized server consolidation, i.e., benchmarks measuring aggregated server performance when physical resources are virtualized and shared, include vConsolidate [49], VMMark [430] and recently SPECvirt_sc2010 [841]. Benchmarks for virtualized servers are still a subject of discussion, as there is a lack of consensus for a metric describing consolidated server performance [143, 373]. The authors of [372] propose new metrics taking into account per-VM performance along with total system throughput. The authors of [143] emphasize that particularly for benchmarking database performance, the consolidation of resource-intensive workloads is of crucial importance. None of the virtual benchmarks available today measure database-centric properties adequately [143].

Unlike virtualized server consolidation, Cloud Computing lacks well-established benchmark suites [398], although benchmarks such as TeraSort, Cloudstone or MalStone exist, and traditional high-performance computing benchmarks have been used (e.g., [322, 481, 705]). In [114], it is argued that the established TPC-W benchmark [838] is not appropriate for Cloud Computing, because Cloud scalability invalidates its metrics, TPC-W relies on database properties often not supported in the Cloud, and because TPC-W does not provide metrics for important Cloud properties such as scalability, pay-per-use pricing, and fault-tolerance. The authors of [114] propose desirable properties of a Cloud benchmark; similarly, [238] proposes a benchmarking framework specific to Cloud data serving systems. Resilience benchmarking in general is further discussed in Chap. 14.[1]

Several simulation approaches for Cloud systems have been proposed. These methods differ in whether they focus on special applications or allow simulation of Cloud systems in general. The simulation framework MRPerf [926] instruments the discrete-event network simulator NS-2 [693] for studying performance and de-

---

[1] "Resilience Benchmarking"

pendability of MapReduce [275]. The framework models node, network, and disk behavior in high detail and thus allows evaluating the impact of network topology choices and node/network failures, but is limited to applications that use MapReduce. In contrast, the CloudSim toolkit [167] is a discrete-event simulation toolkit for general Cloud systems. The toolkit models, among other aspects, virtual machines and VM scheduling, storage, network, and computing resources.

Analytical approaches for the evaluation of dependability and performance of Cloud systems usually focus on the impact of virtualization.

Reliability block diagrams to model system reliability at the host level have been proposed in [759]. These models do not consider the behavior of the underlying hardware and software components. More detailed models based on CTMCs are presented in [878], but these models still only capture behavior at the VM level. The two-level hierarchical approach in [534] uses fault-trees in the upper level and CTMCs in the lower level in order to capture software failures at the VMM, VM, and application level as well as hardware failures. Finally, combinatorial modeling to analyze design choices with a single physical server hosting multiple VMs was proposed in [759].

Virtualized resources shared between VM instances have a non-trivial impact on performance. Due to this overhead, traditional design-time model-based approaches, as surveyed in e.g., [83, 556] may yield imprecise results when used as-is. A typical approach (e.g., [101, 639]) is to construct traditional queueing network models and apply a slowdown factor to capture the effects of virtualization. Another approach is applied in [560], where artificial neural networks are used to predict performance of virtualized applications from a set of observable or controllable parameters related to CPU, memory, disk and network usage.

Prediction of resource utilization is required for dimensioning and workload placement decisions. The work in [940] focuses on predicting CPU utilization of both a VM and the Dom-0 (which hosts the network and disk drivers). The prediction model is automatically derived from a set of microbenchmarks consisting of synthetic CPU, network and disk workloads, using a robust stepwise linear regression between several metrics obtained in native and virtualized microbenchmark executions. Further parametrizations of the model are based on measurements of the application executed natively. Simple models for core utilization and effective shared space allocation are developed in [476, 881]. The authors also note that for some shared resources (such as the cache space), online measurement and modeling is not possible today, due to a lack of appropriate performance counters.

As has long been accepted for dependability, complex systems can never be perfectly secure. Therefore, only quantitative measures allow comparisons between systems with respect to their security. While quantification of security has long been recognized as an important problem [593, 594] and several approaches have been made in recent years [356, 469, 484, 571, 614, 688, 830], the area is still under-explored and subject to dispute [912]. Still, various security metrics have been proposed [30, 186, 356, 484, 618], and experimental studies have been performed [469].

For security evaluation of Cloud systems, even less work exists. In fact, quantitative security evaluation of Cloud systems is still in its infancy. The analytical approach by [808] exemplifies some of the difficulties in quantitative security assessment. In this approach, risk is computed as a weighted sum of the impact of a security incident and its probability. Both incident probabilities and impacts, however, are hard to measure. While the authors of [809] argue that probabilities can be obtained from published incidence reports and impacts can be estimated based on expert opinions, such data may be invalid due to biased report and subjective opinions. Furthermore, taking the weighted sum assumes that security is a static property, whereas it seems likely that the probability of security incidents and their impact changes over time, as both the system, the attacker, and the value of the system to the user evolve.

4.3.2 Approaches for Managing Dependability and Performance

There are many research challenges with respect to managing dependability and performance in Cloud systems (see Sect. 4.3). On the one hand, virtualization provides opportunities to improve these properties, on the other hand, Cloud Computing poses a complex resource allocation problem.

4.3.2.1 Virtualization for Improving Dependability and Performance

Techniques that take advantage of virtualization to improve system dependability have been the focus of recent research [250, 332, 551, 667, 877, 878]. Two lines of research can be distinguished: i) virtualization-based software rejuvenation and ii) using VM replication as a basis for failure recovery.

Software rejuvenation is a proactive fault management technique aimed at cleaning up the system's internal state to prevent occurrence of severe failures due to the phenomena of software aging or caused by transient failures [895]. A detailed introduction to rejuvenation is given in Chap. 8.[2] The approach has been applied to Cloud Computing and virtualization. In [877], a technique that can increase availability of application servers through the use of virtualization, clustering and software rejuvenation is presented. Analytical models are used to analyze multiple design choices when a single physical server and dual physical servers are used to host multiple VMs. It is shown that by integrating virtualization, clustering and software rejuvenation, it is possible to benefit from increased availability, manageability and savings from server consolidation through virtualization without decreasing uptime of critical services. A similar approach based on automated self-healing techniques claimed to induce zero downtime for most of the cases is proposed in [667]. Software aging and transient failures are detected through continuous monitoring of system data

---

[2] "Software Aging and Rejuvenation for Increased Resilience: Modeling, Analysis and Applications"

and performability metrics of the application server. A further virtualization-based rejuvenation technique for application servers using stochastic models was proposed in [878]. The authors present a stochastic model of a single physical server used to host multiple virtual machines (VMs) configured with the proposed technique. The model is intended as a general model capturing the application server characteristics, failure behavior, and performability measures. Finally, in [551], the authors present a technique called warm-VM reboot for fast rejuvenation of VMMs that enables efficiently rebooting only a VMM by suspending and resuming VMs without accessing the memory images. The technique is based on two mechanisms, on-memory suspend/resume of VMs and quick reload of VMMs. The technique is claimed to reduce downtime and prevent the performance degradation due to file cache misses after the reboot. In [895], stochastic models that help to detect software aging and determine optimal times to perform rejuvenation are proposed. Models are constructed using workload and resource usage data collected from the UNIX operating system over a period of time. The measurement-based models are intended to help development of strategies for software rejuvenation triggered by actual measurements.

Accounting for failures by dynamically creating replicas is a common strategy to improve overall dependability. For instance, [748] uses regeneration of new data objects to account for reduction in redundancy and the Google File System [375] similarly creates new file "chunks" when the number of available copies is reduced below a threshold. Even commercial tools such as VMWare High Availability (HA) [923] allow a virtual machine on a failed host to be reinstantiated on a new machine. However, the placement of replicas becomes especially challenging when they are components in a multi-tier application. Recent work on performance optimization of multi-tier applications (e.g., [102, 252, 498, 901]) addresses the performance impact of resource allocation on such multi-tier applications, but does not combine performance modeling with availability requirements and dynamic regeneration of failed components. The tradeoff between availability and performance is always present in dependability research since increasing availability (by using more redundancy) typically increases response time. Examples of work that explicitly address this issue include [271] and [823], both of which consider the problem of when to invoke a (human) repair process to optimize various metrics of cost and availability defined on the system. In both cases, the "optimal policies" that specify when the repair is to be invoked (as a function of system state) are computed offline through solution of Markov Decision process models of the system.

As far as failure recovery mechanisms are concerned, in [332], the authors introduce an extensible grammar that classifies the states and transitions of VM images and can be used to create rules for recovery and high availability exploiting virtualization for simplified fault tolerance. In [250], a fail-over technique based on asynchronous VM replication is proposed that asynchronously propagates changed state to a backup host at frequencies as high as forty times a second, and uses speculative execution to concurrently run the active VM slightly ahead of the replicated system state. In case of a failure, automatic fail-over with only seconds of downtime is provided while preserving host state such as active network connections. Finally, in [671], a proactive fault tolerance technique for Message Passing Interface(MPI) applications is presented exploiting Xen's live migration mechanism to migrate an MPI task from a health-deteriorating node to a healthy one without stopping the MPI task during most of the migration. Experimental results demonstrate that live migration hides migration costs and limits the overhead to only a few seconds. Some further general approaches for leveraging virtualization to improve system dependability are surveyed in [608, 759]. In [657], a high-level approach for autonomic management of system availability including real-time evaluation, monitoring and management is sketched. The authors suggest using analytical models (non-state space or state space models) parametrized using monitoring data collected during operation. The approach, however, is targeted at static system architectures and assumes that the underlying availability models are built manually at system design time.

4.3.2.2 Self-Adaptive Capacity and Power Management in Virtualized Data Centers Including Trade-Offs.

We first describe general approaches to a self-adaptive capacity and power management. Afterwards, approaches specifically targeted at virtualized environments are reviewed.

A number of self-adaptive approaches have been proposed that automatically adapt resource allocations in response to changes in application workloads in a way that utility is maximized, e.g., QoS requirements are satisfied while resources are used efficiently. Existing work mostly focuses on performance as QoS property and utility functions are based on assigning rewards for satisfied SLAs and penalties for violated SLAs (e.g., [18, 252, 601, 641]). In recent years, given the rising cost of energy, capacity management strategies aiming at improving the power usage effectiveness have received increasing attention (e.g., [193, 496, 913]).

Existing approaches to self-adaptive capacity management are typically based on: i) control theory feedback loops, ii) machine learning techniques or iii) general utility-based optimization techniques. Approaches based on feedback loops and control theory (e.g., [17, 31]), can normally guarantee system stability by capturing the transient system behavior [31]. Machine learning techniques, without a need for an a priori analytical model of the system, base their learning sessions on live systems. Such techniques have been used to tackle resource allocation problems [875] as well as the coordination of multiple autonomic managers [479]. In utility-based approaches, the system is typically modeled by means of a performance model embedded within an optimization framework aiming at optimizing multiple criteria such as different QoS metrics [497, 647, 914].

Utility-based optimization frameworks differ in the way in which they trigger adaptations. There are reactive and proactive approaches. The former react on certain events observed in the system, the latter try to anticipate the future system behavior and thus require forecasting mechanisms: a model that allows utility predictions and a way to forecast model input parameters. For workload forecasting, established time series analysis techniques [147] are used, e.g., Brown's quadratic exponential smoothing or general AutoRegressive-Moving Average (ARMA) models have been implemented in [647] and [194, 499], respectively. Regarding performance modeling, existing work mainly uses predictive performance models that capture the temporal system behavior (e.g., queueing networks) where the platform is normally abstracted as a "black-box" (e.g., [18, 102, 194, 901, 956]). Applications are modeled by a single queue with a single workload class [193] or multiple workload classes [102]. In [956], multi-tier applications are modeled using queueing networks where one queue represents one tier. All these models are solved analytically, e.g., in the latter case based on mean-value analysis (MVA). In [498], layered queueing models (LQNs) are solved by means of simulation. A different approach uses fuzzy-logic models to model the resource needs of an application for a given workload intensity [947]. The fuzzy-logic models need to be trained under dynamically changing workloads.

Resource allocation problems have been studied in the literature, frequently using techniques including bin packing (e.g., [122, 497]), multiple knapsack problems, and multi-dimensional knapsack problems [526]. For dynamic resource allocation applications, previous studies address this problem using linear optimization techniques [519] or non-linear optimization strategies based on simulated annealing [929], fuzzy logic [947], or other heuristics [20]. There are approaches to formulate the optimization problem as a network flow problem [601], to solve it with genetic algorithms [646], or to automatically change deployments using profiles capturing experts' knowledge of scaling different types of applications [950]. The above studies differ in the objective of the optimization and the type of applications on which they focus.

In virtualized environments, due to the introduction of virtual resources, the resource allocation problem is more complex. The studies in [637, 749] validate a performance inference in virtualized environments. There are strategies that explicitly make use of VMM configurations or recent CPU technologies. For instance, the authors of [641, 676] propose to exploit the min, max and shares parameters (respectively CPU priorities) for VM placement and power consolidation in data centers. In [363], the power-to-frequency relationship of dynamic voltage and frequency techniques is leveraged to distribute available power among the servers in order to get maximum performance. Some recent work on capacity management in Cloud infrastructures, based on LQN models, considers both performance and power as well as adaptation costs (incl. live migration costs) [497, 498]. To estimate the power consumption, utilization-based models from previous studies [559] are used. The following adaptation actions are considered: adapt a VM's CPU capacity, add/remove a VM, live-migrate a VM between hosts, and shut down/restart a host [497]. For the optimization there are two algorithms: a bin packing algorithm optimizing the power/performance tradeoff and an A* graph search algorithm that takes adaptation costs as well as search costs into account. The case study shows promising results, however, it is based on a simple multi-tier application with read-only transactions and a fixed web tier.

4.4 Emerging Research Directions

In this section, we outline emerging research directions targeting resilience and dependability management in Cloud infrastructures. At first, we discuss the question of how the flexible allocation mechanisms available in virtualized environments can be used to tackle scalability and consolidation issues. Afterwards, we capture the research challenge of finding representative predictive models and model parameters. Finally, we examine the trade-off decisions between performance and energy consumption and highlight the need for self-aware management techniques that enable a continuous application of management activities during system operation.

4.4.1 A Question of Scale

By 2015, it is predicted that more than 75% of computer infrastructure will be purchased from virtualized service providers [362]. Such services are hosted in Cloud environments with computation and network resources multiplexed between many distinct services. Although functionally, services may not impact each other, there is good evidence to suggest that performance stress from one virtual machine can indeed be noticed by another virtual machine instance [294].

Cloud administrators, like software developers, are increasingly responsible for the reliable and performance-driven provision of these software and hardware services. They face difficult quantitative scalability questions, often focused around service-level response-time goals. Being able to create accurate predictive models of such services is a major challenge in performance engineering and stochastic analysis.

Clearly servers could be over-provisioned in an effort to obtain high throughput, availability or resilience for all services. However, this is not a viable solution. The economics of virtualized service provision dictate that a sufficient level of shared or multiplexed computation is in fact a requirement. The energy consumed for unnecessary servers and extra air-conditioning will render a policy of server over-provisioning unsustainable financially, even if in doing so it was able to satisfy a strict service level requirement.

This is one of the major challenges facing Cloud Computing—how can many services be multiplexed in a virtualized environment and how to guarantee service level agreements imposed upon those services while minimizing the energy costs and maximizing the revenue of the overall cloud environment.

4.4.2 Parameter Sweeping

Here are some examples of the quantitative scalability questions and requirements that a virtualized environment might face. Maintaining a predictive model of a Cloud environment will mean both sustaining an accurate behavioral model of the services and virtualized architecture but also addressing the key scalability and configuration issues, for example:

- How many servers does a Cloud cluster need in order to execute 4000 jobs every minute at least 95% of the time?
- Under the predicted traffic profile, at what rate can a Cloud environment hibernate its servers to save energy, given the time penalty involved in power-cycling a host and relocating virtual instances?
- How many virtual machines can be launched on a host (for the same/different service) while maintaining a service level requirement of 96.7% of service requests actioned within 0.88 s?

These are all examples of performance evaluation questions where the result is contingent on specific model parameters. Potentially, small fluctuations in a set of key parameters in the model will have an enormous effect on the overall performance and even functional behavior of the whole system. Discerning which parameters have the most prominent effect on a given performance goal is a question of sensitivity analysis and can be a highly computationally intensive task even for small models.

Where such questions are not asked, or not rigorously answered, the consequences are very familiar. Systems are delivered which fail to win the trust of users because their performance is too unpredictable. Those systems which do deliver the required level of service often have excessively high running costs because their architects over-provisioned the hardware requirements in an attempt to mask failings due to uncertain software performance. There is a growing understanding that the running costs of a system greatly outweigh the development costs and that it is false economy to buy more hardware to cut software costs.

For these reasons, precise query-driven performance evaluation of computer systems and specifically virtualized computer systems is an important practical concern. In the next section we will highlight some of these energy-computation tradeoffs in the context of a simple multi-client, multi-server environment. Achieving this for a more complex Cloud environment with many possible services will require a step change in modeling and analysis approaches.

4.4.3 Trade-off Between Energy Consumption and Performance

We demonstrate the sort of energy/performance trade-off on a simple massively parallel client-server system. It serves to demonstrate the synergy of several critical issues that will need to be considered in a more complex model of a Cloud environment: scalability analysis via parameter sweeping, energy modeling and server hibernation.

The model consists of a large number of clients and a large number of servers cooperating together. The clients access the servers in two stages: first the client requests some data of the server and then the client receives the data from the server in response; the client goes on to process this data individually before restarting. The servers, in addition to serving clients, can hibernate to save energy and can also break. Broken servers are repaired. The details of this stochastic model and analysis can be found in Stefanek et al. [848]. A reward architecture is deployed to keep track of energy consumption and a fluid analysis technique [420] is used to calculate a service level agreement.

In this client/server model, we might be interested in the optimal number of servers that have to be employed in order to guarantee given performance requirements while minimizing the associated running costs. The performance requirements are often given in terms of a *Service Level Agreement* (SLA) for each client. In the context of this model, a suitable SLA might require that a client finishes its first request cycle within a given time period with a given high probability, for example within 4.0 s with probability being at least 0.9. Considering only the configurations that satisfy such an SLA, the *feasible configurations*, we can look for those that minimize the energy expended over the operation of the system.

Figure 4.1 is generated by the Grouped PEPA Analyzer (GPA) tool [847] and shows an example where we vary the number of servers and the rate with which they are hibernated. For each configuration we calculate the energy used and plot a point on the surface only if that configuration satisfies the SLA requirement mentioned above. We are able to find the configuration (84 servers and a hibernation rate of 0.37) which minimizes the energy consumption in the system. Intuitively, increasing the number of servers and decreasing the hibernation rate increases the probability of a client finishing early, but also raises the energy cost of running the system. Although, at this stage we are not capturing issues such as virtualization, multiple services or server classes in the model, this example illustrates the power of predictive modeling in being able to identify so-called sweet spots in operation.

4.4.4 Self-Aware Systems

As discussed in the previous sections, managing system resources in Cloud environments to ensure acceptable end-to-end application QoS and efficient resource utilization is a challenge. Modern enterprise software systems have highly distributed architectures composed of loosely-coupled services that operate and evolve independently, and are subjected to time-varying workloads.

The presented challenges call for novel systems engineering methodologies enabling the engineering of so-called *self-aware software systems* [547, 549]. The latter should have built-in online QoS prediction and self-adaptation capabilities used to enforce QoS requirements in a cost- and energy-efficient manner. Self-awareness in

Fig. 4.1 Global optimization
of the energy consumption
of the server components
from [847]. Configurations
satisfying the SLA are those
in the *upper plane*

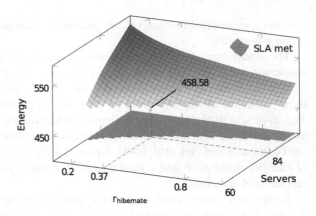

this context is defined by the combination of three properties that systems should
possess:

- *Self-reflective*: aware of their software architecture, execution platform and the
  hardware infrastructure on which they are running as well as of dynamic changes
  that occur during operation,
- *Self-predictive*: able to predict the effect of dynamic changes (e.g., changing user
  workloads) as well as predict the effect of possible adaptation actions,
- *Self-adaptive*: proactively adapting as the environment evolves in order to ensure
  that their non-functional requirements (e.g., availability, performance and reliability) and respective SLAs are continuously satisfied in a cost- and energy-efficient
  manner.

Self-aware systems engineering is a newly emerging research area at the intersection
of several computer science disciplines including software architecture, computer
systems modeling, autonomic computing, distributed systems, and more recently,
Cloud Computing and Green IT [548].

4.5 Conclusion

We have provided an overview of the research challenges and opportunities in providing dependability and resilience in Cloud Computing environments. State-of-the-art
approaches for dependability assessment and for managing dependability, performance and security were presented, including approaches to self-adaptive capacity
and power management in virtualized data centers. The identification of the existing gaps led to an overview of the emerging research directions. It is still an open
question, how a set of services should be multiplexed in a virtualized environment
while SLAs are guaranteed in such a way that the revenue of the overall Cloud
environment is maximized. In particular, modeling the trade-offs between energy
consumption/costs and application QoS remains a challenge.

As shown in this chapter, there are many challenges in assessing and managing dependability and performance in the Cloud. Their solution requires techniques dealing with very large and complex systems, including monitoring, modeling, and online prediction, all discussed in other chapters of this book. An introduction to monitoring and failure diagnosis of complex systems is given in Chap. 12.[3] Modeling of large systems is required for both online and offline management decisions, but highly-complex systems quickly run into the state-space explosion problem. Approaches for solving this problem are presented in (Chap. 6[4]). Online prediction, which is necessary to manage the high dynamicity of Cloud systems, is discussed in Chap. 9.[5] The application of these techniques to Cloud systems is demonstrated on the case-studies in Chap. 19.[6]

Acknowledgments The work of the first author was funded by the German Research Foundation (DFG) under grant No. KO 3445/6-1. Jeremy Bradley and Anton Stefanek are supported by the UK Engineering and Physical Sciences Research Council on the AMPS project (reference EP/G011737/1). Vlastimil Babka is supported by the Czech Science Foundation project GACR P202/10/J042.

---

[3] "Failure Diagnosis of Complex Systems"

[4] "Scalable Stochastic Modelling"

[5] "Online Prediction"

[6] "Providing Dependability and Resilience in the Cloud: Case Studies"

Part II
Modelling Techniques

Chapter 5
Phase-Type Distributions

Philipp Reinecke, Levente Bodrog and Alexandra Danilkina

Abstract Both analytical (Chap. 6) and simulation- and experimentation-based (Chap. 17) approaches to resilience assessment rely on models for the various phenomena that may affect the system under study. These models must be both accurate, in that they reflect the phenomenon well, and suitable for the chosen approach. Analytical methods require models that are analytically tractable, while methods for experimentation, such as fault-injection (see Chap. 13), require the efficient generation of random-variates from the models. Phase-type (PH) distributions are a versatile tool for modelling a wide range of real-world phenomena. These distributions can capture many important aspects of measurement data, while retaining analytical tractability and efficient random-variate generation. This chapter provides an introduction to the use of PH distributions in resilience assessment. The chapter starts with a discussion of the mathematical basics. We then describe tools for fitting PH distributions to measurement data, before illustrating application of PH distributions in analysis and in random-variate generation.

P. Reinecke (✉) · A. Danilkina
Institute of Computer Science,
Free University Berlin,
Takustr. 9, 14195 Berlin, Germany
e-mail: philipp.reinecke@fu-berlin.de

L. Bodrog
Department of Telecommunications,
Budapest University of Technology and Economics,
Budapest 1521, Hungary
e-mail: bodrog@webspn.hit.bme.hu

A. Danilkina
e-mail: danilkin@zedat.fu-berlin.de

K. Wolter et al. (eds.), *Resilience Assessment and Evaluation of Computing Systems*,
DOI: 10.1007/978-3-642-29032-9_5, © Springer-Verlag Berlin Heidelberg 2012

5.1 Introduction

Phase-type (PH) distributions are an often-used type of model for many phenomena in system evaluation, e.g., service-times, delays, and failure times. This chapter provides a gentle introduction to the theory of PH distributions and their application in common evaluation tasks.

As an illustrative example we consider resilience evaluation of a simple system where clients are being served by a faulty server. The server uses multiple threads that require access to shared resources, but resource contention may lead to a deadlock, which manifests as a situation where no service is provided anymore. We assume that we do not have the means to address the root of the problem in the server itself, but might be able to reset the server once it has reached a deadlock. In order to assess resilience of a system relying on this server to work, we want to model the time between server crashes. We use PH distributions for this task, as they provide good approximation of the time-to-failure distribution and are well-suited for both analytical approaches and simulation.

The general workflow for applying phase-type distributions in evaluation tasks is shown in Fig. 5.1: first, a PH distribution describing the phenomenon under study has to be found. This can be achieved both with a white-box and a black-box approach. With the white-box approach (Sect. 5.3), the structure of the system is used to directly infer a PH distribution that describes the behaviour of the system. With the black-box approach (Sect. 5.4), system behaviour is measured and the measurements are fitted by a PH distribution. Both approaches result in a distribution that is a model for the behaviour of the system. This distribution can then be used in analytical approaches such as matrix-analytic methods (Sect. 5.5) and in simulation (Sect. 5.6).

We are going to show how to arrive at a PH distribution for the time-to-failure distribution for our example system using both the white-box and the black-box approach. We will then illustrate application of the distribution both in analytical and in simulation approaches. First, however, we need to introduce the required mathematical background and notation.

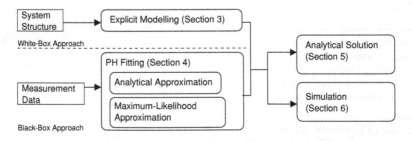

Fig. 5.1 Typical workflow when applying phase-type distributions in system evaluation

5.2 Mathematical Background

Continuous phase-type (PH) distributions represent the time to absorption in a Continuous-Time Markov Chain (CTMC) with one absorbing state [683]. PH distributions are commonly specified by a vector-matrix tuple $(\boldsymbol{\alpha}, \mathbf{A})$, where

$$\boldsymbol{\alpha} = (\alpha_1, \ldots, \alpha_n) \in \mathbf{R}^n \quad \text{and} \quad \mathbf{A} = \begin{pmatrix} \lambda_{11} & \cdots & \lambda_{1n} \\ \vdots & \ddots & \vdots \\ \lambda_{n1} & \cdots & \lambda_{nn} \end{pmatrix} \in \mathbf{R}^{n \times n}.$$

Definition 5.1 The *size* of the $(\boldsymbol{\alpha}, \mathbf{A})$ representation is the size of the vector $\boldsymbol{\alpha}$, which is equal to the size of the square matrix $\mathbf{A}$.

Definition 5.2 The *probability density function (PDF), cumulative distribution function (CDF), Laplace-Stieltjes Transform (LST)* of the CDF and kth *moment*, respectively, are defined as follows [446, 683, 870]:

$$f(x) = \boldsymbol{\alpha} e^{\mathbf{A}x} \mathbf{a}, \tag{5.1}$$

$$F(x) = 1 - \boldsymbol{\alpha} e^{\mathbf{A}x} \mathbb{1}, \tag{5.2}$$

$$\tilde{F}(s) = \alpha_{n+1} + \boldsymbol{\alpha}(s\mathbf{I} - \mathbf{A})^{-1}\mathbf{a}, \tag{5.3}$$

$$E\left[X^k\right] = k! \boldsymbol{\alpha}(-\mathbf{A})^{-k} \mathbb{1}. \tag{5.4}$$

where $\mathbb{1}$ is the column vector of ones of the appropriate size and $\mathbf{a} = -\mathbf{A}\mathbb{1}$. Note that, since $\boldsymbol{\alpha}$ is a row vector and both $\mathbb{1}$ and $\mathbf{a}$ are column vectors, the above equations do indeed specify scalar values. Furthermore, observe that phase-type distributions have rational LST and that the eigenvalues of the transient generator matrix are the poles of the LST of the distribution [697].

The vector-matrix representation of a PH distribution is not unique. In general, there exists another representation $(\boldsymbol{\alpha}', \mathbf{A}')$ of size m that represents the same phase-type distribution. Different representations of a PH distribution may differ both in size $(n \neq m)$ and in the contents of the tuples.

Another representation of the same size can be computed by a similarity transformation, as follows: when $\mathbf{B}$ is invertible and $\mathbf{B}\mathbb{1} = \mathbb{1}$, then $(\boldsymbol{\alpha}\mathbf{B}, \mathbf{B}^{-1}\mathbf{A}\mathbf{B})$ is another representation of the same distribution, since its CDF is

$$1 - \boldsymbol{\alpha}\mathbf{B} e^{\mathbf{B}^{-1}\mathbf{A}\mathbf{B}x} \mathbb{1} = 1 - \boldsymbol{\alpha}\mathbf{B}\mathbf{B}^{-1} e^{\mathbf{A}x} \mathbf{B}\mathbb{1} = 1 - \boldsymbol{\alpha} e^{\mathbf{A}x} \mathbb{1}.$$

It is also possible to generate representations of the same distribution with another size, using a non-square matrix $\mathbf{W}$.

An important property of PH distributions (and in fact one which distinguishes them from larger classes such as the Matrix-Exponential (ME) distributions) is that every PH distribution has a *Markovian* representation $(\boldsymbol{\alpha}, \mathbf{A})$. This representation

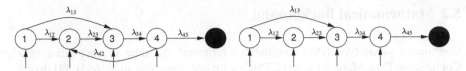

Fig. 5.2 CTMC representations for general and acyclic phase-type distributions

admits an interpretation of the PH distribution as the distribution of absorption-times in a Markov chain. With the Markovian representation, $\mathbf{A}$ describes the transient part of the generator matrix of the associated CTMC,

$$\overline{\mathbf{A}} = \begin{pmatrix} \mathbf{A} & \mathbf{a} \\ \mathbf{0} & 0 \end{pmatrix},$$

and consequently fulfills the required properties: all off-diagonal elements are non-negative ($\lambda_{ij} \geq 0\,(1 \leq i \neq j \leq n)$), all diagonal elements are negative, and the row-sums are non-negative ($\mathbf{a} = -\mathbf{A}\mathbb{1} \geq 0$). The vector $\boldsymbol{\alpha}$ is the vector of initial probabilities of the transient states of the CTMC, and thus $\boldsymbol{\alpha} \geq 0$ and $\boldsymbol{\alpha}\mathbb{1} \leq 1$. In the following, we focus on the Markovian representation of phase-type distributions, where we assume that $\boldsymbol{\alpha}\mathbb{1} = 1$, i.e., there is no probability mass at zero.

5.2.1 PH Classes

Based on the structure of the underlying Markov chain, several classes of phase-type distributions can be distinguished. These classes differ in the statistical properties they can represent. Furthermore, the structure of a PH representation often has an impact on its application, as some structures allow more efficient solutions.

The most important distinction is the one into Acyclic and General Phase-type distributions: every acyclic phase-type (APH) distribution has at least one Markovian representation without cycles in the sub-generator, while for general phase-type distributions cycles are allowed. This is illustrated in Fig. 5.2: the distribution on the left contains a cycle, that is, a backward transition from state 4 to state 2. The distribution on the right does not contain this transition and therefore there are no cycles.

Most approaches in fitting and application of PH distributions focus on the APH class, as this class offers better tractability than the general PH class. Within APH, we distinguish two important sub-classes: the first one is the class of Hyper-Erlang distributions (HErD). Hyper-Erlang distributions are mixtures of Erlang-distributions with different lengths and rates. They can be specified by a tuple $(\boldsymbol{\beta}, m, \mathbf{b}, \boldsymbol{\lambda})$, where $\boldsymbol{\beta}$ is the vector of initial probabilities of each Erlang branch, m is the number of Erlang branches, $\mathbf{b}$ is the vector of the lengths of the Erlang branches, and $\boldsymbol{\lambda}$ is a vector containing the rates. The size of a Hyper-Erlang distribution is given by the

Fig. 5.3 CTMC representations for Hyper-Erlang and hyper-exponential distributions

sum of the lengths of the branches, i.e., $n = \mathbf{b}\,\mathbb{1}$. The general structure is illustrated in Fig. 5.3, where we show a hyper-Erlang distribution with $m = 2$ branches of length $b_1 = 3$ and $b_2 = 2$, respectively. The initial probabilities and the transition rates are given by $\boldsymbol{\beta} = (\beta_1, \beta_2)$ and $\boldsymbol{\lambda} = (\lambda_1, \lambda_2)$. The size of this representation is $n = b_1 + b_2 = 5$. One important example is the Erlang distribution, i.e., a Hyper-Erlang distribution with only one branch and initial probability $\beta_1 = 1$.

The second sub-class of APH we consider is the class of Hyper-Exponential distributions (HEx) of order n, specified by initial probability vector $\boldsymbol{\alpha}$ and rate vector $\boldsymbol{\lambda}$. Figure 5.3 shows an example for a hyper-exponential distribution of size $n = 4$. From this example, it is obvious that the hyper-exponential distributions are a subclass of the hyper-Erlang distributions, as every hyper-exponential distribution is a hyper-Erlang distribution with branch length vector $\mathbf{b} = \mathbb{1}$. Furthermore, setting $n = 1$ and $\alpha_1 = 1$ yields the exponential distribution with rate λ_1.

5.2.2 Canonical Representations

While in general representations for phase-type distributions are not unique, several canonical forms have been defined. For each PH distribution, the canonical form of a given size n is unique in the sense that there exists no representation of the same size n with the structure of the canonical form, but different parameters. Therefore, by comparing canonical forms, we can determine whether PH distributions given by different representations are identical. More important, however, is the use of canonical forms in fitting, analysis, and simulation, where their typically low number of parameters and simple structure enable efficient methods.

In the following we discuss Cumani's Canonical Form 1 (CF-1) [251] and the Monocyclic form introduced in [659], as these are the most common ones.

5.2.2.1 The Canonical Form for APH Distributions

The Canonical Form 1 (CF-1) was defined in [251]. The structure of its underlying CTMC is shown in Fig. 5.4: the Markov chain can be entered at any state $i = 1, \ldots, n$ with probability α_i, but the absorbing state can only be reached by traversing all remaining states. For this structure, the associated generator $\mathbf{A}$ has a bi-diagonal structure, that is, for $i = 1, \ldots, n-1$ and $1 \le j \le n$,

$$-\lambda_{ii} = \lambda_{i,i+1} \quad \text{and} \quad \lambda_{ij} = 0 \text{ for } j \notin i, i+1.$$

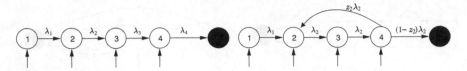

Fig. 5.4 Canonical representations for phase-type distributions

It is often convenient to describe a bi-diagonal generator by the vector

$$\Lambda = (\lambda_1, \ldots, \lambda_n),$$

where $\lambda_i = |\lambda_{ii}|$ for $i = 1, \ldots, n$. The formal definition for the CF-1 form is then

Definition 5.3 [251] The *Canonical Form 1 (CF-1 form)* is a bi-diagonal Markovian representation (α, Λ) where the elements of the diagonal, given in the vector Λ, are ordered by absolute value.

In [251, 698] it has been shown that every acyclic phase-type distribution with a Markovian representation of size n has a unique CF-1 representation of the same size.[1] The CF-1 form for an APH given as $(\alpha, \mathbf{A})$ can be obtained by a similarity transformation. A procedure for constructing the similarity transformation matrix is given in [422].

Note that transforming an APH representation of size n to the CF-1 form considerably reduces the number of parameters: a general APH representation has n initial probabilities $\alpha_1, \ldots, \alpha_n$ and n^2 entries in the subgenerator matrix $\mathbf{A}$, i.e., the number of parameters is $n + n^2$. In the CF-1 form $\mathbf{A}$ is an upper bi-diagonal matrix with $\lambda_{ii} = -\lambda_{i+1,i}$. The CF-1 form therefore only requires the n rates on the diagonal and the n entries in α, resulting in $2n$ parameters.

5.2.2.2 The Monocyclic Form for General PH Distributions

General PH distributions may have complex poles, and the poles of a PH distribution are given by the eigenvalues of the subgenerator matrix $\mathbf{A}$. As the eigenvalues of a bi-diagonal representation $(\alpha, \mathbf{A})$ are equal to entries of the diagonal and $\mathbf{A} \in \mathbf{R}^{n \times n}$ it is easy to see that a bi-diagonal structure like the CF-1 form cannot represent phase-type distributions with complex poles.

For this reason, [659] proposed the Monocyclic form as a chain of Feedback-Erlang (FE) blocks, defined as follows:

Definition 5.4 A *Feedback-Erlang (FE) block* is given by a tuple (b, λ, z) of the length b, transition rate λ, and feedback probability $z \in [0, 1)$. The Feedback-Erlang

---

[1] Smaller CF-1 representations may exist if there is redundancy in the original representation [422, 683, 750].

Fig. 5.5 Structure of a
Feedback-Erlang block

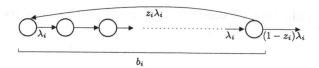

block consists of an Erlang-distribution with length b and rate λ and an additional
(feedback) transition from the last state of the block to the first state.

Figure 5.5 illustrates this concept. Note that the cases $z = 0$ and $b = 1$ are allowed.
For $z = 0$, the Feedback-Erlang is simply an Erlang of order b, while for $b = 1$ it
is an exponential distribution. The importance of this structure lies in the fact that
for $z > 0$ and $b > 1$ the block has a conjugate-complex pair of eigenvalues [659].
Therefore, a chain of FE blocks can be used to represent the complex eigenvalue
pairs of a general phase-type distribution.

Based on this observation, [659] define the Monocyclic representation as a chain
of Feedback-Erlang blocks:

Definition 5.5 A *Monocyclic representation* is given by the tuple $(\alpha, m, \mathbf{b}, \lambda, z)$,
where the vector $\alpha \in \mathbf{R}^{b1}$ specifies the initial state probabilities, and $\mathbf{b}, \lambda$ and z
define the length, rate, and feedback probability of the m Feedback-Erlang blocks.

The FE blocks are positioned such that the absolute values of the dominant eigenvalues r_i are in ascending order: $r_i \leq r_{i+1}$.

Any PH distribution has a Monocyclic representation [659]. If the representation
of the PH distribution is PH-simple [698] and of size n, then the size of the Monocyclic representation is $n' \geq n$. This potential size expansion makes the Monocyclic
representation less efficient in analytical studies, but its simple and still Markovian
structure makes it promising for simulation studies.

The structure of a Monocyclic representation is shown in Fig. 5.4. Note that if
$z_i = 0$ for all FE blocks $i = 1, \ldots, b$ the Monocyclic form is equivalent to the CF-1
form. That is, the CF-1 form is actually a special case of the Monocyclic form.

5.2.3 Properties

Phase-type distributions exhibit a number of properties that make them attractive for
use in resilience evaluation. In particular, the support of PH distributions is the set of
non-negative real numbers. Therefore, they can be used to model typical system properties such as response-times, interarrival times, or inter-failure times. Furthermore,
the PH class is closed under important operations such as minimum, maximum, and
summation, i.e., PH distributed random variables can be combined without losing
the properties of PH distributions.

On the other hand, even though PH distributions are well-suited for fitting data
(see Sect. 5.4), two important limitations may affect their suitability for particular

tasks. First, the density of a phase-type distribution is strictly positive [683]:

$$f(t) > 0, \quad t > 0,$$

i.e., phase-type distributions can only approximate the density if it is close to zero, and may require a large number of phases to do so, rendering models more complex. Second, PH distributions are limited with respect to the moments they can express. In particular, the feasible range of the squared coefficient of variation for the PH of size n (PH(n)) is

$$cv^2 \geq \frac{1}{n}, \tag{5.5}$$

where the equality holds for the Erlang distribution with n phases (Erl(n)) [260]. This bound implies that in order to approximate data with low variation a large number of phases is required.

For higher moments there is no general knowledge, however there are several special cases for which some insights on the moment bounds exist, like e.g., the moment bounds of the APH(2) $\equiv$ PH (2) class [871] and the moment bounds of the PH (3) class implied by the canonical form given in [447]. The bounds of the general APH class within the PH class are known according to the APH canonical form and there also exists a numerical method to determine the general PH bound in [448].

From the fitting perspective the *reduced moment problem* (when a distribution function is determined based on its moments) can also be crucial. This problem is only solved for the larger class of matrix-exponential distributions [907].

5.3 Explicit Modelling

We will now describe how a phase-type distribution can be obtained directly from the structure of the system under study. Returning to our example with the faulty server, recall that the server becomes unavailable due to resource contention between its threads. A very intuitive way of thinking about resource contention is in terms of the Dining Philosopher's problem [292][2]: at least two philosophers sitting around a table want to eat a dish for which they require a fork in each hand. However, there are only as many forks as philosophers. Each philosopher employs the following strategy: if a fork is available either to the left or to the right, they wait a random amount of time before taking it. Once they have one fork, they wait for the other one, before they start eating. They eat for a random amount of time and then drop both forks at the same time. It is immediately obvious that this strategy eventually leads to the situation where each philosopher has one fork in his left (or right) hand

---

[2] Note that we only use the general problem, given in [292], but do not assume a solution.

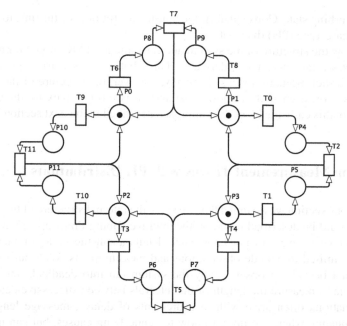

Fig. 5.6 Stochastic Petri Net (SPN) model for the Dining Philosopher's problem with four philosophers

and is waiting for the other one, which, however, is in the corresponding hand of his neighbour.

In terms of the server system, the philosophers represent the threads of the server and the forks are the shared resources. If each thread has access to one of the resources, but not to the other (each philosopher has one fork), the system is in the deadlock situation and unable to serve clients (no philosopher can eat). The only way of leaving this situation is to reset the system to the inital state where all resources are free (all forks are on the table).

Based on the abstraction as a Dining Philospher's problem, we can model our faulty server using a Stochastic Petri Net (SPN), as shown in Fig. 5.6 for $n = 4$: a marking on one of the four innermost places represents the respective fork resting on the table. The eight outermost places represent the hands of the philosophers, i.e., a marking in one of these places models that the philosopher has picked up a fork in this hand. The transitions leading from the inner places to the outer places model the act of picking up a fork with that hand, while the transitions from the outside to the inside model the laying down of both forks. We assume that both the times before picking up a fork and the eating time have exponential distribution, i.e., the transitions are Markovian.

The CTMC underlying this SPN possesses two absorbing states (the two deadlock situations). Since in both cases no philosopher can eat, or, equivalently, the server system cannot serve clients, both states describe the same situation and can be lumped into one absorbing state. Consequently, the time to absorption, i.e., the time to failure, follows a phase-type (PH) distribution.

If we know the structure of the system and can build a CTMC model for it, using e.g., an SPN, we can thus directly derive a PH distribution. This distribution can then be used in further evaluation steps. Typically, however, the structure of the system is not known, or, even if it is known, the system is too complex to allow direct modelling. In this case, the black-box approach described in the next section is more appropriate.

5.4 Fitting Measurement Traces with PH Distributions

In the previous section we assumed that we know the internal structure of the system, and that it could be described based on the intuitive Dining Philosopher's problem. Typically, however, we will not know such details about the system under study, and will be limited to outside observations and measurements. With our example, we might not be able to observe why the system ran into deadlock, but we are certainly able to measure the length of the intervals between successive deadlocks. Similar situations often arise with measurements of delays, message lengths, or other phenomena, where we do not know the underlying causes, but can measure their effects. In such cases we can fit a phase-type distribution to the data in question, and use this model in our evaluation.

Consider Fig. 5.7, where we show both a histogram of some data and the density of a phase-type distribution approximating the data. Our aim is to approximate the data as closely as possible, in order to obtain correct results when using the approximating distribution later on. In this section we provide the basics for fitting data sets with phase-type distributions. We discuss costs, quality metrics, and introduce three established fitting tools.

5.4.1 Costs of Fitting PH Distributions to Data

Since a PH distribution is defined by the tuple $(\boldsymbol{\alpha}, \mathbf{A})$, the problem of fitting translates to finding an initial probability vector $\boldsymbol{\alpha}$ and a sub-generator matrix $\mathbf{A}$ of appropriate size n. While, in general, higher-order PH distributions can provide a better approximation [124], they are more expensive in both analysis and simulation. Furthermore, the time required for fitting a distribution increases with n, as more parameters have to be fitted. Consequently, careful choice of n is important.

As will be shown in Sects. 5.5 and 5.6, the cost of using a PH distribution depends not only on the size n, but also on the structure of the representation. The same holds for the fitting problem. Here, the number of free parameters to be fitted can be reduced significantly by choosing an appropriate representation: if we assume the size n of the representation to be constant, then general phase-type distributions in an arbitrary Markovian representation have $n + n^2$ free parameters, as $\boldsymbol{\alpha}$ is a row vector

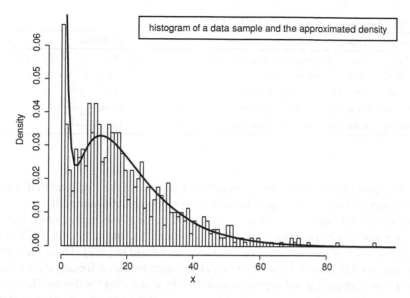

Fig. 5.7 Example data and its approximation with a phase-type distribution

of length n, and $\mathbf{A}$ is a matrix of size $n \times n$. If we assume that the representation is Monocyclic, we have a chain of m Feedback-Erlang blocks, each with a length parameter b_j, rate parameter λ_j and feedback probability z_j, and an initial probability vector of size n. As $m \leq n$, the upper limit for the number of free parameters is $3n + n$. Limiting ourselves to the APH class, we can utilise the CF-1 canonical form, which has only $2n$ free parameters: n transition rates and n initial probabilities. Finally, if we consider only HErD distributions in representations as shown in Fig. 5.3, the number of free parameters reduces to $3m$: m initial probabilities for the m Erlang branches, m lengths for the Erlang branches, and m transition rates.

5.4.2 Quality Measures

Fitting a phase-type distribution to data requires careful choice of the right fitting tool, as well as of fitting parameters such as sub-class and size. As just discussed, the approximation problem becomes less complex if data is fitted with subclasses of phase-type distributions, however, fitting quality may decrease as well, as subclasses cannot represent all properties of the general PH class. For example, hyperexponential distributions cannot approximate distributions with oscillating densities [880].

In order to assess the quality of data approximation, quality measures are required. An intuitive method consists in simply comparing the shape of the empirical PDF or CDF to that of the approximating PH distribution. This gives a visual impression how

Table 5.1 Performance measures defined in [572]

Performance measure	Definition
Area difference between distribution functions ΔF	$\Delta F = \int_0^\infty |\hat{F}(x) - F(x)| dt$
Area difference between densities Δf	$\Delta f = \int_0^\infty |\hat{f}(x) - f(x)| dt$
Relative error in the first moment (mean c_1)	$e_1 = \frac{|\hat{c}_1 - c_1|}{c_1}$
Relative error in the second central moment (variance c_2)	$e_2 = \frac{|\hat{c}_2 - c_2|}{c_1}$
Relative error in the third central moment (skewness c_3)	$e_3 = \frac{|\hat{c}_3 - c_3|}{c_3}$

well the approximating PH distribution reflects the shape of the empirical PDF/CDF. For instance, in Fig. 5.7 the approximated density fits the data quite well.

While a visual impression often yields a good initial assessment, a more formal approach requires exact definitions of quality measures. In the area of PH fitting there exists a set of standard quality measures, as defined in [572]. These measures are summarised in Table 5.1: the first two performance measures formalise the visual comparison of empirical and approximated data, by computing the distance between both curves. The last three measures capture how well the fitted distribution approximates the empirical moments of the data. Based on these performance measures we can decide which tool to use, and which fitting is most appropriate for the requirements and future application of approximation results. For instance, for use in a stochastic model whose behaviour primarily depends on the first three moments, one would aim to get small relative moment errors, while in other applications fitting the shape of the density may be more important.

5.4.3 Introduction to PH-Fitting Tools

Here we outline three tools for data approximation with phase-type distributions: Moment Matching, G-FIT and PhFit. They mainly differ with respect to the algorithms they employ and the subclass of PH distributions they support. There are two general and relevant classes of algorithms: analytical and statistical methods, where the former relies on direct computation of the parameters and the latter is based on iterative procedures for parameter estimation.

5.4.3.1 Analytic Approximation: Moment Matching

Analytic moment-matching methods have the advantage of being fast, easy to implement, and giving low errors in the moments. On the other hand, accuracy of the fitting may be limited by the representation. We illustrate this using the method proposed in [870], which can fit an APH(2) distribution to the first three moments of a data set.

The approach proceeds by computing the approximation parameters directly from the moments, as follows: An APH(2) in CF-1 form with $\boldsymbol{\alpha} = (\alpha_1, 1 - \alpha_1)$ and

$$\mathbf{A} = \begin{pmatrix} -\lambda_1 & \lambda_1 \\ 0 & -\lambda_2 \end{pmatrix},$$

is defined by three parameters, λ_1, λ_2, and α_1. Recall from Definition 5.4 the general moments-generating function for a PH distribution. Writing the first three moments explicitly:

$$E[X] = m_1 = \frac{\lambda_1 + \alpha_1 \lambda_2}{\lambda_1 \lambda_2},$$

$$E[X^2] = m_2 = \frac{2(\lambda_1^2 + \alpha_1 \lambda_1 \lambda_2 + \alpha_1 \lambda_2^2)}{\lambda_1^2 \lambda_2^2},$$

$$E[X^3] = m_3 = \frac{6(\lambda_1^3 + \alpha_1 \lambda_1^2 \lambda_2 + \alpha_1 \lambda_1 \lambda_2^2 + \alpha_1 \lambda_2^3)}{\lambda_1^3 \lambda_2^3},$$

[870] obtain a system of three linear equations. Solving this system for $\lambda_1, \lambda_2, \alpha_1$ yields an APH(2) that matches the first three moments. However, possible solutions are limited by the moment bounds for the APH(2) class (cf. Sect. 5.2.3). For combinations of moments outside the moment bounds, the system has no solution, i.e., data sets with these moments cannot be fitted exactly by an APH(2). For instance, as follows from (5.5), the smallest SCV cv^2 that can be represented by an APH(2) is

$$cv^2 = \frac{1}{2},$$

which puts constraints on the relation of the mean and variance. Data sets with $cv^2 < \frac{1}{2}$ require PH distributions of higher order. Similar constraints exist for the third moment, although in some cases the third moment can be approximated even when no exact fitting is possible.

   Iterative procedures for PH fitting have the advantage of providing more flexibility than analytical moment-matching methods. On the other hand, they are usually slower than the analytical approach. In the following we discuss two important tools of this class.

5.4.3.2 G-FIT for Fitting Hyper-Erlang Distributions

The G-FIT tool [880] approximates data using Hyper-Erlang distributions. Recall that the number of transition rates and the size of the initial vector of a Hyper-Erlang distribution only depend on the number of Erlang branches. This enables an efficient fitting method: once the number m and length b of the Erlang branches have been set, the parameters are

$$\Theta = (\beta, \lambda).$$

In each iteration the Expectation Maximisation (EM) algorithm (cf. [281]) computes parameters β and λ which maximise the likelihood of the parameters, given the data. G-FIT provides convergence checks based on the maximal change in Θ and on the relative differences of the log-likelihood between successive iterations.

The user may specify the number and length of Erlang branches prior to fitting or let G-FIT determine an optimal size. In the first case the user has to set a number of Erlang branches and their length. The second option is more general and is useful for the unexperienced user. It requires as input only a number of phases for the resulting distribution. G-FIT will then estimate optimal number of Erlang branches and their parameters, by trying all possible combinations.

G-FIT expects an input as a text file containing the data set. The first line should be a number of data points in the data set followed by data points themselves, which are given one per line. The output is also a text file, containing the number of Erlang branches, number of phases, initial probabilities and transition rates for each Erlang branch.

5.4.3.3 PhFit for Fitting APH Distributions

The PhFit tool [446] approximates data using acyclic phase-type distributions in CF-1 form. It applies a variant of the Frank/Wolfe algorithm [355, 361] for constrained non-linear minimisation of the distance between the PH distribution and the data. One major advantage is that the user can choose between different distance measures, in order to obtain an optimal fitting. The distance measures supported by PhFit are the relative entropy, PDF area distance, and CDF area distance, defined as

$$\int_0^\infty f(t) \log(\frac{f(t)}{\hat{f}(t)}) dt, \quad \int_0^\infty |\hat{F}(x) - F(x)| dt, \quad \text{and} \quad \int_0^\infty |\hat{f}(x) - f(x)| dt, \text{ respectively,}$$

where $f(t)$ denotes the probability density function (PDF) of the original distribution and $\hat{f}(t)$ the PDF of the approximating distribution, $F(t)$ the cummulative distribution function (CDF) of the original distribution and $\hat{F}(t)$ the CDF of the fitted distribution. Among the fitting tools we discuss, PhFit is the only one with a graphical user interface. This feature is beneficial for finding appropriate fitting parameters and evaluation of results.

PhFit computes optimal values for the distribution parameter (α, A) starting with special initial values $(\alpha^{(0)}, A^{(0)})$ according to the distance measure. PhFit picks optimal values from 1,000 randomly generated pairs of vectors. The distance measure defines the optimality criterion. The optimisation problem is solved by using the iterative linearisation method. After linearisation in a local neighbourhood of the current distribution parameters, the direction for optimisation of the distance measure is determined by the Simplex algorithm. The algorithm stops computation once the relative difference between

$$(\alpha^{(i-1)}, \mathbf{A}^{(i-1)})$$

and

$$(\alpha^{(i)}, \mathbf{A}^{(i)})$$

for iteration i is less than the predefined value, or if a maximum number of iterations is reached.

PhFit provides separate fitting for body and tail. The body is the part of distribution with the most mass, whereas the tail represents rare data points. The user can choose the boundary where the tail begins. The tail is approximated with a heuristic method that determines parameters for a hyper-exponential distribution. The body is then approximated as described before. The resulting distribution is then given by the CF-1 form and the hyper-exponential distribution.

PhFit requires as input a text file containing the data in ascending order. The output consists of the initial probability vector α and the diagonal of the subgenerator matrix. Note, however, that in contrast to the definition we gave in Definition 5.3, PhFit considers the 0th state to be absorbing, instead of state $(n + 1)$. That is, the output of PhFit is reversed, compared to the notation used throughout this chapter.

5.5 Phase-Type Distributions in Model Analysis: Matrix Analytic Methods

In the previous sections we discussed how a phase-type distribution modelling the phenomenon of interest can be obtained either explicitly or by fitting measurement traces. We will now illustrate how such a model can be used in analytical approaches. Referring to our example with the faulty server, we may want to analyse the effect of deadlocks on job processing in a queueing system. We assume that the server is reset after a deadlock, but that the fault leading to the deadlock persists. Then, the instances of deadlock can be fitted by a PH renewal process or a Markovian Arrival Process (MAP). If the process starts from the same initial state each time the process of deadlocks will be uncorrelated and forms a PH renewal process. In case of correlated initial states the time between the deadlocks forms a MAP. In this section we discuss matrix-analytic methods [579] for analysing complex models using phase-type distributions.

Matrix-analytic methods utilize the structure of the Markov chain which, in this chapter, is two-dimensional. Both dimensions have their own characteristics. The first dimension represents the—usually finite—number of phases $J(t)$ of the process. The second dimension, denoted by $N(t)$, is the infinite counting process. This approach results in an infinite, but well-structured, Markov chain on the block level where the blocks describe the phase either with or without arrival. The same block structure appears also in the generator matrix of the Markov chain which can be upper block-bidiagonal or tridiagonal in our cases.

The examples of this section show how the matrix-analytic methods utilize the analytic PH properties during the solution of complex Markov models. The result can be either the short-term or the steady-state behavior. The methods also allow to find the solution of infinite models by solving finite problems.

5.5.1 Processes with PH Marginal Distribution

A sequence of random variables—according to a given (marginal) distribution—defines a stochastic process or simply process. Processes play an important role in stochastic modeling thus it comes naturally to propose the process with PH marginal distribution. Here we investigate both the independent identical distributed (iid) and the correlated arrival process with PH marginal distribution. These are the PH renewal process and the Markov arrival process (MAP) respectively.

Referring to the example in Sect. 5.3 a faulty server has PH distributed time to deadlock if the relevant times are exponentially distributed. Furthermore if the system restarts at deadlock situations then the resulting sequence of times to deadlock is a stochastic process with PH marginal distribution.

5.5.1.1 PH Renewal Process

Given a phase-type distribution represented by the initial vector α and subgenerator matrix $\mathbf{A}$, the generator matrix

$$\mathbf{Q} = \begin{pmatrix} \mathbf{A} & \mathbf{a}\alpha & 0 & \cdots \cdots \\ 0 & \mathbf{A} & \mathbf{a}\alpha & 0 & \cdot \\ 0 & 0 & \mathbf{A} & \mathbf{a}\alpha & 0 \\ \cdots \cdots \cdots \cdots \cdots \end{pmatrix}, \tag{5.6}$$

defines the PH renewal process for which $(\alpha, \mathbf{A})$ is the marginal distribution. $\mathbf{a} = -\mathbf{A}\mathbb{1}$ is the vector of absorption rates of the marginal distribution. The blocks on the diagonal describe the phase transitions of the PH marginal, and the blocks in the upper co-diagonal describes the phase transitions belonging to the renewal instances. The graph of the corresponding continuous time Markov chain (CTMC) is depicted in Fig. 5.8.

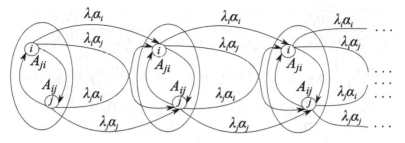

Fig. 5.8 The graph of the PH renewal process

The product in the upper co-diagonal blocks expresses that the initial distribution of the next interarrival is always the same (α) after arrival ("absorption" in the PH marginal) regardless of any of the other interarrivals, i.e., the process is uncorrelated.

The generator matrix of the phase process is $\mathbf{H} = \mathbf{A} + \mathbf{a}\alpha$. The steady state phase distribution (π) is the solution of the linear system of equations

$$\pi \mathbf{H} = 0$$
$$\pi \, \mathbb{1} = 1. \qquad (5.7)$$

The transient phase distribution is

$$\pi(t) = \pi(0)e^{\mathbf{H}t} \qquad (5.8)$$

which is a vector of elements $\pi_i(t) = \Pr(J(t) = i)$ giving the probability that the process is in phase i at time t. Using the transient phase behavior, at time t the remaining time to the next arrival is distributed according to the phase-type distribution $(\pi(t), \mathbf{A})$.

Let the entries of the vector $\pi(n, t) = (\Pr(N(t) = n, J(t) = j))$ give the probabilities that at time t the number of arrivals is equal to n and the level process is in phase j. With initial conditions $\pi(0, 0) = \alpha$ and $\pi(i, 0) = 0$ $(i > 0)$, the transient number of arrivals is given by the differential equation

$$\frac{d\pi(i, t)}{dt} = \pi(i, t)\mathbf{A} + \pi(i - 1, t)\mathbf{a}\alpha, \qquad (5.9)$$

whose z-transform, with initial condition $\pi(z, 0) = \alpha$, is

$$\frac{d\pi(z, t)}{dt} = \pi(z, t)\mathbf{A} + z\pi(z, t)\mathbf{a}\alpha = \pi(z, t)(\mathbf{A} + z\mathbf{a}\alpha). \qquad (5.10)$$

The solution of the differential equation, i.e., the transient distribution of the number of arrivals, is

$$\pi(z, t) = \alpha e^{(\mathbf{A}+z\mathbf{a}\alpha)t}. \qquad (5.11)$$

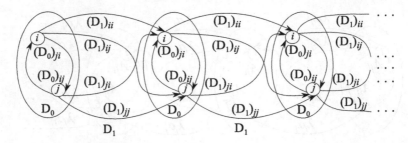

Fig. 5.9 The graph of the Markov arrival process

Regarding to the example in Sect. 5.3, the results on the PH renewal processes can be used to model the faulty server if the system restarts in the same (initial) state after each deadlock situation.

5.5.1.2 Markov Arrival Process

In the PH renewal process the phase distribution is the same after every arrival. In contrast, in the Markov Arrival Process (MAP) after each arrival an arbitrary phase distribution may hold. This allows the modelling of correlated arrival processes. The two-dimensional CTMC of the MAP process is also defined by the phase process $J(t)$, describing the phase of the marginal distribution, and by the counting process $N(t)$, giving the number of arrivals. Its graph is depicted in Fig. 5.9 and its generator matrix is

$$\mathbf{Q} = \begin{pmatrix} \mathbf{D}_0 & \mathbf{D}_1 & 0 & \dots \dots \\ 0 & \mathbf{D}_0 & \mathbf{D}_1 & 0 & . \\ 0 & 0 & \mathbf{D}_0 & \mathbf{D}_1 & 0 \\ \dots \dots \dots \dots \dots \end{pmatrix}, \tag{5.12}$$

where the Markov arrival process is represented by $\mathbf{D}_0$—the phase transitions without arrival—and $\mathbf{D}_1$—the phase transitions with one arrival. Such a MAP is denoted as MAP $(\mathbf{D}_0, \mathbf{D}_1)$.

The interarrival times of the MAP $(\mathbf{D}_0, \mathbf{D}_1)$ are PH $(\alpha_0, \mathbf{D}_0)$, PH $(\alpha_1, \mathbf{D}_0)$ ... The—correlated—phase distribution embedded at arrival instances forms a discrete time Markov chain (DTMC) with state transition probability matrix $\mathbf{P} = (-\mathbf{D}_0)^{-1} \mathbf{D}_1$.

The joint probability density function of the interarrival times, X_0 and X_k, is

$$f_{X_0, X_k}(x_0, x_k) = \boldsymbol{\pi} e^{\mathbf{D}_0 x_0} \mathbf{D}_1 \mathbf{P}^{k-1} e^{\mathbf{D}_0 x_k} \mathbf{D}_1 \mathbb{1}, \tag{5.13}$$

where π is the embedded stationary phase distribution at arrival instances, i.e., it is the solution of the linear system of equations

$$\pi P = \pi$$
$$\pi \mathbb{1} = 1. \tag{5.14}$$

The stationary interarrival time distribution is PH $(\pi, \mathbf{D}_0)$ with nth moment

$$E\left[X^n\right] = n!\pi\left(-\mathbf{D}_0\right)^{-n} \mathbb{1} \tag{5.15}$$

and the joint moment of two interarrivals is

$$E[X_0 X_k] = \int\limits_{x_0} \int\limits_{x_k} x_0 x_k \pi e^{\mathbf{D}_0 x_0} \mathbf{D}_1 \mathbf{P}^{k-1} e^{\mathbf{D}_0 x_k} \mathbf{D}_1 \mathbb{1} dx_0 dx_k$$
$$= \pi\left(\mathbf{D}_0\right)^{-1} \mathbf{P}^k\left(\mathbf{D}_0\right)^{-1} \mathbb{1}. \tag{5.16}$$

The covariance of two interarrivals is

$$\text{cov}\left(X_0, X_k\right) = E[X_0 X_k] - E^2[X] \tag{5.17}$$

and using (5.15)–(5.17) the lag k correlation of the MAP is

$$\text{corr}\left(X_0, X_k\right) = \frac{\text{cov}\left(X_0, X_k\right)}{E\left[X^2\right] - E^2[X]}. \tag{5.18}$$

The MAP can help to model the faulty server of Sect. 5.3 if the initial states of the system (after restart) are correlated.

5.5.2 The Quasi Birth-Death Process

The quasi birth-death (QBD) process [579, 683] is also defined by the phase process $(J(t))$ and the counting process $(N(t))$. But in case of the QBD process the counting, or the "level", process is allowed to be decreased by one as well as to stay on the same level or to be increased by one. It is thus the "multiphase" extension of the birth-death process which is for example the solution of the M/M/1 queueing system. The generator matrix of the QBD process has block-tridiagonal form

$$\mathbf{Q} = \begin{pmatrix} \mathbf{L}' & \mathbf{F} & 0 & .. & . \\ \mathbf{B} & \mathbf{L} & \mathbf{F} & 0 & . \\ 0 & \mathbf{B} & \mathbf{L} & \mathbf{F} & 0 \\ . & . & . & . & . & .. \end{pmatrix}, \tag{5.19}$$

Fig. 5.10 The graph of the
quasi birth-death process

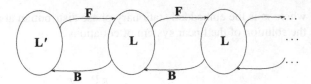

where the blocks or level transition matrices are

$\mathbf{L}'$ local state transitions inside the first—irregular—block,
$\mathbf{B}$ backward (level) state transitions,
$\mathbf{L}$ local state transitions on the regular levels and
$\mathbf{F}$ forward (level) state transitions.

The graph of the QBD is depicted in Fig. 5.10.

We give the solution method of the QBD through the analysis of the MAP/PH/1 queueing system with arrival process MAP $(\mathbf{D}_0, \mathbf{D}_1)$ and service time PH $(\boldsymbol{\alpha}, \mathbf{A})$. The level transition matrices are

$$\mathbf{L}' = \mathbf{D}_0 \otimes \mathbf{I}$$
$$\mathbf{B} = \mathbf{I} \otimes \mathbf{a}\boldsymbol{\alpha}$$
$$\mathbf{L} = \mathbf{D}_0 \oplus \mathbf{A}$$
$$\mathbf{F} = \mathbf{D}_1 \otimes \mathbf{I},$$

where $\mathbf{a} = -\mathbf{A}\mathbb{1}$ and $\mathbf{I}$ is the identity matrix of appropriate size. The operators $\otimes$ and $\oplus$ are the Kronecker product and sum, respectively.

The generator matrix of the phase process is $\mathbf{H} = \mathbf{B} + \mathbf{L} + \mathbf{F}$ and if it is irreducible then the steady state phase distribution is the solution of the linear system of equations

$$\boldsymbol{\pi}\mathbf{H} = 0$$
$$\boldsymbol{\pi}\mathbb{1} = 1. \tag{5.20}$$

The QBD process is stable if its stationary drift is less than zero

$$d = \boldsymbol{\pi}\mathbf{F}\mathbb{1} - \boldsymbol{\pi}\mathbf{B}\mathbb{1} < 0. \tag{5.21}$$

The steady state solution of the QBD is the solution of the infinite system of linear equations

$$\boldsymbol{\nu}\mathbf{Q} = 0$$
$$\boldsymbol{\nu}\mathbb{1} = 1. \tag{5.22}$$

Partitioning v according to the blocks of $\mathbf{Q}$ is

$$v = \begin{pmatrix} v_0 & v_1 & v_2 & \ldots \end{pmatrix}$$

and substituting the partitions into (5.22) we get

$$v_0 \mathbf{L}' + v_1 \mathbf{B} = 0 \tag{5.23}$$

and

$$v_{i-1}\mathbf{F} + v_i \mathbf{L} + v_{i+1}\mathbf{B} = 0 \quad \forall i \geq 1. \tag{5.24}$$

Assuming that the Markov chain is irreducible $v_i = v_{i-1}\mathbf{R} = v_0 \mathbf{R}^i$ ($\forall i$), i.e., its solution is the matrix geometric distribution, the general Eq. (5.24) can be rewritten as

$$v_0 \mathbf{R}^{i-1}\mathbf{F} + v_0 \mathbf{R}^i \mathbf{L} + v_0 \mathbf{R}^{i+1}\mathbf{B} = 0$$
$$v_0 \mathbf{R}^{i-1}\left(\mathbf{F} + \mathbf{R}\mathbf{L} + \mathbf{R}^2\mathbf{B}\right) = 0$$

with a solution determined by

$$\mathbf{F} + \mathbf{R}\mathbf{L} + \mathbf{R}^2\mathbf{B} = 0. \tag{5.25}$$

If the QBD is stable there is one of the solutions of $\mathbf{R}$ whose eigenvalues are within the unit circle on the complex plane.

As all the eigenvalues of the relevant $\mathbf{R}$ are within the unit circle there exists the limit of the sum $\sum_{i=0}^{\infty} \mathbf{R}^i = (\mathbf{I} - \mathbf{R})^{-1}$. Using the convergence the normalizing condition of v can be expressed as

$$v \mathbb{1} = \sum_{i=0}^{\infty} v_i \mathbb{1} = \sum_{i=1}^{\infty} v_0 \mathbf{R}^i \mathbb{1} = v_0 \sum_{i=1}^{\infty} \mathbf{R}^i \mathbb{1} = v_0 (\mathbf{I} - \mathbf{R})^{-1} \mathbb{1} = 1. \tag{5.26}$$

Now substituting $\mathbf{R}$ into (5.23) and using (5.26) we have a linear system of equations

$$v_0 \left(\mathbf{L}' + \mathbf{R}\mathbf{B}\right) = 0$$
$$v_0 (\mathbf{I} - \mathbf{R})^{-1} \mathbb{1} = 1 \tag{5.27}$$

for the zeroth block of v. All the other blocks can be calculated using v_0 and $\mathbf{R}$ as

$$v_i = v_0 \mathbf{R}^i, \quad \forall i. \tag{5.28}$$

By these considerations the infinite problem of solving the QBD in (5.22) is reduced to be the solution of the finite problems in (5.25), (5.27) and (5.28).

5.6 Phase-Type Distributions in Random-Variate Generation

While phase-type distributions enable efficient solutions for analytical models, they have applications beyond analytical approaches. Their ability to provide good models for many different empirical distributions makes them attractive in evaluation techniques where observed phenomena must be represented accurately and efficiently. In particular, they can be used both in discrete-event simulation of models that cannot be solved by analytical methods, and in fault-injection-driven experiments in testbeds. Referring back to our example, in addition to considering an analytical solution we might want to explore the effect of the faulty server on the resilience of our system by running a simulation or performing measurements in a testbed. Then, we need to generate random variates from a PH distribution describing the times to deadlock.

Phase-type distributed samples may be generated by playing the CTMC until absorption, and by numerical inversion of the distribution function [157]. In the following we focus on methods that 'play' the CTMC. Note that these methods require the Markovian representation.

The methods discussed in the following utilise random variates from the uniform, exponential, Erlang, and geometric distributions. We assume that random variates with uniform distribution on $(0, 1)$ are given, and denote these by U. Using the inversion method, a sample with exponential distribution with rate λ is then drawn by

$$\mathrm{Exp}(\lambda) = -\frac{1}{\lambda}\ln(U).$$

A sample from the Erlang distribution with degree b and rate λ is generated by

$$\mathrm{Erl}(b, \lambda) = -\frac{1}{\lambda}\ln\left(\prod_{i=1}^{b} U_i\right).$$

Note that this way of sampling $\mathrm{Erl}(b, \lambda)$ is more efficient than the functional equivalent of drawing b exponentially distributed samples and summing them up, because the ln operation is applied only once. Finally, a sample from the geometric distribution (starting from 0) with parameter p is obtained by

$$\mathrm{Geo}(p) = \left\lfloor \frac{\ln(U)}{\ln(p)} \right\rfloor.$$

The most natural way to generate a PH-distributed sample by playing the CTMC proceeds as follows: first, we select a state i by drawing an integer sample distributed according to the initial probability vector $\boldsymbol{\alpha}$. Afterwards, in each step the next state is selected according to the next-state probability vector, which is given by the ith row of the embedded Markov chain of $\overline{\mathbf{A}}$,

$$\mathbf{S} = \mathbf{I} - diag(\overline{\mathbf{A}})^{-1}\overline{\mathbf{A}}.$$

In the following, let $\mathbf{S}_i$ denote the ith row vector of $\mathbf{S}$. The sojourn time for state i is obtained as a sample from the exponential distribution with rate $-\lambda_{ii}$. Letting e_i denote the row vector with 1 at position i, and 0 everywhere else, the Play method can be given in pseudocode as follows:

Procedure Play:

1) $x := 0$. Draw an $\boldsymbol{\alpha}$-distributed discrete sample i for the initial state.
2) The chain is in state i

   - draw an $\mathbf{S}_i$-distributed discrete sample for the next state,
   - $x\mathrel{+}= \text{Exp}(-\lambda_{ii})$,
   - if the next state is the absorbing one ($i = n + 1$) go to 3), otherwise go to 2)

3) Return x.

In [684], Neuts and Pagano point out that when traversing a state more than once, the Play method adds up multiple samples from the same exponential distribution. The sum of k_i exponential distributions of the same rate $-\lambda_{ii}$, however, is the Erlang distribution with length k_i and rate $-\lambda_{ii}$. As shown above, drawing a sample from the Erlang distribution of length k_i requires only one logarithm operation, as opposed to k_i logarithms when drawing individual exponential samples. Thus, Neuts and Pagano propose the following method, which, instead of drawing exponential samples for each visit to a state i, counts the number of visits and then draws one Erlang-distributed sample for each state:

Procedure Count:

1) $x := 0$, $k_i := 0$, $(i = 1, \ldots, n)$, Draw an $\boldsymbol{\alpha}$-distributed discrete sample i for the initial state.
2) The chain is in state i

   - $k_i \mathrel{+}= 1$,
   - draw an $\mathbf{S}_i$-distributed discrete sample for the next state,
   - if the next state is the absorbing one go to 3) otherwise to 2)

3) for $i = 1, \ldots, n$; do x $\mathrel{+}=$ Erl(k_i, $-\lambda_{ii}$); done
4) Return x.

If the distribution is in Monocyclic form, we can derive another method from the structural properties of the Monocyclic representation. Recall that this representation consists of a chain of Feedback-Erlang blocks. With such a chain, possible state transitions are predetermined by the structure in two ways: First, when we leave a Feedback-Erlang block j, the next state will be the first state of the next Feedback-Erlang block $j + 1$. This implies that no new sample is required for choosing the successor block. Second, recall from Fig. 5.5 that each FE block consists of a chain of $m_j - 1$ states with exactly one outgoing transition (to the next state), and only one state with two outgoing transitions (the feedback state). Thus, within each FE block the only state where the next state is not determined by the structure is the last one. Furthermore, as the last state has only two outgoing transitions, the choice of staying within block j or entering the next block $j + 1$ corresponds to a Bernoulli experiment with parameter z_j. Consequently, the number of 'loops' in each block follows a geometric distribution with parameter z_j. Therefore, in order to generate the sample corresponding to the jth Feedback-Erlang block, we add a geometrically distributed number of exponentially distributed random variates with the same rate λ_j. As discussed when introducing the Count method, an efficient way of doing this is to draw a sample from an Erlang distribution of the appropriate length. These considerations lead to the following method:

Procedure Monocyclic:

1) $x := 0$. Draw an α-distributed discrete sample for the initial state,
2) the chain is in state l of block i (for the left-most state of the block, $l = b_i$)

   - $c = \mathrm{Geo}(z_i)$,
   - $x+ = \mathrm{Erl}(cb_i + l, \lambda_i)$
   - if the next block is the absorbing state go to 3), otherwise $l = b_{i+1}$, $i = i + 1$ and go to 2)

3) Return x.

The first three methods are applicable to general PH distributions. If we restrict our attention to sub-classes, more efficient methods can be designed. First, consider the APH class in CF-1 form. As a special case of the Monocyclic form, the CF-1 form is a chain of states, where each state has exactly one successor state (cf. Fig. 5.4), and thus the next state is not chosen randomly. Hence, once an initial state has been selected, the random variate is simply the sum of exponentially distributed samples from each of the successor states[3]:

---

[3] Note that the transition rates in the CF-1 form are usually not identical, hence we cannot simply draw an Erlang-distributed sample.

Procedure Simpleplay:

1) $x := 0$. Draw an α-distributed discrete sample for the initial state.
2) The chain is in state i.

    – $x+ = \mathrm{Exp}(-\lambda_{ii})$,
    – $i+ = 1$,
    – if the next state is the absorbing state go to 3), otherwise go to 2).

3) Return x.

If we assume a Hyper-Erlang distribution, represented as shown in Fig. 5.3, we can simplify the procedure Count, by using our knowledge that each of the branches is an Erlang distribution:

Procedure SimpleCount:

1) Draw a β-distributed discrete sample to choose an Erlang branch i.
2) Return $\mathrm{Erl}(b_i, \lambda_i)$.

5.6.1 Costs of Generating PH-Distributed Numbers

In the previous section we argue that the methods for generating random variates differ in their efficiency. We will now treat the costs of random number generation from phase-type distributions in a more formal way. All of the algorithms use exponential random variates for the sojourn times and uniform random variates for choosing the initial state. Play and Count additionally use uniform random variates for choosing successor states, while the Monocyclic algorithm needs geometrically distributed numbers for the number of loops in each Feedback-Erlang block. In order to draw from an exponential or geometric distribution, we need uniform random variates and logarithm operations. Therefore, we define the following two metrics for measuring algorithm complexity:

Definition 5.6 Let #*uni* be the number of uniform random variates that need to be generated and let # ln be the number of logarithm operations that must be performed for generating one PH-distributed random variate from a given PH distribution $(\alpha, \mathbf{A})$.

Using these metrics, we can compare the complexity of the algorithms. We consider both worst-case and average costs.

Table 5.2 Theoretical costs of generating PH distributed random variates from different PH classes and using different PH representations (where $\boldsymbol{v} = (n, n-1, \ldots, 1)$, $n^* = \boldsymbol{\alpha}(\text{diag}(\mathbf{A})^{-1}\mathbf{QB})^{-1}\mathbb{1}$)

PH Class	Worst case		Average case	
	#uni	#ln	#uni	#ln
HEx(n) `SimpleCount`	2	1	2	1
HErD(n) `SimpleCount`	$\max\{b_i + 1\}$	1	$\boldsymbol{\beta}\mathbf{b}^\mathsf{T} + 1$	1
APH(n) `SimplePlay`	$n + 1$	n	$\boldsymbol{\alpha}\boldsymbol{v}^\mathsf{T} + 1$	$\boldsymbol{\alpha}\boldsymbol{v}^\mathsf{T}$
PH(n) `Play`	∞	∞	$2\bar{n} + 1$	$\bar{n}$
PH(n) `Count`	∞	n	$2\bar{n} + 1$	n
`Monocyclic`	∞	$3m$	$\boldsymbol{\omega}\boldsymbol{\varphi}^\mathsf{T} + \boldsymbol{\alpha}\boldsymbol{\psi}^\mathsf{T}$	$\boldsymbol{\omega}\boldsymbol{\vartheta}^\mathsf{T}$

5.6.1.1 Worst-Case Costs

Let $\tilde{n}$ denote the length of the longest possible path through the CTMC. For the `Play` method, we draw one exponentially distributed random variate for each traversed state, and hence need one logarithm and one uniform random variate per step, as well as an additional uniform for choosing the next state. For this method, #*uni* and #*ln* are proportional to $\tilde{n}$. However, $\tilde{n}$ is not defined if there are cycles in the CTMC. Therefore, worst-case costs are not defined for `Play`.

The same problem with the unknown maximum number of state traversals occurs with the `Count` method. However, in this case we only draw Erlang-distributed samples (one for each state). Therefore, the maximum number of logarithm operations is bounded by the number of states: #*ln* $= n$. Similarly, for the `Monocyclic` method we draw one Erlang-distributed and one geometrically-distributed sample for each Feedback-Erlang block. The latter requires another two logarithm operations, in addition to the one for generating the Erlang sample. As the worst case occurs when we start in the first block, the worst-case number of traversed FE blocks is m, and thus #*ln* $= 3m$.

For APH in CF-1 form and using the `SimplePlay` method, the worst case is if the chain is entered at state $i = 1$, since in that case we have to traverse the whole chain. Thus, $\tilde{n} = n$. Obviously, for a Hyper-Erlang distribution in CF-1 form, $\tilde{n} = n$ holds as well. However, if we consider the Hyper-Erlang form and simulation using the `SimpleCount` method, the worst case is equivalent to choosing the longest Erlang branch. In that case, $\tilde{n} = \max\{b_i\} \leq n$. The worst-case costs can be computed as follows: With every class, we need one uniform random variate to choose the initial state. When using the APH(n) class in CF-1 form we need $\tilde{n} = n$ uniforms and $\tilde{n} = n$ logarithms for the consecutive phases. With the HErD class and the `SimpleCount` method we need $\tilde{n} = \max\{b_i\}$ additional random variates and one logarithm to obtain an Erlang-distributed random number. We summarise these results in the left half of Table 5.2.

5.6.1.2 Average Costs

In general, we do not expect to have worst-case behaviour, but are more interested in average costs. This measure is based on the average number of state transitions up to absorption,

$$\bar{n} = \alpha (\text{diag}(\mathbf{A})^{-1}\mathbf{A})^{-1}\mathbb{1}.$$

Applying the `Play` method for the general PH class, in each step we need two uniform random variates (one for the exponential sample and one for choosing the next state, see above), and one logarithm operation. As before, applying the `Count` procedure instead, the number of logarithms is $\#ln = n$, while the number of uniforms stays $\#uni = \bar{n}$.

Canonical forms enable explicit expressions for $\bar{n}$. For $\text{Mono}(\alpha, m, \mathbf{b}, \lambda, \mathbf{z})$ we introduce vector ω of size m, whose ith element is the probability of starting from Feedback-Erlang block i (e.g., $\omega_1 = \sum_{j=1}^{b_1} \alpha_j$), vector φ of size m, whose ith element is $\varphi_i = \frac{z_i b_i}{1-z_i} + \sum_{j=i+1}^{m} \frac{b_j}{1-z_j}$ (the mean number of steps spent in a Feedback-Erlang block from the first feedback, i.e., excluding the steps from the initial state to the feedback state in the first passage through the initial block), vector ψ of size n whose ith element indicates how many phases are needed to reach the next Feedback-Erlang block (e.g., if $b_1 \geq 2$ then $\psi_1 = b_1$, $\psi_2 = b_1 - 1$).

Using these notations the mean number of steps till absorption is

$$\bar{n} = \omega\varphi^{\mathsf{T}} + \alpha\psi^{\mathsf{T}},$$

where $\alpha\psi^{\mathsf{T}}$ contains the number of steps if there is no feedback (i.e., if $z_i = 0$, for $i = 1, \ldots, m$) and $\omega\varphi^{\mathsf{T}}$ contains the additional number of steps due to the loops in the Feedback-Erlang block.

The mean number of ln operations is

$$\ell^* = \omega\vartheta^{\mathsf{T}},$$

where ϑ is a row vector of size m whose ith element indicates the number of required ln operations starting from block i. $\vartheta_i = \sum_{j=i}^{m}(1 + 2\,\text{sgn}(z_j))$, since a degenerate Feedback-Erlang block with $z_i = 0$ is $\text{Erlang}(l, \lambda_i)$ distributed which requires one ln operation and a non degenerate ($z_i > 0$) Feedback-Erlang block requires three ln operations, two ln operations for $c = \text{Geo}(z_i)$ and one for $\text{Erl}(cb_i + l, \lambda_i)$.

For the APH class in CF-1 form, there exists an even simpler expression, as the number of traversed states depends only on the initial state, which in turn is determined by the initial probability vector α. Thus, for APH in CF-1 form,

$$\bar{n} = \alpha\nu^{\mathsf{T}}, \quad \text{where} \quad \nu = (n, n-1, \ldots, 1).$$

Equivalently, for the HErD class, $\bar{n}$ is a weighted sum of the lengths of the Erlang branches:

$$\bar{n} = \alpha \mathbf{b}^{\mathsf{T}}.$$

5.6.2 Optimisation

Considering the costs for the different methods discussed in the previous sections, it becomes clear that both the representation of the distribution and the method have an impact on the efficiency of PH random variate generation. One immediate question is then: what is the optimal representation to generate random variates efficiently? While the answer to this question is not yet available for the general PH case, [764] presents the following result for APH in CF-1 form:

Lemma 1 *[764] Given a Markovian representation (α, Λ) in CF-1 form, the representation (α^*, Λ^*) that reverses the order of the rates is optimal with respect to $\bar{n}$ if α^* is a stochastic vector. In this case, all bi-diagonal representations are Markovian.*

The proof given in [764] relies on the observation that swapping two adjacent rates λ_i, λ_{i+1} moves probability mass towards the end of the chain only if $\lambda_i < \lambda_{i+1}$. Thus, reversing the CF-1 order (where $\lambda_i \leq \lambda_{i+1}$ for all i) gives an initial probability vector α where probability mass is concentrated at the higher indices. Recalling from above, $\bar{n} = \alpha v^{\mathsf{T}}$ for APH, i.e., high probability for states close to absorption implies low average costs.

Note, however, that reversing the CF-1 form may result in α with negative entries [764]. In this case, the tuple $(\alpha^*, \mathbf{A}^*)$ still represents the same distribution, but the representation does not have a Markovian interpretation anymore, and thus $\bar{n}$ is not defined, nor can `SimplePlay` be applied. The optimal ordering can then be found by exhaustive search over all $n!$ possible orderings, or by heuristics that try to find a Markovian representation that is as similar as possible to the reversed CF-1. The heuristics presented in [764] either start from the CF-1 form and apply pair-wise swappings until the result would be non-Markovian, or start from the reversed CF-1 and try to reach a Markovian representation.

5.7 Conclusion

In this chapter we introduced the basics of using phase-type distributions as tools in resilience evaluation, discussing the complete workflow from explicit derivation and fitting to application in both analytical and simulation methods. One application of the methods described here is illustrated in Chap. 6, where PH distributions are used to reduce the size of a stochastic model by modelling the behaviour of components based on their delay distributions.

Our goal was to provide our readers with the fundamentals to apply PH distributions in their own work. Of course, this means that we could only scratch the surface of the vast amount of work available on phase-type distributions. We would like to give a few pointers for further study to the reader interested in different aspects of the topic. For readers interested in the mathematical background, we recommend [683], where PH distributions were introduced, and the fundamental work in [251, 697, 698], which provides the basics for many of the closure properties and canonical forms used in the field. More information on PH fitting is available in the papers introducing the fitting tools we discussed here [446, 870, 880], and in [59], where a fitting tool for general PH distributions is described. [572] gives a survey of PH fitting tools that were available at the time and have not been considered here. An in-depth discussion of fitting heavy-tailed data is provided in [48]. Both [579, 683] provide good introductions to matrix-analytic methods. Lastly, with respect to random-variate generation we would like to also point out the approach in [157], where the authors present a method of generating random variates from Matrix-Exponential distributions, of which the PH distributions are but a subclass. The approach is particularly interesting because it differs from the methods discussed in this chapter in that it applies numerical inversion of the distribution function instead of playing the Markov chain.

Chapter 6
Scalable Stochastic Modelling for Resilience

Jeremy T. Bradley, Lucia Cloth, Richard A. Hayden, Leïla Kloul, Philipp Reinecke, Markus Siegle, Nigel Thomas and Katinka Wolter

Abstract This chapter summarises techniques that are suitable for performance and resilience modelling and analysis of massive stochastic systems. We will introduce scalable techniques that can be applied to models constructed using DTMCs and CTMCs as well as compositional formalisms such as stochastic automata networks, stochastic process algebras and queueing networks. We will briefly show how tech-

J. T. Bradley and R. A. Hayden (✉)
Imperial College, London, UK
e-mail: jb@doc.ic.ac.uk

R. A. Hayden
e-mail: rh@doc.ic.ac.uk

L. Cloth
Department of Applied Information Technology,
GU Tech, Oman, Muscat
e-mail: lucia.cloth@gutech.edu.om

L. Kloul
Laboratoire PRiSM, Université de Versailles, Versailles France
e-mail: Leila.Kloul@prism.uvsq.fr

P. Reinecke and K. Wolter
Institute of Computer Science, Free University Berlin,
Takustr. 9, 14195 Berlin, Germany
e-mail: philipp.reinecke@fu-berlin.de

K. Wolter
e-mail: katinka.wolter@fu-berlin.de

M. Siegle
Department of Computer Science,
Universität der Bundeswehr München, Neubiberg, Germany
e-mail: markus.siegle@unibw.de

N. Thomas
School of Computing Science, Newcastle University,
Newcastle, UK
e-mail: nigel.thomas@ncl.ac.uk

K. Wolter et al. (eds.), *Resilience Assessment and Evaluation of Computing Systems*,
DOI: 10.1007/978-3-642-29032-9_6, © Springer-Verlag Berlin Heidelberg 2012

niques such as mean value analysis, mean-field analysis, symbolic data structures and fluid analysis can be used to analyse massive models specifically for resilience in networks, communication and computer architectures.

6.1 Introduction

The techniques presented in this chapter represent the state of the art in performance and resilience analysis when it comes to coping with massive state-space models. Many existing analysis techniques rely on generating underlying stochastic models, such as continuous-time Markov chains. Where there is too close a correspondence between the state space of the model and that of the underlying stochastic process, the state-space explosion in the former can lead to intractability in the latter. The presented techniques in this chapter were chosen as they represent instances of the main approaches to state-space reduction in stochastic systems: aggregation, decomposition, symbolic representation and continuum approximation.

We realise that accurate resilience analysis relies on a detailed and complex model. This kind of model generates huge state spaces and computation time if handled naïvely. In this chapter, we are specifically interested in analysis techniques that side-step the state space explosion problem by making use of efficient representation mechanisms. This is necessary if we are to make headway in directly analysing problems in mobile networks, critical infrastructures and Cloud systems.

We summarise the techniques presented below. In each case, we indicate what type of analysis result can be expected to be obtained using this method. It is important to understand how these techniques are relevant and useful to the understanding of resilience of a system. The types of analysis that can be tackled using these techniques can broadly be put into three categories. In the context of resilience analysis: *steady state distributions* are useful for calculating the probability that a fault state is ever reached; *transient distributions* are useful for calculating the probability that a fault state is entered in a particular time window; and *response* or *passage time analysis* is used to specify service level agreements, along the lines of *The round-trip response of a service request to a virtualised environment should be less that 1.8 s, with probability 0.95.*

> **Decomposition and phase-type representation** We start with a technique which combines simulation and decomposition, to generate simple approximate models based on phase-type representation of key portions of a system's operation (Sect. 6.2). A grey-box technique such as this avoids explicit investigation or representation of individual states in the system. This will be a useful technique for deriving some aggregate steady-state and transient measures but will probably be most widely used for response-time results.
>
> **Product forms and MVA** In Sect. 6.3, we explore two powerful techniques from queueing theory, product form and mean value analysis (MVA). These powerful techniques can be applied to very large or infinite state systems. They explore

balanced flows of traffic between components and in doing so permit compositional rather than system-wide analysis. Product forms will tend to lead to rapid steady-state distribution results, while MVA can calculate mean throughput, response time and job occupancy measures in a system.

Tensor representation Sect. 6.4 presents a decomposition technique which avoids the construction of the underlying explicit state space of a continuous-time Markov chain. Instead smaller submatrices are constructed which broadly reflect the components of the stochastic automata network. Analysis techniques exist which maintain the decomposed tensor representation while producing accurate steady-state distribution results.

Symbolic representation The symbolic representation of a stochastic model is described in Sect. 6.5. Multi-terminal Binary Decision Diagrams are used to encode efficiently the real rate values of a variety of stochastic models, including CTMCs, DTMCs and MDPs (Markov Decision Processes). Steady-state analysis, transient analysis and passage-time analysis are all possible using exclusively MTBDD-based algorithms.

Mean-field analysis In Sect. 6.6, we present developments in mean-field analysis as applied to massively distributed systems that are made up of identical communicating components. Applied to discrete-time systems, mean-field techniques generate sets of deterministic mean-difference equations (MDEs). Solving these MDEs gives access to both transient and passage-time measures. They provide an interesting comparison to the continuous counterpart, fluid analysis, described in the next section.

Fluid analysis Finally, in Sect. 6.7, we show how a continuum approximation of the dynamics of a stochastic system can be encoded in a system of ordinary differential equations (ODEs). Fluid analysis can be applied to large distributed systems that are comprised of groups of components of different types. Usually generated from a higher level formalism such as a stochastic process algebra, the ODEs can encode steady-state, transient and passage time questions. Higher moments of various model quantities are also available to capture measure accuracy.

6.2 Efficient Model Representation and Simulation

Discrete-event simulation is a widely-used approach for evaluating system properties such as response-time. Compared to analytical closed-form approaches, discrete-event simulation has the advantage of allowing the construction and evaluation of highly-detailed models where the modelled behaviour is not limited by the constraints of the formalism.

Unfortunately, the ability to easily build highly-detailed models often leads to scalability problems. While simulations modelling simple scenarios can be evaluated efficiently, increasing the complexity of the scenario quickly results in simulation models whose evaluation takes too long for the results to be useful. In this case, one obvious solution would be to start from scratch with a less detailed model with

Fig. 6.1 Example decomposition of a simulation

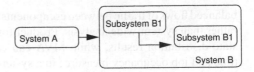

better scalability. However, this solution is often not desirable, since the results from a less detailed model may not be sufficiently accurate, and because of time and cost constraints on the modelling process itself.

Increasing model complexity is a well-known problem with analytical approaches as well. These approaches often suffer from state-space explosion as the system to be studied becomes more complex. Various solutions for this problem have been proposed [854]. In particular, decomposition/aggregation methods allow the solution of highly-complex models by splitting the model into submodels, solving the submodels independently, and aggregating the submodel solutions into a solution for the whole system. This approach may be used to reduce both processing time and memory requirements.

The general idea of decomposition and aggregation can also be applied to discrete-event simulations. In this section we describe a hybrid approach that combines decomposition and approximation using stochastic models to allow faster evaluation of discrete-event simulation models. With this approach, the simulation model is split into blocks whose behaviour can be approximated by phase-type (PH) distributions (Chapter 5). The phase-type distributions replace the approximated blocks in the simulation. The behaviour of the approximated blocks can then be reproduced by drawing random variates from the approximating phase-type distributions. Since this is more efficient than detailed discrete-event simulation, simulation of the whole system becomes much faster. This method can be applied when delay metrics are to be determined. Whether the method also applies for other metrics, such as throughput, availability or state probabilities in general, is as of yet unknown. The approach consists of the following steps:

1. **Decomposition of the simulation model into blocks that affect the metrics.**
   Starting from the complete simulation, one must identify parts that can be approximated by phase-type distributions. These blocks should satisfy a number of criteria: it should be simple to separate the simulation into the blocks, and the blocks should be chosen such that increasing the complexity of the simulation corresponds to multiple application of identical blocks. This step is shown in Fig. 6.1, where block A of the simulation is affected by block B: The effect of block B on the metrics depends on its internal block B1. Increasing the size of system B corresponds to using multiple instances of B1 within B.
2. **Evaluation of simulation blocks.** In order to approximate the behaviour of the simulation blocks identified in step 1 by phase-type distributions, data is needed that shows the effect of the blocks. Typically, this data will be obtained from detailed simulations of the individual blocks, but if a block corresponds to a system where data from measurements is available, such data can also be used.

Fig. 6.2 Data stream in packet-switched network with tree topology

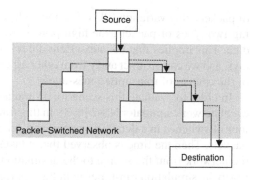

3. **Approximation using phase-type distributions.** The data obtained in step 2 is approximated by a phase-type distribution. The general process and tools for fitting PH distributions to data sets is described in detail in Chap. 5. For the application in the hybrid approach, the focus should be on capturing those properties well that strongly affect the metric. Furthermore, one should choose phase-type distributions that enable efficient random-variate generation, so as to be able to reproduce the behaviour of the approximated building-blocks efficiently.

4. **Integration of the phase-type distributions into the simulation model.** The phase-type models must be integrated into the simulation. For delay metrics, this requires drawing random variates from the distribution and delaying events caused by the approximated system blocks according to these variates. For common discrete-event simulation toolkits such as NS-2 [693] and OMNeT++ [911], the libphprng [765] library provides the necessary routines for generating PH-distributed random numbers.

5. **System evaluation.** The whole system can now be evaluated. In order to scale system size/complexity, the number of blocks containing the approximating PH distribution is increased. As random-variate generation from a PH distribution is typically more efficient than detailed simulation of the blocks, the model can be expected to scale much better to higher numbers of building-blocks. On the other hand, the approximation process introduces an error, since the PH distribution does not represent all behaviour of the detailed model.

6.2.1 Illustrative Example

In [938] we have investigated timing behaviour in a tree-structured network across a variable number of identical switches. The topology is shown in Fig. 6.2 and each data stream is transmitted in a straight feed-forward fashion.

Let us illustrate the approximation technique for the given example. In this example, the transmission delays encountered on the path between the source and the destination are investigated. More precisely, the metric of interest is the 1% quantile of packet delay variation (PDV)[1] of the high-priority packets in a network transporting two types of packets, the high priority traffic and the low priority background data. This implies that the subsystem models must only represent transmission delays, and we can abstract away from other aspects of the network, such as correctness of the content of transmitted messages.

In order to evaluate this scenario, a simulation has been built using the discrete-event network simulator NS-2 [693]. In this simulation, we model the internal behaviour of switches in a detailed manner, and generate streams of high and low priority packets. Then the time is observed that it takes for high-priority packets to traverse the network from the source to the destination. As illustrated by the upper curve in Fig. 6.4a, simulation run-times with this approach increase quickly as the number of switches is increased. Therefore the approximation method is applied as follows:

1. **Decomposition of the simulation model into blocks that affect the metrics.** In this scenario, the obvious block is the individual switch. For simplicity, we assume that the network consists of switches whose behaviour with respect to transmission delay is identical.
2. **Evaluation of simulation blocks.** The transmission delays incurred by each switch must be assessed. As a network consisting of only one switch can still be simulated within an acceptable time, transmission delays for a long simulation run with only one such switch between source and destination are obtained.
3. **Approximation using phase-type distributions.** We fit an acyclic phase-type distribution using the PhFIT tool [446] to the transmission delay of the high-priority stream such that the resulting distribution represents the 1% quantile well. See Chap. 5 for more details on phase-type distributions. The phase-type distribution obtained here represents the transmission delay distribution of a single switch. Figure 6.3a shows the packet delay distribution from step 2 and the cumulative density function of the phase-type distribution fitted to the data. Note that the distribution fits the lower quantiles well and tends to diverge only on the higher quantiles.
4. **Integration of the phase-type distributions into the simulation model.** Now, the phase-type distribution for the packet delay of the switch can be used to simulate the behaviour of the switch. To this end, we build a simple queueing station whose service-time distribution is given by the phase-type distribution fitted to the data. Packets entering the station system are delayed according to the service-time distribution. Note that the presence of a queue is only dictated by the fact that we cannot drop packets, and that this queue does not correspond to any part of the original model. In the considered scenario high-priority packets are sent very infrequently, and therefore queueing is highly unlikely. If the arrival rate of high-priority packets was higher, queueing might occur. The resulting queueing delay would increase the error caused by the approximation, since this queueing is not part of the original simulation.

---

[1] Packet delay variation is defined as the difference between the shortest and the longest transmission time, where lost packets are ignored.

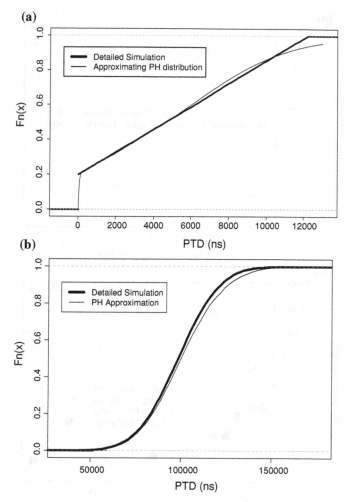

Fig. 6.3 Comparison of simulation data and PH approximation. **a** CDF of simulation data for 1 switch and associated PH approximation. **b** CDF of simulation data and PH approximation with 20 links using full simulation and PH approximation

5. **System evaluation.** The system can now be simulated by transmitting high-priority packets from the source to the destination. Note that it is not necessary to simulate the low-priority stream anymore, since its effect on the packet delay variation is already captured in the service-time distribution.

The advantages of the hybrid approach can be evaluated using a scenario where the number of switches that the data stream has to traverse is increased from 1 to 20.

Figure 6.4a shows run-times for the detailed and approximating simulations for increasing number n of links. Note that the run-times for the detailed simulation increase sharply and reach values in the order of days, while run-times with PH

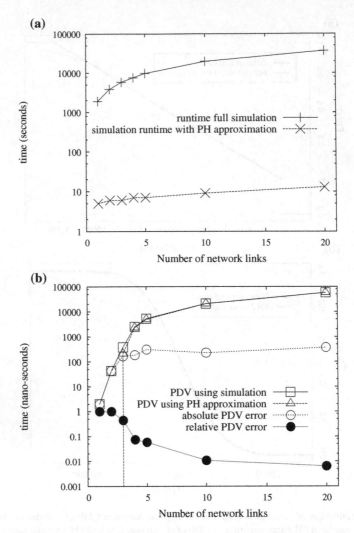

Fig. 6.4 Evaluation of simulation time and the error introduced by approximation. **a** Simulation runtimes for full simulation and with PH approximation. **b** PDV error for full simulation and with PH approximation

approximation stay in the order of hours, even for a high number of links. Figure 6.4b illustrates the error incurred by using PH approximations for the switches instead of simulating them in detail. As expected, the error does increase with n, but it seems to converge, and the relative error even decreases.

6.2.2 Outlook

The decomposition and approximation approach should be extended in various directions. This has not been done yet and there might arise problems that cannot be foreseen. First, the system model need not be a simulation model. Any stochastic discrete event model, such as a CTMC, should be applicable too. Second, the approximation technique need not be a PH distribution. Simpler approximations could consist in Bernoulli variables, and more sophisticated approximation may use correlated stochastic processes, that are still smaller than the original model. Third, metrics other than timing metrics should be considered. Often of interest are system or service availability, throughput, or failure characteristics.

6.3 Product-Form Solution and Mean Value Analysis

One approach to tackling the state space explosion problem common to all compositional modelling techniques is to break the model into smaller parts that can then be solved separately and the solutions combined in some way to give measures for the whole system. This approach is generally known as model decomposition.

One of the most powerful model decomposition techniques are so called, *product-form solutions*. Essentially, a product-form is a decomposed solution where the steady state distribution of a whole system can be found by multiplying the marginal distributions of its components. Thus, for a system described by the pair $\{S_1, S_2\}$, where S_i is the local state of component C_i a product form solution would have the form

$$\pi(S_1, S_2) = \frac{1}{B}\pi(S_1)\pi(S_2)$$

where $1/B$ is the *normalising constant* $(B \leq 1)$, which is necessary if there are combinations of possible local states which are prohibited in the system evolution.

The conditions for such a solution to exist are clearly going to be restrictive, and as such product-form solutions are applicable only in a relatively small number of cases. Despite the restrictions, product-forms are extremely powerful and so the quest for new solutions in stochastic networks has been a major research area in performance modelling for over 30 years, giving rise to a number of seminal results, such as Jackson queueing networks [480] and the BCMP result [93] for closed queueing networks.

A Jackson network is a simple open queueing network with Poisson arrivals, negative exponentially distributed service times and unbounded FCFS queues, where the routing of jobs from one node to another is strictly a priori. For stability it is also required that the utilisation at each node is strictly less than one. In such a network each node can be considered as an independent FCFS $M/M/k$ queue and the steady state probability of being in any given system state can be found simply by multiplying together the marginal probabilities of each node's local state.

Jackson's result does not apply to closed networks of $M/M/k$ queues (where no jobs may enter or leave the network) because the population of a closed network is bounded. This restriction was overcome by Gordon and Newell [386], with the introduction of the normalising constant $1/G(K)$, where K is the population size. If the state is described by the tuple $\mathbf{S} = \{S_1, \ldots, S_n\}$ then $G(K)$ is given by,

$$G(K) = \sum_{\mathbf{S}} \prod_{i=1}^{n} \pi(S_i)$$

The results of Jackson and Gordon and Newell were subsequently generalised by Baskett, Chandy, Muntz and Palacios [93] to allow four possible classes of network for which the product-form solution holds. The first class is FCFS queues with negative exponentially distributed service times. The service rate may be state dependent, which allows the network to be open or closed. The subsequent classes slightly relax the condition of negative exponentially distributed service times, as long as the queueing discipline is either processor sharing, infinite server or LCFS with pre-emptive resume.

The BCMP characterisation of product-form networks greatly extended the potential for applying product-form solutions to real world problems. Subsequently there have been many further results extending the class of queueing networks amenable to product-form solution under specific conditions, for different kinds of queueing network and for different properties. One case worthy of special mention is the work on product-form G-networks by Gelenbe [372]. G-networks are a variant of queueing networks that allow so-called *negative customers*. A negative customer may potentially remove a job from the queue into which it arrives. As well as having external arrivals of negative customers, a normal (or *positive*) customer may become a negative with some probability following completion of service. Subject to similar conditions as Jackson's result described above, a G-network can be shown to exhibit a product form solution. This result is more surprising than it might naively appear, as previously it was assumed that there was a relationship between the product-form solution and the existence of *partial balance*. Partial balance does not hold in G-networks, and so the proof of the product form solution changed the understanding of the conditions for product-form solution.

Most attention has been given to queueing networks and their variants (such as G-networks), but there have also been other significant examples, for example [82, 144, 425]. Many of the approaches to efficiently solving stochastic process algebra models have been based on concepts of decomposition originally derived for queueing networks [437]. Applying such approaches to stochastic process algebra allows the concepts to be understood in a more general modelling framework and applied to non-queueing models. More recently the Reversed Compound Agent Theorem (RCAT) has been developed [412]. RCAT is a compositional result that finds the reversed stationary Markov process of a cooperation between two interacting components, under syntactically checkable conditions [412, 413]. From this a product-form follows simply. RCAT thereby provides an alternative methodology that unifies many product-forms, far beyond those for queueing networks.

Even when a product-form solution exists, or an approximation to a product-form can be derived, obtaining a numerical solution may still be computationally expensive. In particular, finding the normalising constant can be costly in general. Mean value analysis (MVA) [766] depends on the application of the *arrival theorem*, first derived independently by Sevcik and Mitrani [821] and Lavenberg and Reiser [580]. This theorem states that, subject to certain conditions, a job arriving in a queue will, on average, observe the queue to be in its steady state average behaviour. Combining the arrival theorem with Little's law gives rise to a method for deriving average performance metrics based on steady state averages directly from the queueing network specification, without the need to derive any of the underlying Markov chain.

Stated simply, the basic MVA algorithm, when the population size is N, is as follows:

1. Set the population size, n, to be 1.
2. Compute the delay this single job will experience at each node as it traverses the otherwise empty network (i.e. the average service time for one job).
3. Hence compute the probability that this job will be at any given node at a randomly observed instant. This gives the average queue length at each node when the total population consists of one job.
4. Increment the population size, n.
5. Compute the delay an arriving job will experience at each node (i.e. the average service time for one job plus average service time for the average number of jobs in the queue when the population is $n - 1$).
6. Hence compute the average queue length at each node when the total population consists of n jobs.
7. If $n < N$ then go back to step 4.

As such it is relatively computationally efficient as long as the population size is not excessively large. The computational cost for solving a network with a given structure grows linearly with N.

There are many generalisations and approximate solutions of the original mean value analysis algorithm. MVA has been applied to many classes of queueing network, as well as other formalisms, including stochastic process algebra [879]. The key observation in applying MVA to stochastic process algebra is that repeated instances of a component in parallel may be treated as jobs in a queueing network. If the interaction of these components with other components conforms to some simple set of restrictions, then a version of the arrival theorem can be derived relating a component evolving between derivatives (or behaviours) to the steady state solution of a system with one fewer instance of the component.

6.4 Tensor Representation

The *tensor representation* or *compact representation* has been used for some time as a means to address the problem of state space explosion to which state-based performance modelling formalisms are prone. This technique which aims to keep the size of the model representation as small as possible falls into the *largeness avoidance* category of techniques which also includes techniques such as decomposition, aggregation and symbolic encodings [891]. However, unlike the aggregation which can result in a significant reduction in the size of the state space, the tensor representation is not a state space reduction method, but rather an alternative approach to state space explosion which handles the model solution in a decomposed form. The matrix representation of the Markov process underlying a performance model may be decomposed so that the state space of the model, and its dynamics, are not represented by a single matrix but by a number of smaller matrices. Nevertheless the model is solved as a single entity and the solution is exact, unlike the decomposition technique which, generally, gives rise to approximate solution of the original model [243].

The tensor representation has been developed for several state-based modelling formalisms. The pioneering work in this area was carried out, in 1984, by Plateau on Stochastic Automata Networks (SANs) [733]. Using a technique based on tensor or *Kronecker* algebra, it has been proved [732, 733, 735] that this method automatically provides an analytic derivation of a decomposed form of the generator matrix called the *descriptor*. Compared to a monolithic description of the generator, the structure of this descriptor leads to a considerable reduction in memory requirements during the model solution. Moreover, solution techniques have been adapted to this representation [103, 105, 337].

In the following, we present the tensor representation in the context of the SAN approach and show how to derive the descriptor expression for the SAN model of the leaky bucket, an admission control mechanism in ATM networks. We finally discuss the impact of the tensor representation on both the memory requirements and the computation time of the matrix-vector multiplications when solving the model, and this regardless of the modelling formalism used.

6.4.1 Stochastic Automata Networks

In the SAN approach, a system is represented by a number of *automata*, each automaton capturing the dynamic behaviour of a component of the system. Within an individual automaton, the behaviour of a component is captured as a set of *states* and *events* causing transition from one state to another. A label associated with each transition allows us to specify the type of the event, and its occurrence date and probability [736]. The transitions in the network can be of two types: *local* or *synchronised*. A local transition occurs only in an automaton whereas a synchronised transition occurs in several automata at the same time.

More formally, a SAN is a set of N automata in which each automaton A_i, $1 \le i \le N$, is defined by the tuple (S_i, L, Q_i) where

- S_i is the set of states of the automaton,
- L is the set of labels. Each label $l \in L$ is a list that may contain either a function τ, or a list of tuples (e, τ_e, p_e), or both of them such that:

  - e is the name of a synchronising event or synchronisation,
  - τ and τ_e are the transition rates, functions defined from $\Pi_{i=1}^N S_i$ to $\mathbb{R}^+$,
  - p_e is the probability transition function defining a conditional routing probability on e between local states.

- Q_i is the transition function which associates a label from L with every arc of automaton A_i.

A label on an edge allows us to specify the type and the rate of the transition as follows:

- If $Q_i(x_i, y_i)$ contains a function τ, then we have a local transition to A_i between states x_i and y_i. If τ is not a constant, the transition is still local to automaton A_i, but its rate depends on the state of other automata of the network.
- If $(e, \tau_e, p_e(x_i, y_i)) \in Q_i(x_i, y_i)$, then the transition between states x_i and y_i is a synchronised transition, e and τ_e being the name and the rate of the transition of the synchronising event. $p_e(x_i, y_i)$ is the routing probability between local states x_i and y_i. The distinction between the rate and the probability is required because the first must be unique for a given synchronising event, thus the same on all concerned automata, whilst the probabilities may, and generally will, differ. The rate of synchronising events are determined at the global level.

6.4.2 The Descriptor

In the seminal paper [732], Plateau proved that the generator matrix of the Markov process underlying a SAN model can be analytically represented using Kronecker algebra. This matrix is automatically derived from the SAN description, using the individual automata to generate the sub-matrices in the tensor expression. It has been proved in [732, 733, 736] that, if the states are in a lexicographic order, then the generator matrix Q of the Markov process associated with a continuous-time SAN model is given by:

$$Q = \bigoplus_{i=1}^N F_i + \sum_{e \in \varepsilon} \tau_e \left(\bigotimes_{i=1}^N R_{i,e} - \bigotimes_{i=1}^N \overline{R}_{i,e} \right) \tag{6.1}$$

where

- N is the total number of automata in the network.
- ε is the set of synchronisations.

Fig. 6.5 The leaky bucket
mechanism

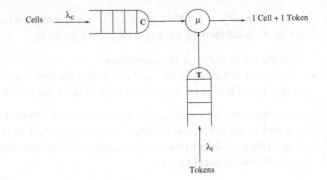

- F_i is the local transition matrix of automaton $\mathcal{A}_i$ without synchronisations.
- $R_{i,e}$ is the transition matrix of automaton $\mathcal{A}_i$ due to synchronisation e whose rate is τ_e.
- $\overline{R}_{i,e}$ is a matrix representing the normalisation associated with the synchronisation e on automaton $\mathcal{A}_i$.
- $\bigoplus$ and $\bigotimes$ denote the tensor sum and product, respectively.

Unlike the local transition matrices F_i, the synchronising matrices $R_{i,e}$ are not generators, that is their rows do not sum to zero. The diagonal corrector matrices $\overline{R}_{i,e}$ have been introduced to normalise these synchronising matrices.

In discrete-time, the transition matrix of the Markov process underlying a SAN model is given by an expression similar to equation (6.1) where the tensor sum $\bigoplus$ is replaced by the tensor product $\bigotimes$. Applying the tensor product on the local transition matrices F_i allows us to catch the phenomenon characterising the discrete-time, that is the occurrence of several events at the same time.

In both the continuous and discrete-time, the solution of the model, that is the steady-state distribution, can then be achieved via the corresponding tensor expression of sub-matrices; the complete generator or transition matrix does not need to be generated.

6.4.3 Application

The *leaky bucket* is the admission control mechanism developed for ATM networks [222]. Its simplest version [538] consists of two buffers B_c and B_t (see Fig. 6.5). Whilst the former is used to store the user's data cells (packets), the latter is dedicated to the tokens. At its arrival to the access buffer, a cell is either lost if the buffer is full or stored before being served. The service of a cell consists of assigning to it a token taken from buffer B_t. For the cell, this token constitutes its access permit to the network.

The generation rate of the tokens is equal to either the average throughput or the peak cell rate characterising the user's stream. Therefore, if there are no tokens in B_t while data cells are still arriving to B_c, then the user's throughput does not conform to the throughput he has initially specified. The cells in B_c will have to wait until new tokens are generated and all the cells arriving while B_c is full are lost.

6.4.3.1 The SAN Model

As the data packets (cells), in ATM networks, have the same size, a discrete-time performance analysis of the leacky-bucket would be more appropriate. However, in order to keep the global automata simple, the SAN model is built in continuous-time. The model parameters are the following:

- cell arrivals to B_c according to a Poisson process of parameter λ_c,
- token arrivals to B_t according to a constant distribution of parameter λ_t,
- cell service times exponentially distributed with rate μ.

The SAN modelling the leaky bucket mechanism consists of two automata, A_1 and A_2 [538]. Automaton A_1 models the number of cells in B_c and A_2 models the number of tokens in B_t. We assume the buffer size limited to two cells. This size remains however sufficient to represent all possible transitions. Thus both A_1 and A_2 have three states, noted s_0, s_1 and s_2.

In the modelled system, the possible events are of three types: the cell arrival with rate λ_c, the token arrival with rate λ_t and the simultaneous departure of a cell and a token with rate μ. Whilst the two first types of events are local events to automaton A_1, and A_2 respectively, the third type of events, noted e_t, is a synchronising event between A_1 and A_2, since it has an impact on both the number of cells in B_c and the number of tokens in B_t. The SAN model $\{A_1, A_2\}$ is depicted in Fig. 6.6.

6.4.3.2 The Descriptor Matrices

We first build F_1 and F_2, the matrices of the local transitions associated with automaton A_1 and A_2, respectively. These matrices are the following:

$$F_1 = \begin{pmatrix} -\lambda_c & \lambda_c & 0 \\ 0 & -\lambda_c & \lambda_c \\ 0 & 0 & 0 \end{pmatrix} \qquad F_2 = \begin{pmatrix} -\lambda_t & \lambda_t & 0 \\ 0 & -\lambda_t & \lambda_t \\ 0 & 0 & 0 \end{pmatrix}$$

As only one synchronising event has an impact on automaton A_1, we have only a single matrix of synchronised transitions, noted R_{1,e_t}, for this automation. By considering the associated normalisation matrix $\overline{R}_{1,e_t}$, we have:

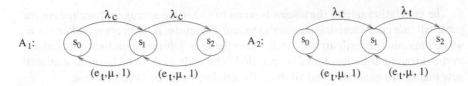

Fig. 6.6 The SAN model

$$R_{1,e_t} = \begin{pmatrix} 0 & 0 & 0 \\ 1 & 0 & 0 \\ 0 & 1 & 0 \end{pmatrix} \qquad \overline{R}_{1,e_t} = \begin{pmatrix} 0 & 0 & 0 \\ 0 & -1 & 0 \\ 0 & 0 & -1 \end{pmatrix}$$

Similarly, only one synchronising event has an impact on automaton A_2. Thus we have a single matrix of synchronised transitions, noted R_{2,e_t} for this automation.

$$R_{2,e_t} = \begin{pmatrix} 0 & 0 & 0 \\ 1 & 0 & 0 \\ 0 & 1 & 0 \end{pmatrix} \qquad \overline{R}_{2,e_t} = \begin{pmatrix} 0 & 0 & 0 \\ 0 & -1 & 0 \\ 0 & 0 & -1 \end{pmatrix}$$

Note that the transition rate of the synchronising event e_t, that is $\tau_{e_t} = \mu$ in the descriptor equation 6.1, is not reported in matrices $R_{i,e_t}, i = 1, 2$; only the transition probabilities are reported. As in continuous-time, the rate of a synchronising event is unique, and thus the same on all automata involved in the synchronisation, this rate appears only once, when the tensor product of matrices $R_{i,e_t}, i = 1, 2$, is performed.

Once all the matrices built, the elements of the generator associated with the SAN model $\{A_1, A_2\}$ can be computed, using equation 6.1. Thus the complete generator, which is the 9×9 matrix given below, does not need to be generated.

$$Q = \begin{array}{c} \\ 0,0 \\ 0,1 \\ 0,2 \\ 1,0 \\ 1,1 \\ 1,2 \\ 2,0 \\ 2,1 \\ 2,2 \end{array} \begin{pmatrix} 0,0 & 0,1 & 0,2 & 1,0 & 1,1 & 1,2 & 2,0 & 2,1 & 2,2 \\ -\lambda & \lambda_t & \lambda_c & & & & & & \\ & -\lambda & \lambda_t & & \lambda_c & & & & \\ & & -\lambda_c & & & \lambda_c & & & \\ & & & -\lambda & \lambda_t & & \lambda_c & & \\ \mu & & & & -(\lambda+\mu) & \lambda_t & & \lambda_c & \\ & \mu & & & & -(\lambda_c+\mu) & & & \lambda_c \\ & & & & & & -\lambda_t & \lambda_t & \\ & & & \mu & & & & -(\lambda_t+\mu) & \lambda_t \\ & & & & \mu & & & & -\mu \end{pmatrix}$$

In this representation, where $\lambda = \lambda_c + \lambda_t$, a global state (C, T) consists of the number of cells C and the number of tokens T. Thus, state $(1, 2)$, for example, refers to the system global state when there are 1 cell in B_c and 2 tokens in B_t. Note that all the global states are reachable.

6.4.4 Memory Requirements

Following the development of the tensor representation for the SAN models, Kronecker representation techniques have been proposed for several other state-based
performance modelling formalisms such as Petri net based-formalisms [163, 170,
215, 219, 306, 307, 531] and stochastic process algebra [162, 439]. In all these
models, the *size* of the state space is open to several interpretations:

- the physical space T needed to store the model using the tensor representation;
- the size of the state space $\hat{S}$ of the cartesian product of the model components;
- the size of the reachable state space S.

In general, in the Kronecker representation, the cartesian product space $\hat{S}$ is repre-
sented, not the reachable state space S. When $|S| = |\hat{S}|$, the benefit of using the tensor
representation may be enormous compared to an explicit saving of the generator as
a sparse matrix. Consider, for example, a model which consists of N components
and where n_i is the size of component i, $i = 1, \ldots, N$. If the generator is full (no
zero elements), the memory needs of the tensor representation are given by $\sum_{i=1}^{N} n_i^2$
whereas the memory requirements of the sparse matrix representation are of the order
of $(\Pi_{i=1}^{N} n_i)^2$.

When $|S| \ll |\hat{S}|$, the benefit of the tensor representation may be lost because
of the unreachable states. If the probability vectors used in the vector-descriptor
multiplications are the extended vectors $\hat{\pi}$, that is with an entry for each unreachable
state in $\hat{S}$, the benefit of the tensor representation is lost not memory-wise only, but
also because of unnecessary computations when solving the model. Therefore, the
probability vectors used in the vector-descriptor multiplications must be reduced to
the reachable states entries only (π). While the sparse matrix representation avoids
unnecessary computations when solving the model, it remains a memory consuming
representation, specially when dealing with big models.

In all the performance modelling formalisms for which a tensor representation
has been developed (for instance, SAN, GSPN, PEPA), the model components are
connected by synchronisations and/or functions. From previous work on SAN, we
know that the use of functions has a positive effect on both the size of the tensor
representation and the size of the product state space. In particular, if we remove a
function it is generally necessary to introduce an additional component. If the new
component has two or more states then we increase both spaces T and $\hat{S}$. However
this should not change the reachable state space S. To do so fundamental changes
have to be considered in the model.

Sometimes, the use of functions, like in process algebra PEPA [440], allows the
tensor representation to be more direct and similar to the one obtained by Plateau for
the SANs. The functional dependency on the state of a component can capture the
different apparent rates that the component may express with respect to an action type.
There is an implicit assumption that an action type uniquely defines a synchronisation
event at the transition system level. This will not generally be the case without
restrictions on the use of types within cooperation sets [439, 440].

6.4.5 The Model Solution

The space efficiency of the tensor representation is obtained at the expense of an increased computation time. Moreover, the presence of functional rates in the components of a model introduces an extra computing time during the matrix-vector multiplications. Indeed when a model contains functional rates, an appropriate numerical value has to be recomputed each time a functional rate is needed. Thus the presence of the functional rates may constitute a determinant factor in the computing time requirements.

In [336, 337, 732], an efficient vector-descriptor multiplication algorithm, known as the *shuffle* algorithm has been developed to be used when solving the stationary distribution. This algorithm is the basic step in iterative methods such as the Power method and Generalised Minimum Residual (GMRES) method [853]. However, the algorithm, which is very efficient when $|S| = |\hat{S}|$, requires the use of probability vectors $\hat{\pi}$. In [219], the reachability states are stored using MDD (Multi-valued Decision Diagrams) while matrix diagrams are used to store the Kronecker representation. The numerical results show that solving the model using a technique such as Gauss–Seidel requires less iterations and less time per iteration than the shuffle algorithm. However, the matrix diagram solution requires twice the memory size required by the basic Kronecker representation. Alternative approaches have been developed [530, 531, 713]. In these approaches, the reachable state space S is first computed and the model is solved using the reduced probability vectors π. The algorithm proposed in [530, 531] is based on a permutation which reorders the states according to their reachability, and the use of π.

Recently, a new version of the shuffle algorithm called *FR-Sh* (*Fully Reduced Shuffle*) has been proposed in [103, 105]. This algorithm, which uses the probability vectors π, improves the memory needs and the computation time when there are a lot of unreachable states. It has been proved that the algorithm allows an important reduction in the memory requirements, in particular when using iterative methods such as Arnoldi and GMRES.

In [267], iterative methods based on splittings, such as Jacobi and Gauss–Seidel, are proved to be better than the Power method.

6.4.6 Outlook

Currently there is a great need for a comparison between the different algorithms. In [219], a comparison between matrix diagrams and the original shuffle algorithm showed a substantial advantage of the matrix diagrams in terms of computation time. Several versions of the shuffle algorithms have been investigated in [103, 105], among which the *PR-Sh* (*Partially Reduced Shuffle*) and the *FR-Sh* algorithms. These new versions, in particular the *FR-Sh* algorithm, improve considerably the original shuffle algorithm. These results will be fully validated if the *FR-Sh* algorithm, for example, is compared to the alternative approaches in the literature. Moreover, it will be interesting to consider in the future a combination of the shuffle algorithms and elaborated data structures such as decision diagrams [105]. Such approaches may allow the analysis of larger systems.

6.5 Symbolic Data Structures in CTMCs and SPAs

This section summarises the state-of-the-art of symbolic approaches to state space representation. In this context, the term "symbolic"—which was originally coined in the context of model checking [636]—refers to the use of decision diagrams as a graph-based data structure for compactly encoding sets of states or transition systems of various kinds. This approach has the potential to handle very large models efficiently while utilising only small amounts of memory.

6.5.1 Introduction to BDDs

A Binary Decision Diagram (BDD) [160] is a symbolic representation of a Boolean function $f : \mathbb{B}^n \mapsto \mathbb{B}$. Its graphical interpretation is a rooted directed acyclic graph with one or two terminal vertices, marked 1 and 0 (for "true" and "false"). Each non-terminal vertex x is associated with a Boolean variable $\mathsf{var}(x)$ and has two successor vertices, denoted by $\mathsf{then}(x)$ and $\mathsf{else}(x)$. The graph is ordered in the sense that on each path from the root to a terminal vertex, the variables are visited in the same order. A reduced BDD is essentially a collapsed binary decision tree in which isomorphic subtrees are merged and "don't care" vertices are skipped (a vertex is called "don't care" if the truth value of the corresponding variable is irrelevant for the truth value of the overall function). Reduced ordered BDDs are known to be a canonical representation of Boolean functions.

As a simple example, Fig. 6.7a shows the full binary decision tree for the function $(\overline{\mathsf{a}} \wedge \mathsf{t}) \vee (\mathsf{a} \wedge \mathsf{s} \wedge \overline{\mathsf{t}})$, where all vertices drawn at one level are labelled by the same Boolean variable, as indicated at the left of the graph. The edge from vertex x to $\mathsf{then}(x)$ represents the case where $\mathsf{var}(x)$ is true; conversely, the edge from x to $\mathsf{else}(x)$ the case where $\mathsf{var}(x)$ is false. (In the graphical representation, then-edges are drawn solid, else-edges dashed.) Part (b) of the figure shows the corresponding reduced BDD which can be obtained from the decision tree by merging isomorphic subgraphs and leaving out don't care vertices. For instance, in the diagrams shown in Fig. 6.7, if $\mathsf{a} = 0$ then s is a don't care variable. As shown in Fig. 6.7c, in the graphical representation of a BDD, for reasons of simplicity, the terminal vertex 0 and its adjacent edges are usually omitted. In all three graphs shown in Fig. 6.7, the function value for a given truth assignment can be determined by following the corresponding edges from the root until a terminal vertex is reached.

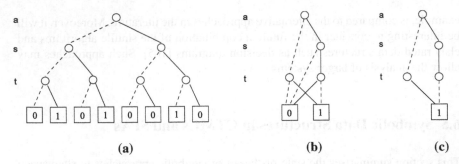

Fig. 6.7 a Binary decision tree, **b** reduced BDD and **c** simplified graphical representation for the Boolean function $(\overline{a} \wedge t) \vee (a \wedge s \wedge \overline{t})$

A finite set, e.g. the reachability set of a state-based model, can be represented by a BDD via its characteristic functions, i.e. a function yielding one or zero, depending on whether the corresponding state—encoded as a bitstring—is in the set or not. Similarly, a finite transition system can be represented by a BDD, as illustrated by the following example:

Example: Fig. 6.8 (left) shows the labelled transition system (LTS) of a simple finite-buffer queueing process. The middle of the figure shows the way transitions are encoded, and the resulting BDD is depicted on the right. Action labels *enq* and *deq* are encoded with the help of two Boolean variables[2] a_1 and a_2. In particular, the encodings of action *enq* resp. *deq* is set to $(0,1)$ resp. $(1,0)$. The LTS has four states, therefore two bits are needed to represent the state number. Note that this BDD uses a special "interleaved" ordering of the Boolean variables encoding the source and target state. This interleaved ordering is a proven heuristics for obtaining small BDD sizes in the context of compositional model construction [327].

6.5.2 Related Decision Diagram Data Structures

Over the years, several variants of the basic BDD data structure have been developed, mostly initiated by the wish to find a data structure perfectly suited to a particular verification or analysis problem. In this sections, the most prominent ones are discussed briefly.

A Multi-terminal BDD (MTBDD) is a symbolic representation of a pseudo-Boolean function $f : \mathbb{B}^n \mapsto \mathbb{D}$, where $\mathbb{D}$ is an arbitrary domain [227]. MTBDDs are constructed similarly to BDDs, but—as the name implies—there may be more than two terminal vertices carrying the function values. If one wishes to encode labelled Markov chains symbolically, MTBDDs can be used, where the transition

---

[2] Since there are only two distinct actions in the LTS, one bit would be enough to encode the action. However, the encoding 0 is often reserved for the special internal action τ, and in any case it is not mandatory to use the smallest possible number of bits.

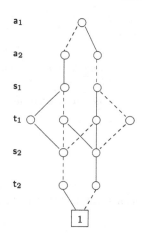

	$(a_1, a_2, s_1, t_1, s_2, t_2)$
$0 \xrightarrow{enq} 1$	$(0, 1, 0, 0, 0, 1)$
$1 \xrightarrow{enq} 2$	$(0, 1, 0, 1, 1, 0)$
$2 \xrightarrow{enq} 3$	$(0, 1, 1, 1, 0, 1)$
$1 \xrightarrow{deq} 0$	$(1, 0, 0, 0, 1, 0)$
$2 \xrightarrow{deq} 1$	$(1, 0, 1, 0, 0, 1)$
$3 \xrightarrow{deq} 2$	$(1, 0, 1, 1, 1, 0)$

Fig. 6.8 Queue LTS, transition encoding and corresponding BDD

rate of each encoded transition is stored in the corresponding terminal vertex of the MTBDD.

Example: As an example, consider the queueing process from Fig. 6.8, now with arrival rate λ and service rate μ, as depicted in Fig. 6.9 (left). Its MTBDD representation is shown in Fig. 6.9 (right). The set of paths leading to a non-zero vertex is of course the same as in Fig. 6.8, but the graph now has two branches, one for the *enq*- and one for the *deq*- transitions. We now consider a scaling of this model: The queueing system shown in Fig. 6.8 has a capacity of 3 customers. It can be generalised to an M/M/1 queue with capacity $c = 2^k - 1$, which means that the labelled Markov chain has 2^k states. One can show that the MTBDD representation of this Markov chain only requires $10k - 2$ MTBDD vertices (by the same argument as the one used in [428]), which means that for a family of models whose state space grows exponentially, MTBDDs provide a representation which only grows linearly! This nice result is of course related to the perfect regularity of the M/M/1 model (in particular if the state space is a power of 2), but many case studies have demonstrated the space efficiency of MTBDD representations for a large class of models, especially if used in a compositional context (see Sect. 6.5.3 below).

An orthogonal strand of research has led to the class of zero-suppressed binary decision diagrams (ZBDD). It is based on the observation that some Boolean functions whose set of minterms contains many negated variables do not have very compact BDD representations. However, if one changes the reduction rules for the decision diagram, more compact representations can be obtained. So, instead of eliminating don't care vertices (as in BDDs), in ZBDDs those vertices are eliminated whose then-successor is the terminal 0-vertex [651]. In other words, if a variable level is skipped from the root to the terminal 1-vertex, this means that the corresponding variable carries the value 0 (in a BDD setting this situation would mean that the corresponding variable is don't care). The use of ZBDDs and their multi-valued variants

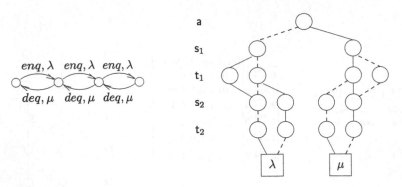

Fig. 6.9 Labelled Markov chain and corresponding MTBDD

has been shown to be beneficial for the analysis of Markov reward models [568]. A recent overview of zero-suppressed decision diagrams can be found in [570].

Working with decision diagrams, the branching decision taken at every vertex of the graph does not necessarily have to be a binary decision. This observation led to a large class of multi-valued (or multiway) decision diagrams (MDD) [505], originally employed for logic synthesis and verification. Multiway decision diagrams are also very well suited to encode the reachability set of a decomposed Petri net (i.e. a Petri net whose set of places is partitioned into subsets). The local marking of a subnet is hereby encoded as an integer, and there is a one-to-one correspondence between the subnets and the levels of the decision diagram. The maximum branching factor at a certain level is thus given by the number of reachable markings of that subnet [653].

6.5.3 Model Generation and Manipulation

Starting from a high-level model description, such as a Generalized Stochastic Petri Net (GSPN) or a stochastic process algebra model, efficient procedures are needed for generating the symbolic representation of the underlying labelled Markov chain. For Markovian stochastic process algebra, symbolic semantics have been developed which map a given process algebraic model directly to its underlying MTBDD-representation, without generating an intermediate labelled transition system in explicit form [561]. The key point of this mapping consists of the exploitation of the compositional structure of the SPA model at hand: Given M_1 and M_2, the MTBDD representations of two SPA processes P_1 and P_2, the MTBDD representation of their parallel composition $P_1|[S]|P_2$ is obtained as

$$
\begin{aligned}
M = \ &(M_1 \cdot S) \cdot (M_2 \cdot S) \\
&+ M_1 \cdot (1 - S) \cdot \mathrm{Id}_2 \\
&+ M_2 \cdot (1 - S) \cdot \mathrm{Id}_1
\end{aligned}
\tag{6.2}
$$

Fig. 6.10 A tandem queueing network

k	c	reachable states	transitions	MTBDD size	
				monolithic	compositional
3	7	128	378	723	148
4	15	512	1,650	1,575	197
7	127	32,768	113,538	11,480	341
10	1,023	2,097,152	7.3308e+06	–	485
14	16,383	5.36871e+08	1.8789e+09	–	677

Fig. 6.11 Statistics for the tandem queueing network

where S is the BDD-encoding of the synchronisation set S, and the Id_i are BDDs denoting stability of processes P_i. This construction guarantees that the size of the resulting MTBDD M is linear in the sizes of the operand MTBDDs and the number of action labels, which is a major source for the compactness of the symbolic representation [427].

Example: As an extension of our previous example, consider the tandem queueing network shown in Fig. 6.10, where an upstream queue with Coxian service is connected to a downstream queue with exponential service (the downstream queue is actually the one already considered in Sect. 6.5.2). Each of the queues has finite capacity of $c = 2^k - 1$, yielding a scalable model with altogether $2^k \cdot 2^k \cdot 2 = 2^{2k+1}$ states (the last factor of 2 is due to the two Coxian phases). Fig. 6.11, cited from [427], shows the growth of this model and of the associated MTBDDs. The last two columns give the numbers of MTBDD vertices for two variants of the symbolic representation. The numbers in column "monolithic" were obtained by directly encoding the labelled Markov chain of the overall model. Clearly, this approach does not lead to compact MTBDDs. In contrast, the numbers in column "compositional" were obtained by constructing the overall MTBDD in a compositional fashion, which means that two MTBDDs (one for the Coxian queue and one for the Markovian queue) were composed according to (6.2). This yields MTBDDs which grow only linearly with the parameter k, although the state space grows exponentially!

It is important to note that the above construction (6.2) yields an encoding of the so-called *potential* transition system which may also include transitions emanating from non-reachable states of the product state space. Symbolic reachability algorithms are employed to determine the set of reachable states. A mapping similar to (6.2) from the high-level model description to the symbolic representation of the underlying labelled Markov chain is employed in the probabilistic model checker PRISM [564], where users specify their models with the help of a guarded command language, based on Reactive Modules, which also features synchronisation between modules. In PRISM, not only CTMCs but also DTMCs and Markov decision processes can be specified, all of which are internally represented using MTBDDs.

In contrast to these structure-oriented approaches, the activity-local state graph generation scheme [569] does not need any a priori structure information and is therefore applicable to a general class of Markovian models. It creates its own, very fine structure, by considering the local effect of every activity (i.e. event) within the model. Since it is a round-based scheme, where reachability analysis needs to be performed in every round, an efficient variant of symbolic reachability analysis was developed as part of this approach. The activity-local approach is implemented (using zero-suppressed multi-terminal BDDs) in the framework of the Moebius modelling environment [276].

The saturation algorithm, first described in [217] also uses a fixpoint iteration scheme. Several variants of it have been described in the literature which work on different types of decomposition of the high-level model, see e.g. [925]. As the underlying data structures, these algorithms use extensible versions of multi-valued decision diagrams and matrix diagrams [219, 652].

For models with both Markovian and immediate transitions, elimination of the so-called vanishing states is a prerequisite for numerical analysis. An efficient symbolic elimination algorithm, which is implemented in the tool CASPA, has been described in [76]. It consists of three steps: 1. Some precomputations (including a realisation of the maximal progress assumption, i.e. the priority of an immediate transition over any timed transition). 2. The main fully symbolic round-based elimination algorithm. 3. A semi-symbolic post-processing for eliminating transitions that form an immediate loop or cycle (of which there are usually very few).

Since numerical analysis of large models is expensive (in terms of processing time and memory, see Sect. 6.5.4), it is desirable to reduce the size of the state space, if anyway possible. Bisimulation minimisation is a fundamental concept on which such a reduction can be based, whereby exact performance and dependability measures of the modelled system are preserved (in the context of Markov chains, bisimulation is known as lumpability). An early approach to symbolic bisimulation minimisation was described in [430], and more recently some very efficient symbolic bisimulation algorithms have been developed [289, 935]. All of these algorithms follow the basic principle of partition-refinement, where initially all states are considered equivalent, and at every step the current state space partitioning is refined according to some lumpability criterion, until stability is reached. However, the representation techniques used by these algorithms for encoding the state space partitions as BDDs are very different, resulting in a runtime-memory tradeoff between the different algorithms.

6.5.4 Numerical Analysis Based on the Structure of the Decision Diagram

For computing the desired performance or dependability measures of the modelled system, numerical analysis of the underlying stochastic process needs to be performed. In the case of CTMCs, this means that the vector of stationary probabilities has to be computed, which is typically done using iterative numerical methods such as Jacobi or Gauss-Seidel or their overrelaxed variants (or the well-known uniformization algorithm in case of transient state probabilities). In principle, such numerical calculations—which involve matrix-vector calculations as their basic operations—could work exclusively on symbolic data structures [360]. However, storing vectors of state probabilities symbolically (e.g. as an MTBDD) has proved to be neither memory-efficient nor time-efficient. Therefore, a hybrid scheme was developed [718], where only the matrix of transition rates is stored symbolically as an MTBDD while the vector of probabilities (of only the reachable states) is stored in explicit form as an array. Even with this approach, the probability vector (and not the storage of the matrix) is still the memory-bottleneck for large models! For speeding up the traversal of the MTBDD (which is done for looking up the transition rates), parts of it are sometimes replaced by sparse matrix data structures, which yields a typical time-space tradeoff. Parallel versions of symbolic numerical algorithms have also been developed [565]. In [814], a symbolic version of the multilevel algorithm (a recursive aggregation/disaggregation scheme) was described. Since the MTBDD possesses a recursive block-structure (due to the nature of its composition) this type of algorithm matches very well with the structure of the MTBDD. The multilevel scheme has also been combined with sparse representations of both terminal and intermediate blocks of the matrix, and some further accelerations of the calculations have been developed [815]. Numerical solution algorithms based on different types of matrix-diagrams have been implemented in the tool SMART [219, 652], as well as approximate algorithms for stationary solution [654].

6.5.5 Outlook

The "symbolic" approach described in this section is now a mature method implemented in several successful tools (e.g. PRISM [564], SMART [652] and CASPA [76]). It supports all phases of modelling, from state space generation to various forms of (qualitative and quantitative) analysis. Decision-diagram-based techniques are capable of dealing with very large state spaces, thus alleviating the problem of state space explosion. They are therefore among the methods of choice for constructing and analysing detailed and scalable resilience models. However, there are still many remaining research problems, for instance the question of how to further improve the numerical analysis of very large models with the help of approximations or bounding methods.

6.6 Mean-Field Approximation

Mean-field methods were first used in Physics to describe the interaction of particles in systems like plasma or dense gases. Instead of providing a detailed model, the influence of the mean environment on a single particle is studied. Subsequently mean-field methods were introduced to many other topics, for an overview see the introduction of [80].

The idea of aggregating the influence of the environment can also help in dealing with the state space explosion problem. We consider large networks of identical components, for example, computers in the Internet running the same piece of protocol software. Modelling each component and its interaction with the other components explicitly results in an intractable state space. Instead we focus on an approximating model where only the average impact of the complete system on the evolution of a component is considered. Results for this mean-field approximation model are cheap to compute (matrix-vector multiplications). It allows for statements about the average behaviour of the underlying original model, especially when the number of components is large.

6.6.1 Computing the Mean-Field

In the following we describe the process of mean-field approximation for discrete-time Markov models. We then illustrate this process with an example.

6.6.1.1 Discrete-Time Model for Single Component

At the beginning of the process we have to determine a discrete-time probabilistic model for a single component. It will typically have a relatively small set of states S. The transition probabilities are allowed to depend on N, the total number of components in the system, and on the so-called *occupancy measure* $\mathbf{m}$, a vector containing the fraction of components in each state. The model is then determined by a local probability matrix $P^N(\mathbf{m})$.

6.6.1.2 The Underlying Stochastic Process

Even though we are never going to explicitly construct it, we have to consider some properties of the stochastic process for the complete system. It consists of the parallel composition of the models for all N components. If the model for a single component has K states, the state space for the composed system would have K^N states. Since all components behave identically, we can aggregate the state space to the occupancy measure where we only keep track of the fraction of components in each state. The state space still would consist of $\binom{K+N-1}{K-1}$ states. The mean-field method gives us an approximation for the *transient occupancy measure*, that is, the occupancy measure at a given point in time t.

6.6.1.3 The Deterministic Limit Process

Under certain convergence requirements [146] the local probability matrix has a limit if N goes to infinity:

$$P(\mathbf{m}) = \lim_{N \to \infty} P^N(\mathbf{m})$$

For a given initial occupancy measure $\mu(0)$, the matrix $P(\mathbf{m})$ defines a *deterministic* process

$$\mu(t+1) = \mu(t) \cdot P(\mu(t))$$

This deterministic process approximates the occupancy measure for large N. The values of $\mu(t)$ for $t \in \mathbb{N}$ are easily determined by simple matrix-vector multiplications. Note that the matrix $P(\mu(t))$ has to be recalculated in each step.

6.6.1.4 Interpretation of Results

The mean-field method gives us a deterministic approximation of the occupancy measure. For each point in time it predicts a distribution of components over the different possible states. If we evaluated the underlying stochastic process we would get a different type of result: it would result in a distribution over all possible occupancy measures, that is, a distribution of possible distributions. However, the larger N is, the more deterministic the occupancy measure becomes (Central Limit Theorem). It is therefore justified to use the deterministic mean-field approximation of the number of components N is large. Even for small N the approximation gives valuable insight into the average behaviour of the system.

6.6.2 Illustrating Example

We consider a very simple information dissemination protocol. A large number of network nodes communicates in order to distribute a small but important piece of information. The communication protocol is completely decentralised which makes it robust even if some of the involved network nodes fail. To increase security, a node only accepts and redistributes the information after it has received it twice. This prevents the distribution of incorrect data, resulting from transmission errors or malicious insertion. Figure 6.12 shows the state-transition diagram for nodes. Each node can be in one of three states: it can be ignorant (state 0), it can be waiting for confirmation (state 1), or it can be knowing (state 2). A node can move from ignorant to waiting state with a certain probability. From here it either moves on to knowing, or, if it does not receive the piece of information a second time, it returns to ignorant. Once knowing, it will never get ignorant again—it never forgets.

Fig. 6.12 Data stream in
packet-switched network with
tree topology

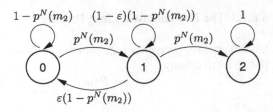

The probabilities in the model depend on two parameters that can change over
the evolution of the system: firstly, it depends on the total number N of nodes in
the network, secondly, it depends on m_2, the fraction of already knowing nodes.
Additionally there is a *gossip* parameter $g \in [0, 1]$ that reflects the probability that
an informed node is going to pass on the information, and a parameter $\varepsilon \in [0, 1]$ that
governs the return to state 0 from state 1.

The probability for a single ignorant node to move from state 0 to state 1 then is

$$p^N(m_2) = g \cdot \frac{m_2 \cdot N}{N-1}$$

which is proportional to the fraction of knowing nodes among the remaining nodes,
and g. The higher the fraction of already knowing nodes is in the network, the higher
is the probability, that a yet ignorant node gets the information.

A waiting node (state 1) moves to the knowing state again with probability
$p^N(m_2)$. It returns to state 0 with probability $\varepsilon(1 - p^N(m_2))$. The parameter ε
can be interpreted as the probability for discarding the first receipt of the data.

To represent all transition probabilities in the model, we state a probability
matrix that is *local* to each node. It depends on N and the *occupancy measure*
$\mathbf{m} = (m_0, m_1, m_2)$.

$$P^N(\mathbf{m}) = \begin{pmatrix} 1 - p^N(m_2) & p^N(m_2) & 0 \\ \varepsilon(1 - p^N(m_2)) & (1 - \varepsilon)(1 - p^N(m_2)) & p^N(m_2) \\ 0 & 0 & 1 \end{pmatrix}$$

Considering all N network nodes in parallel would result in a state space with 3^N
states. Even if we only recorded the occupancy measure, the state space would still
have $\frac{N(N+1)(N+2)}{2}$ states—for three-state components!

Two avoid this blowup, we consider the mean-field limit for the occupancy measure. The limit of the local probability matrix is

$$P(\mathbf{m}) = \lim_{N \to \infty} P^N(\mathbf{m}) = \begin{pmatrix} 1 - g \cdot m_2 & g \cdot m_2 & 0 \\ \varepsilon(1 - g \cdot m_2) & (1 - \varepsilon)(1 - g \cdot m_2) & g \cdot m_2 \\ 0 & 0 & 1 \end{pmatrix}$$

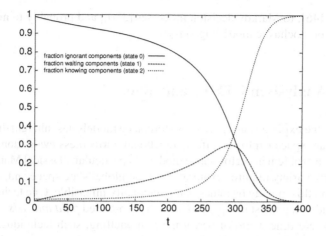

Fig. 6.13 Mean-field approximation for the illustrating example

For a given *initial* occupancy measure $\mu(0)$ we can then approximate the transient evolution of the occupancy measure by the deterministic process [79, 146]

$$\mu(t+1) = \mu(t) \cdot P(\mu(t))$$

Figure 6.13 shows the deterministic occupancy measure $\mu(t)$ for $t \in [0, 400]$ when at the beginning the fraction of ignorant nodes is 99 % and the fraction of knowing nodes is 1%, that is, $\mu(0) = (0.99, 0, 0.01)$. The figure represents the behaviour that could be expected: the fraction of ignorant nodes decreases continually while the number of knowing nodes increases. Since all nodes have to move through the intermediate waiting state, the fraction of nodes in this state first increases and then decreases again.

The mean-field approximation does not give us information about the possible deviations from the computed deterministic value. However, the results depicted in Fig. 6.13 allow a statement about the approximate time at which we expect all nodes to be knowing with high probability.

6.6.3 Outlook

Very often we encounter real-world systems that consist of a large number of identical replicas of the same component. In this section we have shown how the mean-field approximation method can be employed if the components are represented by discrete-time models. The core computation for transient measures then boils down to a series of cheap matrix-vector multiplications. Mean-field analysis is not restricted to discrete-time models. It can also be applied to continuous-time Markov chains [118, 146], to Markov decision processes [370], and probably to many other probabilistic or stochastic modelling classes.

6.7 Fluid Analysis in CTMCs and SPAs

Representing the explicit state space of performance models has inherent difficulties. Just as the state-space explosion effects functional correctness evaluation, so it can also be easily a problem in performance models. In particular, classical Markov chain analysis of any variety requires exploration of the global state space and, even for a simple system, this quickly becomes computationally infeasible. One technique that attempts to side-step the state-space explosion is so-called *fluid analysis*.

In the discrete-time world of performance modelling, such techniques have already been used by Benaïm and Le Boudec [100] to good effect in *mean-field* analysis of performance models. Similarly, Bakhshi et al. [79, 80] have developed some discrete-time model-specific analysis techniques for gossip protocols using the mean-field technique.

In the field of stochastic process algebras, Hillston developed *fluid-flow* analysis [438] to make first-order approximations of massively parallel PEPA models. Bortolussi [142] presented a formulation for the stochastic constraint programming language, sCCP. Cardelli has a first-order fluid analysis translation to ordinary differential equations (ODEs) for the π-calculus [172]. Petriu and Woodside [724] have successfully used Mean Value Analysis of hierarchical queueing models such as Layered Queueing Networks (LQNs) to obtain mean response time results over large component-based systems.

In this section, we briefly outline a fluid analysis technique that is applied to a variety of the PEPA language known as Grouped PEPA (or GPEPA) [420]. It has the advantage over the original fluid approximation for PEPA or π-calculus of having higher moment results available. These can be used to calculate variance information and give a notion of accuracy of the first-order prediction. Additionally, higher moments can also be used to create bounding approximations on some varieties of passage-time distribution [419].

6.7.1 GPEPA

Grouped PEPA (GPEPA) [420] is a simple syntactic extension to PEPA which allows the straightforward identification of models that can be analysed using fluid techniques. Specifically, the component grouping in GPEPA identifies the abstraction level at which the fluid analysis will take place. By adding component group labels to these structures, we also benefit from being able to identify uniquely components in particular states in particular parts of the model structure.

For a detailed summary of PEPA process algebra syntax and operational meaning we refer to the reader to the many papers that have been published discussing and using PEPA. The following is a small selection of such work [226, 352, 381, 436].

A GPEPA model is formed by composing multiple labelled component groups together. The grammar for a GPEPA model G is:

$$G ::= G \underset{L}{\bowtie} G \mid Y\{D\} \tag{6.3}$$

where Y is a group label, unique to each component group. The term $G \underset{L}{\bowtie} G$ represents synchronisation over action types in the set L, where $L \subseteq \mathcal{A}$ is a set of possible action types in the model.

In this context, a component group is a parallel cooperation of a normally large number of *fluid* components. A fluid component is one whose state changes can be captured by a random variable and ultimately a set of differential equations. Parallel, in this setting, means that there is no synchronisation between individual members of the component group. A component group D is specified as follows:

$$D ::= D \parallel D \mid P \tag{6.4}$$

where P is a *fluid component*, a standard PEPA process algebra term. The combinator $\parallel$ represents parallel, unsynchronised cooperation between fluid components.

6.7.2 A Client–Server Example of a Fluid model

To illustrate more clearly how component groups and fluid components are used together to construct GPEPA models, we consider a GPEPA model of a simple client/server system from [419]. The type of model we wish to consider in this section is one which exhibits massive parallelism. We present a simple system with n clients and m servers. The system uses a 2-stage fetch mechanism: a client requests data from the pool of servers; one of the servers receives the request, another server may then fetch the data for the client. At any stage, a server in the pool may fail. Clients may also timeout when waiting for data after their initial request. We could capture this scenario of n clients cooperating on the *request* and *data* actions with m resources with the following GPEPA system equation:

$$\mathbf{CS}(n, m) \overset{def}{=} \mathbf{Clients}\{\mathbf{Client}[n]\} \underset{L}{\bowtie} \mathbf{Servers}\{\mathbf{Server}[m]\} \tag{6.5}$$

where $L = \{request, data\}$ and $C[n]$ represents n parallel cooperating copies of component C. Each client is represented as a **Client** component and each server as a **Server** component. Each client operates forever in a loop, completing three tasks in sequence: *request*, *data* and then *think*; and they may also perform a *timeout* action when waiting for data:

$$\mathbf{Client} \stackrel{\mathrm{def}}{=} (request, r_r).\mathbf{Client\_waiting}$$

$$\mathbf{Client\_waiting} \stackrel{\mathrm{def}}{=} (data, r_d).\mathbf{Client\_think} + (timeout, r_{tmt}).\mathbf{Client}$$

$$\mathbf{Client\_think} \stackrel{\mathrm{def}}{=} (think, r_t).\mathbf{Client}$$

The servers on the other hand first complete a *request* action followed by a *data* action in cooperation with the clients but at either stage they may perform a *break* action and enter a broken state in which a *reset* action is required before the server can be used again:

$$\mathbf{Server} \stackrel{\mathrm{def}}{=} (request, r_r).\mathbf{Server\_get} + (break, r_b).\mathbf{Server\_broken}$$

$$\mathbf{Server\_get} \stackrel{\mathrm{def}}{=} (data, r_d).\mathbf{Server} + (break, r_b).\mathbf{Server\_broken}$$

$$\mathbf{Server\_broken} \stackrel{\mathrm{def}}{=} (reset, r_{rst}).\mathbf{Server}$$

The *request* and *data* actions are shared actions between the clients and servers in order to model the fact that clients must perform these actions by interacting with a server. The actions *timeout*, *think*, *break* and *reset* on the other hand are completed independently.

6.7.3 Generating ODEs for Fluid Analysis

Fluid analysis is used to approximate the mean dynamics of a GPEPA model. That is, at a time t, fluid analysis will tell you how many components in each named component group there are on average. As the size of the component groups increases, so the approximation can be shown to improve [419]. This convergence can also be demonstrated for higher moments of component counts [849].

In this section, we analyse the client/server model of Sect. 6.7.2 to show mean component counts for particular client and server states, Figs. 6.14a and 6.14c. We show the standard deviation of the component counts for the client/server model in Figs. 6.14b and 6.14d. Where there is a local deviation from the simulated results, this is due to the synchronisation model used in PEPA rather than fluid analysis in particular [420, 849].

We will skip any further preamble and write down the generating equation for a set of ODEs that gives a first-order fluid analysis of a GPEPA model. The higher order fluid analysis (that gives quantities can such as variance, standard deviation and skewness) can be found in [420].

Let G be a GPEPA model. We define the evolution of the number of components of type P in group H over time, by $v_{H,P}(t)$. This quantity is defined by a system of first-order coupled ODEs:

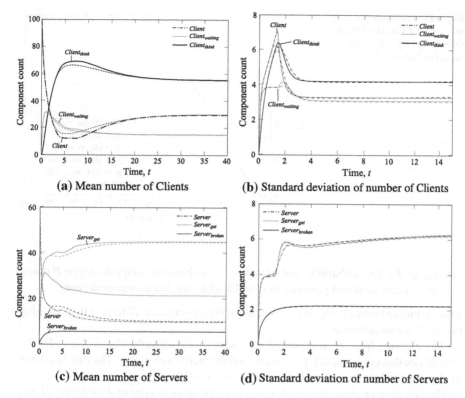

Fig. 6.14 Fluid analysis of the number of components in a Client/Server model. Simulations are given as dashed plots. **a** Mean number of Clients **b** Standard deviation of number of Clients **c** Mean number of Servers **d** Standard deviation of number of Servers

$$\dot{v}_{H,P}(t) = \sum_{\alpha \in \mathcal{A}} \underbrace{\left(\sum_{Q \in \mathcal{B}(G,H)} p_\alpha(Q, P) \, \mathcal{R}_\alpha(G, V(t), H, Q) \right)}_{\text{incoming rate}} - \underbrace{\mathcal{R}_\alpha(G, V(t), H, P)}_{\text{exit rate}}$$

(6.6)

for all component group/component state pairs (H, P) in the GPEPA model G. That is, we generate one ODE for each local state in the compositional model. Hence we avoid any association with the global state space.

This can broadly be explained by slitting (6.6) into two parts, the *incoming rate* and the *exit rate*. The incoming rate part describes the total rate of the model that increases the number of components of type P in group H for a given action α. This is calculated from the product of two quantities:

1. $\mathcal{R}_\alpha(G, V(t), H, Q)$, the overall rate that component type Q enables doing actions of type α, where Q is also in group H, and;

Fig. 6.15 Fluid passage time
analysis of a Client response
time in the Client/Server
model of Sect. 6.7.2

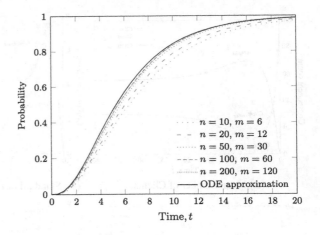

2. $p_\alpha(Q, P)$, the probability that Q can evolve to become component type P doing
an action α (it could perform an α and lead to another component type).

This is summed over all possible component types Q in group H to give the incoming
rate to P for an action α.

The exit rate from P in group H for an action α is simply $\mathcal{R}_\alpha(G, V(t), H, P)$,
overall rate that component type P enables doing actions of type α. The full definition
of the component rate function $\mathcal{R}_\alpha(\cdot)$ is given in [420].

Finally, the aggregate rate of change of components of type of P in group H can
be shown to equal to the total incoming rate minus the total exit rate, summed over
all possible action types α.

We have shown how first order fluid analysis of large stochastic processes can be
used to create passage and response time distributions and related measures [419,
421]. This is done by performing the sort of fluid analysis described here on an
appropriately modified model. Figure 6.15 shows a single Client response time distribution (from steady-state) given all the synchronisations in the system. As with all
fluid analysis, as the model gets larger, the distribution predicted by the fluid analysis
(black line) becomes more accurate.

6.7.4 Outlook

Further developments in fluid analysis have shown how other types of passage time
distribution can be extracted from massively parallel systems [421]. Additionally,
real-valued rewards can be accumulated over the lifetime of a process model to capture quantities such as cost and energy consumption. For smaller scale models, these
can be analysed using discrete-state Markov Reward Models [872]; fluid analysis can

be used here also to analyse larger scale reward models [848] which would otherwise
have a prohibitive state space size.

6.8 Conclusion

In this chapter we have presented a number of efficient representation and analysis
techniques that can aid resilience and performance analysis of large and complex
systems. It is worth saying that there is no one best technique in all cases. As with
many modelling and analysis problems, the best tool for the job will often depend
on the problem being tackled and the type of answer being sought. Some of the
techniques discussed can be used to tackle many types of performance and resilience
problem, while some are better suited to a particular type of analysis. Some of the
techniques will allow the user to recover an explicit state representation, while others
consider many model states in aggregation.

For a particular problem a modeller will wish to select those techniques that can
produce the desired analysis. However, beyond that, it will be worth the modeller's
while to attempt a few of the remaining techniques to see which scales best to their
needs. They will find dramatic differences in operation and those differences (in
execution time and memory usage) will reflect many issues, such as how the model
is originally represented, how many and indeed whether parameters are available to
the modeller and the level of complexity that the modeller wishes to capture in their
model.

Finally, we would like to emphasise that these descriptions are by necessity brief
summaries of much more involved techniques. These topic introductions should be
treated as such and many references have been provided for the reader to follow up
if more information is required.

Acknowledgments Jeremy Bradley, Richard Hayden and Nigel Thomas are supported by the
UK Engineering and Physical Sciences Research Council on the AMPS project (reference EP/
G011737/1). Leïla Kloul is supported by the European Celtic project HOMESNET [8], Philipp
Reinecke and Katinka Wolter are supported by the German Research Council under grant Wo
898/3-1.

Part III
Model-Driven Prediction

Chapter 7
Modelling and Model-Based Assessment

Andrea Bondavalli, Paolo Lollini, István Majzik and Leonardo Montecchi

Abstract This chapter provides an overview of the state of knowledge related to stochastic model-based assessment approaches, which are most commonly used for resiliency evaluation of current computing systems. The chapter first introduces a set of representative surveys developed in recent European projects, and then it provides a deeper description of common techniques used in model-based assessment of resilient systems. The most widely used modelling formalisms are reviewed, with a particular focus on state-based formalisms like Stochastic Petri Nets and its extensions. Techniques used in model construction and solution are also discussed, as well as the different classes of analysis tools and frameworks. The techniques analyzed in the chapter span from largeness avoidance and largeness tolerance techniques to more comprehensive modelling approaches that are integrated in the system's development and assessment process. Some of these techniques try to cope with system's complexity by automatically deriving the analysis models from engineering models like UML or AADL. Other approaches attack the complexity issue combining different evaluation methods, exploiting their possible complementarities and synergies. A discussion on the open research challenges in model-based resilience assessment is finally provided in the last part of the chapter, based on the reviewed techniques and on the activities carried out within the AMBER Coordination Action.

A. Bondavalli (✉) · P. Lollini · L. Montecchi
University of Firenze,
Viale Morgagni 65, 50134 Firenze, Italy
e-mail: bondavalli@unifi.it

P. Lollini
e-mail: lollini@unifi.it

L. Montecchi
e-mail: lmontecchi@unifi.it

I. Majzik
Budapest University of Technology and Economics,
Magyar Tudósok krt. 2, Budapest, Hungary
e-mail: majzik@mit.bme.hu

K. Wolter et al. (eds.), *Resilience Assessment and Evaluation of Computing Systems*,
DOI: 10.1007/978-3-642-29032-9_7, © Springer-Verlag Berlin Heidelberg 2012

7.1 Introduction

As defined in [81], a model is an abstraction of a system "that highlights the important features of the system organization and provides ways of quantifying its properties neglecting all those details that are relevant for the actual implementation, but that are marginal for the objective of the study". There exist several types of models, and the choice of a proper model depends on many factors, like the complexity of the system, the specific aspects to be studied, the attributes to be evaluated, the required accuracy, the questions to be answered about the system, and the available resources for the study.

Models play a primary role in the resilience assessment process of modern computing systems. First of all, they allow an "a posteriori" resiliency analysis, to understand and learn about specific aspects, to detect possible design weak points or bottlenecks, to perform a late validation of the dependability requirements and to suggest sound solutions for future releases or modifications of the systems. Furthermore, assessing the resilience of composite systems, often including a dynamic mixture of components built by different parties and for different purposes, is a difficult task that may require the combination of several assessment methods and approaches. In this perspective, models can be profitably used as support for experimentation and vice-versa. On one side, modelling can help in selecting the features and measures of interest to be evaluated experimentally, as well as the right inputs to be provided for experimentation. On the other side, the measures assessed experimentally can be used as parameters in the models, and the features identified during the experimentation may impact the semantics of the dependability model.

This chapter provides an overview of the state of knowledge related to stochastic model-based assessment approaches, which are most commonly used for resiliency evaluation of current computing systems. An overview of the history in model-based assessment research, as well as current directions, is sketched in Fig. 7.1.

The approaches presented in this figure will be explained and discussed in the next sections. The introduction of Fault Tree Analysis (FTA) and Petri Nets (PN) in the early 1960s had a great impact on the formalization of model-based assessment practice. For several years, model-based assessment was based on these formalisms, as well as on the early theories of Markov Chains and Queuing Networks. Later these formalisms, Petri Nets and Markov Chains in particular, have inspired many other higher-level formalisms (Stochastic PN, Generalized Stochastic PN, etc.) which are currently widely used for model-based analyses. The largeness and complexity of the models rapidly became a challenging issue to be addressed, and in the beginning of the 1980s researchers started focusing on the development of methodologies, techniques and tools to avoid or tolerate model complexity, also (from the early 1990s) exploiting the synergies and complementarities among several evaluation methods. In the same years, stochastic extensions to process algebras started to appear in the literature, most notably with the introduction of Performance Evaluation Process Algebra (PEPA) in the mid 1990s. Finally, in the last ten years, another research direction has focused on the use of engineering languages (UML, AADL, etc.) to facilitate the construction

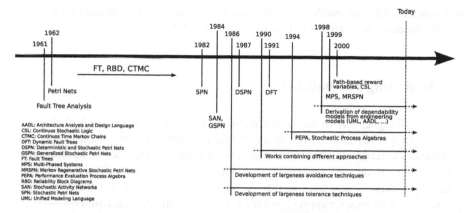

Fig. 7.1 Historical timeline of advances in modelling and model-based assessment

of the models by designers, and on the development of transformation techniques and tools to translate such high-level models to analysis models for dependability evaluation.

Detailed surveys giving a comprehensive overview of model-based assessment can be found in publications of recent European projects, discussed from different perspectives and applied to different system contexts. In HIDENETS [435], a state of the art of evaluation techniques, methodologies and tools is presented, focusing on distributed applications and mobility-aware services, in ubiquitous communication scenarios (see [603]). Individual quantitative evaluation techniques that are suitable to analyze HIDENETS-like systems are reported, as well as existing work describing combinations of them. In the context of the ReSIST NoE [773], the research challenge required for providing scalable resilience of policies, algorithms and mechanisms has been identified. In particular, the state of knowledge and ongoing research on methods and techniques for resilience evaluation are summarized in [774], taking into account the scaling challenges of large and evolving systems considered in the project. In CRUTIAL [246], a state of knowledge [500] was presented, focusing on the existing model-based methodologies, techniques and tools that can be useful to address the challenges raised in the context of interdependent critical infrastructures in general, and electric power system infrastructures in particular. Finally, an overview on the model-based methodologies elaborated for the quantitative evaluation of safety critical applications, in particular of safety critical railway systems, has been produced in the SAFEDMI [796] project, and reported in [797].

The remainder of the chapter is structured as follows. In Sect. 7.2 the available and commonly used modelling formalisms are outlined. Strategies to build and solve models in a time and space-efficient way are briefly discussed in Sect. 7.3, while Sect. 7.4 gives an overview of existing modelling and solution tools supporting model-based assessment. Section 7.5 discusses the methodologies used to derive dependability models based upon engineering models, and Sect. 7.6 surveys works that combine different modelling approaches. Open research challenges for model-based resilience assessment are finally discussed in Sect. 7.7.

7.2 Modelling Formalisms

A system designer has in his or her possession a wide range of analytical modelling techniques to choose from. Each of these techniques has its own strengths and weaknesses in terms of accessibility, ease of construction, efficiency and accuracy of solution algorithms, and availability of supporting software tools. The most appropriate type of model depends upon the complexity of the system, the questions to be answered, the accuracy required, and the resources available to the study.

Analytical models can be broadly classified into non-state space (combinatorial) models and state space models.

7.2.1 Non-State-Space Models

Reliability block diagrams (RBD), fault trees (FT) and reliability graphs (RG) are non-state-space models commonly used to study the dependability of systems. They are concise, easy to understand, and have efficient solution methods. However, realistic features such as interrelated behaviour of components, imperfect coverage, nonzero reconfiguration delays, and combination with performance can not be captured by these models. These arguments led to the development of new formalisms, such as dynamic fault trees (DFT) and dynamic reliability block diagrams (DRBD), to model reliability interactions among components or subsystems. A brief overview of traditional non-state-space models can be found in [688], while some of their 'dynamic' extensions are outlined in [298].

7.2.2 State-Space Models

State-space models, in particular homogeneous continuous time Markov chains (e.g., see [893] for full details) are commonly used for dependability modelling of computing systems. They are able to capture various functional and stochastic dependencies among components and allow evaluation of various measures related to dependability and performance (performability) based on the same model, when a reward structure is associated to them. Unfortunately, not all the existing systems and their features can be captured properly by Markov processes, since these processes require the strong assumption that the holding time in any state of the system is exponentially distributed. In some cases this assumption may be very unrealistic and to properly represent the system behaviour more general processes (e.g., semi-Markov, Markov

Regenerative or even non-Markovian processes) must be used. When dealing with such processes, complex and costly analytical solution techniques may have to be used. If analytic solution methods do not exist, discrete-event simulation must be used to solve the models thus providing only estimates of the measures of interest. Especially for dependability metrics such as reliability and availability, simulation may however be time consuming because of the rare event problem: events of interest occur so rarely that very lengthy simulations are necessary to obtain reliable results. Alternatively, one can approximate an underlying non-Markovian process with a Markov process, and thus represent a non-exponential transition with an appropriate network of exponential ones (phased-type approach). The price to pay following this approach is a significant increase in the number of states of the resulting Markov model. The work in [894] reviews the existing state space models (as well as combinatorial ones) and it discusses the benefits and the limitations of each.

To facilitate the generation of state-space models based on Markov chains and their extensions, higher-level modelling formalisms like Stochastic Petri Nets (SPN) are commonly used. These formalisms allow a more compact model representation because they support concurrency. In [216], the authors explore and discuss a hierarchy of SPN classes where modelling power is reduced in exchange for an increasingly efficient solution, focusing on Generalized Stochastic Petri Nets (GSPN), Deterministic and Stochastic Petri Nets (DSPN), Semi-Markovian Stochastic Petri Nets (SMSPN), Timed Petri Nets (TPN), and Generalized Timed Petri Nets (GTPN). Other widely used modelling formalisms are Stochastic Reward Nets [119], Stochastic Activity Networks (SAN) [805] and Markov Regenerative Stochastic Petri Nets (MRSPN) [210].

Other modelling formalisms exist that allow to provide a high-level representation of Markov Chain models, e.g., Stochastic Automata Networks (SAN) [734] or models based on Stochastic Process Algebras. Such formalisms are extensions of basic Process Algebras, which are enriched with the ability to associate probabilities and/or time delays to the execution of actions. These extensions allow performing quantitative analysis on the model. Several stochastic process algebra languages have been introduced, the most influencing one in dependability analysis has been Performance Evaluation Process Algebra (PEPA). Similarly to Petri Net extensions, some of these formalisms are Markovian, e.g., PEPA, or Markovian Time Processes for Performance Evaluation (MTIPP) [429], therefore having evaluation techniques that rely on Markov chains. Other formalisms allow more general probability distributions, e.g., SPADES [414] or MoDeSt [257], and therefore have evaluation techniques that rely on more general stochastic processes, or discrete-event simulation.

7.3 Model Construction and Solution Approaches

The main problem in using state-based models to realistically represent the behaviour of a complex system is the explosion in the number of states (often referred to as state-space explosion problem). Significant progress has been made in addressing the challenges raised by the large size of models both in the model construction and model solution phase, using a combination of techniques that can be categorized with respect to their purpose: largeness avoidance and largeness tolerance, see [501, 603, 688] for three comprehensive surveys.

Largeness avoidance techniques try to circumvent the generation of large models using, for example, state truncation methods [189], state lumping techniques [529], hierarchical model solution methods [798], fixed point iterations [623], hybrid models that judiciously combine different model types [678] and the fluid flow approximation [438, 444].

However, these techniques may not be sufficient as the resulting model may still be large. Thus, *largeness tolerance* techniques are needed to facilitate the generation and the solution of large state space models.

Largeness tolerance techniques propose new algorithms and/or data structures to reduce the space and time requirements of the model. This is usually achieved through the use of structured model composition approaches, where the basic idea is to build the system model from the composition of sub-models describing system components and their interactions. Generic rules are then defined for the elaboration of the sub-models and their interconnection. Following the approach proposed in [732], for example, the generator matrix of a CTMC is not entirely stored, but it is implicitly represented as Kronecker product of a number of smaller matrices. In [220] largeness is tolerated using Multivalued Decision Diagram (MDD) data structures to efficiently explore large state spaces.

Other techniques try to reduce the complexity of the model making use of concepts borrowed from the model checking theory. A first approach combines process algebras with Markov chains, to take advantage of their powerful and well-defined composition operators, leading to the Input/Output Interactive Markov Chains (I/O-IMC) formalism [145].

Rather than focusing on model composition, another approach concentrates on the definition of the dependability measures of interest to be evaluated. In fact, many sophisticated formalisms exist for specifying complex system behaviours, but methods for specifying performance and dependability variables remain quite primitive. To cope with this problem, modellers often must augment system models with extra state information and event types to support particular variables. To address this problem the so-called path-based reward variables have been introduced [695]. Numerical methods to compute such reward variables, defined with the Continuous Stochastic Logic (CSL), are given in [78], while in [417] the model checking approach is illustrated through a workstations cluster example.

Other approaches try to tolerate model largeness using model decomposition and aggregation of the partial results. The basic idea is to decouple the model into simpler and more tractable sub-models, and the measures obtained from the solution of the sub-models are then aggregated to compute those concerning the overall model. A survey on decomposition/aggregation approaches can be found in [602]. In the same paper, the authors also propose a general modelling framework that adopts three different types of decomposition techniques to deal with model complexity: at functional, temporal, and model-level. The key point is that the approach is non-domain-specific, i.e., not specifically developed for a particular class of systems or tailored for a specific modelling formalism or solution technique. Other largeness tolerance techniques also exist, such as disk-based approaches [278], where the model structure is stored in the disk thus allowing larger models to be solved, or on-the-fly approaches [277] which completely avoid the storage of structures in memory, generating them iteratively while computing the solution.

Even if these techniques are used, solving large state-space models is still a difficult task. Moreover, under certain conditions model solution may be a challenge even for models having only a few states. In particular, a large difference between the rates of occurrences of events leads to the *stiffness* problem. Stiff models cause problems in the numerical solution, because they require the use of an integration step of the order of the smallest time constants even though the analysis is to be carried out for an interval consistent with the largest time constants. Stiffness usually arises when dependability and performance models are mixed into a single model, but stiffness may also arise in a simple failure/repair model because of the different orders of magnitude between failure and repair rates. Stiffness may be avoided using aggregation and decomposition techniques in which the resulting sub-problems are non-stiff (e.g., see [121]), or it may be tolerated using special numerical solvers (e.g., see [668, 910]).

It is important to note that all the above techniques (largeness avoidance/tolerance and stiffness avoidance/tolerance) are complementary and all may be needed at the model construction and model solution levels, when detailed and large dependability models need to be generated and processed to evaluate metrics characterizing the resilience of real life systems.

7.4 Modelling and Solution Tools

Several software tools developed over the last thirty years address dependability and performability modelling and evaluation. Surveys of the problems related to techniques and tools for dependability and performance evaluation can be found for example in [176, 418, 763, 802]. Tools can be grouped in two main classes:

- *Single-formalism/multi-solution tools*, which are built around a single formalism and one or more solution techniques. They are very useful inside a specific domain, but their major limitation is that all parts of a model must be built in the single formalism supported by the tool. In the following we cite two sets of tools. The first set of tools is based on Stochastic Petri Nets formalism and its extensions. They all provide analytic/numerical solution of a generated state-level representation and, in some cases, support simulation-based solution as well. This set includes DSPNexpress [591], GreatSPN [208], SURF-2 [107], DEEM [133], TimeNET [374], UltraSAN [806]. Other tools are based on Stochastic Process Algebra models; they provide numerical solutions and in some cases simulation-based results as well. This set includes for example the PEPA Eclipse Plugin [888], CASPA [562],

PEPS [104], and PRISM [564]. Another set of tools uses other model specification approaches, sometimes tailored to a particular application domain, and includes HIMAP [844] and TANGRAM-II [173].

- *Multi-formalism/multi-solution tools*, which support multiple modelling formalisms, multiple model solution methods, and several ways to combine the models, also expressed in different formalisms. They can be distinguished with respect to the level of integration between formalisms and solution methods they provide. In particular, some tools try to unify several different single-formalism modelling tools into a unique software environment. Examples are the following: IMSE [738], IDEAS [357], FREUD [909], DRAWNET++ [354]. In other tools, new formalisms, composition operators and solvers are actually implemented within a unique comprehensive tool. Though more difficult than building a software environment out of existing tools, this approach has the potential to much more closely integrate models expressed in different modelling formalisms. To the best of our knowledge, there are five main tools having these attributes: SHARPE [890], SMART [218], DEDS [95], POEMS [22] and MÖBIUS [256].

The solution is considered as a computation of a measure by using one of the following classes of techniques:

- *Closed form solutions*, which yield exact measures but can be obtained for only a limited class of models.
- *Direct analytical techniques*, like matrix inversion, which still yield exact measures but can be obtained for only a limited class of models.
- *Iterative numerical techniques* (it is important to note that no general guarantees of convergence of iterative methods do exist for some problems, and the determination of a suitable error bound for termination is not easy).
- *Simulation techniques*, which provide an estimate with a confidence interval for the result, but may be costly in terms of run time.

In a large number of scenarios, steady-state solutions by themselves are insufficient. Two prominent examples for that are (1) scenarios in which steady-state does not exist at all, and (2) systems that can only be described by a homogeneous model for a very limited time during which steady-state behaviour cannot be observed. Transient analysis is frequently performed in simulation models although there exist a number of methods (supported by several tools) for analytical models as well.

7.5 Deriving Dependability Models from Engineering Models

The emergence of model-based engineering methodologies and the elaboration of automated model transformation techniques have opened up new ways to integrate model-based assessment into the development process. Model-Driven Engineering (MDE) refers to the systematic use of models as primary artefacts throughout the engineering lifecycle [811]. Precise, albeit informal or semi-formal engineering languages (like UML—the Unified Modeling Language, BPEL—the Business Process

Execution Language, AADL—the Architecture Analysis and Design Language, etc.) allow not only a reasonable unambiguous specification and design but also serve as the input for subsequent development steps like code generation, formal verification, evaluation, and testing. One of the core technologies supporting model-based engineering is *model transformation* [253]. Transformations can be used to refine models, apply design patterns, and project design models to various mathematical analysis domains in a precise and automated way.

These initiatives and technologies influenced model-based assessment as well, since they offered an efficient and integrated approach to *derive dependability analysis models from engineering models*. Resilience assessment requires specific support for the specification and description of non-functional aspects of the system (like reliability, safety), which are not properly covered by the common engineering languages, as these focus primarily on functional aspects. Recently, significant effort has been spent in the definition of standard languages that support the high-level specification of non-functional properties of systems; the UML profile for QoS and fault tolerance [703], the UML profile for Modeling and Analysis of Real-Time and Embedded systems (MARTE) [704], the Error Model Annex for AADL [795], and the Dependability Analysis Modeling (DAM) profile [110] are the most notable examples. However, there are not comprehensive high-level languages that support MDE dependability evaluation yet, since most of the existing approaches are tailored to a specific analysis method, or to a specific application domain. Properly addressing dependability concerns in this context is still a challenge (e.g., see [662] for further details) and it is actually one of the objectives of the ongoing ARTEMIS-JU CHESS project [199].

Different approaches for the automated derivation of dependability models have appeared in literature, often using ad-hoc language extensions:

- *Direct modelling of dependability related behaviour*: system designers use the extended engineering language to directly describe failure and repair/recovery processes (e.g., occurrence of different failure modes, error propagation) and also the corresponding properties of components (e.g., error rates, propagation probabilities). A good example is the usage of the AADL Error Model Annex: the behaviour of the components can be described in presence of internal faults and repair events, as well as in presence of external propagations. The dependability evaluation toolset constructs the analysis models by mapping the dependability related behaviour to the analysis formalism and then computes system-level dependability measures. A stepwise approach for GSPN dependability modelling on the basis of AADL is presented in [789]. As another example, in [714] UML is used as a language to describe error propagation and module substitution, that is then mapped to dynamic fault trees.

- *Modular construction of system-level models using predefined generic sub-models*: dependability experts construct *analysis sub-models* that represent the generic structure of both the failure/recovery processes of the different types of components and the error propagation among them. System designers use the language extensions just to identify the component types and assign local dependability parameters to hardware and software artefacts in the engineering model. These dependability parameters (typically available from component handbooks or from component level evaluation) are used to parameterize the generic sub-models. The dependability model construction tools (1) apply pattern matching and model transformation to assemble the relevant parameterized sub-models in a modular way on the basis of the architecture design, and then (2) invoke solution algorithms to solve the system level model. In a UML based approach [134], language extensions are defined as a UML profile (stereotypes and tagged values), analysis sub-models are assigned to architectural components and relations, and then composed as a system level Stochastic Reward Net (SRN). Modular model construction is supported by automated tools [619]. In case of web service based process models [385], web service language extensions are utilized, the services are mapped to DSPN sub-models, and then integrated into a Multiple Phased System model. An MDE transformation workflow for the quantitative evaluation of dependability-related metrics has been recently presented in [663]. The workflow is integrated in a more comprehensive modelling framework that is currently developed within the CHESS project, which combines MDE philosophy with component-based development techniques.

- *Integration of various aspects from different models*: in complex, dynamic distributed systems the dependability model shall be constructed from several engineering models that capture various aspects of the system at different hierarchy levels. Typically user, application, architecture and network levels are distinguished. For example, in case of large, critical mobile systems and infrastructures [135], the construction of the dependability model for computing user-level dependability attributes is based on (1) the workflow model of the user activities, (2) the topology models of the network connections in the various phases of the user activities (also constructed automatically from user mobility traces), and (3) the application-service-resource dependency models. This way a complex evaluation tool-chain is required to integrate the different mapping, abstraction, model transformation, and solution steps [555].

The automated derivation of dependability analysis models from the engineering models (that were created during the model based development process) has the advantage that—besides the application of certain model extensions—there is no need to learn and use specific dependability analysis formalisms, and modelling efforts can be saved. This is definitely a benefit if dependability analysis necessitates the creation of state-based dependability models in complex systems, as these models require higher learning and modelling effort than traditional combinatorial methods.

7.6 Works Combining Different Evaluation Approaches

The largeness and complexity of current real-life systems call for a *composite verification and validation* (V&V) framework, where the synergies and complementarities among several evaluation methods can be fruitfully exploited. For example, it is well established and widely recognized that modelling and experimentation complement each other, at least at the conceptual level, but the two approaches are infrequently combined in the literature to evaluate real-life systems.

An interesting area of research is the construction of analysis models *on the basis of measurements* performed in a running prototype or in a full deployment. The most comprehensive method was developed for performance and performability analysis: software performance models of distributed applications are extracted from traces recorded during execution [474]. A similar approach is recording error propagation traces induced by fault injection experiments [54] to support the construction of error propagation models [207]. Other works (e.g., [53]) derive high-level behavioural models using experimental measurements obtained from fault injection experiments, while in other papers (e.g., [229]) the values obtained from field data are used to setup parameters of analytical models.

Other attempts in exploiting the potential interactions among different evaluation approaches were reported in the context of recent European projects. In DBench [268], a framework for dependability benchmarking tailored for on-line transactional systems was established (see [513] for more details). One of the benchmarks developed in the project was based on both modelling and experimentation; the two final measures evaluated from such benchmark are the stationary system availability and the total cost of failures. The measures are evaluated by combining measures obtained from experimentation on the target system (e.g., the percentages of the various failure modes) and information from outside the benchmark experimentation (e.g., the failure rate, the repair rate and the cost of each failure mode).

A major research line of the European project CRUTIAL [246] focused on the development of a model-based methodology for the dependability and security analysis of the power grid information infrastructures. Within electric power systems, changes in the system's state (e.g., component failures) may propagate in a cascading fashion, following complex power flow equations. To accurately represent this behaviour, the modelling framework developed within the project combined a Stochastic Activity Networks (SAN) model with ad-hoc external mathematical functions (see [95, 202, 203]). The overall model that represents the organization and topology of the power grid was built using the SAN formalism, and it interacted with external functions to evaluate the effects on the complete power grid of environment variations.

HIDENETS [435] addressed the provisioning of available and resilient distributed applications and mobile services in highly dynamic environments characterized by unreliable communications and components, mostly concerning the field of car-to-car and car-to-infrastructure communications. One of its main achievements was the definition of a holistic evaluation framework (see [135]) where the synergies and complementarities of the different evaluation approaches could be fruitfully exploited. In the quantitative assessment of complex systems, like those targeted by the HIDENETS project, a single evaluation technique is not capable of tackling the whole problem, i.e., the dependability evaluation of end-to-end scenarios. To master complexity, the application of the holistic approach allows defining a "common strategy" using different evaluation techniques applied to the different components and sub-systems, thus exploiting their potential interactions. The idea underlying the holistic approach follows a "divide and conquer" philosophy: the original problem is decomposed into simpler sub-problems that can be solved using appropriate evaluation techniques. Then the solution of the original problem is obtained from the partial solutions of the sub-problems, exploiting their interactions. The different evaluation techniques may interact in different ways, including (1) *cross validation*, if a partial solution validates some assumptions introduced to solve another sub-problem, or validates another partial solution; (2) *cross-fertilization*, if a partial solution obtained solving a sub-problem is used as input to solve another sub-problem, possibly using a different technique; and (3) *problem refinement*, if a partial solution gives some additional knowledge that leads to a problem refinement. In [135, 136], for example, a highway scenario has been modelled, and the impact of user mobility on the QoS of UMTS communication has been evaluated combining a Stochastic Activity Network (SAN) model with a mobility simulator, exploiting this way the cross-fertilization of the techniques. The model of the users' behaviour and of the UMTS network has been provided using the SAN formalism, while the mobility of the users within the scenario has been accurately represented by the mobility simulator. A specific SAN submodel took charge of progressively reading the traces produced as output by the simulator, and synchronizing the states of the two models. The integration of the output produced by the mobility simulator into the modelling process itself allowed to capture system characteristics at a more detailed level, thus enabling a more refined analysis that could be hardly obtained using a single technique.

7.7 Open Research Challenges

As discussed in this chapter, a wide literature exists on the use of models for the assessment of dependability-related indicators of a system and, in general, for fault-forecasting, that is to probabilistically estimate the occurrence of faults and their impact on the ability of the system to provide a proper service. Nevertheless, the modelling and analysis of complex (large, dynamic, heterogeneous, ubiquitous) systems still needs continued research, both in model construction and in model solution. A crucial point in this context is also to assess the *approximations* introduced in the modelling and solution process to manage the system complexity, as well as their impact on the final results.

The role of modelling in a more comprehensive assessment process is, on the contrary, not well addressed in the literature. The largeness, dynamicity, heterogeneity and ubiquity of current computing systems actually calls for the development of a *composite and trustable assessment framework* including complementary evaluation techniques, covering modelling and experimental measurements. Mechanisms are needed to ensure the cooperation and the integration of these techniques, in order to provide realistic assessments of architectural solutions and of systems in their operational environments.

Of increased significance is also the use of quantitative evaluation methods to support the effective use of *adaptation mechanisms* in current systems. Efficient on-line mechanisms are needed to monitor the environment conditions of the system and to dynamically adapt to their changes.

Besides assessing the impact of accidental threats, extensions are also needed to *quantify the impact of malicious threats*. Quantitative evaluation techniques have been mainly used to evaluate the impact of accidental faults on systems dependability, while the evaluation of security has been mainly based on qualitative evaluation criteria. Therefore, there is a need for a comprehensive modelling framework that can be used to assess the impact of accidental faults as well as malicious threats in an integrated way.

Acknowledgments The authors acknowledge the support given by the European Commission to the AMBER Coordination Action [38]. This work has been partially supported by the Italian Ministry for Education, University, and Research (MIUR) in the framework of the Project of National Research Interest (PRIN) "DOTS-LCCI: Dependable Off-The-Shelf based middleware systems for Large-scale Complex Critical Infrastructures".

Chapter 8
Software Aging and Rejuvenation for Increased Resilience: Modeling, Analysis and Applications

Alberto Avritzer, Ricardo M. Czekster, Salvatore Distefano
and Kishor S. Trivedi

Abstract Software aging and rejuvenation research has shown that the application of approaches for software aging modeling, monitoring, and rejuvenation has the potential to significantly increase software resilience. In this chapter, we present an overview of important analytical models and measurement approaches for software aging and rejuvenation. We start by describing the Markov based approaches and renewal process based approaches for software aging and rejuvenation modeling. In addition, we present measurement based approaches using both online and offline methods for software rejuvenation. We conclude by presenting a categorization of the approaches and by presenting a brief overview of applicability of each of the approaches presented in this chapter.

A. Avritzer (✉)
Siemens Corporate Research and Technology, 755 College Road East,
Princeton NJ 08540, USA
e-mail: alberto.avritzer@siemens.com

R. M. Czekster
PUCRS/Faculdade de Informatica, Avenida Ipiranga, 6681, Predio 32,
Sala 505, CEP 90619-900 Porto Alegre, Brazil
e-mail: ricardo.czekster@pucrs.br

S. Distefano
Dipartimento di Elettronica e Informazione (DEI),
Politecnico di Milano, Piazza L. da Vinci, 32,
20133 Milan, Italy
e-mail: distefano@elet.polimi.it

K. S. Trivedi
Department of Electrical and Computer Engineering, Duke University, Durham
NC 27708, USA
e-mail: kst@ee.duke.edu

K. Wolter et al. (eds.), *Resilience Assessment and Evaluation of Computing Systems*,
DOI: 10.1007/978-3-642-29032-9_8, © Springer-Verlag Berlin Heidelberg 2012

8.1 Introduction

The introduction of software for monitoring and control of mission-critical systems has created a need for the validation of the resilience and safety of these systems. The activities required for the assessment and enforcement of these systems reliability and availability include requirements, architecture, modeling, testing, online monitoring, and software rejuvenation.

In this chapter we present models, algorithms and applications of software rejuvenation to increase software resilience. This chapter is closely related to the chapter on resilience assessment based on performance testing, where performance measurement results of smoothly degrading systems were presented. Smooth performance degradation has been also called software aging and is a consequence of the exhaustion of system resources, such as system memory or kernel structures, invalid pointers, the accumulation of round off errors, database deadlocks, and the contention for a pool of limited software resources. Therefore, transient application faults and operating system faults can be a major source of system performance degradation. Examples of operating system related faults are invalid allocation or deallocation of memory, kernel data corruption, and incorrect or sub-optimal kernel resource management [369, 895, 905].

Software aging research was initially directed towards the implementation of data collection tools for monitoring of application and operating system resources [72, 73, 181, 369, 895]. The development of stochastic models of software aging and the parameterization of these models with the time to failure distribution, the input workload, and its influence on software aging were presented in [303, 585, 905]. The *xSeries Software Rejuvenation Agent* (SRA) [181] is a tool introduced by IBM and Duke University to monitor system resources and to calculate the expected time to resource exhaustion. Approaches to monitor a customer-affecting metric, such as response time, to detect software aging due to resource exhaustion or security intrusions were introduced in [65–68].

The types of software faults that cause software aging have been shown to be very difficult to test, reproduce, and correct [392]. Some examples of major software outages that were attributed to software aging were reported in [111, 400]. The *Patriot* anti-missile software aging event allowed a *Scud* missile to penetrate US defenses, when the *Patriot* software started to miscalculate routes. This software aging event led to the death of U.S. soldiers during the Gulf War [401, 625]. The event investigation concluded that the problem was caused by a numerical accumulation error that was never caught during testing. Therefore, the system was deployed in production with the faulty software. The investigation report recommendation was to periodically restart the guiding system every eight hours of continuous operation to reset the accumulated variables to their initial valid states.

Software Rejuvenation is a mechanism to proactively and efficiently counteract the effects of software aging [72, 74, 449, 799, 895, 906]. Software rejuvenation architecture artifacts are a good match for complex industrial mission-critical applications that are susceptible to software aging. For example, the process of quickly shutting down and restarting a given process is a successful strategy to clear internal data structures and replenish system resources to their original specification. The main purpose of introducing software rejuvenation into the architecture of mission-critical systems is to proactively restore the critical system resources to full capacity before a customer impacting failure occurs. Software rejuvenation functions as preventive maintenance to ensure high availability. Software rejuvenation cost effectiveness depends on the state of the environment [74], the state of the mission [68], and the extent of system degradation. The addition of monitoring for software aging, and software rejuvenation as architecture artifacts, have been shown to be a cost-effective approach to increase the resilience of large industrial mission-critical systems [41, 86, 304, 305]. These systems are susceptible to software aging because of their complexity and the high cost of finding and correcting transient software faults. Software rejuvenation architecture artifacts have been applied to telecommunication billing and provisioning data [74, 449], transaction processing system [368], operating systems [904, 905], cluster systems [903], cable modem termination systems [599], web Servers [585], worm mitigation in tactical MANETs [67], and virtualization [828]. Dynamic software rejuvenation algorithms that are based on online monitoring of the environment and of the real-time system performance, can outperform static algorithms, for systems where mission success is dependent on real-time performance [66]. In addition, several empirical studies have identified relevant customer affecting metrics and the best software rejuvenation trigger interval for different applications [41, 86, 304, 305].

Examples of software rejuvenation approaches are the rebooting of a process, releasing of memory, clearing of a deadlock, or performing any other fast action that would prevent software aging from manifesting itself as a system wide failure that could lead to a system crash. These system wide crashes can cause significant damage to the mission the software is controlling and to the infrastructure that is being used to support the software system. For example, a database corruption could take significant time to recover from.

In this chapter, we present models of software aging and different algorithms that were developed to counter aging and security intrusions by applying the so called software rejuvenation techniques.

The outline of the chapter is as follows. Section 8.2 presents a review of the analytical models that were developed for capturing the effects of software aging and for providing recommendations for the best times to trigger software rejuvenation. Section 8.3 presents measurement based studies of software aging and rejuvenation. Section 8.4 presents our conclusions.

8.2 Analytical Models

One of the aims of developing analytic models of software aging is to determine optimal times to perform software rejuvenation to maximize software availability, to minimize the probability of loss, to minimize the mean response time of a transaction

Fig. 8.1 Basic two-step
software rejuvenation model
proposed by Huang et al. [449]

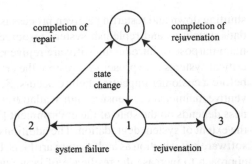

(e.g., transaction processing system), or to minimize maintenance costs. Performance optimization is particularly important for business-critical applications for which adequate response times can be as important as system uptime. Modeling and analysis of software aging is done for different kinds of software systems exhibiting varied failure/aging characteristics.

8.2.1 Markov Models

8.2.1.1 Continuous-Time Markov Chain

Markov models have been often used in the representation and investigation of software aging and rejuvenation policies [449]. Although the software aging phenomenon is characterized in analytical terms by increasing failure rate (IFR), the first attempts at representing software aging and rejuvenation were based on homogeneous Markov chains [366, 449, 552, 808].

The Markov model used to represent software aging and software rejuvenation is based on a phase-type expansion, where software aging is discretized into a finite number of states of the Markov chain, each characterized by a specific degradation level and a transition rate to the next state. For example, the Markov chain models proposed in [366, 449] restrict the time to failure to be hypo-exponentially distributed. This approach was initially introduced in [449], where a two-step failure model was used with only one degraded state between the initial state (State 0) and the failed state (State 2), as shown in Fig. 8.1. State 1 represents the failure probable state, where a failure would take the system to state 2 and a software rejuvenation trigger would take the system to state 3. The authors solved the model to compute the costs that would be accrued by software rejuvenation and by downtime after a hard failure event. The authors concluded that for the parameters evaluated, software rejuvenation costs would have to be less than 2% of the costs of a hard failure, for software rejuvenation to be cost effective. From this Markov model, the system availability and subsequently the optimal rejuvenation trigger interval was computed.

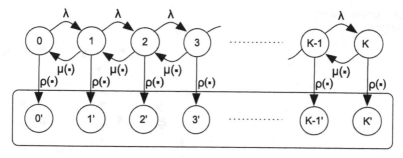

Fig. 8.2 Subordinated non-homogeneous continuous-time Markov chain of [368]

In [808] a single node system was introduced to investigate the effects of software aging on two different operating systems. The authors compared the effectiveness of several software rejuvenation policies, considering a few alternatives and different system degradation levels (till 10). In this work, system performance degradation was assessed in terms of the system resource utilizations, as for example, processor, memory, and thread utilizations.

8.2.1.2 Non-Homogeneous Continuous-Time Markov Chain

An alternative modeling approach to represent the increased failure rate of the software aging process employs non-homogeneous continuous-time Markov chain model. Non-homogeneous continuous-time Markov chains have been traditionally used for software reliability modeling, and have also been successfully applied to solve software aging and rejuvenation problems [85, 86, 368, 554]. In [368] Garg et al. analyzed a queueing system with preventive maintenance as a mathematical model for a transaction-based software system. The proposed non-homogeneous continuous-time Markov chain model [368] used a time-to-failure function that was generally distributed and a time-varying failure rate to capture the effects of load on software aging. Two software rejuvenation schemes based on the cumulative operation time (before or after the idle time) were investigated. The authors were able to derive the optimal rejuvenation interval T^* under the two policies so as to maximize the steady-state availability, minimize the transactions probability loss, and/or minimize the upper bound on the mean response time. The non-homogeneous continuous-time Markov chain with K states is shown in Fig. 8.2, where the state definition represents the number of transactions queued including the one in service. $K > 1$ is the maximum capacity of the transaction buffer. In Fig. 8.2, λ represents the transaction arrival rate, $\mu(\cdot)$ represents the software service rate as an arbitrary function, as it can be constant, or a function of time, load dependent, or a combination of these factors. $\rho(\cdot)$ represents the software failure rate, which is also an arbitrary function. The model is able to capture aging and performance degradation of systems that lose transactions due to software failures.

Fig. 8.3 Markov reward
model of [958]

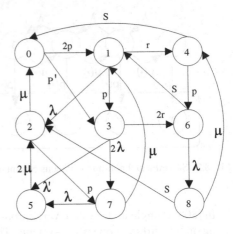

The main goal of [554] was to study the overall behaviour of a software system
by modeling the time-dependent rejuvenation rates and deriving an optimal rejuvenation policy using a cyclic non-homogeneous Markov chain. A non-homogeneous
continuous-time Markov chain was built to assess the tradeoff between system degradation and software rejuvenation cost.

8.2.1.3 Markov Decision Process and Reward Model

Another Markov modelling framework that was applied to assess software aging and
rejuvenation is the Markov decision process. In [726] the authors developed a Markov
decision process based framework to compute optimal rejuvenation schedules. The
optimal rejuvenation schedule solved the optimal stopping problem, as applied to
software aging and rejuvenation, by the use of a gradual decrease of the failure rate.
The authors have also considered the impact of using realistic cost assumptions and
simple rules that could yield an optimal software rejuvenation schedule.

In [958] the authors extended the mathematical characterization of common
software-aging-related faults introduced in [449] with a Markov reward model representing a redundant fault-tolerant software system, which is modeled using the
software aging and rejuvenation approach introduced in [449]. The proposed model
is shown in Fig. 8.3 and represents the states of the joint state of the two parallel
software systems, where each individual software system can be in the states defined
in [449]. For example, in state 0 both systems are operating correctly. In state 1 one
software system is operating correctly and the other one is in the failure probable
state. In state 3 both software systems are in the failure probable state, while in state
2 one of the software systems is operating correctly and the other has failed. The
other states are derived similarly.

8.2.2 Renewal Processes

Non-Markovian processes shall be employed when the exponential distribution for the time to failure is not sufficient to model the system under study. Renewal theory provides the tools to adequately represent more complex aging processes or rejuvenation policies. Semi-Markov and Markov regenerative processes have been widely used for representing software aging and software rejuvenation.

8.2.2.1 Semi-Markov Process

In [304] the authors developed a semi-Markov model by generalizing the continuous-time Markov chain approach introduced in [449]. The optimal software rejuvenation schedules were analytically derived to optimize the steady-state availability objective and the average cost expended by the software rejuvenation approach. In addition, non-parametric statistical algorithms to estimate the optimal software rejuvenation schedules were also developed.

In [85, 86] the authors used a semi-Markov model to represent a high-level proactive fault management approach. The main contribution of this work was the development of an hierarchical modeling approach composed of a lower level non-homogeneous continuous-time Markov chain and an upper level semi-Markov model. The hierarchical software rejuvenation model triggers different software rejuvenation actions depending on the degradation level that the system has experienced, the time elapsed since the last software rejuvenation event, or any other specific criterion that needs to be modeled [85, 86]. The first-level software rejuvenation, or partial rejuvenation, consists of stopping and rejuvenating certain applications, while the second-level software rejuvenation, or full rejuvenation, consists of stopping all the running applications and restarting the system. Therefore, the first-level software rejuvenation incurs lower cost as measured by system downtime than the second-level software rejuvenation. The hierarchical model proposed in [85, 86] allows for decomposition of the analysis by first evaluating the impact of resource leakage and then assessing the effectiveness of software rejuvenation on the metric of interest.

The tradeoff between using the partial or the full software rejuvenation approach is studied in [553], where a computer system with one standby redundant node was evaluated. The authors considered five different software rejuvenation models for the redundant system, evaluating the steady-state behaviour and the asymptotic availability for each rejuvenation model.

A semi-Markov (SMP) process has been used in [717] to model the availability of personal computer-based active/standby cluster system with software rejuvenation to handle software related system failures. Software rejuvenation and switchover states were mapped into a semi-Markov model whose analysis provided the steady-state availability.

Fig. 8.4 The MRGP proposed in [423]

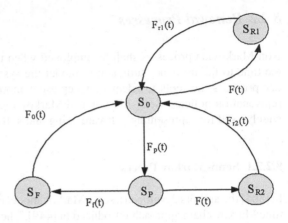

8.2.2.2 Markov Regenerative Process

Markov regenerative processes have been successfully applied in several software rejuvenation studies. For example, the focus of [423] was to solve the problem of system unavailability caused by the restart operation in the rejuvenation phase. The authors proposed a software rejuvenation model of a hot standby architecture, implementing the software restart by switching between the active copy and the backup copy. The Markov regenerative process availability model was created by changing the Markov chain model of [449] by using two states to represent two different software rejuvenation policies, as shown in Fig. 8.4. The authors have validated their model by comparing the system availability obtained using the model of [449] with the system availability with the proposed hot standby architecture.

Examples of the application of Markov regenerative process software aging and rejuvenation are [309, 368, 926]. As discussed above, in [368] the Markov regenerative process is used on top of a non-homogeneous continuous time Markov chain thus composing a hierarchical model. In [926] three time-based rejuvenation policies used to improve the performability measures of a cluster system under varying workload were evaluated. Similarly, in [309] Markov regenerative process has been used to evaluate the software aging process that was studied by the authors.

In [120] a fine grained software degradation model was proposed, where the current software degradation level could be observed based on the monitoring of a system parameter. In this work, the degradation process consists of a sequence of additive random shocks. The system is considered out of service as soon as the appropriate parameter reaches an assigned threshold level. The system model is a complex reward-renewal processes that is analyzed using the theory of renewal processes with cost/rewards. The approach was used to analyze the impact of system parameters and two alternative rejuvenation policies on a redundant database management system unavailability.

8.2.3 Petri Nets

Petri nets are one of the modeling frameworks used to evaluate software aging and software rejuvenation approaches, because Petri net models can accurately incorporate the most common characteristics of computer systems like concurrency, synchronization, sequencing, and queueing for multiple resources.

Petri nets have been mainly used as a modeling notation. The underlining stochastic processes are derived by using specific techniques. Markov chains, renewal theory, phase type expansions, simulation and similar solution techniques have been used in the evaluation of the Petri nets underlying processes.

Several different Petri net variations have been used to model software aging characteristics and software rejuvenation approaches. One of the specific requirements of software aging modeling is the ability to represent non-Markovian behaviors. Modeling of software aging processes have to take into account the age/history of the software, which can be approximated using Markov models [400].

8.2.3.1 Stochastic Petri Nets

One of the first attempts to apply Petri nets in software rejuvenation is reported in [367]. The authors used the Markov regenerative stochastic Petri net of Fig. 8.5a to deal with a deterministic software rejuvenation trigger interval. The system is fully operational in the place P_{up}. When the T_{fprob} transition fires, which represents software aging, a token reaches the place P_{fprob}, where the system is the failure probable state. The system is in the crash state after the firing of transition T_{down}. While the system is restarting, all transactions are suspended, as shown by the inhibitor arc from P_{down} to the T_{clock} transition, which models the periodic software rejuvenation trigger. T_{clock} fires when the clock expires, if it has not been inhibited. The other transitions are understood similarly.

Another interesting model was described in [117], where the fluid stochastic Petri net shown in Fig. 8.5a was used. The Petri net formalism allows the modeler to represent software aging and software rejuvenation in systems that use specific techniques for software rejuvenation, restoration, and checkpointing. Specifically, a fluid flow approximation approach can be used to model the software aging process also taking into account the workload condition, where the fluid level at a certain time t represents the extent of system degradation that has occurred up to t.

Another type of Petri net often used to model software aging and rejuvenation is the deterministic stochastic Petri net, used for representing the cluster system described in [926]. The performability metric was evaluated by the numerical analysis of the underlying subordinated Markov chain. The software rejuvenation policies was there evaluated by considering both the historical data and the current running state of the system.

(a)

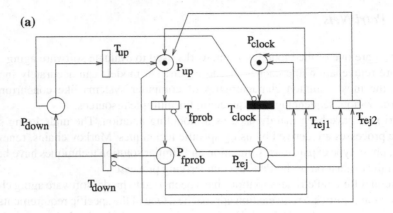

(b)

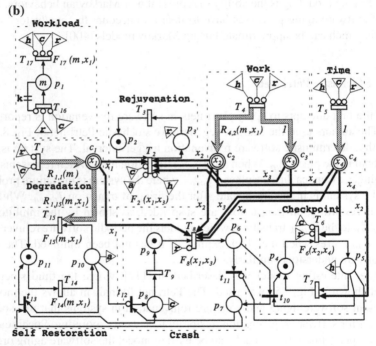

Fig. 8.5 MRSPN (**a**) and FSPN (**b**) software-aging/rejuvenation models proposed in [117, 367], respectively

8.2.3.2 Stochastic Reward Nets

Another Petri net notation used in the software aging and software rejuvenation context is the stochastic reward notation. Stochastic reward nets are particularly suitable to model software aging and rejuvenation approaches, since this formalism allows for modeling the software aging process by using reward rates and guards,

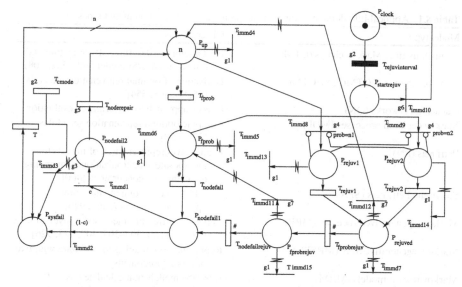

Fig. 8.6 The SRN proposed in [903] for evaluating cluster systems software rejuvenation policies

among other features. Therefore, stochastic reward nets can be used to represent complex aging processes rejuvenation policies and other quantities such as, for example, the cost [181, 489, 598, 903, 946, 951].

In [903] an evaluation of the application of software rejuvenation to cluster systems was performed. The stochastic reward net introduced in [903] is shown in Fig. 8.6. Software rejuvenation was shown to significantly improve the evaluated system availability and productivity. Both time-based and prediction-based software rejuvenation techniques were there evaluated by modeling, using stochastic reward nets. In [181] the application of software rejuvenation to cluster systems using a real case study related to the xSeries IBM cluster was presented.

Stochastic reward net models were applied to study the cable modem termination system cluster in [598]. Capacity-oriented availability and downtime cost with and without deploying software rejuvenation were evaluated showing significant availability improvement and downtime cost reduction when time-based and condition-based (a prediction-based) rejuvenation strategies were used.

Fan Xin-yuan et al. [946] analyzed the dispatch-worker based cluster system and proposed a Stochastic Reward net model for dispatch-worker based architectures with prediction-based software rejuvenation.

In [951], dependencies among cluster nodes were taken into account in the evaluation of software aging and time/prediction-based software rejuvenation strategies by using a stochastic reward net. The authors evaluated the impact of these rejuvenation strategies on the cluster system reliability by minimizing the software rejuvenation cost, while considering the comprehensive relations between nodes in the entire system.

Table 8.1 Applicability domain of the software aging and rejuvenation model types

Model type	Observations
Continuous time Markov chains (CTMC)	First attempt to describe software aging and rejuvenation, cost model [449, 552, 808]
Non-homogeneous CTMC (NHCTMC)	Derivation of optimal rejuvenation intervals [85, 86, 368, 554]
Queueing networks (QN)	Classic approach to performance evaluation
Markov decision processes (MDP)	Compute optimal rejuvenation schedules [726, 958]
Petri nets (PN)	Higher level representation of models that incorporate synchronization, sequencing, queueing and concurrency [400]
Stochastic Petri nets (SPN)	Could be used to compute fluid flow approximations [117, 367, 926]
Markov regenerative processes (MRP)	Represent different time-based rejuvenation policies [120, 303–305, 309, 423, 926]
Markov regenerative SPN (MRSPN)	Representation of aging for software with two or more components
Markov reward models (MRM)	Describe models that calculate costs of different rejuvenation policies [958]
Stochastic reward networks (SRN)	[181, 489, 599, 903, 946, 951]
Semi-Markov process (SMP)	Hierarchical modeling of rejuvenation and switchover states; allows specification of renewal states in contrast to NHCTMCs; SMPs overcome the NHCTMC models lack in representing rejuvenation, allowing to specify renewal states [717]
Semi-Markov reward models (SMRM)	Ability to model aging with rewards [85, 86]

In [489] a mixed time and prediction based software rejuvenation policy was evaluated using a stochastic reward net. The model was used to evaluate the system availability and downtime cost. The authors were able to show that under the same conditions, a mixed time and prediction based software rejuvenation policy could achieve higher availability and lower downtime cost than either one of the time-based and prediction-based software reliability policies.

Table 8.1 presents a summary of different models presented in this section with a brief description of the domain of applicability of each model.

8.3 Measurement Based Approaches

The software aging and rejuvenation analytical models either assume that the time to failure distribution of the software is known (in case of time-based software reju-venation) or that the degradation level of the software system is known (in case of inspection-based software rejuvenation). To facilitate the latter approach, measurement based approaches monitor and collect data on the attributes responsible for determining the health of the executing software. The data is then analyzed to obtain predictions about possible impending failures due to resource exhaustion. The data analysis can be executed online or offline.

In this section we describe measurement-based approaches for detection and validation of the existence of software aging.

Garg et al. [369] introduced an approach for the detection and the estimation of aging in the UNIX operating system. An SNMP-based distributed resource monitoring tool was used to collect operating system resource usage and system activity data from nine heterogeneous UNIX workstations connected by an Ethernet LAN at Duke University. A central monitoring station was used to run the manager program, which was used to send get requests periodically to each of the agent programs running on the monitored workstations. The agent programs in turn obtained data for the manager from their respective machines by executing various standard UNIX utility programs like pstat, iostat and vmstat. For quantifying the effect of aging in operating system resources the metric *Estimated time to exhaustion* was proposed. The objective of the study was to detect aging or a long term trend (increasing or decreasing) in the measured values. This approach assumed that the accumulated depletion of a resource over a time period depended only on the elapsed time. However, it is intuitive that the rate at which a resource is depleted is also dependent on the current workload. An approach to estimate the rate of exhaustion of operating system resources as a function of both time and the system workload was presented in [905, 906].

A methodology based on time-series analysis was used to detect and estimate resource exhaustion times due to software aging in a web server while subjecting it to an artificial workload [585]. The experiments were conducted on an Apache web server running on the Linux platform. The analysis was done using two different approaches: (1) building a univariate model for each of the outputs or, (2) building only one multivariate model with seven outputs. Seven univariate models were built and then combined into a single multivariate model. First, the parameters were analyzed and incorporated into the model with one output and four inputs for each parameter as follows: connection rate, linear trend, periodic series with a period of one week, and periodic series with a period of one day. The autocorrelation function (ACF) and the partial autocorrelation function (PACF) for the output were computed. The ACF and the PACF were used to select the appropriate model for the data [825]. The autoregressive multiple input single output (MISO) model of order 1 (AR(1)) is considered for the single multivariate model, also taking into account the inputs above identified, an autoregressive model with the exogenous input of order 1 (ARX(1)) is specified for each of them, and obtaining seven ARX(1) models. In summary, the models have been combined into a single multiple input multiple output (MIMO) ARX(1) model. The next step after determination of the model orders is to estimate the coefficients of the model by using the least squares method. The first half of the data is used to estimate the parameters and the rest of the data is then used to verify the model. The obtained results show that the predicted values are very close to the measured values.

In [85] a model was developed to account for the gradual loss of system resources, specifically, the memory resource. The model is able to represent both the correct system operation with no memory leakage and the faulty system operation when a memory leak fault is present. The model relates system degradation to resource request, resource release, or resource holding intervals, and memory leaks. These quantities can be monitored and modeled directly from the system data measurements [585].

Cassidy et al. [180] have developed an approach for software rejuvenation of large online transaction processing servers. The authors monitored various system parameters over a period of time and were able to determine that 13 of these parameters deviate from normal behavior just prior to a crash, thus providing sufficient warning to warrant the initiation of software rejuvenation. A feedback control loop approach for software rejuvenation in a web server was presented in [487].

Machine learning [34], Support Vector Machines (SVM) and similar techniques have also been applied to analyze software aging data [443]. Accelerated life testing and accelerated degradation testing techniques have been applied to reduce the time needed for aging approximations [627].

Algorithms for online monitoring of a defined customer-affecting metric have been applied to the security domain. In [68] the effectiveness of the basic bucket-based online monitoring algorithm introduced in [67] was assessed for mission-critical systems by computing the probability of mission success. The analysis results showed that online monitoring and software rejuvenation are very effective in ensuring a high probability of mission success, when the mission-critical system is under attack by a worm infection.

In [71] an application of the basic bucket-based online monitoring algorithm using known system performance signatures was used to detect security intrusions. The research presented in [71] uncovered a significant difference between the performance signatures associated with failure events and the performance signatures associated with security attacks. The performance signatures obtained from the analysis of system failures showed significant system degradation, i.e., CPU values of up to 100%. In contrast, the performance signatures obtained from the analysis of the execution of security test suites, showed that the observed CPU usage values were constrained to a narrow band. A new version of the bucket-based online monitoring algorithm, which was introduced in [71], was able to successfully distinguish between software aging that results from failure events and software aging that results from security intrusion events.

8.4 Conclusions

We have presented in this chapter an overview of several software rejuvenation approaches that can be used to increase software resilience by using analytical modeling, offline and online system measurements.

Different analytical models have been applied to the evaluation of software aging and rejuvenation. We have categorized these models according to the stochastic process and the technique used in the analysis. Software can be modeled as a degrading system that is characterized by the software age. The selection of the stochastic process used to represent software aging and rejuvenation policies is driven by the need to incorporate the software age into the model. Markov models are used in software aging as an approximation where software age is represented by using different degradation states. Approximated models that implement a phase type-like discretization of the software degradation into two or more degradation states have been introduced in [449, 552, 808]. One of the benefits of using Markov models to represent software aging is the reduced model solution cost for simple models. However, a large number of degradation states may be required to model the problem with high accuracy, thus increasing the Markov model solution cost.

Non-homogeneous Markov chains have been used to model software aging [85, 86, 368, 554] but to adequately model software rejuvenation more complex models are required as a single global clock will not provide accurate results. Therefore, semi-Markov and the Markov regenerative processes are required to adequately represent software rejuvenation [85, 86, 120, 303–305, 309, 368, 553, 717, 926]. Specifically, semi-Markov processes can be used to specify renewal states. However, semi-Markov processes cannot adequately represent the aging that occurs between regeneration epochs. As a consequence, Markov regenerative processes are used to model software aging in software hierarchies composed of two or more components, by separately modeling each individual component age.

Markov reward models and variants have been often used when different rejuvenation policies have to be evaluated and compared to establish the optimal software rejuvenation policy [958]. The rewards associated with the Markov model states are used to represent and quantify the cost of the software rejuvenation policies being evaluated.

The Petri net formalism can be considered as a higher level of modeling representation, which can be used to provide clarity to the software modeling process. Petri nets have been used to model software aging and rejuvenation problems, because of their compactness and expressiveness [367, 926]. Specifically, stochastic reward nets have been used when software rejuvenation policies have to be compared in terms of their costs to investigate the optimal software rejuvenation policy [181, 489, 598, 903, 946, 951].

Measurement based approaches rely on data collection and analysis for determining system health and the best time to trigger the software rejuvenation routines. Online monitoring of system resources and/or of a customer affecting metric were shown to be a cost-effective approach to ensure software resilience.

Software aging has been originally observed in the telecommunications domain [70, 72, 111, 449], because of strict engineering efforts that were conducted by telecommunication companies to assess service reliability. The detection of smooth degradation of the available resources or software aging, and the periodic workload characteristics of telecommunication applications, led to the development of techniques to counteract aging that were called software rejuvenation. These software

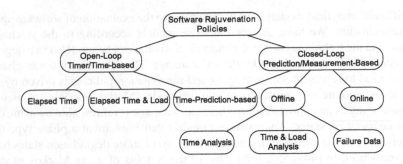

Fig. 8.7 Software rejuvenation techniques classification

Table 8.2 Bibliography classification based on Fig. 8.7

Open-loop	
Elapsed time	[55, 85, 117, 120, 304, 367, 449, 553, 554, 717, 726, 808, 958]
Elapsed time and load	[86, 309, 366, 368, 423, 926]
Closed-loop	
Offline	
Time analysis	[85, 120, 180, 304, 369, 627]
Time and load analysis	[585, 905, 906]
Failure data	[34, 85, 443]
Online	[66–68, 71, 487, 649, 825]
Open/closed-loop	
Time/prediction based	[181, 489, 598, 903, 946, 951]

rejuvenation techniques were shown to be cost effective to increase system resilience, when the system contains soft faults or is a victim of security attacks. Therefore, we expect that the software aging and rejuvenation approaches described in this chapter will see increased deployment in systems that are designed for resilience.

Figure 8.7 presents a classification of software aging and rejuvenation approaches. The software aging and rejuvenation approaches can be divided broadly into *time-based* (*open loop*) and *prediction-based (closed loop)* approaches. Time-based techniques are usually used in the early stages of software development process and are often implemented by analytical models, as discussed in Sect. 8.2.

In the classification shown in Fig. 8.7, we further characterized the time-based class in terms of the quantities and the parameters taken into account in the model, such as time and load.

On the other hand, in the prediction-based approach, the objective is to monitor and collect data on the attributes responsible for determining the health of the executing software. The data is then analyzed to obtain predictions about possible impending failures due to resource exhaustion. The data analysis can be executed online or offline. Offline techniques have been further characterized according to the type of data available and used in the evaluation:

- *time analysis*—time parameters,
- *time and load analysis*—time and load data,
- *failure data*—for reliability analysis.

Statistical techniques are used to collect and process the data as discussed in Sect. 8.3.
Table 8.2 presents a summary of some of the techniques discussed in the chapter.

Acknowledgments   We like to thank Dr. Fumio Machida, Ermeson Andrade and Dr. Jing Zhao for their useful comments.

Chapter 9
Online Prediction: Four Case Studies

Katja Gilly, Fabian Brosig, Ramon Nou, Samuel Kounev and Carlos Juiz

Abstract Current computing systems are becoming increasingly complex in nature and exhibit large variations in workloads. These changing environments create challenges to the design of systems that can adapt themselves while maintaining desired Quality of Service (QoS), security, dependability, availability and other non-functional requirements. The next generation of resilient systems will be highly distributed, component-based and service-oriented. They will need to operate in unattended mode and possibly in hostile environments, will be composed of a large number of interchangeable components discoverable at run-time, and will have to run on a multitude of unknown and heterogeneous hardware and network platforms. These computer systems will adapt themselves to cope with changes in the operating conditions and to meet the service-level agreements with a minimum of resources. Changes in operating conditions include hardware and software failures, load variation and variations in user interaction with the system, including security attacks and overwhelming situations. This self adaptation of next resilient systems can be achieved by first online predicting how these situations would be by observation of the current environment. This chapter focuses on the use of online predicting

K. Gilly (✉)
Universidad Miguel Hernandez, 03202 Elche, Spain
e-mail: katya@umh.es

F. Brosig · S. Kounev
Karlsruhe Institute of Technology, 76131 Karlsruhe, Germany
e-mail: fabian.brosig@kit.edu

S. Kounev
e-mail: kounev@kit.edu

R. Nou
Barcelona Supercomputing Center, 08034 Barcelona, Spain
e-mail: ramon.nou@bsc.es

C. Juiz
Universitat de les Illes Balears, 07004 Palma, Spain
e-mail: cjuiz@uib.es

K. Wolter et al. (eds.), *Resilience Assessment and Evaluation of Computing Systems*,
DOI: 10.1007/978-3-642-29032-9_9, © Springer-Verlag Berlin Heidelberg 2012

methods, techniques and tools for resilient systems. Thus, we survey online QoS adaptive models in several environments as grid environments, service-oriented architectures and ambient intelligence using different approaches based on queueing networks, model checking, ontology engineering among others.

9.1 Introduction

New resilient systems have to consider QoS variations that occur and then react to these changes online acting accordingly to maintain a certain Service Level Agreement (SLA). Consequently, these systems need to predict these variations found even at the risk of being wrong on a certain value.

Predictions are based on a model that has to be representative in the sense that it reflects the system's QoS-relevant behaviour. Typically, the user behaviour is an input of such a model. Thus, the user behaviour has to be predicted as well when obtaining model predictions to anticipate QoS problems. In the context of performance predictions, user behaviour prediction is often referred to as workload forecasting. For workload forecasting, established time series analysis techniques [147] are often used. For instance, Brown's quadratic exponential smoothing or general AutoRegressive—Moving Average (ARMA) models have been implemented in [194, 499].

Concerning online performance prediction, in [640, 642], the authors describe a framework using analytic performance models in the design of self configurable and self-managing computer systems. An general overview on performance models that can be evaluated efficiently, is provided in, e.g., [125]. Typically, these models are based on queuing networks and markov chains. A different approach is applied in [588], where the online performance prediction is based on a machine-learning approach.

In this chapter, we consider four different case studies in order to show how online prediction could help in this way to the resilience of systems. The first case study shows how detailed architecture-level performance models can be extracted and maintained automatically at run-time based on on-line monitoring data. Even though the current version of the extraction method is not 100 % automated, and there are some prediction error yet, the case study demonstrated that the existing gap between low-level monitoring data and high-level performance models can be closed. In the second case, we augmented the Grid middleware with an online performance prediction mechanism that can be called at run-time to predict the Grid performance for a given resource allocation and load-balancing strategy, demonstrating the benefits of online performance prediction for run-time performance management. In the third example, we include an adaptive time slot scheduling based on a burstiness metric, that permits to control the monitoring frequency of the system depending on the burstiness levels detected by the algorithm. This means a considerable decrease of the overhead of the monitoring process, whose frequency can be adapted to the stress detected at the entry point of the system. This technique is used in the fourth

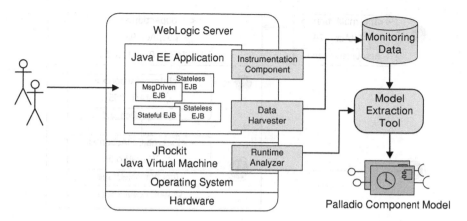

Fig. 9.1 Model extraction tool architecture

case to build an admission control and load balancing algorithm that is based on throughput prediction for a Web system. These cases studies are just four individual examples, but they illustrate how on-line predictions increase the resilience of any kind of system performance problem. All of them are related one to the others, in several ways that coincide with three general questions to face off during their design: first, the necessity of gathering data from either monitoring or measurements in order to predict the future; second, the dynamicity of the on-line decisions based on partial temporal information and finally, the overhead of doing both procedures is the price to be paid in order to get the on-line predictions. The challenge in all cases is how to reduce overhead time as the QoS problem permits.

9.2 Automatic Model Extraction at Run-Time

As a proof-of-concept for automatic model extraction at run-time, we conducted a case study with a complex Java EE application. The case study shows how detailed architecture-level performance models can be extracted and maintained automatically at run-time based on online monitoring data [150], The Java EE application we considered was a beta version of the new SPECjEnterprise2010 standard benchmark. We deployed the benchmark on Oracle WebLogic Server (WLS) and used the WebLogic Diagnostics Framework (WLDF) as a monitoring and instrumentation tool (Fig. 9.1). The considered architecture-level performance model was the Palladio Component Model (PCM).

The PCM is a domain-specific modelling language for describing performance-relevant aspects of component-based software architectures [96]. In PCM, a component specification normally includes a definition of which interfaces the component provides and requires together with a set of *Resource Demanding Service Effect Specifications* (RDSEFFs). Each RDSEFF describes the performance-relevant

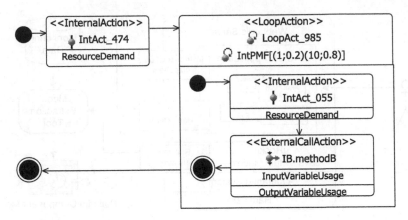

Fig. 9.2 Example: extracted RDSEFF

internal behaviour of a provided component service in an abstract manner. The control flow and the resource consumption of the service can be modelled probabilistically as well as depending on the input parameters. Figure 9.2 shows a component service's RDSEFF in a notation similar to the notation of UML activity diagrams. The RDSEFF consists of an internal action abstracting component-internal resource demanding instructions, followed by a loop action containing a further internal action and an external call action to a required service. The loop iteration number of LoopAct_985 is specified as a probability mass function (PMF). The PMF states that the loop iterates one time with a probability of 20 % and ten times with a probability of 80 %.

The extraction method for Java EE applications was implemented using the WLDF monitoring tool that is provided with Oracle WebLogic Server (WLS). The three main steps of the model extraction process are: i) the extraction of the application architecture, ii) the extraction of performance-relevant control flow and iii) the extraction of resource demands.

In the first step, the effective application architecture is extracted. The latter refers to the set of components and connections between components that are effectively used during operation. The components and connections are identified on the basis of trace data reflecting the observed call paths during execution. Based on the call paths, the effective connections among components can be determined, i.e., required interfaces of components can be bound to components providing the respective services. In the second extraction step, the tracing technique is applied to extract the performance-relevant control flow inside the components. We focus on monitoring the effective control flow and therefore extract probabilities of different call paths in contrast to extracting explicit parametric dependencies. Figure 9.2 shows an RDSEFF that has been extracted from trace data generated by WLDF. To estimate the resource demands of individual internal actions, we investigated two approaches: i) approximate resource demands with measured response times, ii) estimate resource demands based on measured utilization and throughput data. While the first approach is only

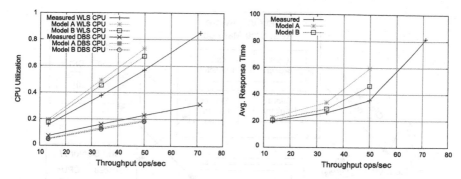

Fig. 9.3   Validation of the scheduleworkorder performance models

applicable during phases of low resource utilization, i.e., <20 %, the second approach can be applied during an observation period with medium to high load.

We applied the model extraction method to a beta version of the new SPECjEnterprise2010 benchmark. The benchmark workload is generated by an application that is modelled after a real-world business scenario. We deployed the benchmark in a system environment consisting of three machines. The JavaEE application was deployed on an Oracle WebLogic Server (WLS) instance. As a database server (DBS), Oracle Database 11 g was installed on the second machine. The benchmark driver was running on the third machine. The machines all have Intel Pentium Dual Core E2180 CPUs (2x2.0 GHz), 3 GB of RAM and are connected using a 1 GBit Ethernet.

To validate the extraction method, we compared predictions derived from the extracted PCM models with measurements on the real system. We considered two different models: i) *Model A*—PCM model in which resource demands were approximated with measured response times, ii) *Model B*—PCM model in which resource demands were estimated based on utilization and throughput data. We analysed the extracted models my means of simulation [96]. As performance metrics, we considered the average response times of business operations as well as the average utilization of the WLS CPU and the DBS CPU. We analysed scenarios under low load conditions, medium load conditions and high load conditions.

In the scenario we consider here, the workload consisted of the business operation ScheduleWorkOrder. Figure 9.3 shows the results. Predictions based on Model B perform slightly better than predictions based on Model A. For the highest considered throughput level, both models deliver no performance predictions. This is because the system as represented by the models is not able to sustain the injected load since the WLS CPU utilization is overestimated to be 100 %. Both models overestimate the WLS CPU utilization while underestimating the DBS CPU utilization. The modelling prediction error for CPU utilization is mostly about 20 %. The modelling prediction error for response times increases with the throughput level. The higher the CPU utilization, the bigger the impact of the overestimated WLS CPU demands on the predicted response times. We assume that the overestimation of the WLS CPU demands is due to the instrumentation overhead during resource demand extraction.

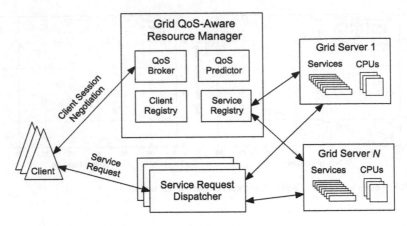

Fig. 9.4 Grid QoS-aware resource manager architecture

We considered a number of different scenarios, on the one hand, varying the operation mix and throughput level under which the PCM models were extracted, and on the other hand, varying the operation mix and throughput level for which performance predictions were made. The extracted models consisted of up to six components, eight RDSEFFs and 13 internal actions annotated with WLS CPU demand or DBS CPU demand estimations [150]. The results were similar to the ones presented here. The prediction error was between 20 and 30 %. Even though the current version of the extraction method is not 100 % automated, the case study demonstrated that the existing gap between low-level monitoring data and high-level performance models can be closed.

9.3 Autonomic QoS-Aware Grid Resource Managers

As a second proof-of-concept demonstrating the benefits of online performance prediction for run-time performance management, we conducted a case study of a SOA application running in a service-oriented Grid computing environment [550, 691, 692]. The latter was implemented using the Globus Toolkit [349] middleware which is based on open Web Services standards and can be seen as an incarnation of SOA. We augmented the Grid middleware with an online performance prediction mechanism that can be called at run-time to predict the Grid performance for a given resource allocation and load-balancing strategy. The online performance prediction mechanism was used as a basis for building a novel QoS-aware Grid resource manager architecture depicted in Fig. 9.4. A *resource manager* is responsible for managing access to a set of Grid servers each offering some Grid services. The resource manager keeps track of the available Grid resources and mediates between clients and servers to make sure that SLAs are continuously satisfied. Before a Grid server can be used, it must register with the resource manager providing information on the services it offers, their resource requirements and the server capacity made available to the Grid. The Grid server must provide an architecture-level performance model that captures the information relevant to predicting the performance of the services it offers. For a client to be able to use a service, it must first send a *session request* to the resource manager. The session request specifies the type of service addressed, the frequency with which the client will send requests for the service, and the required average response time (SLA). The resource manager tries to find a distribution of the workload among the available servers that would provide the requested QoS. For each client session, a certain number of threads (from 0 to unlimited) is allocated on each Grid server offering the respective service. Incoming service requests are then load-balanced across the servers according to thread availability.

The resource manager considers different configurations in terms of thread allocation and for each of them it generates a predictive performance model (more specifically, a queueing Petri net model [546]) based on the architecture-level performance models of the involved services. The generated model reflects the current system environment in terms of available server resources and active client sessions. The model is analysed through simulation and used to predict the performance of the system in order to ensure that the client SLAs are satisfied. If no configuration can be found that satisfies the client SLAs, the session request is rejected or a counter offer with lower throughput or higher response time is sent back to the client.

We now present some experimental results that demonstrate the effectiveness of the above approach. Three sample services each with different behaviour and resource demands were run as part of the experiments. The services use the Grid to execute some business logic requiring a given amount of CPU time. The business logic includes calls to external (third-party) service providers which are not part of the Grid environment. Figure 9.5 shows the results from an experiment in which 99 session requests were sent to the resource manager over a period of 2 h. The average session duration was 18 min in which 92 service requests were sent on average. We compare the behaviour of the system in two different configurations: i) with basic overload control and ii) with QoS control. In the first configuration, the resource manager simply load-balances the incoming requests over the Grid servers without considering SLAs, however, requests that arrive during periods in which both Grid servers are saturated are automatically rejected. In the second configuration, the resource manager uses its online performance prediction mechanism as described above to ensure that SLAs are satisfied. As we can see, without QoS control, the SLAs of the majority of accepted sessions were not fulfilled, whereas with QoS control, the response times of accepted sessions were much lower and all SLAs were fulfilled. The experiment was repeated for a number of different workload configurations varying the transaction mix, the average session length and the server utilization. The results were of similar quality as the ones presented here and they confirmed the effectiveness of our online performance prediction mechanism.

So far we have assumed that when a Grid server is registered with the resource manager, information on the service resource demands (i.e., CPU service times) is

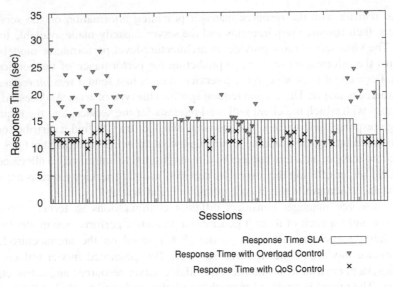

Response Time SLA ▭
Response Time with Overload Control   ▽
Response Time with QoS Control   ×

Fig. 9.5 Response time results for 99 sessions over a period of 2 h

provided as part of the supplied architecture-level performance models. In case the resource demands are not known in advance, a simple method for estimating them on-the-fly based on monitoring data can be used. The method, described in detail in [691], is applicable for services with no internal parallelism. The method is conservative in that it starts with conservative estimates of the resource demands and refines them iteratively as requests are processed. We consider three different configurations in an experiment with 85 sessions over a period of 2 h: i) Basic overload control; ii) QoS control with resource demands available in advance; iii) QoS control with resource demands estimated on-the-fly.

The experiment was conducted in a virtualised setup with 9 Grid servers. Table 9.1 presents a break down of the client sessions into: i) sessions for which the client SLA was observed, ii) sessions for which the client SLA was violated and iii) sessions that were rejected by the resource manager. Without QoS control, 96 % of the requested sessions were admitted, however, the client SLAs were observed in only 22 % of them. In contrast to this, in all configurations with QoS control, the SLAs were observed for nearly 100 % of the accepted sessions. Indeed, only two sessions had their SLAs violated and the violation was by a tiny margin. The price for estimating resource demands on-the-fly was that 14 sessions more were rejected which amounts to 16 % of the total number of sessions.

Finally, we extended the resource manager architecture to support adding Grid servers on demand as well as dynamically reconfiguring the system after a server failure. Whenever the QoS requested by a client cannot be provided using the currently available server resources, the extended algorithm considers to launch an additional server to accommodate the new session. At the same time, each time a

Table 9.1 Summary of session SLA compliance

Configuration	SLA fulfilled	SLA violated	Rejected
1	19	63	3
2	46	2	37
3	34	0	51

server failure is detected, the resource manager reconfigures all sessions that had threads allocated on the failed server. Existing sessions might have to be cancelled in case there are not enough resources available to provide adequate QoS. The extended algorithm was subjected to an extensive experimental evaluation the results of which are available in [691]. The results showed that adding servers on demand does not have a significant impact on the performance of the resource manager despite of the decreased flexibility in distributing the workload.

9.4 Adaptive Time Slot Scheduling

The advantages of predicting the performance of a system online can also be applied to generic distributed algorithms. As a third proof-of-concept we include an adaptive time slot scheduling based on a burstiness metric, that permits to control the monitoring frequency of the system depending on the burstiness levels detected by the algorithm. This means a considerable decrease of the overhead of the monitoring process, whose frequency can be adapted to the stress detected at the entry point of the system.

Considering a locally distributed cluster-based Web information system, a fundamental aim is the monitoring of some Web servers' parameters in an adaptive way in order to reduce the algorithm overhead. Some of the Web servers' parameters likely to be monitored are the arrival rate, the CPU/disk utilization, I/O performance, etc. The performance of the nodes that compound the Web system have to be monitored continuously in order to know their status and make the appropriate decisions in case of overload to avoid a possible congestion situation. This can be done in several ways: *(i)* each time a request arrives at the front-end of the Web system; (ii) at fixed times by using static time slot scheduling; or *(iii)* at non-fixed times by using dynamic time slot scheduling. The overhead introduced by option *(i)* is the biggest because each time a request arrives at the Web system, Web node parameters are monitored. While option *(ii)* introduces a constant overhead, option *(iii)* monitors the system at non-fixed intervals, hence, its overhead will depend on the frequency of those intervals. The drawback of defining monitoring in a constant duration interval schedule (option *(ii)*) is the choice of monitoring time interval. It is very difficult to set a duration interval that fits with all possible Internet arrival rates at the Web system due to its heavy tailed pattern.

We have considered six different approaches to define burstiness factors in order to compare their behaviour and detect their benefits or drawbacks under the same

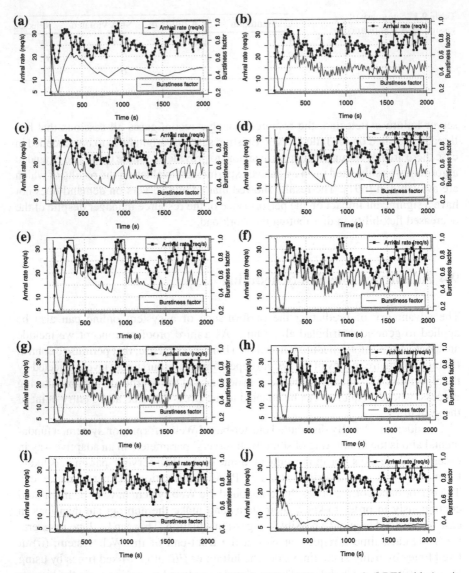

Fig. 9.6 Arrival rate and burstiness factors: **a** BF1; **b** BF2; **c** BF3 with $j = 3$; **d** BF3 with $j = 4$; **e** BF3 with $j = 10$; **f** BF4 with $j = 3$; **g** BF4 with $j = 4$; **h** BF4 with $j = 10$; **i** BF5; **j** BF6

circumstances. All the burstiness factor values are defined in [0, 1]. The precise definition of the burstiness factors can be found in [379]. Instead of defining them formally, let us describe them visually in Fig. 9.6, where the arrival rate to the system is also shown.

Burstiness Factor 1 (BF1) smooths the arrival rate curve. Figure 9.6a illustrates that it follows the arrival rate but does not accurately represent its quick variations.

We consider that the burstiness factor should alert the system as quickly as possible of an increase in the arrival rate, and this factor increases or decreases along with the increasing or decreasing arrival rate trend but very slowly and delayed.

We propose the direct inclusion of the arrival rate value in the burstiness factor in the next proposal, as a way to modify it quantitatively. Figure 9.6b shows that, in this case, BF2 also varies with the variations of the arrival rate. Nevertheless, there are some peaks in the arrival rate that are not followed by the factor. In the next proposal we introduce a penalisation when detecting a consecutive number of *bursty* slots.

Figure 9.6c–e represent the results obtained with BF3 and a record of 3, 4 and 10 slots, respectively. It can be observed that as the number of slots considered increases, the burstiness factor penalisation also increases. We need to check if this penalisation leads to an increase in the system performance or otherwise, decreases its performance because of an overreaction to the arrival rate.

The BF4 values are shown in Fig. 9.6f–h, representing the results obtained with a maximum record of 3, 4 and 10 slots. We can observe that the resulting curves of BF4 are similar to the BF3 curves, but in this case the burstiness factor is also sensitive to changes in the arrival rate.

Figure 9.6i shows the results obtained with this burstiness factor and the resulting curve can be observed as being even smoother than the one obtained from the original BF1.

In Fig. 9.6j, it can be observed that the BF6 curve does not accurately follow the arrival rate changes. The BF6 curve decreases in some points of Fig. 9.6j when the arrival rate curve increases. The main drawback of this burstiness factor is the fact that its calculation is made for each incoming HTTP request and then it needs a huge computational effort, which leads to a considerable overhead compared to the other proposals.

In order to define the adaptive time slot scheduling, we divide the total observation time T of the experiment in several slots of variable duration. While the experiment is simulated, the duration of the slot changes based on the value obtained by the burstiness factor. Hence, the duration of the slot $k + 1$ is dependent on the burstiness of the two previous slots, $b(k)$ and $b(k - 1)$, as follows:

$$d(k + 1) = \frac{d(k)}{1 + b(k) + b(k - 1)}, \quad \text{if} \quad b(k) \geq b(k - 1) \tag{9.1}$$

$$d(k + 1) = \frac{d(k)}{1 + b(k) - b(k - 1)}, \quad \text{if} \quad b(k) < b(k - 1)$$

Therefore, the number of slots defined during the simulation time is also variable. We can calculate the total number of slots that divide the observation time T during each slot. Considering the duration of the slot $k + 1$, the frequency of slots is defined as:

$$e(k + 1) = \frac{T}{d(k + 1)}$$

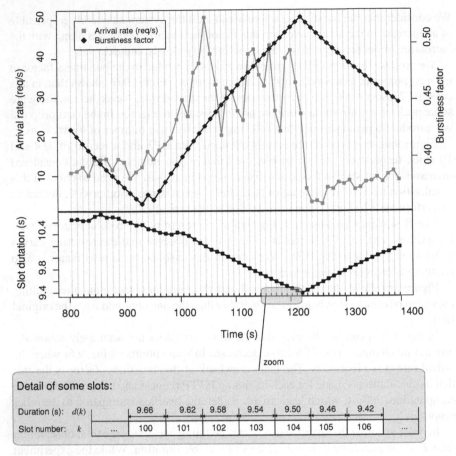

Fig. 9.7 Arrival rate monitored following adaptive time slot scheduling and detail of some of the slots using the BF1

As the duration of the following slot is defined by the value of the burstiness factor on the current slot, when a burstiness increase is detected, the following testing time is brought nearer in order to check the incoming arrival rate early enough and then tune again the algorithm parameters. If a decrease in burstiness is perceived, the duration of the following slot is enlarged to reduce the overhead. By controlling the burstiness in the arrival rate, and then the duration of testing slots, a sudden reduction in the future performance of the Web servers may be forecasted.

An example of adaptive time slot scheduling is depicted in Fig. 9.7. In the upper part of the figure the arrival rate and the burstiness factor curve are drawn following adaptive time slot scheduling. As the arrival rate increases from time instant 910 s, the burstiness factor also increases. We have used BF1 to illustrate burstiness factor behaviour in this case. Below this figure, the slot duration is represented in another

Fig. 9.8 Adaptive admission control and load balancing algorithm overview

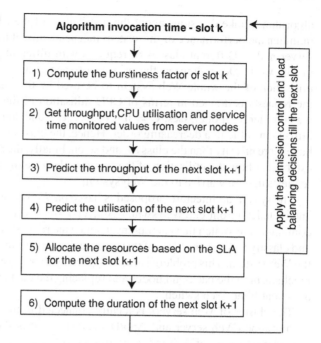

scale. It can be observed how the duration of the slots decreases when the arrival rate increases. Some slots have been zoomed in to detail the decrease of their durations.

The adaptive time slot scheduling has been implemented in an OPNET Modeler scenario and the complete simulation results can be found in [379].

9.5 Admission Control and Load Balancing Algorithm

In this section we want to describe an admission control and load balancing algorithm that is based on throughput prediction for a Web system as a fourth case study of online prediction. The invocation times of the algorithm are planned based on the adaptive time slot scheduling described in previous section.

This algorithm is adaptive because it is invoked adaptively depending on the arrival rate. Each time it is invoked, some computations need to be done in order to take the admission control and load balancing decisions that will remain till the next invocation. The period of time between invocations is considered a slot.

Figure 9.8 depicts the general steps that are taken by the algorithm. Once the burstiness factor has been computed and the monitored throughput, CPU utilisation and service time values obtained from the server nodes, the throughput that the nodes will get during the next slot is predicted. Five throughput predictors are defined in order to give us the trend of the system behaviour. These predictors permit the algorithm to take decisions about the distribution of the load in the Web system to maintain the performance of the system independently of the congestion level of the server nodes. Different classes of requests with different priorities are considered in this work. Depending on the priority of each request, we set a fraction of the utilisation of the whole Web system to be used by that request class. The SLA of the requests is defined in terms of CPU utilisation of the Web servers. Therefore, we consider a set of classes, $C = \{c_1, c_2, \ldots, c_r\}$, and define for them a normalised utilisation value in a decreasing order. Hence, the class of requests that represent c_1 have more priority than the class c_2, and so on. Finally, the resource allocation policy establishes how the utilisation of the server nodes is assigned to attend each class of requests that may arrive to the Web system.

The system architecture proposed is based on Web cluster-based network servers and includes a front-end Web switch. A layer-7 Web switch is normally described as a content-aware switch that can de-encapsulate the requests up to the application level and classify them on the basis of this information, but it can easily be the bottleneck of the Web system. This problem is easily solved by transferring the request distribution mechanism to the back-end nodes and replacing the content-aware Web switch with a content-blind Web switch.

The cluster of Web servers is locally connected to the Web switch in a two-tier organisation (Web server and App/DB server), as it is shown in Fig. 9.9. We have considered five sets of Web and App/DB servers. Each Web server attends the requests that ask for static files, namely static requests and the App/DB server is accessed when the request asks for a Web page that needs to retrieve dynamic content (dynamic requests).

The six throughput predictors (P1–P6) are defined and completely detailed in [380]. We have implemented our algorithm in the simulation tool OPNET Modeler which facilitates accurate simulation of the layers of the TCP/IP stack. We consider two different service classes, named c_1 and c_2, in all the simulations. Each service class contains two types of applications: one that asks for dynamic content and another that asks for static content. Static requests are attended by the Web servers while dynamic requests require access to the App/DB server.

As an admission control algorithm is going to be tested, we need to overload the system. The workload is generated in the Web system by 30, 40, 50, 60, 70, 80, 90 and 100 Web clients, as we are interested in stressing the system to test the algorithm with an increasingly high workload. So, the Web system starts rejecting requests when it is overloaded.

We configure two workloads in order to test the algorithm more accurately. Both are basically the same, the only difference is in the user think time. In Fig. 9.10, we can observe that the arrival rate increases up to 350 Web requests per second for 100 clients during these 30 s periods.

The response time of dynamic requests, represented in Fig. 9.11, is more meaningful than the one obtained by static requests because the App/DB servers are more congested with the increase of traffic. If we analyse the case of *Workload 1* in Fig. 9.11a, we can note some differences among the response time obtained by the predictors. Focusing on the last case, 100 clients, we can detect that the predictors

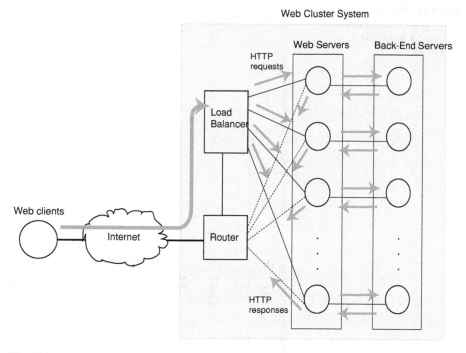

Web Cluster System

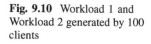

Fig. 9.9 The Web architecture is made up of several mirrored Web servers and their corresponding database servers. The model architecture is one-way, which means that the incoming HTTP requests go through the front–end node but their HTTP responses use a different way to prevent a system bottleneck in this node

Fig. 9.10 Workload 1 and Workload 2 generated by 100 clients

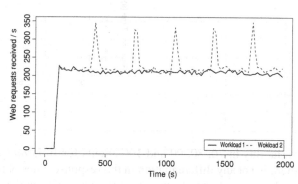

P1, P2 and P3 obtain a higher response time than predictors P4 and P5. This is also depicted in Fig. 9.11b, which represents the response time for *Workload 2*. We can also observe that the maximum response time for both workloads is around 2.5 s, that means that our algorithm achieves an extra goal, that is the limitation of the response time regardless of the amount of traffic arriving to the system. The predictor that shows a good response time and the most stable behaviour is P4, as P5 shows some

Fig. 9.11 95th percentile of the response time for dynamic requests: **a** Workload 1; **b** Workload 2

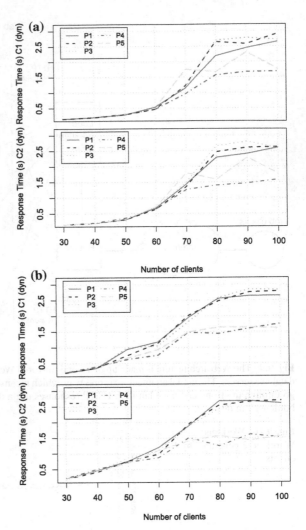

variability in 70, 80, 90 and 100 clients for *Workload 1*. We can also observe that there is not any differentiation in the response times obtained by class-1 and class-2 traffic, as we do not distinguish different queues in the Web and App/DB servers in order to keep the approach simple.

The response time of dynamic Web pages obtained from the simulations leads us to the conclusion that P4 is the most suitable predictor for our admission control and load balancing algorithm. However, we would like to remark that the predictors P1, P2 and P3 do also obtain good performance results and that have an important advantage: they are easily obtained from the throughput of the two previous slots and that do not need a record of more previous slot throughput values as predictors P4 and P5, which are more complicated to compute (please, see [380] for more information).

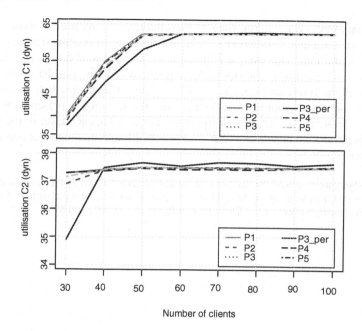

Fig. 9.12 95th percentile of the App/DB server utilisation for dynamic requests

In order to show the benefits of the adaptive time slot scheduling (described in previous Section), the algorithm has been configured to be executed on a fixed time slot scheduling. The predictor chosen for these simulations is P3. The workload chosen for this comparison is *Workload 2*.

The 95th percentile of the App/DB server utilisation is represented in Fig. 9.12. The results obtained when invoking the algorithm periodically are named as "P3_per" in the figure. Here we observe that the utilisation level of the App/DB servers is lower for P3_per in the first points of the x-axis of the graph. In the case of class-1 traffic, the servers seem to be less loaded for 30, 40 and 50 clients with P3_per. The case of 30 clients also reaches a lower utilisation level for class-2 traffic.

However, if we analyse the P3_per utilisation level of class-2 traffic after 40 clients, we can also observe that it is slightly greater that the rest of the simulations. In fact, this indicates to us that the fixed time slot scheduling introduces some errors in the utilisation level reached for each traffic class. That also means that the algorithm is less accurate in its reservations and that the SLA is less guaranteed.

9.6 Conclusion

Although the use of online prediction methods and techniques are not yet generalised to all systems, it is clear that resilient systems should consider different strategies to ensure a certain QoS, despite failures, overwhelming services and other inconveniences that usually occur at run-time. In this chapter, we have tried to show through four practical examples how to use simple tools to bring interesting benefits, thanks to online predictions. However, it is much research in this direction, especially in finding common methodologies for building resilient systems considering the online prediction of the future and react accordingly by adjusting the prediction over time. These methodologies should consider not only the techniques presented here but other appropriate to each level of design abstraction and at each layer of the system during operation. We hope these four case studies illuminate the reader about these possibilities.

Acknowledgments The work of Samuel Kounev was partially funded by the German Research Foundation (DFG) under grant No. KO 3445/6-1. The work of Carlos Juiz was partially funded by the Spanish Ministry of Science and Technology under grant TIN2007-60440.

Part IV
Measurement and Metrics

Chapter 10
Foundations of Metrology in the Observation of Critical Systems

Andrea Bondavalli, Andrea Ceccarelli, Lorenzo Falai
and Michele Vadursi

Abstract The scientific literature as well as the industrial practice shows that the observation of a system when it operates in its real environment is a common and attractive option to obtain highly accurate measurements of the system's performance and monitor its behavior. Methodology and tools for the performance evaluation and monitoring of distributed systems can successfully take advantage of the methodological approach and the mathematical tools and techniques which are typical of metrology, the science of measurement. We approach metrology from the perspective of an expert in evaluation and monitoring of critical systems, and we address the issues of the importance and of the concrete applicability of the main practices and notions from metrology to the observations of dependability properties of critical systems. We review the foundations of measurement theory and analyze the awareness of such concepts in the observations of critical systems, and the main open research challenges in this area.

A. Bondavalli (✉) · A. Ceccarelli
University of Firenze, Viale Morgagni 65, I-50134
Firenze, Italy
e-mail: bondavalli@unifi.it

A. Ceccarelli
e-mail: andrea.ceccarelli@unifi.it

L. Falai
Resiltech S.R.L. Piazza Nilde Iotti 25,
I-56025 Pontedera (Pisa), Italy
e-mail: lorenzo.falai@resiltech.com

M. Vadursi
University of Naples, "Parthenope" Centro Direz. Is. C4,
I-80143 Naples, Italy
e-mail: michele.vadursi@uniparthenope.it

K. Wolter et al. (eds.), *Resilience Assessment and Evaluation of Computing Systems*,
DOI: 10.1007/978-3-642-29032-9_10, © Springer-Verlag Berlin Heidelberg 2012

10.1 Introduction

The past years have seen a growing interest in methods for studying the behavior of computer-based systems. The scientific literature, as well as the industrial practice, shows that measuring resilience and dependability attributes is a key issue: in particular, experimental measurement is an attractive option for evaluating an existing system or prototype, because it allows observing the real execution of the system in its real environment to obtain highly accurate measurements of the system.

Performing measurements on computing systems in their real operating environment is useful in several contexts. It is done mainly to achieve two goals:

- to monitor their actual behavior (e.g., to on-line react to events);
- to evaluate their behavior (e.g., to quantitatively assess a system).

In both these contexts, the more accurate the observation of the system is, the more reliable the collected results are, and the more effective the decisions that can be consequently taken. In fact,

- in on-line monitoring, it is possible to react in a more appropriate way;
- in the performance evaluation, it is possible to take (off-line) decisions, on the basis of more accurate information.

The observation of systems is typically performed relying on *tools*. Tools that are used to experimentally assess and monitor resilience and dependability properties of critical systems should be treated for what they really are: measurement instruments. Methodology and tools for evaluation and monitoring of systems, and especially distributed systems (which respect to the centralized ones they typically suffer of additional challenges for their observation as increased complexity and absence of central control), can thus successfully take advantage of the methodological approach and the mathematical tools and techniques which are typical of metrology, the science of measurement.

First of all, since measuring a quantity (the *measurand*) consists in quantitatively characterizing it, a clear and univocal definition of the measurands is of uttermost importance. Metrology has developed theories and good practice rules to make measurements, to evaluate measurements results and to characterize measuring instruments. The main metrological properties (uncertainty, repeatability, resolution and intrusiveness) should be precisely identified in each methodology (and supporting tool) for experimental assessment and monitoring of resilience and dependability properties of computing system.

In a similar way, the results obtained using a tool should include uncertainty evaluation and, when comparing results achieved through different measurement methods, compatibility of measurement results should be assessed.

Dependability and resilience measurements on computing systems involve a wide variety of measures, from *discrete* measures, such as number of source code lines, packet size in packet-switched networks, to *continuous* measures which include delays experienced in an end-to-end connection, quality of clock synchronization, quality of service metrics, etc. A closer look at this class highlights the crucial role of time measurements: dependability-related measurements are very often based on time measurements, for example because the measurand is a time interval, or because the measurement result is obtained through indirect measurements based on timestamps.

Issues with the way measurement is applied in assessing computer dependability, and the need for giving practice a better theoretical basis, were first raised with respect to software reliability assessment. Problems were identified separately in two communities of research and practice: software reliability [149] and software metrics [334]. There were three sets of inter-related issues: confusion about the meaning of a measure (leading for instance to redefining software "reliability" as a count of bugs in a piece of code, or to seeking scalar measures for inherently multi-dimensional attributes), confusion between problems of measurement and of prediction (leading for instance to naïve methods for inference from observed failures to future reliability), and insufficient fitness for purpose of the metrics [335].

More recently, the problem of awareness of measurement theory in evaluating dependability attributes of computing systems has been raised [128]. In the paper, a set of well-known tools for experimental assessment of dependability, and papers describing results of experimental evaluations are analyzed, identifying whether and to what extent the most important metrological properties and attributes, which will be explained in the next section, namely *uncertainty, repeatability, resolution, intrusiveness* and results *compatibility*, are taken into account. Up to now, it is the only document that presents a deep state-of-the-art in this area.

Considering the metrology research community, some works on the performance assessment of distributed measurement systems and computer networks, mainly in terms of the evaluation of some specific network parameters (e.g., one-way delay, packet loss ratio, jitter) have been published [43, 44, 88]. Some papers concerning fault diagnosis in electronic and automotive systems and some others related to the design of distributed measurement systems for specific applications, such as power quality measurements, are also available in the literature [415, 558, 898]. Nevertheless, the problem of measuring and assessing resilience of critical systems with a strict and systematic metrological approach has never been faced.

A methodological approach for the observation of critical systems is thus needed. In distributed systems things are even more complex, for the lack of central control, and for the difficulties in obtaining a precise global time and an accurate view of the global state of the system. Dependability issues are very rarely addressed in the major conferences and scientific journals in the area, and resilience is practically not considered at all.

In the following we investigate the awareness and correct application of the main metrological concepts in literature when designing tools and experiments for the assessment of *dependability* and *resilience* properties.

A complete presentation of the basic concepts and definitions in metrology science can be found in [493].

The rest of the work is organized as follows. In Sect. 10.2 we focus on the main results in the dependability community regarding the problem of awareness and applications of measurement theory; results of this section greatly overlap with the ones presented in [128]. In Sect. 10.3 some papers involving dependability and resilience issues, included in the major Instrumentation and Measurement conference proceedings and journals are enlisted, and the way dependability and resilience are defined and studied is discussed. In Sect. 10.4 we conclude the paper highlighting which are the main open research challenges of this area.

10.2 Awareness of Metrology in the Academic Dependability Research Community

In a recent paper, the authors catalogue tools and experiments on the basis of the characteristics of the systems on which the tools were executed or the experiments were performed [128]. The identified classification is the following: *real-time/non real-time* systems, *centralized/distributed* systems, *safety critical/non-safety critical* systems.

Before discussing the results of such classification, we briefly introduce the main metrological properties considered there (uncertainty, intrusiveness, resolution, repeatability):

- It is well known that any measurement system perturbs the measurand, determining a modification of its value. Minimizing such perturbation, that is minimizing the system's *intrusiveness*, is therefore desirable when designing a measurement system.
- *Resolution* is the ability of a measuring system to resolve among different states of a measurand. It is the smallest variation of the measurand that can be appreciated, i.e., that determines a perceptible variation of the instrument's output.
- *Repeatability* is the property of a measuring system to provide closely similar indications in the short period, for replicated measurements performed i) independently on the same measurand through the same measurement procedure, ii) by the same operator, and iii)in the same place and environmental conditions.
- *Uncertainty* provides quantitative information on the dispersion of the quantity values that could be reasonably attributed to the measurand and it has to be included as part of the measurement result. Ref. [492] contains a comprehensive guideline for the evaluation and the expression of uncertainty in measurements.

It is important to observe that uncertainty is not just something that should be evaluated *a posteriori*. On the contrary, it is advisable (if possible) to design the measurement system and procedure to keep uncertainty below a given threshold (*target uncertainty*) [613].

Table 10.1 describes the *importance* of several metrological properties for the different categories of systems introduced above. Accordingly, it also describes the most important metrological properties that should be considered when measurement tools that shall provide reliable results are designed. A rank of the importance of

Table 10.1 Relevance of main metrological properties in measuring different kinds of systems

System Topology			Uncertainty	Intrusiveness	Resolution	Repeatability
Centralized	Non R-T	Non-Crtitical	Recomm.	Recomm.	Recomm.	
Centralized	Non R-T	Non-Crtitical	Recomm.	Recomm.	Recomm.	Recomm.
Centralized	Real Time	Non-Crtitical	Mandatory	Mandatory	Recomm.	
Centralized	Real Time	Non-Crtitical	Mandatory	Mandatory	Mandatory	Recomm.
Distributed	Non R-T	Non-Crtitical	Recomm.	Recomm.		
Distributed	Non R-T	Crtitical	Recomm.	Recomm.		Recomm.
Distributed	Real Time	Non-Crtitical	Mandatory	Mandatory	Recomm.	
Distributed	Real Time	Non-Crtitical	Mandatory	Mandatory	Recomm.	Recomm.

assessing metrological properties in each configuration is provided by classifying properties as "*Recommended*" or "*Mandatory*".

Looking at Table 10.1, *intrusiveness* is considered a parameter of fundamental importance in the evaluation of computing systems, in particular for real-time systems: a tool being able to collect sufficiently reliable data in a non real-time environment, may collect unreliable data in a hard real-time environment.

Intrusiveness is thus particularly critical in hard real-time systems, where timing predictability may be altered by the additional overhead of monitoring tasks, or other mechanisms e.g., fault injection probes.

Intrusiveness and uncertainty are related to each other since intrusiveness has consequences on uncertainty. This explains why in Table 10.1 all the rows in which intrusiveness is mandatory exhibit the same importance for uncertainty.

Time resolution may be critical in real-time systems since it needs to be much lower than the imposed time deadline to allow useful quantitative evaluations of time or dependability metrics. In computing systems, it can be generally assumed that the resolution of the time interval measurements is equal to that of the clock used in the experiment. In a centralized context it can happen that resolution is of the same order of magnitude of the measure, and it is thus of great importance to evaluate it. On the other hand, when experiments are performed on distributed systems, uncertainty is usually far greater than resolution; in such cases, the evaluation and the control of resolution may be less crucial.

Repeatability is often not achievable when measurements are carried out on computer systems. In fact it can be hard to guarantee the same environmental conditions, especially in distributed systems, where differences among local clocks, in addition to the problems of thread scheduling and timing of events, enormously increase the difficulty of designing repeatable experiments.

The difficulty of reaching a satisfactory level for repeatability has been taken into account in some of the surveyed papers, but the ability to design repeatable experiments is generally occluded by limits on accurate time stamping.

Conclusive remarks from the work surveyed in [128] are that in general a full awareness of all metrological properties is lacking, and an exhaustive analysis of measurement parameters is performed only in a few cases. There is a wide-spread consciousness of the importance of intrusiveness, but there are few efforts that try to evaluate it with a rigorous approach. Regarding resolution, it is frequently not considered, even if it is usually the easiest parameter to estimate: the reason is probably that it is often considered not important, at least if compared with intrusiveness and uncertainty.

Moreover, the approach followed to quantitatively assess algorithms and systems is not univocal, but generally varies from a paper to another, making the comparison among different results quite difficult, if not meaningless (in the surveyed tools and experiments, results are never dealt with in terms of compatibility).

10.3 Resilience Measurement, Assessment and Benchmarking in the Field of Academic Metrology Research

A survey of the most important and representative world congresses and international journals in the metrology field shows that the concept of dependability and resilience are not commonly addressed in the metrology research community.

Some years ago, in the most important conference of the field, IEEE Instrumentation and Measurement Technology Conference (IMTC), a paper was presented, which analyzed some possible failures of a distributed measurement or control architecture [88]. The paper was still preliminary, as it only took into consideration reliability of the measurement system, and seems to have had no continuation. In 2005, the editor of the IEEE Instrumentation & Measurement Magazine, wrote a short article, entitled "Dependability", introducing some basic dependability concepts to the instrumentation and measurement community [353]. He focused on dependable (electronic) system design, and stressed the role of testing. The fact that such a "general" article is so recent and practically has had no evolution gives an idea of how little is the penetration of dependability concepts in the metrology research community.

Even less spread is the concept of *resilience*. In the very few cases in which the word resilience is mentioned, it either refers to a very specific concept [415] or it is not well defined [558, 898].

In particular, the work in [415] deals with the *error resilience* of some compressed codes used for telecommunication device testing. Error resilience is defined as the capability of a compressed test data stream, which is transferred from automatic test equipment to the device under test, to tolerate errors. The effects of errors, such as bit flips, on the test data are analyzed and error resilience is evaluated.

In [898], the authors analyze the resilience of hard disks to vibrations. In this case, resilience of the hard disk was not clearly defined, but they simply checked if vibrations prevent the hard disk from transferring data; in other words, by hard disk resilience it is meant the ability to complete data transfer. What is interesting, the authors pay much attention to be as less intrusive as possible when experimentally measuring resilience.

Also in [558], the authors present a theoretical model to calculate the performance/resilience of optimized multipath routing in variable network topology. They deal with resilience of routing algorithms, but again do not clearly define what they mean by resilience.

10.4 Conclusions: Recent Advances and Open Research Problems

The previous analysis has focused mainly on *time-related* dependability and resilience attributes. While time appears the most critical factor, many algorithms may have additional continuous attributes other than time that may suffer from the unawareness of the metrological properties previously described (e.g., location algorithms, which suffer from space uncertainty).

The results of the survey presented in [128] show that some general awareness about some metrological issues is indeed present, but the followed approaches are quite intuitive, and usually quite incomplete, as well. In particular, there is a lack of (i) a common systematic approach, (ii) wide-spread studies on feasibility of repeatable experiments, and (iii) diffusion of comparable results.

A preliminary result in the field of designing tools for measuring dependability properties of distributed systems is [329], where the author defines a conceptual framework for experimental evaluation and monitoring activities that supports a rigorous (from a metrological point of view) observation of distributed systems.

Most recently, a paper has appeared in the most relevant journal of the Instrumentation and Measurement community [129]. It presents a tool for dependability measurements in distributed systems that is capable of evaluating the uncertainty of measurement results based on distributed time measurements, and can consequently discard the results characterized by the largest uncertainty. After synthesizing the results of the survey presented in detail in [128], which has constituted the starting point for the development of the new tool, the paper describes the tool and includes the results of two case studies.

In conclusion, we can say that some awareness on the importance of making measurements and assessments on critical systems and comparing their results in a rigorous and fair way, according to the best practices and the theoretical findings of metrology *is* spreading, but still not as largely as it should. There is still much to do in this way, and probably we may succeed if we put the accent on the usefulness and fairness of such an approach.

Acknowledgments This work has been partially supported by the European Community through the Coordination Action FP7-ICT-216295 (AMBER - Assessing, Measuring and Benchmarking Resilience).

Chapter 11
Field Studies on Resilience: Measurements and Repositories

Joao Duraes, José Fonseca, Henrique Madeira and Marco Vieira

Abstract This chapter is devoted to field studies and the aspects related to this kind of measurements. The importance of measurements collected from the operational scenarios is discussed, and two case studies are presented. Field measurements are closely tied to data repositories, and this chapter presents an overview of some field data repositories available to the public.

11.1 Introduction

Field measurements refer to observations of systems in the operational phase, i.e., systems that are actually in use. The results obtained from these observations have the very important characteristic of being realistic: the operation conditions and environment, and the workload are not mere experimental approximations. Very often, field studies are not representative as there is no guarantee that all possible, important system configurations have been observed. Nevertheless, field measurements and field data are a unique and very important source of information for researchers when studying resilience properties, such as availability, reliability and robustness.

J. Duraes (✉)
DEI/CISUC, Polytechnic Institute of Coimbra, 3030-290 Coimbra, Portugal
e-mail: jduraes@isec.pt

J. Fonseca
DEI/CISUC, University of Coimbra & UDI, Polytechnic Institute of Guarda,
3030-290 Coimbra, Portugal
e-mail: josefonseca@ipg.pt

H. Madeira · M. Vieira
DEI/CISUC, University of Coimbra, 3030-290 Coimbra, Portugal
e-mail: henrique@dei.uc.pt

M. Vieira
e-mail: mvieira@dei.uc.pt

K. Wolter et al. (eds.), *Resilience Assessment and Evaluation of Computing Systems*,
DOI: 10.1007/978-3-642-29032-9_11, © Springer-Verlag Berlin Heidelberg 2012

There are basically two main driving forces behind the collection of field data: development and research. The first is committed to the improvement of specific systems and to solve problems on those specific systems that are discovered during the operational phase. The second driving force aims to understand the issues related to systems reliability and dependability and to propose new techniques to increase the reliability of non-specific (non vendor-specific) systems. A third driving force is a market-driven one, to promote awareness of a given product (e.g., network providers, such as sprint and AT&T, publish their performance and dependability data to promote the company and attract new customers). However, the first two driving forces are those more relevant to research works.

The research driving force, although not tied to specific vendors or industry goals, is necessarily dependent on the existence of data. These data is mainly that which was collected by users or operators and is not related to any research goal. Thus, so far, the main origin of field data is the occurrence of incidents. This fact has an overwhelming impact on the nature of the available data, which is mainly related to computer failures and security incidents. To demonstrate the importance of field measurement and what can be achieved, in this chapter we present two case studies: the first on software faults and the second related to security vulnerabilities.

The most complex and error prone components of computer-based systems are the software. Understanding software faults is essential to devise mechanisms to mitigate faults existing in software. Thus, the first case study presented in this chapter is a field study on software faults aimed at the characterization of software faults for emulation and fault injection purposes.

Security issues are currently one of the major concerns surrounding software systems. Networking is one of the scenarios that most exposes a system to the general public and potential malicious users and attacks, representing a high relation with security-related incidents. Web-based systems are currently the basis of the majority of network-enabled systems. The second case study presented is thus related to security vulnerabilities.

Although field data (field measurements) are highly relevant to the research community to understand and improve computer-based systems robustness, reliability and security, the availability of such data remains hard to guarantee. The few data available are based on open-source projects and published research works. The importance of field data is widely recognized among researchers as shown in workshops such as RAF07: Reliability Analysis of System Failure Data organized by Microsoft Research in Cambridge and Darmstadt University in 2007. Each open-source development team or research team presents its own data and its own view. One important initiative to mitigate the scarcity and fragmented view of field data is the development of public repositories, to store data and results based on that data originating from many sources and teams. We include in this chapter a brief overview of available data repositories.

The outline of the chapter is as follows. Section 11.2 presents a field study on software faults. Section 11.3 presents a field study on security vulnerabilities. An overview of field data repositories is presented in Sect. 11.4. Section 11.5 concludes this chapter.

11.2 Case Study 1: Field Data on Software Faults

This section presents a field study on real software faults. This case study was conducted to understand the nature of faults, and to obtain a classification scheme usable for fault injection. Injecting faults is a time-proved method of validating fault tolerant mechanisms and assess system robustness. Given the relevance of software faults, it is very relevant to be able to inject software faults. The usefulness of fault injection is tied to the representativeness of the faults injected. To that aim, we need to understand what exactly is a software fault (a clear, but detailed description usable for automated fault injection), and obtain information on the types of faults that represent the faults more common in the operational scenario. The case study presented here is a summarized description of that field study. More details can be found in [315]. A technique to emulate software faults at the binary executable was proposed based on the findings of this study (G-SQFIT, see [315]), however, the details of such technique do not fit in a field study description and it is not presented here.

Section 11.2.1 presents the source of the software faults used in this case study and details the methodology used for the classification of the faults. Section 11.2.2 presents a first overview of the fault distribution and makes a comparative analysis with the field study done by Christmansson and Chillarege in [213] using the ODC classification [205, 206] scheme. Section 11.2.3 presents an overview of the details classification of the collected faults. Some conclusions about this field study are presented in Sect. 11.2.4.

11.2.1 Sources of Real Software Faults and Classification Methodology

To address the representativeness issue of our study, we collected a large set of real software faults from software used in the field. The goal was to improve the knowledge about the exact nature of faults and their occurrence distribution using data from the real operational scenario. More specifically, the software faults that are pertinent to emulate by fault injection are those that originated in the coding phase and eluded the testing procedures and go with the deployed product.

The information source used in our work was a set of diff/patch files for several open source programs. The diff/patch files contain source code corrections for faults discovered after the software was released. By manual inspection of those files we were able to extract information to understand and classify software faults. From those diff/patch files, a total of 668 faults were analyzed. Table 11.1 presents a summary of the programs used in this study. It is worth noting that these programs encompass a broad range of program types: both user programs (including interactive and command line programs) and operating system (Linux kernels) were used.

The total number of faults collected for each program is dependent on the program age, maturity and the user community size. Some of the programs (e.g. Bash) are

Table 11.1 Source of the field data

Programs	Source location	Description	# faults
CDEX	http://sourceforge.net/projects/cdexos/	CD Digital audio data extractor	11
Vim	http://www.vim.org	Improved version of the UNIX vi	249
FreeCiv	http://www.freeciv.org	Multiplayer strategy game	53
pdf2h	http://sourceforge.net/projects/pdftohtml/	pdf to html format translator	20
GAIM	http://sourceforge.net/projects/gaim/	All-in-one multi-protocol IM client	23
Joe	http://sourceforge.net/projects/joe-editor/	Text editor similar to Wordstar®	78
ZSNES	http://sourceforge.net/projects/zsnes/	SNES/Super Famicom emulator	3
Bash	http://cnswww.cns.cwru.edu/~chet/bash/bas	GNU Project's Bourne Again	2
LKernel	http://www.kernel.org	Linux kernels 2.0.39 and 2.2.22	93
Firebird	http://sourceforge.net/projects/firebird/	Cross-platform RDBMS engine	2
MingW	http://www.mingw.org/	Minimalist GNU for Windows	60
ScummV	http://sourceforge.net/projects/scummvm	Iterpreter for adventure engines	74
Total faults collected			668

in a mature phase and have few recent faults; other programs (e.g. VIM) are still in the maturation phase and have a large user community that provides many fault reports. The notion of fault requires the notion of correctness. Generally speaking, the software is correct if it conforms to the user needs, as specified in the software requirements. However those might be wrong. For the purpose of this work, it was assumed that the requirements and specification are correct. Thus, a software fault means that the code is not correct somehow (i.e., it does not implement the specification in some particular aspect) because the code does not contain the instructions that should have.

The approach used to analyze and classify the faults was the following:

1. First we classified the faults according to the Orthogonal Defect Classification scheme (ODC) [205, 206]. The use of general and well accepted fault classification is the best way to make our results available for the research community and it allows us to compare our results with previous field studies.
2. In a second step we grouped the faults in each ODC class according to the nature of the defect, defined from a building block programming point of view. That is, for each ODC class a software fault is further characterized by one or more programming language constructs that is either missing, wrong or in excess. Programming language constructs may be statements, expressions, function calls, etc. A fault may then fall in one of three possible types: missing construct, Wrong construct, and Extraneous construct. This is very relevant to fault emulation/injection since emulating an omission (missing construct) is substantially different from emulating a wrong construct (e.g., erroneous expression).
3. In a last step, faults were further described and grouped into specific types. Each type is defined according to the language construct and program context surrounding the fault location. This description refinement is also particularly relevant for fault injection purposes since it helps (a) the identification of suitable

locations in the target code, and (b) the code modifications necessary to emulate a given fault type.

The resulting final classification can be viewed as an extension to ODC and is used to define fault emulation operators (each operator emulates one specific type of faults).

11.2.2 ODC Classification and General Analysis

According to the Orthogonal Defect Classification, a software fault is characterized by the change in the code that is necessary to correct it, i.e., to put the code consistent with the specification, which is assumed to be correct in our case. From the list of ODC types, the following are directly related to the code and relevant to our work:

- Assignment: value(s) assigned incorrectly or not assigned at all.
- Checking: missing or incorrect validation of data and conditional statements, wherever these checks and conditions may appear (e.g., an incorrect loop condition).
- Interface: errors in the interaction among components, modules, device drivers, functions calls, and similar.
- Timing/serialization: missing or incorrect serialization of shared resources.
- Algorithm: incorrect or missing implementation that can be fixed only by (re)implementing an algorithm or data structure without the need of a design change.
- Function: affects a sizeable amount of code and refers to capability that is either implemented incorrectly or not implemented at all.

As field data available to us did not include any information on timing or serialization properties, we did not consider the Timing/serialization ODC type. The mapping of the faults into one of the remaining ODC types was straightforward with the exception of the Function type which required a more detailed analysis of the code in order to figure out whether the correction of the fault has required a design change or not. Due to the decentralized nature of the software development methodology of open source projects, we didn't have direct information on redesign decisions, which forced us to a more detailed analysis of the faults identified as candidates for the Function ODC type. Table 11.2 presents the distribution of faults across the five ODC fault types addressed in this work.

One interesting topic to both the theme of field-based works and to the theme of software faults is the comparison of our results with other available field studies that also used ODC to classify field-discovered faults. We compared our fault distribution with the one presented in [213] as that work is the one most closely related to our own. Because that work included Time/Serialization faults, we removed that particular type from the comparison and normalize all the percentages leaving so that a direct comparison could be made. Table 11.2 presents this comparison (values shown in parenthesis are those from [213] after normalization.

It is relevant that both our data and that presented in [213] show the same trend in the fault distribution across ODC fault types: assignment faults have approximately

Table 11.2 Fault distribution across ODC types

ODC type	# faults	ODC distribution (%)	
Assignment	143	21.4	(21.98)
Checking	167	25.0	(17.48)
Interface	49	7.3	(8.17)
Algorithm	268	40.1	(43.41)
Function	41	6.1	(8.74)

the same weight as Checking faults; Interface and Function faults are clearly the less frequent ones; and Algorithm are the dominant faults. All ODC classes have approximately the same weight in both works. The fact that independent research works obtained a similar fault distribution suggests that this distribution is representative of programs in general and gives us confidence in our results. Also, the programs analyzed in [213] (large database and operating system code) were quite different from the ones used in our study, suggesting that this fault distribution across ODC types is reasonably independent from the nature of the program. Although more field studies should be conducted to consolidate this conclusion, these results suggest that fault injection experiments should take this fault distribution trend into consideration to improve representativeness.

Table 11.3 presents the fault distribution observed for each individual program used in this study. To observe a trend in fault distribution across programs, only those programs with a significant number of faults should be considered (the number of faults is presented in the first row). Nevertheless, we decided to show the results for all the programs. We observed that the programs with a higher number of faults show a similar ODC fault distribution; the only observed deviation was presented by "Joe" program, which had more checking faults than the global trend. This trend existing across programs reinforces the suggestion that software faults do follow a clear pattern of distribution across ODC types.

11.2.3 Extended Classification and Discussion

For the purpose of fault injection the fault types provided by ODC are not practical as they are too broad, meaning that many different faults fall in the same type and the types themselves lack the fine details required by an automated tool to be able to reproduce the fault in the target code. Clearly, further refining is needed, not in the sense of an alternative classification but as an additional detail layer to ODC. As explained in Sect. 11.2.1, we propose to achieve this extra layer by analyzing faults from the point of view of the (program) context in which fault occur and relate the faults with programming language constructs. Using this notion, a defect is then one or more programming language constructs that are either missing, wrong or in excess. A construct is any building block of the traditional programming languages:

Table 11.3 Fault distribution across ODC types by individual programs

Programs	CDEX	Vim	FCiv	Pdf2h	GAIM	Joe	ZSNES	Bash	LKernel	FireBird	MingW	M	Total (%)
# faults	11	249	53	20	23	78	3	2	93	2	60	74	668
ODC type													
Assignment (%)	18.2	21.3	11.3	55	4.3	25.6	66.7	100	22.6	50	10	24.3	21.4
Checking (%)	18.2	22.5	13.2	5	52.2	44.9	0	0	25.8	50	38.3	8.1	25
Interface (%)	54.5	6.4	7.5	0	4.3	14.1	0	0	5.4	0	5	4.1	7.3
Algorithm (%)	9.1	44.6	52.8	40	26.1	15.4	33.3	0	33.3	0	46.6	56.8	40.1
Function (%)	0	5.2	15.1	0	13	0	0	0	12.9	0	0	6.8	6.1

statements, expressions, function calls, etc. Following this idea, we classified each fault according to its nature which can be one of these: missing construct, Wrong construct, or Extraneous construct. Although this classification is orthogonal to ODC and can be used alone (as is in Table 11.4), we used it as an extension to ODC fault types to provide a refined view of the faults specifically aimed at emulation by fault injection.

As we can see in Table 11.4, faults of the extraneous nature are clearly less frequent than the other two. This was an expected result, as programmers are more prone to forget to put something in the program, or to put it in a wrong way, than to insert surplus code. We can also see that missing programming constructs seem to be the dominant type of software fault. From the point of view of representativeness for fault injection experiments, this information is valuable.

Table 11.5 presents the total number of missing, wrong or extraneous faults for each of the five ODC fault types addressed in this study. We also provide some examples of fault to help the reader understand what kind of fault is included in each type (this will be detailed further on). As we can see from Table 11.5, there are once again trends that we can use to achieve representativeness in the injection of software faults, e.g., for the assignment and interface types, missing program construct faults are less frequent than the wrong construct faults.

We then further detailed the description of faults describing exactly what constructs were missing, wrong or extraneous. We did this for all ODC types and obtained a reasonable small list of fault types (for each ODC type). This is an interesting result, as we do not want a small list of generically-described faults where many faults fit and for which no practical tool can emulate those faults due to lack of details, and we also do not want a long list of over-detailed description where each fault fits into and only into its own type, rendering any effort of representativeness useless. The complete list of fault types for all ODC types is outside the goal of this section and chapter. We present here in Table 11.6 the list of faults for the ODC type algorithm and refer the user to [315] for a detailed description of this work.

The faults listed in Table 11.6 are now described with a level of detail that is useful for practical fault injection. For example, the type MFC—missing function call refers to the omission of a call to a routine in the program. This is an easy understandable description that can be easily emulated into the target code. Another important issue is the identification of suitable location where a given fault can be injected. Using the MFC fault type again, it is relatively easily to identify occurrences of function call in the target, even in the binary code. It is worth noting that this study was part of an effort to devise and implement a fault injection technique able to inject realistic software fault directly into the binary code of the target, without requiring source code (goal that was achieved). This scenario is relevant because many fault injection applications involve common-of-the-shelf components for which there is no source code available.

To help readers understand the level of details that is now used to describe faults, we use another example from Table 11.6 . Fault MIFS—Missing if construct plus statements. This fault refers to the omission of a conditional statement deciding if a givel (small) block of statements is executed. In C language it is something like

Table 11.4 Fault distribution by fault nature

Fault nature	CDEX	Vim	FCiv	Pdf2h	GAIM	Joe	ZSNES	Bash	LKernel	Firebird	MingW	ScumVM	Total	(%)
Missing construct	3	157	35	11	17	34	1	0	63	2	45	61	429	(64.2)
Wrong construct	8	85	18	9	6	41	2	2	24	0	14	12	221	(33.1)
Extraneous construct	0	7	0	0	0	3	0	0	6	0	1	1	18	(2.7)

Table 11.5 Fault nature totals across ODC types

ODCtype	Nature	Examples	# faults	% of total
Assignm.	Missing	A variable was not assigned a value, a variable was not initialized	62	9.3
	Wrong	A wrong value (or expression result, etc) was assigned to a variable	70	10.5
	Extraneous	A variable should not have been subject of an assignment	11	1.6
Checking	Missing	An "if" construct is missing, part of a logical condition is missing,etc	113	16.9
	Wrong	Wrong logical expression used in a condition in brach and loop on struct (if, while, etc.)	53	7.9
	Extraneous	An "if" construct is superfluous and should not be present	1	0.1
Interface	Missing	A parameter in a function call was missing; incomplete expression was used as param.	11	1.6
	Wrong	Wrong information was passed to a function call (value, expression result etc)	38	5.7
	Extraneous	Surplus data is passed to a function (e.g. one parameter too many in function call)	0	0
Algorithm	Missing	Some part of the algorithm is missing (e.g. function call, a iteration construct, etc)	222	33.2
	Wrong	Algorithm is wrongly coded or ill-formed	40	6
	Extraneous	The algorithm has surplus steps; A function was being called	6	0.9
Function	Missing	New program modules were required	21	3.1
	Wrong	The code structure has to be redefined to correct functionality	20	3
	Extraneous	Portions of code were completely superfluous	0	0

If (cond) {statement1; statement2; .... }

Once again the identification of this type of construct is easily identifiable in the target code and easily emulated through modification in said code. One very important aspect of the information in Table 11.6 is the number of occurrences for each fault type. The two fault types described here are much more common than other types (e.g., MIEA). This is a very important information to build representative faultloads for fault injection experiments. Table 11.7 presents a global view of all the occurrences for all fault types (all ODC types and programs).

The information summarized in Table 11.7 is very relevant. It offers two conclusions about software faults:

Table 11.6 Detailed analysis of algorithm faults

Fault nature	Fault specific types	CDEX	Vim	FCiv	pdf2h	GAIM	Joe	ZSNES	Bash	L.Kernel	FireBird	MinGW	ScumVM	Total
Missing construct	Missing function call (MFC)	28	7	1	1	5				4		2	23	71
	Missing If construct plus statements (MIFS)	27	10			1				15		15	12	80
	Missing if-else construct plus statements (MIES)											4	3	7
	Missing if construct plus statements plus else before statements (MIEB)	10	4		2							1	1	18
	Missing if construct plus else plus statements around statements (MIEA)									2		1		3
	Missing iteration construct around statement(s) (MCA)	1												1
	Missing case: statement(s) inside a switch construct (MCS)	1												1
	Missing break in case (MBC)	3				1								4
	Missing small and localized part of the algorithm (MLPA)	9	4	4	1	2		1		1		5		23
	Missing sparsely spaced parts of the algorithm (MLPS)	5			1	1								6
	Missing large part of the algorithm (MLPL)	3			1			1		3				8
Wrong construct	Wrong function called with same parameters (WFCS)	1			1			1		6				9
	Wrong function called with different parameters (WFCD)	9	1										3	13
	Wrong branch construct—goto instead break (WBC1)	1	1											2
	Wrong algorithm—small sparse modifications (WALD)	4	1	1										6
	Wrong algorithm—code was misplaced (WALR)	5	3	1										9
	Wrong conditional compilation definitions (WSUC)			1										1
Extraneous	Extraneous function call (EFC)	1	1			2	1	1						6
Total faults found		108	28	8	6	12	1	4	0	31	0	28	42	268

Table 11.7 The "Top-N" fault in this study by occurrence frequency

Fault types	Description	Fault coverage (%)	ODC types
MIFS	Missing "If (cond) { statement(s) }"	9.96	Algorithm
MFC	Missing function call	8.64	Algorithm
MLAC	Missing "AND EXPR" in expression used as branch condition	7.89	Checking
MIA	Missing "if (cond)" surrounding statement(s)	4.32	Checking
MLPC	Missing small and localized part of the algorithm	3.19	Algorithm
MVAE	Missing variable assignment using an expression	3	Assignment
WLEC	Wrong logical expression used as branch condition	3	Checking
WVAV	Wrong value assigned to a value	2.44	Assignment
MVAV	Missing variable assignment using a value	2.25	Assignment
MVI	Missing variable initialization	2.25	Assignment
WAEP	Wrong arithmetic expression used in parameter of function call	2.25	Interface
WPFV	Wrong variable used in parameter of function call	1.50	Interface
	Total faults coverage (field data)	50.69	

- There is a relatively small set of fault types that is responsible for a large portion of all the fault occurrences. The 12 fault types in Table 11.7 put together are responsible for 50 % of all the faults discovered in this field study.
- There are faults that are clearly more frequent than others, and this information is important to build representative faultloads for fault injection scenarios.

The results of this field study are very interesting for research on software faults and for the injection of software faults. It offers insight on fault details aimed at the realistic emulation of faults, it offers information about the distribution of the most common type of faults in the operational scenario aimed at generating representative faultloads, and is the basis of the G-SWFIT technique for fault injection. These results and this technique have been used on several research works (e.g., [315, 664, 665]), and the classification scheme is used as basis for different application areas (still related to software faults), such as security (e.g., relate vulnerabilities with its root cause faults).

To conclude the presentation of this field study we present here one example of a software fault as classified and described in this field study (Fig. 11.1), and one example of a fault emulation operator of the G-SWFIT fault injection technique developed in the sequence of this field study (Fig. 11.2). We refer the reader to [315] for more details.

```
*** 4858,4864 ****
        for (p=name; isalpha(*p) || isdigit(*p) || *p == '_'; ++p)
            ;
  !     if (p == name)
        {
            EMSG("Function name required");
            return;
        }
--- 4858,4864 ----
        name = eap->arg;
        for (p=name; isalpha(*p) || isdigit(*p) || *p == '_'; ++p)
            ;
  !     if (p == name && !eap->skip)        Was missing
        {
            EMSG("Function name required");
            return;
        }
```

Fig. 11.1 Example of a diff/patch file (excerpt). In this example, the patch applies a "&& !eap->skip" that was missing. The fault type is MLAC—Missing "AND EXPR" in expression used as branch cond

Operator	Example	Example with fault	Search pattern	Code change

OMIEB	if (expression)
	{
	  statements-IF
	}
	else
	{
	  statements-ELSE
	}
	... remaining code
		if (expression)
		{
		  statements-IF
		}
		else
		{
		  statements-ELSE
		}
		... remaining code
			flag-affecting instr.
			jcond elsecode
			... instrs (IF)
			
			jmp after
			elsecode:
			... instrs (ELSE)
			after:
			... remaining code
				- All the conditional jumps to the address loc01 are changed into unconditional jumps
				- Call instructions and stores to memory existing between the cond jumps are removed

Notes				
There may be several cond. jumps to *elsecode* if expressions is composed of several sub-expressions The side-effects (if any) of the first sub-expression are not ommited				

Fig. 11.2 Operator to emulate a fault OMIEB—missing if construct and the statements surrounded by it plus an else statement. It is not one of the most common fault types, but it serves to illustrate the changes at the high level code and its related modification at low level to emulate the fault, as well as search pattern used to identify suitable fault locations

11.2.4 Considerations on the Case Study

In this case study a large number of software faults were analyzed to improve the knowledge about the nature of software faults: its nature, the frequency of its occurrence frequency by fault types, and how they can be emulated through fault injection. The contributions of this case study were a fault classification scheme allowing practical injection of software faults and the knowledge about the fault distribution across fault type as they occur in the operational scenario. The source of the data was a set of open-source programs, without which this study would have been much harder if not impossible: in closed-source projects, the information regarding faults and their correction is kept within the development team. As the correction of faults (patch code) was directly used to conduct this field study, we stress the importance of having data available for research purposes, even in closed-source projects. This data can hardly be used for commercial purposes, and, excepting issues related to security, a concerted effort should be made by academia to try and obtain data such as the one used for this study. This effort should be articulated with the creation of data repositories to help spreading the data and results of field data studies.

11.3 Case Study 2: Field data on Security Vulnerabilities

In this section we present the results of a field study on the most common vulnerabilities, which provides a truthful body of knowledge on real security vulnerabilities that accurately emulate real world security problems. The data was obtained by analyzing past versions of representative web applications with known vulnerabilities that have already been corrected. The main idea is to compare the piece of defective code with the corrections made to secure it. This code change (or the lack of it in the vulnerable application) can be viewed as the reason for the presence of the vulnerability. Note that this methodology can generically be used in other field studies to obtain the characterization and distribution of the source code defects that originate vulnerabilities in web applications.

The field study uses data from 655 SQL Injection and XSS security patches of six widely used web applications. The detailed analysis of the code of the patches shows that web application vulnerabilities result from software bugs affecting only a restricted collection of statements, which greatly facilitates the emulation of vulnerabilities through fault injection, as the effort can be concentrated on the emulation of vulnerabilities in a small number of types of statements.

Sections 11.3.1 and 11.3.2 describe the methodology used to collect the field data in this field study. Section 11.3.3 presents the systems addressed in the study, and the vulnerabilities addressed are presented in Sect. 11.3.4. Section 11.3.5 details the information gathered in the study and the results are presented in Sect. 11.3.6. Section 11.3.7 summarizes this case study.

11.3.1 Vulnerability Analysis and Classification Methodology

When web application vulnerabilities are discovered, software developers correct the problem releasing application updates or patches. In our study, we used these patches to understand which code is responsible for security problems in web applications.

With this approach, we can classify the code structures that cause real security flaws and identify the most frequent types of vulnerabilities observed in the web applications considered in our field study.

For each web application under test, the methodology to classify the security patches is the following:

1. Verification of the patch to obtain the right version of the web application where it applies. We need to confirm the availability of the specific version of the web application and obtain it for the rest of the process. It is mandatory to have both the patch and the vulnerable source code to be able to analyze what code was fixed and how, unless the patch file has all this information (which we found to be unusual).
2. Analysis of the code with the vulnerability and compare it with the code after being patched. The difference between the vulnerable and the secure piece of code is what is needed to correct the vulnerability. This is what the software developer should have done when he first wrote the program and this is what we have to classify.
3. Classification of each code fix that is found in the patch. The absence of the actions programmed in the patch represents what causes the vulnerability. For example, if the patch replaces the variable $id with intval($id)[1], we consider that the vulnerability is caused by the absence of the intval function in the original code. To be accurate, we followed the patch code analysis guidelines described in the next section.
4. Loop through the previous steps until all available patches of the web application have been analyzed.

11.3.2 Patch Code Analysis Guidelines

Web applications are developed using different coding practices and during the classification of the security patches we face different scenarios and have to make some decisions that need to be clarified. To avoid classification mistakes and misinterpretations the following guidelines are followed:

1. We assume that the information publicly disclosed in specialized sites is accurate and that the fix developed by the programmer of the patch and made available by the company that supports the web application solved the stated problem. We do not test the presence of the vulnerability nor confirm its correction.
2. To correct a single vulnerability several code changes may be necessary. This way, each code change was considered as a singular fix. For example, suppose that two functions are needed to properly sanitize a variable. Missing any of these functions makes the application vulnerable, so both of them must be taken into

---

[1] The intval is a PHP function that returns the numeric value of a variable, or 0 on error.

account. In this case, if we want to simulate the vulnerability, we may remove any of the singular fault type fixes.

3. When a patch can fix several vulnerability types simultaneously, each one is accounted separately. This occurred naturally because we analyzed each vulnerability independently, as if we were doing several unrelated analyses, one for each vulnerability type. For example, this occurs when a not properly sanitized variable is used in a query (e.g. allowing SQL Injection) and later on is displayed on the screen (e.g. allowing XSS). When this variable is properly sanitized, both vulnerabilities are mitigated simultaneously, however this situation accounts for the statistics of both XSS and SQL Injection vulnerabilities.

4. When a particular code change corrects several vulnerabilities of the same type, each one is considered as a singular fix. For example, suppose that the value assigned to a specific variable comes from two sources of external inputs; and the variable is displayed in one place without ever being sanitized. We consider that the application has two security vulnerabilities because it can be attacked from two different inputs. However, to correct the problem all that is needed is to sanitize the variable just before it is displayed. In this example we consider that two security problems have been fixed, although only one code change was needed.

5. A security vulnerability may affect several versions of the application. This happens when the code is not changed for a long time, but it is vulnerable. The patch to fix the problem is the same for all versions, and therefore it is considered to be only one fix.

By following the previous guidelines, it was possible to classify almost all the code fixes analyzed. However, in some situations, patching one or more vulnerabilities may involve so many changes, including the creation of new functions or a change in the structure of the overall piece of code, that it is too difficult to classify it properly. These situations are usually associated with major code changes involving simultaneously security and other bug fixes related to functional aspects. These occurrences were quite marginal (5.4 %) and were not considered in our study because they are too complex and difficult to analyze due to the lack of source code documentation.

11.3.3 Web Applications Analyzed

One mandatory condition for our field study is to have access to the source code of the web applications under analysis. The code of previous versions and the associated security patches must also be accessible. The other mandatory condition is the availability of information correlating the security fix and the specific version of the web application.

The goal is to be sure that it is possible to access the source code (including the code of older versions) in order to be able to analyze and understand the security vulnerability and how it was fixed. Actually, the way a given vulnerability is fixed is a key aspect in the classification of the fault type originating the vulnerability.

For the present study we have selected six LAMP (Linux, Apache, MySQL and PHP) web applications: PHP-Nuke [726], Drupal [307], PHP-Fusion [493] , Word-Press [943], phpMyAdmin [728] and phpBB [727]. These are open source web applications that represent a large community of users and, fortunately, there is enough information available about them to be researched. Additionally, they represent a large slice of the web application market and have a large community of users:

- Drupal (winner of the first place at the 2007 and 2008 Open Source CMS Award), PHP-Fusion (one of the five winner finalists at the 2007 Open Source CMS Award) and phpBB (the most widely used Open Source forum solution and the winner of the 2007 SourceForge Community Choice Awards for Best Project for Communications) are Web Content Management Systems (CMS). A CMS is an application that allows an individual or a community of users to easily create and administrate web sites that publish a variety of contents.
- PHP-Nuke is a well-known web based news automation system built as a community portal. PHP-Nuke is one of the most notorious CMS and it has been downloaded from the official site over 8 and half million times.
- WordPress is a personal blog publishing platform that also supports the creation of easy to administrate web sites. It is one of the most used blog platforms in the World.
- phpMyAdmin is a web based MySQL administration tool. It is one of the most popular PHP applications, is included in many Linux distributions, and was the winner of the 2007 SourceForge Community Choice Awards for Best Tool or Utility for SysAdmins.

The six web applications analyzed are so broadly used since several years ago that they have a large number of vulnerabilities disclosed from previous versions, which were the subject of analysis of the field study. It is important to emphasize that a single vulnerability opens a door for hackers to successfully attack any of the millions of web sites developed with a specific version of the web application. Furthermore, it is common to find a single vulnerability in a specific version that also affects a large number of previous versions. The overall situation is even worse because web site administrators do not always update the software in due time when new patches and releases are available.

11.3.4 Security Vulnerabilities Studied

In the present work we focus on two of the most critical vulnerabilities in web applications: XSS and SQL Injection. A Cross Site scripting (XSS, but also known as CSS) vulnerability allows the attacker to inject HTML and/or a scripting language (usually JavaScript) into a vulnerable web page [710] . A SQL Injection vulnerability allows the attacker to tweak the input fields of the web page in order to alter the query sent to the back-end database [709].

Exploits of these vulnerabilities take advantage of unchecked input fields at user interface, which allows the attacker to change the SQL commands that are sent to the database server (SQL Injection), or allows the attacker to input HTML and a scripting language (XSS). Two main points account for the popularity of these attacks:

- The easiness in finding and exploiting such vulnerabilities. They are very common in web applications and within a web browser the attacker can probe for these vulnerabilities tweaking GET and POST variables that are available in the HTML page. The building of an exploit for fun or profit can be a bit more time consuming, but there are plenty information and guides on how to do it (e.g. look at [409, 708] for XSS and [408, 708, 720] for SQL Injection, just to mention a few).
- The importance of the assets they can disclose and the level of damage they may inflict. In fact, SQL Injection and XSS allow attackers to access unauthorized data (read, insert, change or delete), gain access to privileged database accounts, impersonate another user (such as the administrator), mimicry web applications, deface web pages, get access to the web server, malware injection, etc. [347].

11.3.5 Patch Code Sources

For all the applications analyzed, we collected the source code of both the vulnerable and the patched versions. By comparing these two versions, we could understand the characteristics of the vulnerability and classify what code was changed to correct it.

Software houses and developers follow their own policies in what concerns the public availability of older versions of the software, particularly when they have security problems. In some cases, they can be hard to find and even the access to the past collection of vulnerability patches can be a cumbersome task. Furthermore, most security announcements publicly available are so vague that it is too difficult (or even impossible) to know which source files of the application are affected by a particular vulnerability. Moreover, some of the disclosed information about security problems is too generic and groups together several types of security vulnerabilities (e.g., using the same document to refer to directory traversal, remote file inclusion and COOKIE poisoning vulnerabilities), which makes it more difficult to map our target vulnerabilities to the code fixing them.

In order to gather the actual code of security patches, we have to use several sources of data, such as mirror web sites, other sites that provide the source code (mainly on blogs or forums), online reviews, news sites, sites related to security, hacker sites, change log files of the application, the version control system repository, etc.

For the purpose of this study, we just need the changes made to the code of the application correcting the vulnerability problem (i.e., the source code of the entire application is not required). However, as there is no standard way of providing the data about the security vulnerability fix, different sources of information have to be considered, each one following its own specific format. The four main source types used in the current work are the following:

1. Security patch files with information about the target version of the application. In this case, we have the reference to the buggy version of the web application and to the patch file that must be applied to mitigate the target vulnerability.
2. Updated version of the web application. Actually, this is a completely new version of the application containing new features and bug fixes (including security ones). This is the most common source of information we have found, but it is also the one that needs more exploration work to be done.
3. Available security diff file. In this case, there is a diff file, which is a file containing only the code differences between two other files with information about what lines of the original file have been removed, added or changed. It has, therefore, the precise code changes needed to fix a referenced vulnerability.
4. Version control system repository. Almost all relevant open source applications are developed using a version control system to administer the contributions of the large community of developers from around the world. This is the most complete source of information we can have about the application, although it may be difficult to find what we are looking for in such a vast collection of files and versions.

Once the vulnerable code and the respective patch are obtained using one of the previous sources of information, a differential analysis is performed to identify the locations in the code where the defects are fixed. This operation is done mainly through the use of diff utility. The Unix diff utility is a file comparison tool that highlights the differences between two files using the algorithm to solve the longest common subsequence problem [455]. A manual analysis of the code can be also performed when the output of the diff utility is too complex due to a large number of changes between the two versions of the source code, or when many corrections are done in the same file. The manual analysis also helps grouping several security corrections and discarding the code changes not related to security issues.

11.3.6 Field Study Results and Discussion

In the field study we classified 655 XSS and SQL Injection security fixes found in the six web applications analyzed (PHP-Nuke, Drupal, PHP-Fusion, WordPress, phpMyAdmin and phpBB). We followed a classification scheme based on the software fault classification proposed in [314] and adapted the fault types specific to XSS/SQL injection (e.g., MFC to MFCext).

The overall distribution of the fault types found in the six web applications analyzed is shown in Table 11.8. In this table we can see the individual results for each fault type allowing us to understand how they are distributed along the web applications analyzed.

A common belief is that vulnerabilities related to input validation are mainly due to missing if constructs or even missing conditions in the if construct. However, our

Table 11.8 Detailed results of the field study on the most common software faults generating vulnerabilities

Web app.	PHP-Nuke		Drupal		PHP-Fusion		WordPress		phpMyAdmin		phpBB		
Fault type	SQL	XSS	SQL	XSS	SQL	XSS	SQL	XSS	SQL	XSS	SQL	XSS	%
MFCext	120	133	4	39	6	13	6	94	1	51	3	27	76
WPFV	31			3	2	5				4		1	7
MIFS	5	2		2	7	6				10		2	5
WVAV	2			3				2		4		17	4
EFC				1						1		4	1
WFCS				3	1	1		13					3
MVIV		1			1	3						4	1
MLAC				1	2	4				2			1
MFC				2	1					1			1
MIA				1	1								0
MLOC		1											0
ELOC				1									0
Total faults	158	137	4	55	21	33	6	109	1	73	3	55	100

field study shows that this is not the case, as the overall "missing IF..." fault types (MIFS and MIA: see Table 11.8) only have a weight of 5.5%. As for the "missing <condition>..." fault types (MLAC and MLOC), they represent only 1.52% of all the fault types. This suggests that programmers typically do not use if constructs to validate the input data, and this may occur due to the complexity of the validation procedures needed to avoid XSS and SQL Injection.

The typical approach we found in the field is the use of a function to clean the input data and let it go through, instead of stopping the program and raise an exception (or show an error page). This may be understood as a design goal trying to prevent the disruption of the interaction of users to the least possible. In what concerns security, it would be better to allow only inputs known as correct (white list) as this prevents any input with suspicious characters to go any further and is more secure than just cleaning the input from malicious characters and let the operation continue normally.

Analyzing the global distribution of web applications vulnerabilities we found 70.53% of XSS and 29.47% of SQL Injection showing that XSS is the most frequent type by far. As shown, all the fault types account for XSS vulnerabilities but only eight fault types report to SQL Injection, which might help justify the fact that XSS is more prevalent than SQL Injection, confirming the results of the IBM X-Force®2008 Trend and Risk Report [819]. This trend is also confirmed by vulnerability reports disclosed in CVE [657, 707]. However, the four fault types that do not contribute to SQL Injection (MFC, MIA, MLOC and ELOC) only account for 1.22% of all the fault types. Obviously, we do not have enough sample values to conclude that SQL Injection may not be derived from one of these fault types. We can only say that we did not find them in our field study.

There are several factors that contribute to the prevalence of XSS. XSS is easier to discover because it manifests directly in the tester web browser window. Every input variable of the application is a potential attack entry point for XSS, which is not the case for SQL Injection, where only variables used in SQL queries matter. Another factor that contributes to the prevalence of XSS is that SQL Injection alters the database records and this cannot be always seen in the interface, at least so explicitly as XSS. Moreover, the knowledge needed to test for XSS [409, 708] is not as complex as for SQL Injection, for which the attacker needs to have deep knowledge about the SQL language. Although the SQL language is usually based on the SQL-92 standard [290], every database management system (DBMS) has its own extensions and particularities [408, 708, 720], that need to be taken into account when searching for SQL Injection.

The most representative and widespread fault type is the "Missing function call extended (MFCext.)". It represents 75.87 % (140 SQL Injection + 357 XSS out of 655 vulnerabilities studied) of all the fault types found. The high value observed for the MFCext fault type comes from the massive use of specific functions to validate or clean data that comes from the outside of the application (user inputs, database records, files, etc.). In many cases, functions are also used to cast a variable to a numeric value, therefore preventing string injection in numeric fields.

The next three most common fault types are "wrong variable used in parameter of function call (WPFV)", "missing IF construct plus statements (MIFS)", and "wrong value assigned to variable (WVAV)".

A recurring problem is that, looking at several versions of the same program, we frequently found the same regex string being slightly updated as new attacks are discovered. These situations were found in WPFV and WVAV faults.

Excluding the faults types already discussed (MFCext., WPFV, MIFS and WVAV), the remaining fault types correspond to only 7.63 % of the security vulnerabilities found. These fault types are EFC, WFCS, MVIV, MLAC, MFC, MIA, MLOC and ELOC.

11.3.7 Considerations on the Case Study

In this case study we presented a methodology for characterizing the most frequent fault types associated with the most common web application vulnerabilities based on a field study. We focused on XSS and SQL Injection vulnerabilities of six widely used web applications, using 655 security fixes as the field data. Results show that only a small subset of 12 generic software faults is responsible for all the XSS and SQL Injection vulnerabilities analyzed.

One relevant outcome of the field study performed is referred to the distribution of vulnerabilities by a reduced number of fault types. In fact, we observed that a single fault type, the MFCext. (missing the function responsible for cleaning the input variable), is responsible for about 76 % of all the security problems analyzed. Previous studies on software fault types [212, 312] also show this large dependency on a few bug types. Furthermore, this trend is not new in the security area: Microsoft has already stated that fixing the top 20 % of the reported bugs eliminates around 80 % of errors [785] and the Gartner Group reported that 20 % of security test rules uncover

80 % of errors [574]. This concentration of the responsibility of most vulnerabilities on just a few fault types can be very important to address the web applications security and makes it feasible to emulate vulnerabilities by means of fault injection, which has already been started to be addressed by the research community [322, 342, 344, 815].

11.4 Overview of Data Repositories

Data repositories are an excellent resource to store and share information for research purposes. One type of valuable information that can be shared through data repositories is the result from field data studies. Although data repositories to store failure data and dependability experiments results are relatively rare (especially considering the huge value of real failure data to help designers in improving computer systems), several initiatives have been proposed and are currently available.

The Data & Analysis Center for Software (DACS) is a Department of the US Defense Information Center supporting research on software reliability and quality. It serves as centralized source for data related to software metrics. The DACS maintains the Software Life Cycle Experience Database (SLED). This repository is intended to support the improvement of the software development process. The SLED is organized into nine data sets covering all phases and aspects of the software lifecycle ([253] and [477]). Examples of these datasets are:

- The DACS Productivity Dataset (collected from government and private industry sources). This dataset consists of data on over 500 software projects and is mainly oriented to software cost modelling and productivity analysis [678]. The data represents software from early 60s to early 80s and includes software projects ranging from avionics to off-the-shelf packages. The information in this dataset includes the following: size of project, effort, language, schedule, errors.
- The NASA/SEL Dataset (contributed by the Software Engineering Laboratory (SEL) at NASA Goddard Space Flight Center). This repository maintains data on avionic applications since 1976. The dataset is available by request on disk and it can be accessed through web browser. Using the latter, users have access to analytical summaries including linear regression, scatter plots and histograms. The analytical results are created dynamically per request during the HTTP session and served to the user browser. The repository information is stored in a relational database and the link between the data repository and the web server is supported through Perl applications.
- The Software Reliability Dataset (collected at Bell Laboratories) [669]. This repository describes failures in a wide range of application domains including real time, control, office, and military applications. This dataset was primarily aimed at the validation of software reliability models and to assist software managers to monitor and predict software tests. As in the NASA/SEL dataset, the information can be obtained by request, and it can also be accessed through web interface.

The Metrics Data Program (MDP) Repository is a database maintained by the NASA Independent Verification and Validation facility [674]. The repository is aimed at the dissemination of non-specific data to the software community and it is made available to the general public at no cost. All the data available in the repository are sanitized by the projects representatives, and all the necessary clearances are provided. Users of the repository are free to analyze the data for their specific research goals.

The MDP repository is part of the MDP on-going effort to improve the ability to predict error in software by improving the quality of the problem data related to software (e.g., improve the quality of the information about the relationship of the error and the development phase). To this effort, the MDP recruits the participations of private-sector and public-sector projects. Recruited projects maintain complete control of data release and the level of participation in the program. The effort required by the participating projects is minimal. The repository contains data on the software projects that were collected and validated by the MDP program, spanning more than 8 years and including more than 2700 error reports. The information stored in the repository consists of error data, software metrics data, and error data at the function/method level. The dataset enables data associations between products, metrics, and errors classified according to the Orthogonal Defect Classification (ODC) [204].

The Software Reference Fault and Failure Data Project [689] is maintained by the National Institute of Standards and Technology and is aimed at the development of metrology, taxonomy and repository for reference data for software assurance. The project maintains a repository on software fault data specifically aimed at helping industry protect against releasing software systems with faults and to help assess software systems quality by providing statistical methods and tools. The repository is available to the public upon request. The access to the information online allows users to view data and execute simple queries. Analytical and statistical use of the data is possible through a program developed within the project and available to the public (the EFFTool).

The Computer Failure Data Repository (CFDR) is a public repository on computer failure data ([74] and [182]) supported by USENIX. The repository is aimed at the acceleration of the research on system reliability with the ultimate goal of reducing or avoiding downtime in computer systems. To this goal, the CDFR hopes to remove the main difficulty faced by researchers, which is the lack of reliable and precise information about computer failures. The CDFR repository is open to both obtaining and contributing data. The repository comprises nine independent data-sets focusing mainly on very large storage systems. The repository information covers many aspects, including: software failures, hardware failures, operator errors, network failures, and operational environment problems. The raw data are available to the public [182] through web interface. The project does not offer online capability for analytic and statistical data-processing.

The AMBER Raw Data Repository [32] is a repository of field data and raw results from resilience assessment experiments. Its goal is to grant both the research and IT industry communities with an infrastructure to gather, analyze and share field data resulting from resilience assessments of systems and services, stimulating a better coordination of high quality research in the area, and contributing to the promotion of a standardization of resilience measurement, which will in turn have a positive impact in the industry. While experimental and field data repositories are recognizably fundamental for supporting the advance of research and the dissemination of knowledge, the research community still seems somewhat reluctant in embracing such enterprises. This repository aims to encourage acceptance from the community to share its data and promote the research involving several partners sharing data.

Publicly available vulnerability databases currently play a very important role in making the information on vulnerabilities available to researchers and have completely reshaped the way software vulnerabilities are reported and disseminated in recent years. Examples of popular vulnerability databases are the National Vulnerability Database [693] and The Open Source Vulnerability Database [705], which provide comprehensive reports about discovered software vulnerabilities including the nature of a vulnerability (its type, the component where it was located, the list of vulnerable system versions, its discovery date, and so on) and include examples on how to exploit it, as well as the patch or the workaround provided by system vendor to fix it (when available). Additionally, to alert users about the severity and security risk of reported vulnerabilities, these databases typically provide vulnerability impact and exploitability levels assigned by security advisors. These databases also provide a web-based interface that enables users to search vulnerabilities and browse a list of the vulnerabilities reported for a given system.

11.5 Conclusion

The case studies presented in the chapter allow drawing some conclusions on field measurements and field data studies. Although the focus of the chapter is software faults and security vulnerabilities, these conclusions apply to any type of measurement obtainable in the field. Important aspects that are self-evident are the representativeness of the measurements and results, the classification used to describe them and manipulate data, and the mechanisms to make data and results available to the research community and general public.

Concerning data on the robustness of the computer-based systems, field data is mostly obtained from reports (bug reports, incident reports, security logs, and so on, depending on the nature of the incident). These reports are filed by the users and operators and are typically used by the system developers to solve the incidents and improve the system.

Observations made in closed-source, proprietary systems are typically not available to the public. Observations originating from open-source systems are normally made available to the community (e.g., stored in a repository). However, these repositories are normally not oriented to a systematic storage and classification of the discovered faults and remedies. Instead they are the result of the accumulation of solution to problems resulting in a kind of logfile-like information about which problems were discovered (bug reports, many times repeated), and how were solved. The exception to this are the repositories maintained by researchers in the context of long-term research in large companies, such as IBM. These are good initiatives, but typically are very different from one another. It would be of great value to the research community to have information on software faults available in a systematized and uniform way. Repositories like the ones described in the chapter are good initiatives in that direction.

Concerning security, the information pertinent to research is even harder to find than those about software faults. It is not the case of data availability (as it is for faults in closed-source systems). On the contrary, there is plenty of information. The major problem is that there is too much information, scattered and mostly repeated, and classified using different schemes. A given security issue may have been classified according to in scheme and given one value of severity, for instance, and in another repository, the same vulnerability may appear with a different description and different characterization.

The usefulness of public repositories to the research communities is demonstrated by the existence of studies based on the information stored in publicly available repositories (e.g. [32]). Nevertheless, and in spite of the different repository initiatives already available, the raw data from the vast majority of research works on experimental dependability evaluation and on field failure data, among other examples, is not available in any repository. Hundreds of papers have been published but the raw data that have led to the final results presented in those papers is not available. Data repositories do seem a very promising initiative to provide the means to have a uniform description of raw data and results and make this information available to the public, and perhaps some more concerted effort should be placed towards creating and maintaining said repositories. One example among several is the AMBER repository, which was built specifically to share data among different teams.

Chapter 12
Failure Diagnosis of Complex Systems

Soila P. Kavulya, Kaustubh Joshi, Felicita Di Giandomenico
and Priya Narasimhan

Abstract Failure diagnosis is the process of identifying the causes of impairment in a system's function based on observable symptoms, i.e., determining which fault led to an observed failure. Since multiple faults can often lead to very similar symptoms, failure diagnosis is often the first line of defense when things go wrong - a prerequisite before any corrective actions can be undertaken. The results of diagnosis also provide data about a system's operational fault profile for use in offline resilience evaluation. While diagnosis has historically been a largely manual process requiring significant human input, techniques to automate as much of the process as possible have significantly grown in importance in many industries including telecommunications, Internet services, automotive systems, and aerospace. This chapter presents a survey of automated failure diagnosis techniques including both model-based and model-free approaches. Industrial applications of these techniques in the above domains are presented, and finally, future trends and open challenges in the field are discussed.

S. P. Kavulya (✉) · P. Narasimhan
Carnegie Mellon University, PA, USA
e-mail: spertet@ece.cmu.edu

P. Narasimhan
e-mail: priya@cs.cmu.edu

K. Joshi
e-mail: kaustubh@research.att.com

F. Di Giandomenico
ISTI Department, Italian National Research Council, via Morazzi 1,
I-56124, Pisa Italy
e-mail: felicita.digiandomenico@isti.cnr.it

K. Wolter et al. (eds.), *Resilience Assessment and Evaluation of Computing Systems*,
DOI: 10.1007/978-3-642-29032-9_12, © Springer-Verlag Berlin Heidelberg 2012

12.1 Introduction

The issue of diagnosing hardware and software failures to find the underlying causes has existed for as long as computers have been around. Using the fault, error, and failure nomenclature of [576], failure diagnosis is the process of identifying the fault that has led to an observed failure of a system or its constituent components. In any sufficiently large computing system, many types of faults are often not directly visible for a number of reasons-either due to the characteristics of the fault itself, due to fault-tolerance mechanisms built into the system that hide the expression of the fault, or as is most often the case, the lack of detailed monitoring functionalities that can detect and report on the occurrence of the fault directly. In some cases, monitoring systems may provide only an indication that a fault has occurred, but may not provide sufficient information to precisely locate it.

Failure diagnosis is a technically challenging endeavor because the relationship between faults, failures, and their observable symptoms is a complex one; single faults often produce multiple symptoms in different parts of a system, e.g., a misconfiguration fault in a critical network component such as a Dynamic Host Configuration Protocol (DHCP) server can cause all client computers on the network to fail; conversely, similar symptoms may be caused by many different types of faults, e.g., the failure of a networked computer to receive an IP address can have several causes including, but not limited to, packet loss in the physical network, a client misconfiguration, or a problem with the DHCP server. As operational systems become more mature, the failures they encounter often transition from easy to detect "hard failures" that cause a significant impairment to the system's primary function, to "soft failures" such as those due to performance bottlenecks or transient faults that are much harder to detect. Therefore, the process of diagnosis often also includes the identification of anomalous conditions that are symptoms of the occurrence of faults.

In addition to its essential role as the precursor to any remediation actions for maintaining a system's health at runtime, failure diagnosis also serves several important roles in resilience assessment of complex systems. Since it is only the symptoms of a fault that are usually observed at runtime, diagnosis is essential for the accurate cataloguing of fault occurrences in the field. Conversely, any data that reports on occurrences of actual system faults is by definition the product of a diagnostic process, whether it is a simple one (in case of a one-to-one mapping between faults and symptoms), a complex manual process, or an automatic one. Understanding this process is important for understanding the biases and limitations of the field data. Diagnosis is also important for discovering new fault types that can then be used to drive fault injection campaigns as discussed in Chap. 13 .In fact, diagnosis is the converse process of fault injection. In fault injection, one injects faults into a system according to a predefined fault model in order to analyze the resulting symptoms, or if the system tolerates the fault, the absence of any symptoms. In diagnosis, one infers the faults from the observed symptoms. Finally, diagnosis is also important in the emerging field of online resilience assessment as described in Chap. 15. In this area, diagnosis can be used, under the label of fault localization, to infer the true health of complex distributed systems, including what components have actually failed, by eliminating those failure symptoms that are a result of error propagation to an otherwise operational part of the system.

Due to the complexity of computing systems and difficulty of formalizing the scope of the diagnosis task itself, diagnosis has historically been a largely manual process requiring significant human input. However, techniques to automate as much of the process as possible have significantly grown in importance. In domains such as communication networks and Internet services, the sheer scale of modern systems and the high volumes of impairments they face drive such trends, while in domains such as embedded systems and spacecraft, it is increasing complexity together with the need for autonomic operation (i.e., self-healing) when human expertise is not available, that are the drivers. Due to the diversity of the domains, a variety of failure diagnosis techniques drawing from diverse areas of computing and mathematics such as artificial intelligence, machine learning, statistics, stochastic modeling, Bayesian inference, rule-based inference, information theory, and graph theory have been studied in the literature. Finally, when automated techniques fail, approaches that assist humans perform diagnosis more efficiently via the use of visualization aides have also been widely deployed. While a comprehensive survey of this broad topic can provide sufficient material for a book of its own, in this chapter, we provide a summary of the most important techniques, and provide references to more in-depth surveys where available. This chapter is organized as follows: Sect. 12.2 discusses types of problem diagnosis techniques using illustrative examples; Sect. 12.3 highlights practical uses of these diagnosis techniques in industrial applications; Sect. 12.4 presents future trends and open challenges in diagnosis; and Sect. 12.5 concludes.

12.2 Techniques

Automated problem diagnosis techniques localize the most likely sources of a problem to a set of metrics, e.g., anomalous CPU usage, a set of nodes, e.g., anomalous web server, or a type of problem, e.g., misconfiguration. Operators use the output of automated problem diagnosis to guide root-cause analysis by analyzing source-code, or hardware and software settings at the identified culprits. For example, an examination of the source-code at the web server might show that the anomalous CPU activity at the web server was due to an infinite loop in a scheduling function. Automated diagnosis techniques are not perfect and they can either fail to detect a problem resulting in a false negative, or indict the wrong component resulting in a false positive. These techniques rely on tuning to minimize the number of false negatives and false positives generated. Visualization tools complement automated problem diagnosis tools by allowing operators to visualize anomalies and explore different hypotheses on the root-cause of problems. Table 12.1 provides a summary of the techniques described in this chapter. For each technique, we first use an illustrative example to highlight its application, before delving into the different approaches

Table 12.1 Summary of diagnosis techniques

Technique	Limitations
Rule-based techniques rely on expert knowledge expressed as a set of predefined rules to diagnose problems (Sect. 12.2.1).	Rules are human-interpretable and extensible. However, they cannot diagnose unforeseen problems, and large knowledge bases are difficult to maintain.
Model-based techniques define a mathematical representation of a system, testing the observed state against the model to see if it conforms (Sect. 12.2.2).	Model-based techniques are well suited for diagnosing application-level problems. However, building models requires a deep understanding of the system.
Statistical techniques summarize and interpret empirical data using techniques such as correlation, histogram comparison and probability theory, for diagnosis (Sect. 12.2.3).	Statistical techniques require little expert knowledge or detailed models on system internals. However, they have difficulties distinguishing legitimate changes in behavior such as workload changes, from illegitimate changes such as performance problems.
Machine-learning techniques identify patterns in behavior using clustering, or use training data to determine if the system is unhealthy and the likely cause (Sect. 12.2.4).	Machine-learning techniques automatically learn profiles of system behavior, but can suffer from the curse of dimensionality that reduces accuracy when the number of features is large.
Count-and-threshold techniques allow discrimination between transient and intermittent faults (Sect. 12.2.5).	Diagnosis accuracy strongly depends on proper parameter calibration. However, solutions for parameter tuning based on rigorous mathematical formulations and analytical models are available.
Visualization techniques allow operators to visualize trends in data and spot anomalous behavior (Sect. 12.2.6).	Visualization tools allow operators to explore different hypotheses on the root-cause of problems. However, they do not automatically identify they source of problems.

proposed in the research literature. We conclude each discussion with a critique of the technique that highlights its strengths and limitations.

12.2.1 Rule-Based

Rule-based techniques rely on expert knowledge expressed as a set of predefined directives, i.e. rules, to diagnose problems. The rules are typically formatted as a set of *if-then* statements where the *if-part* of the rule is called the *premise*, and the *then-part* of the rule is the conclusion. An example of a rule used for diagnosis is *"if* CPU utilization exceeds 90% *then* node is overloaded". Rule-based techniques for diagnosis typically rely on forward-chaining inference mechanisms [850] to synthesize results when multiple rules fire. Forward inference processes events, such as high CPU and memory utilization, and uses the triggered rules to draw conclusions on the root-cause of the problem.

Illustrative example Chopstix [113], a lightweight monitoring tool, relies on a small collection of rules to guide diagnosis in production systems. They describe a recurrent problem at a production system that caused nodes to crash every 1–7 days. Shortly before such crashes they observed that ssh sessions to nodes would stall for tens of seconds. They observed that the symptoms of this problem matched the rule "*if* combined value of CPU utilization for processes is low, and scheduling delay is high *then* kernel bottleneck is likely". This rule led them to trace the problem to a tight loop in kernel's scheduler.

12.2.1.1 Types of Rule-Based Techniques

One approach for representing rules is *codebooks* [326, 952] which map each problem to a unique signature consisting of symptoms in both the faulty component where the problem occurs, and related components affected by the original problem. The codebook is instantiated as a dependency matrix where the columns represent the problems, and the rows represent the symptoms. Problems are uniquely diagnosable if all the columns are different. Codebooks diagnose the underlying problem by identifying the closest match to the observed symptoms.

Other diagnosis tools, such as Chopstix [113] and Vertical Profiling [416] rely on a small collection of rules based on the semantics of the application, and the underlying behavior of the operating system to map changes in system performance on individual nodes to known problems. These tools provide an intuitive approach for diagnosing problems on individual nodes, however they currently do not correlate metrics across multiple nodes and do not address problems that can propagate across the network in distributed systems.

Diagnosis tools that analyze large sets of rules require more sophisticated techniques, such as expert systems that rely on forward inferencing to synthesize results and resolve conflicts when multiple rules fire. These expert systems allow administrators to cope with the deluge of alarms generated by large-scale distributed systems. [596] presents a specification language for expressing rules that captures the timing relationship among correlated events. For example, alert operator if a link is down and no corresponding link up event occurs within 2 min. Commercial tools such as HP Operations Manager [949] use an optimized Rete algorithm [347] to perform pattern matching on rules in a scalable manner that is independent of the number of rules.

12.2.1.2 Limitations

Rule-based approaches are prevalent in commercial tools, such as IBM Tivoli Enterprise Console [458] and HP Operations Manager [949], as they offer an intuitive approach for expressing system behavior that allows users to augment the rule-base by developing new rules tailored to their unique operating environments. In addition, rule-based systems do not require profound understanding of the underlying system architectural and operational principles. However, rule-based systems suffer from the inability to learn from experience, and the inability to deal with problems not described within the rule-base. Rule-based systems are also difficult to maintain because the rules frequently contain hard-coded network configuration information [850].

12.2.2 Model-Based

Model-based techniques define a mathematical representation of a system, and test the observed state of the system against the learned model to diagnose problems. Some models represent the normal operation of the system, and detect problems whenever the observed system behavior fails to conform to the learned model. Other techniques generate graphical models of how problems propagate through the system [77, 508, 532, 539], and exploit this knowledge to infer the source of the problem. Alternatively, graphical models [496, 780] can represent how successes propagate through the system. These graphical models then analyze patterns of probe failures and successes to infer the source of the problem. Lastly, graphical models may represent expected communication patterns within a system and flag problems whenever these patterns are violated.

Illustrative example
Sherlock [77] localizes performance problems in large enterprise networks using a graphical model of how errors propagate to infer the source of the problem. Sherlock's inference engine learns service-level dependencies by sniffing packets and detecting which services are likely to be used together, e.g., DNS and web service. Sherlock models three types of components: (i) clients which observe response times delays; (ii) root-cause nodes which are potential sources of faults in the system; and (iii) meta-nodes which model how errors propagate through the system. An example of a meta-node is a fail-over node which requires all nodes in the high-availability group to fail for an error to propagate. If a client observes a high response time, Sherlock uses the fault-propagation model to compute the probability that a client observes a set of symptoms given that a root-cause node is at fault. It outputs a list of root-cause nodes which best explain the observed symptoms at the client.

12.2.2.1 Types of Model-Based Techniques

Model-based techniques can be classified into: (i) *physical model based techniques* which use the physical laws that a system operates under to model constraints on system behavior; (ii) *regression and queuing models* which model relationships between resource consumption and application behavior; and (iii) *graph-theoretic models*

which exploit knowledge on how errors or successes propagate in a system to localize problems.

Physical models use models of the physical world, such as the laws of mechanics,
electomagnetics, or chemical kinetics to model system behavior and to determine
when anomalous behavior is present. They typically model continuous cyber-
physical systems in industrial, automotive and aerospace domains whose physics
are well understood, e.g., powertrain [633] and chassis systems [455] in cars. These
systems run in a closed-loop, where sensors monitor the system output, then feed the
data into a controller that signals actuators to adjust control as necessary to maintain
the desired system output. Problems are diagnosed by executing the physical model
alongside the actual system at run-time to detect when the system fails to conform to
the model. The fault model typically associated with the control-theoretic approach
includes sensor faults, actuator faults, and faults in the mechanical, electromechani-
cal, or hydraulic plant being controlled [574]. Isermann et al. [473] provide a more
detailed discussion of these techniques.

Regression and queuing models are useful for workload characterization, capacity
planning and detecting performance problems. These models represent relationships
between resource consumption and application behavior, and detect anomalies whenever these relationships are violated.

Some techniques model multi-tier Internet applications as queues, and use mean-
value analysis [597, 901] to predict transaction response times. These techniques use
a network of queues to represent how the tiers in the multi-tier application cooperate
to process requests. Mean-value analysis assumes closed queueing models in which
the number of clients in the system remains constant. However, it is often difficult in
practice to obtain the client session information required to calibrate closed models
for real-world production applications [852].

Real-world production workloads are non-stationary, i.e., the relative frequencies
of transaction types changes over time. Queuing approaches which leverage regression to learn the relationship between resource consumption and application behavior
can be used to predict response times for non-stationary workloads [198, 528, 852].
These models assume that the system contains a small number of types of transactions, and that transaction types strongly influence system resource demands. These
models rely on open queues, where clients can join and leave the system model.
Open models facilitate more thorough empirical validation in production systems
than would be possible with closed models as they do not require client session
information [852].

In addition, using queuing theoretic approaches to model transaction mixes allows
these systems to distinguish anomalies from workload changes. Cherkasova et al.
[198] use queues to model the relationship between CPU usage and transaction
response times for a transaction mix. They also exploit regression to define an application performance signature that allows them to detect software upgrades by monitoring changes in the application signature. Stewart et al. [852] model the relationship
between multiple physical resources, namely CPU, disk and network, and response
times for a transaction mix. These models need to be re-trained to cope with new

transaction types. They also ignore interaction effects across transaction types and implicitly assume that queueing is the only manifestation of congestion.

Graph-theoretic models analyze communication patterns across nodes and processes to model the probability that errors, or successes, propagate through the system. The models may also monitor violations in expected communication patterns. Graph-theoretic models are useful for diagnosing both correctness and performance problems in distributed systems. They can be used to detect multiple independent problems - ranking them by likelihood of occurrence.

SCORE [539] and Shrink [508] localize problems in the IP network by modeling error propagation patterns in the wide-area networks. Both Shrink and SCORE model the system as a two-level graph between the IP layer and the underlying wide-area network. Sherlock [77] and Khanna et al. [532] extend on Shrink and SCORE to deal with multi-level dependencies and with more complex operators that capture load-balancing and failover mechanisms.These techniques infer the root-cause by computing the probability that errors propagate from a set of possible root-cause nodes to the observation nodes. They indict the root-cause nodes that best explain the symptoms at the observation nodes, and scale by assuming that there can only be a small number of concurrent problems in the system at a given time.

Rish et al. [780] propose an active probing approach that exploits a dependency-matrix to represent the failed components that each probe, e.g., server ping, detects. Active probing allows probes to be selected and sent on-demand, in response to one's belief about the state of the system. At each step the most informative next probe is computed and sent. As probe results are received, belief about the system state is updated using probabilistic inference. This process continues until the problem is diagnosed. They extend their active probing approach to cope with dynamic systems [779], where problems may occur and disappear, by maintaining two sets of probes: one set for repair detection to monitor nodes that are known to have failed, and another set for failure detection to monitor nodes that are known to be working. Their approach assumes a sequential fault model in which only one fault or repair can occur at a time. Joshi et al. [496] use a Bayesian approach to diagnose problems in systems with different types of monitors, or probes, that have differing coverage and specificity characteristics. They use a dependency matrix to represent the probability that a monitor detects a failure in a component, and incrementally update their belief about the set of failed components based on the observed monitor output.

Khanna et al. [779] address diagnosis in distributed systems where errors can propagate across nodes. They track message exchanges between nodes and detect problems by comparing communication patterns against a rule-base of allowed state transitions. Pip [775] detects application-specific problems in distributed systems by allowing programmers to embed expectations about application behavior in the source code. Pip detects problems by comparing actual behavior against expected behavior. Black-box approaches that track message exchanges are more generic and can be easily applied to new systems, whereas white-box approaches like Pip are able to diagnose application-specific problems but require a deeper understanding of system behavior.

12.2.2.2 Limitations

Model-based techniques are well-suited for diagnosing application-specific problems because they encapsulate semantic knowledge on the expected behavior of the system. The incorporation of semantic knowledge can also help them distinguish legitimate changes in behavior, e.g. workload changes, from illegitimate changes due to failures [198, 528, 852]. However, model-based techniques in general require a deep understanding of system behavior to construct the models. Even in cases where automatic model construction is feasible, there is often a tradeoff between the amount of semantic knowledge the model incorporates and the fidelity of the diagnosis. For example, graph-theoretic models [77] that are automatically constructed by examining a system's communication patterns can localize a problem to a single node or a small neighborhood of nodes, but cannot tell what the deeper root cause is. Another disadvantage of model-based techniques is that they can fail to detect novel problems that were not considered in the model.

12.2.3 Statistical

Statistical techniques for diagnosis summarize and interpret empirical data using techniques such as correlation, histogram comparison and probability theory. These techniques are data-centric and require little expert knowledge or detailed models on system internals. Statistical techniques are either: (i) *parametric* techniques that assume that data is drawn from a known distribution such as a normal distribution, or (ii) *non-parametric* techniques that do not rely on data belonging to a particular distribution. Non-parametric methods make fewer assumptions than parametric methods, making them more robust and giving them wider applicability. However, there is a cost - larger sample sizes are required to draw conclusions with the same degree of confidence as parametric methods.

Illustrative example Multivariate Adaptive Statistical Filtering (MASF) [168] is a parametric technique for detecting and visualizing anomalies in data centers that is similar to the Bucket algorithm by Avritzer et al. described in Chap. 2. MASF detects anomalies by tracking deviations from the mean in performance counters, such as CPU and memory usage. MASF assumes that data is drawn from a normal distribution and flags an anomaly if a metric exceeds 3 standard deviations from the mean. To cater for seasonal variations in behavior, such as heavy load during the day and light load at night, MASF maintains separate behavioral profiles for computing the mean and standard deviation of each metric. MASF alerts operators to suspicious behavior and allows them to visualize anomalies, but it does not automatically localize the problem.

12.2.3.1 Types of Statistical Techniques

Statistical techniques are pervasive in problem diagnosis literature. Some model-based techniques discussed earlier rely on statistical techniques, such as correlation and regression, in conjunction with deep knowledge of the application's behavior to diagnose problems. In contrast, the statistical techniques discussed in this section make fewer assumptions about the application's behavior. Statistical techniques can be classified as parametric or non-parametric techniques.

Parametric techniques assume that data is drawn from a known distribution. Normal distributions are commonly used for anomaly detection and diagnosis because of their tractability, and because normality can sometimes be justified by the *central-limit theorem* which explains why many distributions tend to be close to the normal distribution. These techniques typically detect anomalous behavior by identifying significant deviations from the mean for performance counters, which they assume follow a normal distribution. However, hardware failure rates are better modeled using Weibull distributions which capture the increased failure rates of devices as they age [812, 813].

Agarwal et al. [23] use change-point detection and problem signatures to detect performance problems in enterprise systems. They detect abrupt changes in system behavior by monitoring changes to the mean value of performance counters over consecutive windows of time. This technique does not scale well if the number of nodes and metrics is large. NetMedic [509] diagnoses propagating problems in enterprise systems by analyzing dependencies between nodes, and correlations in state perturbations across processes to localize problems. NetMedic represents state for each system component as a vector that indicates whether each metric was anomalous or normal by assuming that each metric obeys a normal distribution and flagging anomalies based on deviation from the mean. If two components which depend on each other are anomalous, NetMedic searches for time periods where the source component's state is similar to its current state, and searches for destination states that have experienced significant changes in the same period. These destination states are the likely culprits.

Draco [524] performs statistical diagnosis of problems in large Voice-over-IP (VoIP) systems by comparing differences in the distributions of attributes, such as hostnames and customer IP addresses, in successful and failed calls. Draco assumes that these attributes are drawn from a Beta distribution and localizes problems by identifying attributes that are most correlated with failed calls. By comparing successes and failures over the same window of time, Draco avoids the need for separate learning passes, and can thus diagnose problems that have never been seen before.

Non-parametric techniques assume that data is drawn from an unknown distribution. Non-parametric techniques estimate the underlying data distribution using histograms or kernel density estimators, or make generalizations about the populations from which the samples were drawn, e.g., using correlation.

Histogram-based techniques typically diagnose problems by comparing histograms of performance counters before and during an anomalous period to identify the metrics most likely to be associated with the problem. Tan et al. [715, 866] diagnose problems in large clusters using histogram-comparison of performance counters to identify "odd-man-out" behavior. Peer-comparison allows their approach to be robust to workload changes. However, propagating errors such as packet-loss that affect communication across multiple nodes reduce the accuracy of their approach. Shen et al. [822] propose a reference-driven approach to diagnose performance problems due to configuration changes or upgrades. Their approach relies on histogram comparison to identify the collection of single-parameter changes that best explain the performance deviation observed.

Correlation-based techniques analyze historical data to automatically discover relationships between pairs of metrics that are stable over time [488, 490]. Changes in these learned correlations may signal problems. Correlation can also be used to automatically discover causal relationships between metrics in distributed systems. Giza [620] exploits knowledge of the system's topology to identify spatial correlations between events. For example, to detect that customers in Texas are experiencing poor video quality. Next, Giza uses cross correlation to discover causal relationships between the observed symptoms and root-cause events. Oliner et al. [702] also use cross correlation to discover causal relationships between anomaly signals across components. The anomaly signals represent the changes in the behavior of components over time in terms of resource usage, message timing or semantics. Project5 [26] records packet traces at each node and uses message correlation algorithms to automatically extract end-to-end causal traces for requests, and detect high-latency paths. Correlation-based approaches can discover spurious relationships depending on the thresholds used to determine whether a correlation is significant. In addition, correlation-based approaches do not scale well if the number of nodes and metrics is large because they search for metric correlations both locally, and remotely between nodes communicating with each other.

Dimensionality-reduction techniques like Principal Component Analysis can reduce the number of metrics to compare when diagnosing problems by summarizing dominant trends. Xu et al. [948] use source-code analysis to apply structure to console logs and discover dominant historical trends in application state and message counts using Principal Component Analysis. PeerWatch [510] uses peer-comparison to detect anomalies in heterogeneous clusters running different hardware. Their peer-comparison algorithm uses a dimensionality-reduction technique known as canonical correlation analysis to normalize performance differences due to different hardware, and discover correlations between peers.

12.2.3.2 Limitations

Statistical techniques require little expert knowledge or detailed models of system internals. The diagnosis techniques can rely on well-established statistical theories to ground their algorithms, and test that their results are statistically significant, i.e., unlikely to have occurred by chance alone. For example, hypothesis tests such as the t-test, allow us to reject the hypothesis that the observed system behavior is consistent with the expected system behavior with a degree of confidence. When building statistical profiles of behavior, care must be taken to include sufficient data samples and test assumptions on data distributions to ensure validity. For example, incorrectly assuming that the data is drawn from a normal distribution can lead to a high error rate. Since statistical techniques do not incorporate much semantic knowledge about semantic behavior, they can experience difficulties distinguishing legitimate changes in behavior such as workload changes from performance problems.

12.2.4 Machine Learning

Machine learning is a scientific discipline that is concerned with the design and development of algorithms that allow computers to evolve behaviors based on training data. Machine-learning techniques borrow heavily from statistical techniques and probability theory. Machine learning relies on training and cross-validation which involves partitioning a sample of data into complementary subsets, performing the analysis on one subset called the training set, and validating the analysis on the other subset called the validation set or testing set. Cross-validation can provide an estimate of model accuracy.

Illustrative example Cohen et al. [231] describe an approach for automatically extracting signatures of system behavior so that operators can identify and quantify recurrent problems, e.g., slowdowns due to insufficient database connections. They use Service-Level Objective (SLO) violations to identify periods of time where the system was behaving abnormally and use tree augmented Bayesian networks (TANs) to determine which metrics are most correlated with the anomalous periods. They build signatures of the anomalous periods using metric attribution as follows: 1 indicates a metric is selected by model and attributed to failure, -1 indicates a metric is selected by model but not attributed to failure, and 0 indicates a metric was not selected by model (irrelevant). They cluster the signatures based on a purity score which indicates what fraction of signatures in the cluster are associated with failures. Clusters with greater purity provide more confidence in the signature. They found that the metric attribution gives better results than using raw metric values. They also found that they can leverage signatures from different sites to identify or rule out recurrent problems.

Types of machine learning techniques Diagnosis algorithms that rely on machine learning can be categorized into two broad categories namely: (i) unsupervised learning which identifies patterns in unlabeled data typically through clustering, and (ii) supervised learning which infer a function that best classifies successful and failed states from labeled data.

Unsupervised learning identifies patterns in unlabeled data typically through clustering, and detects unexpected outlier data points that might be indicators of failures.

Kiciman and Fox [533] uses probabilistic context-free grammars to model the causal paths in the system. The grammar rules represent the probability that one component calls another. They identify anomalous causal paths by measuring the difference between the probability of the observed transition and the expected probability of the transitions that make up the causal path. Magpie [89] uses a string-edit-distance comparison to group together requests with similar behaviour, from the perspective of request structure, synchronization points and resource consumption. The representative requests from each clusters allow them to construct concise workload models and detect outlier requests.

Supervised learning uses labeled data of successful and failed states to learn which metrics are most correlated with failed states, or to identify signatures of recurrent problems from a database of known problems.

Metric attribution approaches localize problems by identifying resource-usage metrics or components that are highly correlated with failed states. They allow operators to sift through the hundreds or thousands of metrics available in their system and narrow down the handful of metrics that yield insight to the cause of the problem and its location and guide operators in performing more detailed root-cause analysis. Once the operators determine the root-cause, they can then annotate the output of metric attribution with the root-cause and build the database of known problems used by signature-based approaches.

Pinpoint [192] and MinEntropy [193] localize components highly correlated with failed requests using data clustering [192] or decision trees [193]. They represent requests using a matrix where each row is a client request, and columns are components. An additional column indicates whether the request was successful or failed. The matrix serves as input into the machine learning algorithm. These approaches detect problems that result in changes in the causal flow of requests such as exceptions. More recently, Spectroscope [780] categorizes requests based on functionality, e.g., read or write requests, and applies data clustering to requests in each category to identify outliers due to changes in causal flows or request durations. Some limitations of these approaches are that they cannot distinguish between sets of components that are tightly coupled and are always used together, and they require requests to be independent of each other. If a request corrupts state and affects subsequent requests, the non-independence of requests makes it difficult to detect the real faults because the subsequent requests may fail while using a different set of components [192].

Cohen et al. [230] use tree augmented Bayesian networks to determine which resource-usage metrics are most correlated with the anomalous periods. They proposed an extension [957] to their work that uses ensembles of Bayesian models to adapt to changing workloads and infrastructure.

Signature-based approaches allow system administrators to identify recurrent problems from a database of known problems. Signature-based approaches have wide applicability because studies have shown that typically half, and as much as 90% of software failures are due to recurrent problems [310]. Research has centered on how to represent and retrieve signatures of known problems from the database of known problems. However, these approaches do not fare well at automatically identifying problems that have not previously been diagnosed.

Yuan et al. [954] learn signatures of known problems in standalone systems by analyzing sequences of system calls. They target problems that have the same manifestation, e.g., a web page may fail to load due to different underlying root causes such as an invalid IP address or an unplugged network cable. Analyzing system calls allows them to distinguish between problems that might be indistinguishable when analyzing resource usage data. They use multi-class Support Vector Machines to learn signatures of problems. However, their approach does not address distributed systems.

Cohen et al. [231] and Bodik et al. [123] generate signatures of recurrent problems in distributed systems by using the discrete feature vectors obtained through metric attribution. They found that using discrete values to represent signatures performs better than using real-valued metrics. In addition, they found that they can leverage signatures learned at one geographical location to diagnose problems in data centers at a different location.

Duan et al. [310] present an approach that can be used for both known problems, and problems that have not previously been seen. They use a supervised approach (decision trees or signature databases) to identify recurrent problems. If the current failure does not match the annotated failures in the database, they compare it to the healthy data to identify features that are correlated with the failure. They then select multiple instances of the same failure which they can present to the system administrator to annotate.

12.2.4.1 Limitations

Machine-learning techniques automatically learn profiles of system behavior, for example, using clustering to identify signatures of known problems. Machine-learning can also help localize problems by identifying resource-usage metrics or components that are highly correlated with failed states. However, these techniques can suffer from the curse of dimensionality that reduces accuracy when the number of features is large. Additionally, they are also susceptible to *overfitting*, a phenomenon in which the learner learns features of the evidence that are circumstantial rather than those that actually define the relationship between the faults and their effects. Over-fitted models generalize poorly, and can fail when presented with evidence that is only slightly different from the one on which the model was trained. Finally, because machine learning techniques learn a direct mapping between the symptoms and underlying root causes without an intermediate structural model of the system, lengthy retraining is required whenever the system behavior changes significantly. Furthermore, previously learned models often have to be thrown away during the period of retraining, leaving the system vulnerable to any problems. Therefore, machine learning techniques may not be appropriate for systems that are upgraded frequently.

12.2.5 Count-and-Threshold Techniques

Physical faults are distinguished by their nature and duration of impact as being permanent or temporary [63]. Permanent faults may lead to error whenever the component is activated; the only way to handle such faults is to remove the affected component. Temporary faults can be internal (usually known as intermittent) or external (transient). The former are caused by some internal part deviating from its specified behavior. After their first appearance, they usually exhibit a relatively high occurrence rate and, eventually, tend to become permanent. On the other hand, transient faults, often manifesting the encountered interferences as noise-pulses on the communication channels, cannot be easily traced to a defect in a particular part of the system and, normally, their adverse effects tend to disappear. In industries like transportation and telecommunications, where operating with permanently faulty modules would carry high risks or costs, it is common that modules, disconnected because they were considered faulty, are later proved to be free from permanent faults when tested during repair operations. Therefore, treating transient faults as permanent has a high cost for these industries. A good discrimination between transient and intermittent/permanent faults solves two important problems: (i) prevents the undue removal of nodes affected by transient faults, thus avoiding unnecessary depletion of system resources; and (ii) helps to maintain the correct coverage of the system fault hypotheses (i.e., the assumption on the number of faults tolerated by the core system protocols within a given time window) by keeping in operation nodes not permanently faulty. Considering that most perturbations encountered are transient [234, 590], the issue of *proper* diagnosis of transients is a significant issue of interest.

Illustrative example A generic class of online low-overhead count-and-threshold mechanisms, called alpha-count, has been initially proposed in [131] and later enriched with a double threshold in [132]. It is characterized by: a) tunability through internal parameters, to warrant wide adaptability to a variety of system requirements; b) generality with respect to the system in which they are intended to operate, to ensure wide applicability; c) simplicity of operation to allow high analyzability through analytical models and to be implementable as small, low-overhead and low-cost modules, suitable especially for embedded, real-time, dependable systems. In its basic formulation, an error counter is associated to each component, which is incremented when the component fails and decremented when it delivers a correct service. When the value of the counter exceeds a given threshold value , the component is diagnosed as affected by a permanent or an intermittent fault.

Heuristic mechanisms The importance of distinguishing transient faults, so that they can be dealt with specifically, is testified by the wide range of solutions proposed, e.g., [25, 477, 590, 661, 841], although with reference to specific systems. Most of these solutions are based on more or less simple heuristic mechanisms. One commonly used method, for example, in several IBM mainframes [841], is to count the number of error events: too many events in a given time frame would signal that the component needs to be removed. In TMR MODIAC, the architecture proposed in [661], two failures experienced in two consecutive operating cycles by the same hardware component that is part of a redundant structure make the other redundant components consider it as definitively faulty. Another architecture using similar mechanisms, designed for distributed ultra-dependable control systems, is described in [567]. In this case, a combination of diversified design, temporal redundancy and comparison schema is used to obtain a detailed determination of the nature of faults. Counting mechanisms are also used to solve the so called 2-2 splits, i.e., to determine the correct value among four proposals in a quadruple modular redundancy (QMR) system when there is a tie. In [25], a list of *suspect* processors is generated during the redundant executions; a few schemes are then suggested for processing this list including assigning weights to processors that participate in the execution of a job and fail to produce a matching result and taking down for diagnostics those whose weight exceeds a certain threshold. Other approaches do, instead, concentrate on off-line analysis of system error logs, and therefore are not applicable on-line. In [590], some heuristics, collectively named Dispersion Frame Technique, for fault diagnosis and failure prediction are developed and applied to system error logs taken from a large Unix-based file system. The heuristics are based on the inter-arrival patterns of the failures (which may be time-varying). For example, there is the 2-in-1 rule, which warns when the time of inter-arrival of two failures is less than one hour, and the 4-in-1 rule, which fires when four failures occur within a 24-hour period. In [477], an error rate is used to build up error groups and simple probabilistic techniques are then recursively applied to discern similarities (correlations) which may point to common causes (permanent faults) of a possibly large set of errors.

Other count-and-threshold solutions In [130], a methodology and an architectural framework for handling multiple classes of faults (namely, hardware-induced software errors in the application, process and/or host crashes or hangs, and errors in the persistent system stable storage) in a COTS and legacy-based application have been defined. Also, a generic FDIR (Fault Detection followed by Isolation and system Reconfiguration) framework for integrating existing distributed diagnosis approaches with a count-and-threshold algorithm is proposed in [819].

Formulation based on Bayesian inference Another direction of research has addressed a rigorous mathematical formulation of diagnosis based on Bayesian inference [731]. Bayesian inference provides a standard procedure for an observer who needs to update the probability of a conjecture on the basis of new observations. Therefore, after a new observation is provided by the error detection subsystem, the on-line diagnosis procedure produces an updated probability of the module being affected by a permanent fault. This leads to an optimal diagnosis algorithm, in the sense that fault treatment decisions based on its results would yield the best utility among all alternative decision algorithms using the same information. This higher accuracy with respect to simple heuristics comes at the cost of higher computational complexity.

Formulation based on Hidden Markov Models A formalization of the diagnosis process, addressing the whole chain constituted by the monitored component, the deviation detector and the state diagnosis through Hidden Markov Models has been proposed in [255], with the goal of developing high accuracy diagnosis processes based on probabilistic information rather than on merely intuitive criteria-driven heuristics. Because of its high generality and accuracy, the proposed approach could be usefully employed: (i) to evaluate the accuracy of low-cost on-line processes to be adopted as appropriate and effective diagnostic means in real system applications; (ii) for those diagnostic mechanisms equipped with internal tunable parameters, to assist the choice of the most appropriate parameter setting to enhance effectiveness of diagnosis; and (iii) to allow direct comparison of alternative solutions.

12.2.5.1 Limitations

The accuracy of diagnosis performed through threshold-based mechanisms strongly depends on proper calibration of the mechanism parameters, namely the threshold's value and the function adopted to update the counter. Actually, proper setting of the mechanism's parameters is fundamental to trade between accuracy and promptness, which are the typical contrasting requirements to be satisfied in fault discrimination, that is:

- To signal, as quickly as possible, all components affected by permanent or intermittent faults. Gathering information to discriminate between transient and intermittent faults takes time, thus giving rise to a longer fault latency. This increases the chances of catastrophic failure and also increases the requirements on the error processing subsystem in fault tolerant systems.
- To avoid signaling components that are not affected by permanent or intermittent faults. In fact, depriving the system of resources that can still do valuable work may be even worse than relying on a faulty component.

Practitioners have long used expertise and trial-and-error approach to tune their systems. However, solutions based on rigorous mathematical formulations, such as alpha-count and its variants, are amenable to high analyzability of the parameters tuning through analytical models. Therefore, the system designer is equipped with a systematic, predictable, and repeatable way to identify a proper setting, taking into account requirements of the targeted application field.

12.2.6 Visualization

Automated diagnosis tools might not always be available, and when available they occasionally miss the true root-cause and typically reduce the search space to a small number of likely culprits [600]. Visualization tools allows operators to cope with these scenarios by: (i) summarizing data trends, (ii) supporting interactive graphs that allow operators to explore different hypotheses on the root-cause of problems, and (iii) integrating output from automated diagnosis tools.

Visualization tools [364, 827, 843] provide an array of simple graphs, line plots, barcharts, and histograms, to display trends in performance counters such as CPU utilization. They use simple statistical tests such as the deviation from the mean to flag outliers, and use color to highlight these outliers. LiveRAC [635] is a visualization system that supports the analysis of large collections of system management timeseries data consisting of hundreds of parameters across thousands of network devices. LiveRAC provides high information density using a reorderable matrix of charts, with semantic zooming that dynamically adapts different aspects of each chart based on available space.

Magpie [89], X-trace [345], and Dapper [827] are primarily tools for tracing causal request paths, but they also offer support for visualizing requests whose causal structure or duration is anomalous. Artemis [244] provides a pluggable framework for distributed log collection, data analysis, and visualization. Mochi [867] is a log-analysis based debugging tool that visualizes both the flow of data and the flow of control for a large-scale parallel processing framework known as Hadoop. NetClinic [600] visualizes data from computer networks using directed graphs, and presents suggested diagnostics for observed problems by incorporating output from an automated analytic reasoning engine [509].

12.3 Industrial Applications

In this section we summarize the use of the previously described diagnosis techniques in several industrial applications ranging from large scale telecommunications infrastructures, to Internet services, to embedded systems and the automotive industry, to aerospace and unmanned spacecraft.

12.3.1 Telecommunications

The telecommunications industry has long operated some of the largest scale distributed systems in use - from digitally switched phone networks and Internet backbones, to high-speed cellular networks and Internet Protocol Television (IPTV) deployments. The high resilience requirements of these systems have led to widespread deployment of diagnosis techniques by telecom operators. [850] provides a survey of diagnosis techniques for communication systems.

Work in the telecom domain has traditionally revolved around alarms produced by network elements, and trouble ticket systems [587] that track and coordinate troubleshooting. The use of rule-based expert systems for troubleshooting was common - as early as 1990, Wright et al. [945] survey a list of 40 rule-based expert system in use within the telecom industry. More recently, codebook-based approaches [326, 952] have been used to correlate alarms across many network devices to a single "root cause" alarm that the operators can investigate. SCORE [539] uses a model of how IP links are routed over an underlying optical network to localize optical layer failures (e.g., fiber-cuts) based on IP layer loss measurements. rcc [333] uses static analysis to detect faults in BGP router configurations by checking them against a high-level correctness specification.

However, such knowledge-based techniques often fail to capture emergent behaviors that are rife in highly heterogenous telecom networks. Therefore, increasing attention is being devoted to "knowledge-free" techniques such as statistical methods and machine learning. Giza [620] uses spatial (i.e., in the same geographical neighborhood) and temporal correlations between network alarms in a large IPTV network to determine the true root cause of network outages that result in many alarms across different layers (e.g., video, TCP, IP). Draco [524] performs statistical comparisons between successful and dropped calls in a large voice-over-IP (VoIP) service to identify features that discriminate the failures. Mahimkar et al. [621] perform statistical comparisons of various performance metrics such as CPU utilization and loss of network elements before and after upgrades to identify problems that result from upgrades.

12.3.2 Internet Services and Data Centers

Diagnosis in data center applications has centered on interactive applications in Internet Services [192, 231, 533, 800] and enterprise systems [77, 509, 954], and batch applications in data-intensive computing [715, 866, 948]. Interactive applications typically have well-established Service Level Objectives, e.g., 99% of Internet requests should be serviced within 4 s, to ensure high-availability. Some techniques use metric attribution [123, 192, 231, 800] localize problems by identifying resource-usage metrics or components that are highly correlated with failed states. Signature-based techniques [123, 231, 310, 954] have been used to diagnose recurrent problems by generating signatures of known problems using techniques such as metric attribution. Regression and queuing models [198, 528, 852] detect performance problems in Internet services by modeling the relationships between performance counters, e.g., CPU utilization and application response times, and detecting performance anomalies whenever these relationships are violated.

Batch applications in data-intensive computing have more diverse runtimes [523]. Peer-comparison techniques [656, 715, 866] diagnose problems by exploiting the parallelism inherent in these applications to compare behavior across components and detect "odd-man-out" behavior. The distributed nature of data center applications facilitates the use of graphical models to analyze communication patterns across nodes (or processes) to model the probability that errors [532], or successes [780] propagate through the system. Log analysis [701, 702, 948] and rule-based techniques [113, 416, 433] are also widely used in data center applications.

12.3.3 Embedded Systems

Embedded systems are computer systems designed to do one or a few dedicated functions, often with real-time computing constraints. Embedded systems are present in a large variety of systems such as consumer electronics (e.g., mobile phones), and automotive safety-critical systems (e.g., anti-lock braking, and drive-by-wire systems). Lanigan et al. [574] provides a comprehensive survey of failure diagnosis in automative systems.

Preparata et al. [744] proposed the Preparata, Metze, and Chien (PMC) model to identify faulty components by collating results of diagnostic tests across a distributed system. Heuristic mechanisms based on thresholds have been also adopted, such as [661] in railway control systems. Serafini et al. [819] distinguish between healthy nodes from unhealthy nodes in time-triggered automotive systems by applying penalties and rewards to the collated diagnostic tests. The penalty counter is increased when a node's entry in the consistent health vector indicates a fault, otherwise the reward counter is increased according to the criticality of the node. When the reward threshold for a node is crossed, the penalty counter for that node is reset to zero. When the penalty threshold for a node is crossed, the node is diagnosed as faulty. Peti et al. [723] introduce Out-of-norm Assertions (ONAs) as a way to correlate fault effects in the three dimensions of value, time and space. They use ONAs to describe fault patterns that discriminate between different types of faults, i.e., wearouts, massive transient faults, and connector faults in automotive systems. Other diagnosis techniques for embedded systems rely on physical models to diagnose problems in powertrain [633] and chassis systems such as braking [455] in cars.

12.3.4 Aerospace

Stroupe et al. [858] and Patton [719] provide a detailed survey on diagnosis in aerospace systems. Livingstone is a model-based system, developed at NASA Ames, used to autonomously control the New Millennium Deep Space One Probe (DS 1) [858]. Livingstone accepts a model of the components of a complex system such as a spacecraft or chemical plant and infers from it the overall behavior of the system. From this, Livingstone monitors the operation of the system, diagnoses its current state, determines if sensor readings are implausible, and recommends actions to put the system into a desired state even in the face of failures. MARPLE is an expert system that relies on a model-based technique known as constraint suspension to diagnose problems [858]. Constraint suspension views the system to be monitored as a network of black-box components and places constraints on the behavior of each component. When observed behavior violates these constraints, MARPLE suspends the components in the network, one at a time until it finds a component that can account for all the inconsistent values at the nodes. MARPLE has been demonstrated to work for the NASA LRC Space Station Freedom (SSF) power system testbed.

Kalman filtering [158] is a state and parameter estimation technique that fuses data from different sensors together to produce an accurate estimate of the true system state. Jayakumar and Das [485] use a single Kalman filter, driven by the motor shaft velocity sensor, to diagnose problems in a flight control system. They diagnose incipient sensor faults using structured residuals that are generated using the Kalman filter estimates. Patton [719] discusses the use of filters to diagnose faults in flight control systems. At the moment, the analytic redundancy provided by model-based approaches cannot be used to replace hardware redundancy due to the safety-critical nature of aerospace applications. However, analytical redundancy can be used to suppress some levels of replication, e.g., to replace quadruple by triplex schemes [719].

12.4 Future Trends and Challenges

Despite the tremendous progress that has been made in automated fault diagnosis, many open problems remain. Below, we enumerate a few such problems that may serve to inspire new contributions in the field.

12.4.1 Online Recovery and Self Healing

The eventual outcome of any automated diagnosis technique is the identification and removal of any impairments to a system's proper operation. Therefore, a natural evolution of diagnosis is the construction of "self-healing" systems that can automatically perform recovery actions upon the outcome of an online diagnosis procedure to remove faults. Self-healing is relatively risk-free either when the fault detection and diagnosis mechanisms are highly accurate, or when the recovery actions do not impose any penalties if applied wrongly. For example, JAGR [171] presents an autonomous self-recovering Enterprise Java Bean (EJB) application server that allows recovery using quick microreboots of components. The basic philosophy in that work is to make recovery mechanisms cheap enough that they can be liberally applied without consequences even if diagnosis produces poor outcomes. When recovery actions are not cheap, self-healing becomes a risky proposition because wrong diagnosis can lead to poor recovery decisions.

[496] propose a decision theoretic framework using Partially Observable Markov Decision Processes (POMDPs) to reason about recovery decisions of different costs under uncertain fault diagnoses. They combine the decision algorithm with a graph-theoretic diagnosis algorithm to determine when components of a multitier Enterprise system should be rebooted using the results of end-to-end system tests. [595] propose a model-free approach for choosing recovery actions by using reinforcement learning to learn the effectiveness of previously executed actions as a function of the observable symptoms. However, none of these techniques are sufficient when faced with unanticipated problems due to emergent behavior.

12.4.2 Automatic Model Construction

Although model-based techniques have several advantages such as the ability to predict error propagation, the ability to provide semantically meaningful diagnoses, and the ability to cope with structural system changes without the need to relearn, they require detailed and accurate models that have to be constantly updated. There is some literature on automatically constructing system models, primarily those suitable for graph-theoretic approaches, but also some on learning queuing-theoretic models. Examples include work on automatic determination of component dependencies by system perturbation (e.g., [153]), work on dependency generation via passive observation (e.g., [24, 26, 776]), approaches based on statistical clustering (e.g., [193]), and approaches to learn the parameters of queuing models using statistical regression [852]. However, all of these techniques are only suitable for learning models of a system during normal operations. Learning the dependencies of a system that may be exercised during fault modes is an open problem whose solution is likely to require a combination of static analysis (to discover all dependencies) along with runtime measurement (to identify those dependencies which are explained by normal behaviors).

12.4.3 Cross Domain and Cross Layer Diagnosis

In many domains such as Internet services and telecommunications, large systems are increasingly built as a composition of multiple horizontal "technology layers" and vertical "administrative domains". For example, consider a typical Internet application constructed using the Java runtime and its libraries, hosted in a Tomcat application server running on a Linux OS inside a virtual machine that runs on a Windows host running in a rack in a particular data center of a cloud provider. For communication with a backend database, it uses the Simple Object Access Protocol (SOAP) that runs over HTTPS (secure HTTP) that runs over an IP virtual private network that is provisioned over an Ethernet service provided by an Internet Service Provider (ISP) that provisions it as a tunnel over an MPLS (multi-protocol label-switching) backbone network. In addition, this application uses the Bing mapping service from Microsoft, obtains analytics support from Google Analytics, and uses PayPal as a payment service. Each of these services also run on very similar infrastructure layers, and depending on which cloud provider the application users, some of these services may also share a data-center and/or network provider with the application.

In such a highly layered and highly silo'ed setup, faults can occur in each of the technology layers or third party providers the service uses. Furthermore, symptoms of lower layer problems (e.g., packet loss on the MPLS network) can translate into symptoms in higher layers (slow response from database server). Seemingly independent third party providers may have common dependencies - e.g., they use the same cloud provider, resulting in correlated failures. No single layer or administrative domain may have sufficient information to completely determine the root cause of a fault occurring in the system. These complications make diagnosis a challenging task. Although there has been some preliminary work on combining information across technology layers (e.g., [539, 620]), comprehensive approaches that can take a whole system view when performing diagnosis are still elusive.

12.5 Conclusions

Diagnosis of failures occurring in systems in the field is an important aspect of system resilience and its assessment. In this chapter, we provided a broad overview of automated techniques for fault diagnosis ranging from knowledge-based techniques that encode expert knowledge in the form of rules or system models to model-free techniques that rely on statistical correlations, regression, and machine learning to perform some aspects of the diagnosis task without any prior human knowledge. We provided examples of industrial applications in which automated diagnosis has proven to be a valuable tool for ensuring and evaluating resilience. Today, while these mainly include telecommunications and Internet services that have to deal with issues of scale and automotive and aerospace systems that have to deal with the absence of human expertise when problems occur, automated diagnosis is poised to make a foray into an increasingly number of domains ranging from software debugging tools to agents that help troubleshoot configuration problems in personal computer systems. Finally, we review some of the open problems in this area - these include the need to deal with problems that occur due to emergent, unpredictable behaviors, and the need for recovery techniques to automatically act upon the output of diagnosis algorithms.

Chapter 13
Fault Injection

Raul Barbosa, Johan Karlsson, Henrique Madeira and Marco Vieira

Abstract Resilient systems are designed to operate at acceptable levels even in the presence of faults and other adverse events. Assessing the resilience of a given system therefore requires that the effects of such events can be measured and examined in detail, which in turn requires the ability to introduce faults and observe the subsequent behaviour of the system. Fault injection is therefore a fundamental method for resilience assessment. It allows us to study the effect of faults on a system, and can thus be used to identify weaknesses in fault handling, to assess the effectiveness of error detectors and fault tolerance mechanisms, and to quantify the effect of faults on the quality of service achieved by the system. This chapter provides an introduction to fault injection for resilience assessment. We start with a brief overview of the area of fault injection, including the necessary terminology. We then discuss common techniques for the injection of hardware, software, and security faults, aiming to cover the wide spectrum of techniques proposed in the literature.

R. Barbosa (✉) · H. Madeira · M. Vieira
DEI/CISUC, University of Coimbra,
3030-290 Coimbra, Portugal
e-mail: rbarbosa@dei.uc.pt

J. Karlsson
Department of Computer Science and Engineering,
Chalmers University of Technology,
412 96 Gothenburg, Sweden
e-mail: johan@chalmers.se

H. Madeira
e-mail: henrique@dei.uc.pt

M. Vieira
e-mail: mvieira@dei.uc.pt

K. Wolter et al. (eds.), *Resilience Assessment and Evaluation of Computing Systems*,
DOI: 10.1007/978-3-642-29032-9_13, © Springer-Verlag Berlin Heidelberg 2012

263

13.1 Introduction

A critical issue in the development of a resilient computer system is the validation
of its fault-handling mechanisms. We use the term fault-handling mechanism as an
umbrella concept encompassing all mechanisms that handle faults and errors in a
computer system. This includes mechanisms for achieving fault tolerance, preserving data integrity, ensuring safety, etc. Ineffective or unintended operation of these
mechanisms can significantly impair the resilience of a computer system. Assessing the effectiveness and verifying the correctness of fault-handling mechanisms in
computer systems is therefore of vital importance.

Fault injection is an important experimental technique for assessment and verification of fault-handling mechanisms. It may be defined as the process of deliberately
introducing faults or errors in computer systems, allowing researchers and system
designers to study how computer systems react and behave in the presence of faults.
Fault injection is used in many contexts and can serve different purposes, such as:

- Assess the effectiveness, i.e., fault coverage, of software and hardware imple-
  mented fault-handling mechanisms.
- Study error propagation and error latency in order to guide the design of fault-
  handling mechanisms.
- Test the correctness of fault-handling mechanisms.
- Measure the time it takes for a system to detect or to recover from errors.
- Test the correctness of fault-handling protocols in distributed systems.
- Verify failure mode assumptions for components or subsystems.
- Study the impact of faults and fault-tolerance mechanisms on QoS of the system.

In this chapter our goal is to give the reader an introduction to fault injection.
As fault injection has a long history in research, we structure the chapter around a
thorough survey of the state of the art. Over the years, numerous papers on assessment and verification of fault-tolerant systems or individual mechanisms, and on
fault injection tools have been published. This chapter covers existing literature on
injection of hardware faults, injection of software faults, and for test of protocols for
fault-tolerant distributed systems.

Before we discuss and compare different fault injection techniques in detail, we
must first introduce some terms and basic concepts. We use *target system* as a generic
term for the system under assessment or verification. The target system executes a
workload, which is determined by the program executed by the target system and the
data processed by the program. The faults injected during the experiments constitute
the *faultload*.

Fault injection can in principle be carried out in two ways: faults can be injected
either in *a real system* or in *a model of a system*. By a real system we mean a physical
computer system, either a prototype or a commercial product. System models for fault
injection experiments can be built using two basic techniques: *software simulation*
and *hardware emulation*.

We distinguish between a *fault injection experiment* and a *fault injection campaign*. A fault injection experiment corresponds to injecting one fault and observing,

or recording, how the target system behaves in presence of that fault. To gain statistical confidence in the assessment or the verification of a target system, we need to collect data from many fault injection experiments. A series of fault injection experiments conducted on a target system is called a fault injection campaign.

Fault injection techniques can be compared and characterized on the basis of several different properties. The following properties are applicable to all types of fault injection techniques:

- **Controllability**. Ability to control the injection of faults in time and space.
- **Observability**. Ability to observe and record the effects of an injected fault.
- **Repeatability**. Ability to repeat a fault injection experiment and obtain the same result.
- **Reproducibility**. Ability to reproduce the results of a fault injection campaign.
- **Faultload representativeness**. How accurately the faultload represents real faults.
- **Workload representativeness**. How accurately the workload represents real system usage.
- **System representativeness**. How accurately the target system represents the real system.

The main advantage of performing fault injection in a real system is that the actual implementation of the fault-handling mechanisms is assessed and verified. Thus, system representativeness is usually higher when using a physical system compared to using software simulation or hardware emulation. On the other hand, fault models used in simulation-based and emulation-based fault injection can usually imitate real faults more accurately than artificial faults injected into a real system. Fault representativeness is therefore often higher for simulation-based and emulation-based fault injection. Also, controllability, observability, repeatability, reproducibility and reachability are normally higher in simulation-based and emulation-based fault injection compared to fault injection in real systems.

However, there are also several drawbacks and limitations to simulation-based and emulation based fault injection. The development of the simulation/emulation model can increase development cost. Performing software simulations with an accurate simulation models can be very time-consuming. In fact, simulating a large amount of system activity (e.g., the execution of several million lines of source code) may not be feasible using a highly detailed model of the target system.

In simulation-based fault injection, it is therefore essential to make a trade-off between simulation time, on one hand, and the accuracy of the fault model(s) and the system model, on the other hand. Time overhead is much lower in hardware emulation-based fault injection than in software simulation-based fault injection. However, it may still be a concern in hardware emulation-based fault injection, e.g., in the verification of real-time systems.

Commonly used fault injection techniques have specific advantages and drawbacks. Several researchers have therefore proposed *hybrid fault injection* approaches, in which different techniques are combined in order to increase the scope and confidence in the verification or the assessment of a target system.

13.2 Techniques for Injecting Hardware Faults

In this section, we describe techniques for injecting or emulating hardware faults. The first three subsections deal with *fault injection into real systems*, covering *hardware-implemented fault injection*, *software-implemented fault injection*, and *radiation-based fault injection*. Three additional subsections cover *software simulation-based fault injection*, *hardware emulation-based fault injection* and *hybrid fault injection*. Thus, hardware faults can be injected or emulated by all the techniques mentioned in the introduction of this chapter.

13.2.1 Hardware Implemented Fault Injection

Hardware-implemented fault injection includes three techniques: *pin-level fault injection*, *power supply disturbances*, and *test port-based fault injection*.

In pin-level fault injection, faults are injected via probes connected to electrical contacts of integrated circuits or discrete hardware components. This method was used already in the 1950s for generating fault dictionaries for system diagnosis. Many experiments and studies using pin-level fault injection were carried out during the 1980s and early 1990s. Several pin-level fault injection tools were developed at that time, for example, MESSALINE [53] and RIFLE [617]. A key feature of these tools was that they supported fully automated fault injection campaigns. The increasing level of integration of electronic circuits has rendered the pin-level technique obsolete as a general method for evaluating fault-handling mechanisms in computer systems. The method is, however, still valid for assessment of systems where faults in electrical connectors pose major problem, such as automotive and industrial embedded systems.

Power supply disturbances (PSDs) are rarely used for fault injection because of low repeatability. They have been used mainly as a complement to other fault injection techniques in the assessment of error detection mechanisms for small microprocessors [518, 655, 757]. The impact of PSDs is usually much more severe than the impact of other commonly used injection techniques, e.g., those that inject single bit-flips, since PSDs tend to affect many bits and thereby a larger part of the system state. Interestingly, some error detection mechanisms show lower fault coverage for PSDs than for single bit-flip errors [757].

Test port-based fault injection encompasses techniques that use test ports to inject faults in microprocessors. Many modern microprocessors are equipped with built-in debugging and testing features, which can be accessed through special I/O-ports, known as test access ports (TAPs), or just test ports. Test ports are defined by standards such as the IEEE-ISTO 5001-2003 (Nexus) standard [466] for real-time debugging, the IEEE 1149.1 standard test access port and boundary-scan architecture (JTAG) [462], and the background debug mode (BDM) facility. Nexus and JTAG are standardized solutions used by several semiconductor manufacturers, while BDM is a proprietary solution for debugging developed by Freescale, Inc. Tools for test port-based fault injection are usually implemented on top of an existing commercial microprocessor debug tool, since such tools contain all functions and drivers that are needed to access a test port.

The type of faults that can be injected via a test port depends on the debugging and testing features supported by the target microprocessor. Normally, faults can be injected in all registers in the instruction set architecture (ISA) of the microprocessor. BDM and Nexus also allow injection of faults in main memory. Test ports could also be used to access hardware structures in the microarchitecture that are invisible to the programmer. However, information on how to access such hardware structures is usually not disclosed by manufacturers of microprocessors.

Tools that support test port-based fault injection include GOOFI [28, 833] and INERTE [955]. GOOFI supports both JTAG-based and Nexus-based fault injection, while INERTE is specifically designed for Nexus-based fault injection. An environment for BDM-based fault injection is described in [762].

Injecting a fault via a test port involves four major steps: (i) setting a breakpoint via the test port and waiting for the program to reach the breakpoint, (ii) reading the value of the target location (a register or memory word) via the test port, (iii) manipulating this value and then writing the new, faulty value back to the target location, and (iv) resuming the program execution via a command sent to the test port.

The time overhead for injecting a fault depends on the speed of the test port. JTAG and BDM are low-speed ports, whereas Nexus ports can be of four different classes with different speeds. The simplest Nexus port (Class 1) is a JTAG port, which uses serial communication and therefore only needs 4 pins. Ports compliant with Nexus Class 2, 3 or 4 use separate input and output ports, know as auxiliary ports. These are parallel ports that use several pins for data transfer. The actual number of data pins is not fixed by the Nexus standard, but for Class 3 and 4 ports the standard recommends 4 to 16 data pins for the auxiliary output port and 1 to 4 data pins for the auxiliary input port.

The main advantage of test port-based fault injection is that faults can be injected internally in microprocessors without making any alterations of the system's hardware or software. Compared to software-implemented fault injection, it provides better or equal capabilities of emulating real hardware faults. Finally, advanced Nexus ports (Class 3 and 4) provide outstanding possibilities for data collection and observing the impact of injected faults within a microprocessor. Existing tools have not fully exploited these possibilities. Hence, microprocessors with high-speed Nexus ports constitute interesting targets for the development of new fault injection tools, which potentially can achieve much better observability than existing tools do.

13.2.2 Software-Implemented Fault Injection of Hardware Faults

Software-implemented fault injection encompasses techniques that inject faults through software executed on the target system.

There are basically two approaches that we can use to emulate hardware faults by software: *run-time injection* and *pre run-time injection*. In run-time injection, faults are injected while the target system executes a workload. This requires a mechanism that (i) stops the execution of the workload, (ii) invokes a fault injection routine, and (iii) restarts the workload. Thus, run-time injection incurs a significant run-time overhead. In pre run-time injection, faults are introduced by manipulating either the source code or the binary image of the workload before it is loaded into memory. Pre run-time injection usually incurs less run-time overhead than run-time injection, but the total time for conducting a fault injection campaign is usually longer for pre run-time injection since it needs more time for preparing each fault injection experiment.

There are several fault injection tools that can emulate the effects of hardware faults by software, but they use different techniques for injecting faults and support different fault models. Most of these tools use run-time injection, since it provides better opportunities for emulating hardware faults than pre run-time injection.

Software-implemented fault injection relies on the assumption that the effects of real hardware faults can be emulated either by manipulating the state of the target system via run-time injection, or by modifying the target workload through pre run-time injection.

The validity of this approach varies depending of the fault type and where the fault occurs. Consider for example emulation of a *soft error*, i.e. a bit-flip error induced by a strike of a high energy particle. Flipping bits in main memory or processor registers can easily be done by software. On the other hand, the effect of a bit-flip in a processor's internal control logic can be difficult, if not impossible, to emulate accurately by software manipulations.

Emulating a permanent hardware fault requires a more elaborate set of manipulations than emulating a transient fault. For example, the emulation of a stuck-at fault in a memory word or a processor register would require a sequence of manipulations performed every time the designated word or register is read by a machine instruction. On the other hand, a transient fault requires only a single manipulation. The time overhead imposed by fault emulation thus varies for different fault types.

We here describe eight tools that are capable of emulating hardware faults through software. These tools represent important steps in the development of software-implemented fault injection for emulation of hardware faults. The tools are FIAT [92], FERRARI [506], FINE [516], DEFINE [515], FTAPE [896], DOCTOR [408], Xception [175], MAFALDA [56] and Exhaustif [259]. These tools use different approaches to emulating hardware faults and implement partly different fault models. Some of the tools also provide support for emulating software faults, as we describe later in this chapter.

Researchers started to investigate software-implemented fault injection in the late 1980s. In the beginning, the focus was on developing techniques for emulating the effects of hardware faults. Work on emulation of software faults started a few years later.

One of the first tools that used software to emulate hardware faults was FIAT [92], developed at Carnegie Mellon University. FIAT injected faults by corrupting either the code or the data area of a program's memory image during run-time. Three fault types were supported: *zero-a-byte*, *set-a-byte* and *two-bit compensation*. The last fault type involved complementing any 2 bits in a 32 bit word. Injection of single-bit errors was not considered, because the memory of the target system was protected by parity.

More advanced techniques for emulation of hardware faults were included in FERRARI [506], developed at the University of Texas, and in FINE [516], developed at the University of Illinois. Both these tools supported emulation of transient and permanent hardware faults in systems based on SPARC processors from Sun Microsystems. FERRARI could emulate three types of faults: *address line*, *data line*, and *condition code* faults, while FINE emulated faults in *main memory*, *CPU-registers* and the *memory bus*. DEFINE [515], which was an extension of FINE, supported fault injection in distributed systems and introduced two new fault models for intermittent faults and communication faults.

DOCTOR [408] is a fault injection tool developed at the University of Michigan targeting distributed real-time systems. It supports three fault types: *memory faults*, *CPU faults* and *communication faults*. The memory faults can affect a single-bit, two bits, one byte, and multiple bytes. The target bit(s)/byte(s) can be set, reset and toggled. The CPU faults emulate faults in processor registers, the op-code decoding unit, and the arithmetic logic unit. The communication faults can cause messages to be lost, altered, duplicated or delayed. DOCTOR can inject transient, intermittent and permanent faults, and uses run-time injection for the transient and intermittent faults. Permanent faults are emulated using pre run-time injection.

FTAPE [896] is a fault injector aimed at benchmarking of fault tolerant commercial systems. It was used to assess and test several prototypes of fault tolerant computers for online transaction processing. FTAPE emulates the effects of hardware faults in the CPU, main memory and I/O units. The CPU faults include single and multiple bit-flips and zero/set registers in CPU registers. The memory faults include single and multiple bit and zero/set faults in main memory. The I/O faults include SCSI and disk faults. FTAPE was developed at the University of Illinois in cooperation with Tandem Computers.

Xception [175] is a fault injection tool developed at the University of Coimbra. This tool uses the debugging and performance monitoring features available in advanced microprocessors to inject faults. Thus, it injects faults in a way which is similar to test portbased fault injection. The difference is that Xception controls the setting of breakpoints and performs the fault injections via software executed on the target processor rather than sending commands to a test port.

Xception injects faults through exception handlers executing in kernel mode, which can be triggered by the following events: *op-code fetch* from a specified address, *operand load* from a specified address, *operand store* to a specified address, and a *specified time* passed since start-up. These triggers can be used to inject both permanent and transient faults. Xception can emulate hardware faults in various functional units of the processors such as the integer unit, floating point unit and the address bus. It can also emulate memory faults, including *stuck-at-zero*, *stuck-at-one* and *bit-flip* faults. Xception is unique because it is one of very few academic tools that has been commercialised. Xception is sold by Critical Software, Coimbra, which released the first commercial version of the tool in 1999.

MAFALDA is a tool for assessment of commercial off-the-shelf microkernels. It uses software implemented fault injection to inject single or multiple bit-flips in the code and data segments of the microkernel under assessment. In addition, MAFALDA also allows corruption of input parameters during invocation of kernel system calls, and thus supports robustness testing of microkernels.

A more recent commercial tool, similar in functionality to Xception, is called Exhaustif [259]. It instruments the workload with a software component that injects faults at run-time. This component is configured through a communication interface (e.g., serial port, Ethernet) using a graphical user interface. It supports several fault models based on corruption of processor registers and memory, and interception of function calls. The software component that injects faults on the target system requires several kilobytes for code and data, which may be significant in terms of intrusiveness.

The differences between the numerous fault injection techniques raise the issue of metrological compatibility, as uncertainties associated with measurement procedures, instruments and target systems create difficulties in comparing and reproducing results of dependability measurements. An investigation of sources of uncertainty in fault injection [832] examined whether the results obtained by three different fault injection techniques (two software-implemented and one test port-based technique) were metrologically compatible. The three injection techniques (supported by the GOOFI-2 tool [833]) were used to inject a single set of faults defined in a shared database. Focusing on the values produced by the target system (i.e., abstracting the temporal aspect of the outputs), it was observed that the outcome of many individual experiments is different, although the three injection techniques produce similar average results over a large number of experiments. However, if we also take into consideration the temporal aspect of the target system's output, the three techniques may produce very different measurements, due to significant differences in their temporal intrusiveness.

13.2.3 Radiation-Based Fault Injection

Modern electronic integrated circuits and systems are sensitive to various forms of external disturbances such electromagnetic interference and particle radiation. One way of validating a fault tolerant system is thus to expose the system to such disturbances.

A growing reliability concern for computer systems is the increasing susceptibility of integrated circuits to soft errors, i.e., bit-flips caused when highly ionizing particles hits sensitive regions within a circuit. Soft errors have been a concern for electronics used in space applications since the 1970s. In space, soft errors are caused by cosmic rays, i.e., highly energetic heavy-ion particles. Heavy-ions are not a direct threat to electronics at ground-level and airplane flight altitudes, because they are absorbed when they interact with Earth's atmosphere. However, recent circuit technology generations have become increasingly sensitive to high energy neutrons, which are generated in the upper atmosphere when cosmic rays interact with the atmospheric gases. Such high energy neutrons are a major source of soft errors in ground-based and aviation applications using modern integrated circuits. All modern microprocessors manufactured in technologies with feature sizes below 90nm are therefore equipped with fault tolerance mechanisms to cope with soft errors.

Although computer systems often are used in environments where they can be subjected to electromagnetic interference (EMI), it is not common to use such disturbances to validate fault tolerance mechanisms. The main reason for this is that EMI injections are difficult to control and repeat. In [55], EMI was used along with three other fault injection techniques to evaluate error detection mechanisms in a computer node in a distributed real time system. A primary goal of this study was to compare the impact of pin-level fault injection, EMI, heavy-ion radiation and software-implemented fault injection. The study showed that the EMI injections tended to "favour" one particular error detection mechanism. For some of the fault injection campaigns almost all faults were detected by one specific CPU-implemented error detection mechanism, namely spurious interrupt detection. This illustrates the difficulty in using EMI as a fault injection method. Another attempt to use EMI for fault injection is reported in [912].

To assess the efficiency of such fault tolerance mechanisms, semiconductor manufacturers are now regularly testing their circuits by exposing them to ionising particles. The neutron beam facility at the Los Alamos Neutron Science Center (LANSCE) in the United States is often used for such tests as its energy spectrum is very similar to that of natural neutrons at sealevel. Similar neutron beam facilities are Osaka University's RCNP and the ANITA Neutron Source at The Svedberg Laboratory in Uppsala, Sweden. Results of neutron beam testing are reported for Intel's Itanium microprocessor in [235], for the SPARC64 V microprocessor in [40], and for several generations of microprocessors from Sun Microsystems in [299]. Sometimes proton radiation is used as a slightly less expensive alternative to neutron beam testing. Results of proton beam testing of the IBM POWER 6 processor can be found in [527].

The sensitivity of integrated circuits to heavy-ion radiation can be exploited for assessing the efficiency of fault-handling mechanisms. In [402] and [518], results from fault injection experiments conducted by exposing circuits to heavy-ion radiation from a Californium-252 source is reported. This method was also used in the previously mentioned study [55], in which the impact of four different fault injection techniques was compared. In this study, the main processor as well as the communication processor of a node in distributed system was exposed to heavy-ion radiation. The results showed that the impact of the soft errors injected by the heavy-ions varied extensively and that they activated many different error detection mechanisms in the target system.

Finally, we note that radiation-based fault injection has very low, or non-existent, repeatability. Due to low controllability, it is not possible to precisely synchronize the activity of the target system with the time and the location of an injection in radiation-based fault injection. Thus, it is not possible to repeat an individual experiment. However, the ability to statistically reproduce results over many fault injection campaigns is usually high in particle radiation experiments. Both repeatability and reproducibility are low for EMI-based fault injection.

13.2.4 Simulation-Based Fault Injection

As mentioned in the introduction, simulation-based fault injection can be performed at different levels of abstraction, such as the device level, logical level, functional block level, instruction set architecture (ISA) level, and system level. Simulation models at different abstractions layers are often combined in so called mix-mode simulations to overcome limitations imposed by the time overhead incurred by detailed simulations.

FOCUS [209] is an example of a simulation environment that combines device-level and gate-level simulation for fault sensitivity analysis of circuit designs with respect to soft errors. At the logic level and the functional block level, circuits are usually described in a hardware description language (HDL) such as Verilog or very high speed integrated circuit hardware description language (VHDL). Several tools have been developed that support automated fault injection experiments with HDL models, e.g., MEFISTO [486] and the tool described in [283]. There are several different methods for implementing the fault injection process, such as modifying the HDL code [60], modifying the HDL simulator, commanding the simulator through scripts or, in a more recent example [258], using the force and release constructs in Verilog to emulate stuck-at faults.

Recently, several studies aimed at assessing the soft error vulnerability of complex high-performance processors have been conducted using simulation-based fault injection. In [928] a novel low-cost approach for tolerating soft errors in the execution core of a high-performance processor is evaluated by combining simulations in a detailed Verilog model with an ISA-level simulator. This approach allowed the authors to study the impact of soft errors for seven SPEC2000 integer benchmarks through simulation.

DEPEND [387] is a tool for simulation-based fault injection at the functional level aimed at evaluating architectures of fault-tolerant computers. A simulation model in DEPEND consists of number of interconnected modules, or components, such as CPUs, communication channels, disks, software systems, and memory. DEPEND is intended for validating system architectures in early design phases and serves a complement to probabilistic modelling techniques such as Markov and Petri net models. DEPEND provides the user with predefined components and fault models, but also allows the user to create new components and new fault models, e.g., the user can use any probability distribution for the time to failure for a component.

13.2.5 Hardware Emulation-Based Fault Injection

The advent of large field programmable gate arrays (FPGAs) circuits has provided new opportunities for conducting model-based fault injection with hardware circuits. Circuits designed in a HDL are usually tested and verified using software simulation. Even if a powerful computer is used in such simulations, it may take considerable time to verify and test a complex circuit adequately. To speed up the test and verification process, techniques have been developed where HDL-designs are tested by hardware emulation in a large FPGA circuit. This technique also provides excellent opportunities for conducting fault injection experiments. Hardware emulation-based fault injection has all the advantages of simulation based fault injection such as high controllability and high repeatability, but requires less time for conducting a fault injection experiment compared to using software simulation.

The use of hardware emulation for studying the impact of faults was first proposed in [197]. The authors of that paper used the method for fault simulation, i.e., for assessing the fault coverage of test patterns used in production testing.

Fault injection can be performed in hardware emulation models through compile time reconfiguration and run-time reconfiguration. Here reconfiguration refers to the process of adding hardware structures to the model which are necessary to perform the experiments. In compile-time reconfiguration, these hardware structures are added by instrumentation of the HDL models. An approach for compile-time instrumentation for injection of single event upsets (soft errors) is described in [223]. This work presents different instrumentation techniques that allow injection of transient faults in sequential memory element as well as in microprocessor-based systems.

One disadvantage of compile-time reconfiguration is that the circuit must be re-synthesised for each reconfiguration, which can impose a severe overhead on the time it takes to conduct a fault injection campaign. In order to avoid re-synthesizing the target circuit, a technique for run-time reconfiguration is proposed in [46]. This technique relies on directly modifying the bit-stream that is used to program the FPGA-circuit. By exploiting partial reconfiguration capabilities available in some FPGA circuits, this technique achieved substantial time-savings compared to other emulation-based approaches to fault injection.

A tool for conducting hardware emulation-based fault injection called FADES is presented in [269, 270]. This tool uses run-time configuration and can inject several different types of transient faults, including bit-flips, pulse, and delay faults, as wells as faults that cause digital signals to assume voltage levels between "1" and "0".

Although FPGA-based techniques overcome the performance issues present in software simulation, it is often difficult to obtain ideal observability due to the communication required for observing the behaviour of emulated circuits. A recently proposed method [321] reduces this overhead by having a single combinational circuit for both the faulty circuit and the fault-free circuit. The complete circuit repeatedly executes one clock cycle with the fault-free flip-flops followed by one clock cycle with the faulty flip-flops, and the output is multiplexed to a comparator, in order to identify which faults cause errors on the target. This method provides good observability under the bit-flip fault model and avoids duplicating the entire combinational circuit, which is assumed to be unaffected by faults.

There is a concern with representativeness when using hardware emulated circuits, rather than the actual hardware. In [758], the results of fault injection on a hardware-emulated IBM POWER6 processor (the authors call it "hardware accelerated simulation") are compared to radiation-based fault injection. The results show a close match between the two techniques, therefore providing evidence in favour of hardware emulation-based fault injection.

13.2.6 Hybrid Approaches for Injecting Hardware Faults

Hybrid approaches to fault injection combine several fault injection techniques to improve the accuracy and scope of the verification, or the assessment, of a target system.

An approach for combining software-implemented emulation of hardware faults and simulation-based fault injection is presented in [403]. In this approach, the physical target is run until the program execution hits a fault injection trigger, which causes the physical system to halt. The architected state of the physical system is then transferred to the simulation model, in which a fault is injected, e.g., in the non-visible parts of the microarchitecture. The simulator is run until the effects of the fault have stabilized in the architected state of the simulated processor. This state is then transferred back to the physical system, which subsequently is restarted so that the system-level effects of the fault can be determined.

An extension of the FERRARI tool which allows it to control a hardware fault injector is described in [507]. The hardware fault injector can inject logic-0/logic-1 faults into the memory bus lines of a SPARC 1 based workstation. The authors used the hardware fault injector to study the sensitivity of the computer in different operational modes. The results showed that system was more likely to crash from bus faults when the processor operated in kernel mode, compared to when it operated in user mode. This study showed that it is feasible to extend a tool for software-implemented fault injection with other techniques at reasonable cost, since many of the central functions of a tool are independent of the injection technique.

A more recent tool that supports the use of different fault injections techniques is NFTAPE [856], developed at the University of Illinois. This tool is aimed at injecting faults in distributed systems using a technique called LightWeight Fault Injectors. The purpose of this technique is to separate the implementation of the fault injector from the rest of the tool. NFTAPE provides a standardized interface which simplifies the integration and use of different types of fault injectors. NFTAPE has been used with several types of fault injectors using hardware-implemented, software-implemented, and simulation-based fault injection.

13.3 Techniques for Injecting or Emulating Software Faults

Software faults are currently the dominating source of computer system failures. Making computer systems resilient to software faults is therefore highly desirable in many application domains. Much effort has been invested by both academia and industry in the development of techniques that can tolerate and handle software faults. In this context, fault injection plays an important role in assessing the efficiency of these techniques. Hence, several attempts have been made to develop fault injection techniques that can accurately imitate the impact of real software faults.

The current state-of-the-art techniques in this area rely exclusively on software-implemented fault injection. There are two fundamental approaches to injecting software faults into a computer system: fault injection and error injection [315]. Fault injection imitates mistakes of programmers by changing the code executed by the target system, while error injection attempts to emulate the consequences of software faults by manipulating the state of the target system.

Regardless of the injection technique, the main challenge is to find fault sets or error sets that are representative of real software faults. Other important challenges include the development of methods that allow software faults to be injected without access to the source code, and techniques for reducing the time it takes to perform an injection campaign. First, we discuss emulation of software faults by error injection, and then software fault injection.

13.3.1 Emulating Software Faults by Error Injection

There are two common techniques for emulating software faults by error injection: program state manipulation and parameter corruption. Program state manipulation involves changing variables, pointers and other data stored in main memory or CPU-registers. Parameter corruption corresponds to modifying parameters of functions, procedures and system calls. The latter is also known as API parameter corruption and falls under category of robustness testing. Here we discuss techniques for emulating software faults by program state manipulation.

Many of the tools that we described in conjunction with emulation of hardware faults through software-implemented fault injection, e.g., FIAT [92], FERRARI [506], FTAPE [896], DOCTOR [408], Xception [175], and MAFALDA [56] can potentially be used to emulate software faults since they are designed to manipulate the system state. However, none of these tools provide explicit support for defining errors that can emulate software faults and the representativeness of the injected faults is therefore questionable.

An approach for generating representative error sets that emulates real software faults is presented in [214]. This approach was based on a study of software faults encountered in one release of a large IBM operating system product. Based on their knowledge of the observed faults, the authors developed a procedure for generating a representative error set for error injection. The study addressed four important questions related to emulation of software fault by error injection: what error model(s) should be used; where should errors be injected; when should errors be injected; and how a representative operational profile (workload) should be designed. This work shows the feasibility of generating representative error sets when data on software faults is available.

An experimental comparison between fault and error injection is presented in [214]. Fault and error injection experiments were carried on a safety-critical real-time control application. A total of 200 assignment, checking and interface faults were injected by mutating the source code, which was written in C. The failure symptoms produced by these faults were compared with failure symptoms produced by bit flip errors injected in processor registers, and in the data and stack areas of the main memory. A total 675 errors were injected. A comparison of the failure distributions were made for eight different workload activations (test cases).

The authors conclude that the choice of test case caused greater variations in the distribution of the failure symptoms than the choice of fault type, when fault injection was used. On the other hand, for error injection the choice of error type caused greater variations in the failure distribution than the choice of test case.

There were also significant differences between the failure distributions obtained with fault injection and with error injection. The authors claim that these differences occurred because a time-based trigger was used to control the error injections. They also claim that the fault types considered could be emulated more or less perfectly by using a break-point based trigger, although no experimental evidence is presented to support this claim. This study points out that it may be difficult to find error sets that emulate software faults accurately, and that the selection of the test case (workload activation) is as important as the selection of the fault/error model for the outcome of an injection campaign.

13.3.2 Techniques for Injection of Software Faults

An obvious way to inject software faults into a system is to manipulate the source code, object code or machine code. Such manipulations are known as mutations. Mutations have been used extensively in the area of program testing as a method for evaluating the effectiveness of different test methods. They have also been used for the assessment of fault-handling mechanisms.

The studies presented in [686] and [687] inject software faults in an operating system through simple mutation of the object code. The primary goal of the fault models used in these studies was to generate a wide variety of operating system crashes, rather than achieving a high degree of representativeness with respect to real soft faults.

FINE [516] and DEFINE [515] were among the first tools that supported emulation of software faults by mutations. The mutation technique used by these tools requires access to assembly language listings of the target program. FINE and DE-

FINE emulate four types of software faults: initializations, assignment, condition check, and function faults. These fault models were defined based on experience collected from studies of field failure data.

An interesting technique, called Generic Software Fault Injection Technique (G-SWFIT), for emulation of software faults by mutations at the machine-code level is presented in [312]. This technique analyses the machine code to identify locations that corresponds to highlevel language constructs that often results in design faults. The main advantage of G-SWFIT is that software faults can be emulated without access to the source code. A set of operators for injection of representative software faults using G-SWFIT was presented in [313]. These operators were derived from a field failure data study of more than 500 real software faults. These two works jointly represent a unique contribution, since they provide the first fault injection environment that can inject software faults which have been proven to be representative of real software faults. They also constitute the foundation of a methodology for definition of faultloads based on software faults for dependability benchmarking presented in [315].

13.3.3 Techniques for Injecting Security Vulnerabilities

Most information systems and business applications that are built nowadays (e.g., e-commerce, banking, transportation, web mail, blogs, etc.) have a web front-end. They need to be universally accessed by clients, employees and partners around the world as online trading is becoming more and more ubiquitous in the global economy. These applications, which can be used from anywhere, also become so widely exposed that any existing security vulnerability will most probably be uncovered and exploited by hackers. Hence, the security of web applications is a major concern and is receiving more and more attention from the research community. However, in spite of this growing awareness of security aspects at web application level, there is an increase in the number of reported attacks that exploit web application vulnerabilities [851, 855].

The use of fault injection techniques to assess security is a particular case of software fault injection, focused on the software faults that represent security vulnerabilities or may cause the system to fail in avoiding a security problem. Security vulnerabilities are in fact a particular case of software faults, which require adapted injection approaches. In [341] the vulnerabilities of six web applications using their past 655 security fixes as the field data are presented and analyzed. Results show that only a small subset of 12 generic software faults is responsible for all the security problems. In fact, there are considerable differences by comparing the distribution of the fault types related to security with studies of common software faults.

Neves et al. presented a tool (AJECT) focusing on the discovery of vulnerabilities on network servers, specifically on IMAP servers [685]. In their work the fault space is the binomial (attack, vulnerability) creating an intrusion that will cause an error and, possibly, a failure of the target system. To attack the target system they used predefined test classes of attacks and some sort of fuzzing.

A procedure inspired on the fault injection technique (that has been used for decades in the dependability area) targeting security vulnerabilities is proposed in [344]. In this work, the "security vulnerability" plus "attack" represents the space of the "faults" that can be injected in a web application; and the "intrusion" is the "error" [685, 743]. To emulate with accuracy real world web vulnerabilities this work relies on results obtained from a field study on real security vulnerabilities and use them in a novel Vulnerability Injection tool. As proposed in [344], this tool is a key instrument that can be used in several relevant scenarios, namely: building a realistic attack injector, train security teams, evaluate security teams, and estimate the total number of vulnerabilities still present in the code.

13.4 Techniques for Testing Resilient Distributed Systems

Several fault injection tools and frameworks have been developed for testing of fault-handling protocols in distributed systems. The aim of this type of testing is to reveal design and implementation flaws in the tested protocol. The tests are performed by manipulating the content and/or the delivery of messages sent between nodes in the target system. We call this message-based fault injection. It resembles robustness testing in the sense that the faults are injected into the inputs of the target system.

A careful definition of the failure mode assumptions is crucial in the design of distributed fault-handling protocols. The failure mode assumptions provide a model of how faults in different subsystems (computing nodes, communication interfaces, and networks) affect a distributed system. A failure mode thus describes the impact of subsystem failures in a distributed system. Commonly assumed failure modes include Byzantine failures, timing failures, omission failures, crash failures, fail-stop failures and fail-signal failures. At the system-level, these subsystem failures correspond to faults. Hence, tools for message-based fault injection intend to inject faults that correspond to different subsystem failure modes.

The experimental environment for fault tolerance algorithms (EFA) [318] is an early example of a fault injector for message-based fault injection. The EFA environment provides a fault injection language that the protocol tester uses to specify the test cases. The tool inserts fault injectors in each node of the target system and can implement several different fault types, including message omissions, sending a message several times, generating spontaneous messages, changing the timing of messages, and corrupting the contents of messages. A similar environment is provided by the DOCTOR tool [408], which can cause messages to be lost, altered, duplicated or delayed.

Specifying test cases is a key problem in testing of distributed fault-handling protocols. A technique for defining test cases from Petri-net models of protocols in the EFA environment is described in [319]. An approach for defining test cases from an execution tree description of a protocol is described in [64].

A framework for testing distributed applications and communication protocols called ORCHESTRA is described in [264, 266]. This tool inserts a probe/fault injection layer (PFI) between any two consecutive layers in a protocol stack. The PFI layer can inject deterministic and randomly generated faults in both outgoing and incoming messages. ORCHESTRA was used in a comparative study of six commercial implementations of the TCP protocol reported in [265].

Neko [900] is another framework for testing distributed applications. It supports rapid prototyping of distributed algorithms and, with the NekoStat extension [330], it can be used to perform quantitative evaluations of distributed systems. Neko provides supports for fault injection, e.g., injection of faults between consecutive layers in the application stack. The core part of Neko is pure Java (J2SE) and therefore Neko processes are able to work on top of many different operating systems and platforms. Neko provides support for both simulation and experimental execution of distributed applications.

The failure of a distributed protocol often depends on the global state of the distributed system. It is therefore desirable for a human tester to control the global state of the target system. This involves controlling the states of a number of individually executing nodes, which is a challenging problem. Two tools that address this problem are CESIUM [35] and LOKI [187]. An environment for message-based fault injection for assessment of OGSA middleware for grid computing is presented in [607].

A similar tool for testing of Web-services, called WS-FIT, is presented in [606]. This fault injector can decode and inject meaningful faults into SOAP-messages. It uses an instrumented SOAP API that includes hooks allowing manipulation of both incoming and outgoing messages. A comparison of this method and fault injection by code insertion is presented in [605]. Another tool aimed at testing grid services and cluster applications is Fail-FCI [441]. This tool is based on FAIL, a fault injection language that is used to define failure scenarios in a distributed system using a state machine approach. The Fail-FCI tool is extended in [442] to allow fault injection in distributed Java applications. Results from message-based fault injection assessment of several implementations of CORBA middleware are reported in [624].

Another tool for testing the robustness of Web Services is presented in [915]. This tool injects invalid data during the invocation of web service methods to discover both programming and design errors. The parameter values used in the method invocations are modified (i.e., corrupted) based on the data types and the semantics of the method parameters. The Web Services are classified according to the type and number of failures observed during the tests.

13.5 Applications of Fault Injection

As described in the first section of this chapter, fault injection is useful for a number of purposes. It has, for example, been applied successfully to assess diverse properties of operating systems. Fault injection is the technology underlying robustness testing of the system call interface, as applied for instance in Ballista [541] and DeBERT [240]. Faults are injected by corrupting parameters just before function calls to the operating system. This process requires knowledge of what consists, at a low level, a system call and allows one to assess the robustness of this important interface to the operating system. Among other measures, one may assess response time and resilience in terms of avoiding system crashes and hangs.

Another problem area for operating systems is the evaluation of resilience against device driver failures [30]. A study targeting the MINIX operating system [426] applied fault injection by emulating hardware faults (corruption of driver code in memory) and software faults (pointer, assignment, checking, parameter, and omission errors). The operating system executed a workload consisting of several device drivers, in which faults were injected. While drivers were expected to fail, the operating system should be able to maintain correctness. The outcome of the experiments was the discovery of several defects in error-handling code, which were subsequently removed. A comprehensive campaign on the final corrected version of the operating system found it able to handle all injected faults properly, thereby showing the usefulness of the fault injection process.

Another application of fault injection is the study of error coverage provided by software and hardware implemented mechanisms. In [834], a series of experiments targeting a brake-by-wire controller revealed that the vast majority of the errors that led to critical failures (in terms of the behaviour of the controller) affected either the stack pointer or the controller's integrator state. Subsequently, the authors devised two simple mechanisms specifically for protecting those two elements of the system. A second campaign showed that those mechanisms reduced the proportion of critical failures by one order of magnitude. This example shows how fault injection may be used to improve error coverage in a very efficient manner, using the information provided by fault injection experiments to guide the development of new mechanisms. The examples provided in this section, along with the ones briefly described throughout the chapter, make a strong case for the usefulness of fault injection in diverse domains and for various applications.

Fault injection is also a key technique in the dependability benchmarking area [514]. A dependability benchmark is a standard procedure to assess and compare dependability aspects of computer systems and/or components. To obtain dependability metrics, the system under benchmarking is exercised by injecting a representative faultload during the execution of a typical workload. The faultload is applied using typical fault injection techniques, as the ones introduced in this chapter. See chapter on Resilience Benchmarking for examples on dependability benchmarks and the application of fault injection in such context.

13.6 Conclusion

This chapter provided an overview of the state-of-the-art and historical achievements in the field of fault injection. The overview covers tools and techniques for injecting three main fault types: physical hardware faults, software design and implementation faults, and faults affecting messages in distributed systems. Our intention has been to highlight the most important fault injection tools and techniques used for assessment and test of resilient computer systems, although it would not have been possible to cover all works in the field, since the number of publications that deal with fault injection is quite large.

As fault injection techniques and tools have matured, we see that recent developments tend to improve and refine existing methods rather than exploiting new principles for fault injection. Along with this trend, fault injection is reaching mainstream usage, becoming widely adopted not only in academia but also in industry. Consequently, fault injection is regarded as an important experimental technique for assessment and verification of resilient systems.

Chapter 14
Resilience Benchmarking

Marco Vieira, Henrique Madeira, Kai Sachs
and Samuel Kounev

Abstract Computer benchmarks are standard tools that allow evaluating and comparing different systems or components according to specific characteristics (performance, dependability, security, etc). Resilience encompasses all attributes of the quality of 'working well in a changing world that includes faults, failures, errors and attacks'. This way, resilience benchmarking merges concepts from performance, dependability, and security. This chapter presents an overview on the state-of-the-art on benchmarking performance, dependability and security. The goal is to identify the existing approaches, techniques and problems relevant to the resilience-benchmarking problem.

14.1 Introduction

Benchmarks are standard tools that allow evaluating and comparing different systems or components according to specific characteristics such as performance, dependability, and security. While historical benchmarks were only a few hundreds lines

M. Vieira (✉) · H. Madeira
DEI/CISUC, University of Coimbra,
Coimbra 3030-290, Portugal
e-mail: mvieira@dei.uc.pt

H. Madeira
e-mail: henrique@dei.uc.pt

K. Sachs
SAP AG, 69190 Walldorf, Germany
e-mail: kai.sachs@sap.com

S. Kounev
Karlsruhe Institute of Technology,
76131 Karlsruhe, Germany
e-mail: kounev@kit.edu

K. Wolter et al. (eds.), *Resilience Assessment and Evaluation of Computing Systems*,
DOI: 10.1007/978-3-642-29032-9_14, © Springer-Verlag Berlin Heidelberg 2012

long, modern benchmarks are composed of hundreds of thousands or millions of lines of code. Compared to traditional software, the benchmark development process has different goals and challenges. Unfortunately, even if an enormous number of benchmarks exist, only a few contributions focusing on the benchmark concepts and development process were published.

The best-known publication on benchmarking is Gray's *The Benchmark Handbook* [391]. Besides a detailed description of several benchmarks, the author discusses the need for domain specific benchmarks and defines four important criteria, which a domain-specific benchmark has to fulfill:

- *Relevance:* the benchmark result has to measure the performance of the typical operation within the problem domain.
- *Portability:* it should be easy to implement on many different systems and architectures.
- *Scalability:* it should be scalable to cover small and large systems.
- *Simplicity:* the benchmark should be understandable to avoid lack of credibility.

Another work dealing with the criteria that a benchmark should fulfill is [457].The questions, what a 'good' benchmark should look like and which aspects should be kept in mind from the beginning of the development process, are discussed in detail and five key criteria are presented:

- *Relevance:* the benchmark has to reflect something important.
- *Repeatable:* the benchmark result can be reproduced by rerunning the benchmark under similar conditions with the same result.
- *Fair & Portable:* All systems compared can participate equally (e.g., portability, 'fair' design).
- *Verifiable:* There has to be confidence that documented results are real. This can, e.g., be assured by reviewing results by external auditors.
- *Economical:* The cost of running the benchmark should be affordable.

The work on performance benchmarking has started long ago. Ranging from simple benchmarks that target a very specific hardware system or component to very complex benchmarks focusing on complex systems (e.g., database management systems, operating systems), performance benchmarks have contributed to improve successive generations of systems. Research on dependability benchmarking has been boosted in the beginning of the millennium, having already led to the proposal of several dependability benchmarks. Several works have been carried out by different groups and following different approaches (e.g., experimental, modeling, fault injection). Due to the increasing relevance of security aspects, security benchmarking is becoming an important research field.

Resilience encompasses all attributes of the quality of 'working well in a changing world that includes faults, failures, errors and attacks' [127]. This way, resilience benchmarking merges concepts from performance, dependability, and security benchmarking. In practice, resilience benchmarking faces challenges related to the integration of these three concepts and to the adaptive characteristics of the systems under benchmarking. This chapter overviews the state-of-the-art on benchmarking performance, dependability and security, identifying the current approaches, techniques and problems relevant to the resilience benchmarking problem.

The outline of this chapter is as follows. The next section introduces the concept of performance benchmarking. Section 14.3 focuses on dependability benchmarking and presents existing research work. Section 14.4 introduces the security benchmarking problem. Section 14.5 discusses the current needs and challenges on resilience benchmarking. An overview of further research trends is provided in Sect. 14.6. Finally, Sect. 14.7 concludes the chapter and puts forward a potential research path to accomplish existing resilience benchmarking challenges.

14.2 Performance Benchmarking

Performance benchmarks are standard procedures and tools aiming at evaluating and comparing different systems or components in a specific domain (e.g., databases, operating systems, hardware, etc.) according to specific performance measures. Standardization organizations such as the SPEC (Standard Performance Evaluation Corporation) and the TPC (Transaction Processing Performance Council) use internal guidelines covering the development process of such benchmarks. A short summary of the keypoints of the SPEC Benchmark Development Process is provided in [573]. However, these guidelines mostly cover formal requirements, e.g., design of run rules and result submission guidelines, not the benchmark development process itself.

In general, a performance benchmark must fulfill the following fundamental requirements to be useful and reliable [545, 793, 794]:

- It must be based on a workload *representative* of real-world applications.
- It must *exercise all critical services* provided by platforms.
- It must *not be tuned/optimized for a specific product*.
- It must generate *reproducible* results.
- It must not have any inherent *scalability* limitations.

The major goal of a performance benchmark is to provide a standard workload and metrics for measuring and evaluating the performance and scalability of a certain platform. In addition, the benchmark should provide a flexible framework for performance analysis. To achieve this goal, the workload must be designed to meet a number of *workload requirements* that can be grouped according to the following five categories [793]:

1. *Representativeness*
2. *Comprehensiveness*
3. *Focus*
4. *Configurability*
5. *Scalability*

Representativeness No matter how well a benchmark is designed, it would be of little value if the workload it is based on does not reflect the way platform services are exercised in real-life systems. Therefore, the most important requirement for a benchmark is that it is based on a representative workload scenario including a representative set of interactions. The scenario should represent a typical transaction mix. The goal is to allow users to relate the observed behavior to their own applications and environments.

Comprehensiveness Another important requirement is that the workload is comprehensive in that it exercises all platform features typically used in the major classes of applications. The features and services stressed should be weighted according to their usage in real-life systems. There is no need to cover features of the platforms that are used very rarely in practice.

Focus The workload should be focused on measuring the performance and scalability of the platform under test. It should minimize the impact of other components and services that are typically used in the chosen application scenario.

Configurability In addition to providing standard workloads and metrics, a benchmark aims to provide a flexible performance analysis framework which allows users to configure and customize the workload according to their requirements. Many users will be interested in using a benchmark to tune and optimize their platforms or to analyze the performance of certain specific features. Others could use the benchmark for research purposes in academic environments where, for example, one might be interested in evaluating the performance and scalability of novel methods and techniques for building high-performance servers. All these usage scenarios require that the benchmark framework allows the user to precisely configure the workload and transaction mix to be generated. This configurability is a challenge because it requires that interactions are designed and implemented in such a way that one could run them in different combinations depending on the desired workload mix. The ability to switch interactions off implies that interactions should be decoupled from one another. On the other hand, it should be ensured that the benchmark, when run in its standard mode, behaves as if the interactions were interrelated according to their dependencies in the real-life application scenario.

Scalability Scalability should be supported in a manner that preserves the relation to the real-life business scenario modeled. In addition, the user should be offered the possibility to scale the workload in an arbitrary manner by defining an own set of scaling points.

14.2.1 SPEC Benchmarks

The Standard Performance Evaluation Corporation (SPEC) is one of the leading standardization bodies for benchmarks. While the most known benchmarks published by SPEC are the SPEC CPU series for the performance evaluation of CPUs, SPEC published benchmarks in many other areas, such as High Performance Computing, Java or Graphical Applications. Inside the SPEC, four groups exist [846]:

- *Open Systems Group (OSG)* focuses on benchmarks for desktop systems, high-end workstations and servers running open systems environments.
  Example benchmarks: SPEC CPU2006 (CPU performance), SPECjms2007 (message-oriented middleware benchmark, SPECpower_ssj2008 (power and performance benchmark), SPECvirt_sc2010 (virtualization benchmark) and SPEC-jEnterprise2010 (JavaEE benchmark).
- *High-Performance Group (HPG)* published a suite of benchmarks that represent high-performance computing applications for standardized, cross-platform performance evaluation.
  Example benchmarks: OMPM2001 / OMPL2001 (benchmarks for OpenMP applications and shared-memory systems) and MPIM2001 / MPIL2001 (benchmarks focusing on Message-Passing Interface (MPI) across a wide range of cluster and SMP hardware).
- *Graphics and Workstation Performance Group (GWPG)* develops graphics and workstation performance benchmarks.
  Example benchmarks: SPECapc benchmark series (addresses graphics and workstation performance evaluation based on actual software applications) and SPECviewperf 11.
- *Research Group (SPEC RG)* promotes research on benchmarking methodologies and tools facilitating the development of benchmark suites and performance analysis frameworks for established and newly emerging technology domains.

14.2.2 TPC Benchmarks

The benchmarks of the Transaction Processing Performance Council (TPC) became the de-facto standard in the database area [884]. Currently the TPC has three active benchmarks, two in the area of transaction processing (TPC-E/TPC-C) and one for benchmarking decision support. Their currently active benchmarks are based on a static workload mix. Additionally, TPC published the TPC-Energy Specification, which contains the rules and methodology for measuring and reporting an energy metric in TPC Benchmarks. It is important to note that, unlike SPEC, TPC does not provide implementations of its benchmarks. A TPC benchmark is essentially a specification that defines an application and a set of requirements on the workload that has to be run. The user is expected to implement the benchmark application and workload on the platform to be tested.

Further, TPC has released two benchmarks that can be used for benchmarking enterprise software systems. The first one is the TPC Benchmark W (TPC-W) [886], which has been available since 2000. The second one is the TPC Benchmark App (TPC-App) [885], which was released in December, 2004. However, both of these benchmarks are obsolete and there is no active benchmark for enterprise software systems.

14.2.3 EEMBC Benchmarks

The Embedded Microprocessor Benchmark Consortium (EEMBC) is developing performance benchmarks for the hardware and software used in embedded systems [325]. EEMBC microprocessor benchmark suites are targeting telecommunications, networking, digital media, Java, automotive/industrial, consumer, and office equipment products. Further, an additional suite that allows users to observe the energy consumed by the processor when performing these algorithms and applications exists. EEMBC also has a series of multicore-specific benchmarks that span multiple application areas.

14.2.4 Other Performance Benchmarks

Besides industry-standard benchmarks, numerous proprietary performance benchmarks for all kinds of systems have been developed and used in the industry and research. Due to the lack of space and the high number (e.g., we are aware of more than 15 benchmarks and performance tests suits for message-oriented middleware [793]) we will not discuss them here in detail.

14.3 Dependability Benchmarking

The notion of dependability and its terminology have been established by the *International Federation for Information Processing (IFIP) Working Group 10.4*. IFIP WG 10.4 defines dependability as *'the trustworthiness of a computing system which allows reliance to be justifiably placed on the service it delivers'*. Dependability is an integrative concept that includes the following attributes [576]: availability (readiness for correct service), reliability (continuity of correct service), safety (absence of catastrophic consequences on the user(s) and the environment), confidentiality (absence of unauthorized disclosure of information), integrity (absence of improper system state alterations), and maintainability (ability to undergo repairs and modifications).

A dependability benchmark can be defined as a specification of a standard procedure to assess dependability-related measures of a computer system or computer component. The main components of a dependability benchmark are: measures (characterize the performance and dependability of the system), workload (work that the system must perform during the benchmark run), faultload (set of faults that emulate real faults experienced in the field), and benchmark procedure and rules (description of the procedures and rules that must be followed to run the benchmark).

Two classes of measures can be considered when assessing dependability attributes:

- *Conditional measures:* measures that characterize the system in a relative fashion (i.e., measures that are directly related to the conditions disclosed in the benchmark report) and are mainly meant to compare alternative systems (e.g., response time, throughput, up-time, recovery time).
- *Unconditional measures on dependability attributes:* measures that characterize the system in a global fashion taking into account the occurrence of the various events impacting its behavior (i.e., reliability, availability, maintainability, safety, etc.) [576].

The conditional measures are directly obtained as results of the benchmark experiments. The unconditional measures on dependability attributes have to be calculated using modeling techniques with the help of external data, such as fault rates, MTBF, etc. This external data could be provided from field data or based on past experience considering similar systems. However, models of complex systems may be very difficult to define and the external data difficult to obtain.

Dependability benchmarks typically focus on direct measures (conditional measures), following the traditional benchmarking philosophy based on a pure experimental approach. These measures are related to the conditions disclosed in the benchmark report and can be used for comparison or for system/component improvement and tuning. This is similar to what happens with performance benchmark results, as the performance measures do not represent an absolute measure of system performance and cannot be used for capacity planning or to predict the actual performance of the system in field.

The faultload represents a set of faults that emulates real faults experienced by systems in the field. Among the main components needed to define a benchmark, the faultload is clearly the most complex one due to the nature of faults. A faultload can be based on three major classes of faults:

- *Operator faults:* operator faults are human mistakes. The great complexity of administration tasks in some systems and the need of tuning and administration in a daily basis, clearly explains why human faults (i.e., wrong human actions) should be considered in a dependability benchmark.
- *Software faults:* software faults (i.e., program defects or bugs) are recognized as an important source of system outages, and given the huge complexity of today's software the weight of software faults tends to increase.
- *Hardware faults:* includes traditional hardware faults, such as bit-flips and stuck-at, and high-level hardware failures, such as hard disk failures or failures of the interconnection network. Hardware faults are especially relevant in systems prone to electrical interferences.

Concerning the definition of the workload, the job is considerably simplified by the existence of workloads from performance benchmarks. Obviously, these already established workloads are the natural choice for a dependability benchmark. However, when adopting an existing workload some changes may be required in order to target specific system features. An important aspect to keep in mind when choosing a workload is that the goal is not only to evaluate the performance but also assess specific dependability features.

The procedures and rules define the correct steps to run a benchmark and obtain the measures. These rules are, of course, dependent on the specific benchmark but the following points give some guidelines on specific aspects needed in most of the cases:

- Procedures for 'translating' the workload and faultload defined in the benchmark specification into the actual workload and faultload that will apply to the system.
- Uniform conditions to build the setup and run the dependability benchmark.
- Rules related to the collection of the experimental results.
- Rules for the production of the final measures from the direct experimental results.
- Scaling rules to adapt the same benchmark to systems of very different sizes.
- System configuration disclosures required for interpreting the benchmark results.
- Rules to avoid optimistic or biased results.

The awareness of the importance of dependability benchmarks has increased in the recent years and dependability benchmarking is currently the subject of strong research. The following subsections present the recent advances on dependability benchmarking, both at universities and computer industry sites.

14.3.1 Special Interest Group on Dependability Benchmarking (SIGDeB)

The *Special Interest Group on Dependability Benchmarking* (SIGDeB) was created by the International Federation for Information Processing (IFIP) Working Group 10.4 in 1999 to promote the research, practice, and adoption of benchmarks for computer-related systems dependability. The work carried out in the context of the SIGDeB is particularly relevant and merges contributions from both industry and academia.

A preliminary proposal issued by the SIGDeB was in the form of a set of standardized classes for characterizing the dependability of computer systems [934]. The goal of the proposed classification was to allow the comparison among computer systems concerning four different dimensions: availability, data integrity, disaster recovery, and security. The authors have specifically developed the details of the proposal for transaction processing applications. This work proposes that the evaluation of a system should be done by answering a set of standardized questions or performing tests that validate the evaluation criteria.

A very relevant effort in the context of SIGDeB is a book on dependability benchmarking of computer systems [514]. This book presents several relevant benchmarking initiatives carried out by different organizations, ranging from academia to large industrial companies.

14.3.2 DBench Project

The DBench project was funded by the European Commission, under the Information Society Technologies Programme (IST), Fifth Framework Programme (FP5). The main goal of DBench project was to devise benchmarks to evaluate and compare the dependability of COTS and COTS-based systems, in embedded, real time, and transactional systems. Several works on dependability benchmarking have been carried out in the DBench project. The following subsections summarize those works.

General purpose operating systems The works presented in [502, 503, 511, 512] address the problem of dependability benchmarking for general purpose operating systems (OS), focusing mainly on the robustness of the OS (in particular of the OS kernel) with respect to faulty applications.

The measures provided are: 1) OS robustness in the presence of faulty system calls, 2) OS reaction time for faulty system calls and 3) OS restart time after the activation of faulty system calls. Three workloads are considered: 1) a realistic application that implements the experiments control system of the TPC-C performance benchmark [887]. 2) the PostMark [521] file system performance benchmark for operating systems and 3) the Java Virtual Machine (JVM) middleware. The faultload is based on the corruption of systems call parameters.

Another research work on the practical characterization of operating systems behaviour in the presence of software faults in OS components is presented in [312]. The methodology used is based on the emulation of software faults in device drivers and the observation of the behaviour of the overall system regarding a comprehensive set of failure modes analyzed according to different dimensions related to different user perspectives.

Real time kernels in onboard space systems The work presented in [666] is a preliminary proposal of a dependability benchmark for real time kernels for onboard space systems. This benchmark, called DBench-RTK, focuses mainly on the assessment of the predictability of response time of service calls in a Real-Time Kernel (RTK).

The DBench-RTK dependability benchmark provides a single measure that represents the predictability of response time of the service calls of RTKs used in space domain systems. The workload consists in an Onboard Scheduler (OBS) process based on a functional model derived from the Packet Utilization Standard [864]. The faultload consists of a set of faults that are injected into kernel functions calls at the parameter level by corrupting parameter values.

Engine control applications in automotive systems The work presented in [790] represents a preliminary proposal of a dependability benchmark for engine control applications for automotive systems. This benchmark focuses on the robustness of the control applications running inside the Electronic Control Units (ECU) with respect to transient hardware faults.

This dependability benchmark provides a set of measures that allows the comparison of the safety of different engine control systems. The workload is based on the standards used in Europe for the emission certification of light duty vehicles [320].

The faultload consists of transient hardware faults that affect the cells of the memory holding the software used in the engine control.

On-line transaction processing systems The DBench-OLTP dependability benchmark [917, 918] is a dependability benchmark for on-line transaction processing systems. The DBench-OLTP measures are divided in three groups: baseline performance measures, performance measures in the presence of the faultload, and dependability measures. The DBench-OLTP benchmark can be used considering three different faultloads each one based on a different class of faults, namely: operator faults, software faults and high-level hardware failures.

In [161] it is presented a preliminary proposal of another dependability benchmark for on-line transaction processing systems. The measures provided by this dependability benchmark are the system availability and the total cost of failures. These measures are based on both measurements obtained from experimentation (e.g., percentages of the various failure modes) and external data (e.g., the failure rates and the repair rates). The external data used to calculate the measures must be provided by the benchmark user. The workload was adopted from the TPC-C performance benchmark [887] and the faultload includes exclusively hardware faults, such as faults in the storage hardware and in the network.

Web-servers The work presented in [316] proposes a dependability benchmark for web-servers (the WEB-DB dependability benchmark). This dependability benchmark uses the basic experimental setup, the workload, and the performance measures specified in the SPECWeb99 performance benchmark [845].

The measures reported by WEB-DB are grouped into three categories: baseline performance measures, performance measures in the presence of the faultload, and dependability measures. The WEB-DB benchmark uses two different faultloads: one based on software faults that emulate realistic software defects (see [314]) and another based on operational faults that emulate the effects of hardware and operator faults.

14.3.3 Berkeley University

The work developed at Berkeley University has highly contributed to the progress of research on dependability benchmarking in the last few years, principally on what concerns benchmarking the dependability of human-assisted recovery processes.

A general methodology for benchmarking the availability of computer systems is introduced in [155]. The workload and performance measures are adopted from existing performance benchmarks and the measure of availability of the system under test is defined in terms of the service provided by the system. The faultload (called fault workload by the authors) may be composed of a single-fault (single-fault workload) or of several faults (multi-fault workload).

The work presented in [156] addresses human error as an important aspect in system dependability, and proposes that human behaviour must be considered in dependability benchmarks and system designs.

A technique to develop dependability benchmarks that capture the impact of human operators on the tested system is proposed in [154]. The workload and measures are adopted from existing performance benchmarks and the dependability of the system can be characterized by examining how the performance measures deviate from their normal values as the system is perturbed by injected faults. In addition to faults injected using traditional fault injection, perturbations are generated by actions of human operators that actually participate in the benchmarking procedure.

In [151] are presented the first steps towards the development of a dependability benchmark for human assisted recovery processes and tools. This work proposes a methodology to evaluate human-assisted failure recovery tools and processes in server systems. This methodology can be used to both quantify the dependability of single recovery systems and compare different recovery approaches, and combines dependability benchmarking with human user studies.

14.3.4 Carnegie Mellon University

Vajra [674] is a research project whose goal is benchmarking the survivability in distributed systems, focusing on the objective and quantitative comparison of the runtime implementations of different Byzantine fault-tolerant distributed systems. The benchmark uses as the point of injection APIs that are common across various Byzantine fault-tolerant systems. A variety of accidental and malicious faults are injected at various rates across the system.

Although not resulting in a formal benchmark proposal, the research on robustness testing developed at the Carnegie Mellon University [540] has effectively set the basis for robustness benchmarks of operating systems. This will be further discussed in Chap. 16, which includes a survey on robustness testing techniques.

14.3.5 Sun Microsystems

Research at Sun Microsystems has defined a high-level framework [959] specifically dedicated to availability benchmarking of computer systems. The proposed framework follows a hierarchical approach that decomposes availability into three key components: fault/maintenance rate, robustness, and recovery. The goal was to develop a suite of benchmarks, each one measuring an aspect of the availability of the system. Within the framework proposed by [959], two specific benchmarks have already been developed.

In [960] is proposed a benchmark for measuring a system's robustness (degree of protection that exists in a system against outage events) in handling maintenance events, such as the replacement of a failed hardware component or the installation of a software patch.

In [629] is proposed a benchmark for measuring system recovery in a non-clustered standalone system. This benchmark measures three specific system events; clean system shutdown (provides a baseline metric), clean system bootstrap (corresponds to rebooting a system following a clean shutdown), and a system reboot after a fatal fault event (provides a metric that represents the time between the injection of a fault and the moment when the system returns to a useful state).

Another effort at Sun Microsystems are the Analytical RAS Benchmarks [324], which consists of three analytical benchmarks that examine the Reliability, Availability, and Serviceability (RAS) characteristics of computer systems:

- The Fault Robustness Benchmark (FRB-A) allows assessing and comparing the techniques used to enhance resiliency, including redundancy and automatic fault correction.
- The Maintenance Robustness Benchmark (MRB-A) assesses how maintenance activities affect the ability of the system to provide a continuous service.
- The Service Complexity Benchmark (SCB-A) examines the complexity of mechanical components replacement.

14.3.6 Intel Corporation

Work at Intel Corporation has focused on benchmarking semiconductor technology. The work presented in [236] shows the impact of semiconductor technology scaling on neutron induced SER (soft error rate) and presents an experimental methodology and results of accelerated measurements carried out on Intel Itanium microprocessors. The proposed approach can be used as a dependability benchmarking tool and does not require proprietary information about the microprocessor under benchmarking.

Another study [236] presents a set of benchmarks that rely on environmental test tools to benchmark undetected computational errors, also known as silent data corruption (SDC). In this work, a temperature and voltage operating test (known as the four corners test) is performed on several prototype systems.

14.3.7 IBM Autonomic Computing Initiative

At IBM, the Autonomic Computing initiative developed benchmarks to quantify a system's level of autonomic capability, which is defined as the capacity of the system to react autonomously to problems and changes in the environment. The goal was to produce a suite of benchmarks covering the four categories of autonomic capabilities: self-configuration, self-healing, self-optimization, and self-protection.

The first steps towards a benchmark for autonomic computing are described in [589]. The benchmark addresses the four attributes of autonomic computing and is able to test systems at different levels of autonomic maturity.

The work presented in [152] identifies the challenges and pitfalls that must be taken into account in the development of benchmarks for autonomic computing capabilities. This paper proposes the use of the workload and driver system from performance benchmarks and the introduction of changes into benchmarking environment in order to characterize a given autonomic capability of the system. The paper proposes that autonomic benchmarks must quantify the level of the response, the quality of the response, the impact of the response on the users, and the cost of any extra resources needed to support the autonomic response.

14.4 Security Benchmarking

Theoretically, a security benchmark provides a metric (or small set of metrics) able to characterize the degree to which security goals are met in a given piece of code [483], allowing developers and administrators to make informed decisions. However, one of the biggest difficulties in designing such metric is related to the fact that security assessment is, usually, much more dependent on what is unknown about the applications (e.g. unknown bugs, hidden vulnerabilities) than by what is known (e.g., known features, existing security mechanisms).

Security metrics are hard to define and compute [883] because they involve making isolated estimations about the ability of an unknown individual (e.g., a hacker) to discover and maliciously exploit an unknown system characteristic (e.g., a vulnerability). A feasible alternative is to assume that such metrics can be obtained using information about the system itself, without taking into account external factors. In fact, a security benchmark based on such metrics would allow characterizing the *degree to which security goals are met in a given web application or component.* In practice, due to the difficulties of quantifying security, most works on security benchmarking are based on analysis and qualification of configurations/systems.

Several security evaluation methods have been proposed in the past [232, 233, 288, 807]. The Orange Book [288] and the Common Criteria for Information Technology Security Evaluation [233] define a set of generic rules that allow developers to specify the security attributes of their products and evaluators to verify if products actually meet their claims. Another example is the red team strategy [807], which consists of a group of experts trying to hack its own computer systems to evaluate security. However, none of these security evaluation approaches is oriented towards security benchmarking, as comparing security has been largely absent from these security evaluation methods.

The work presented in [630] addresses the problem of determining, in a thorough and consistent way, the reliability and accuracy of anomaly detectors. This work addresses some key aspects that must be taken into consideration when benchmarking the performance of anomaly detection in the cyber-domain.

The set of security configuration benchmarks created by the *Center for Internet Security* (CIS) is a very interesting initiative. CIS is a non-profit organization formed by several well-known academic, commercial, and governmental entities that has created a series of security configuration documents for several commercial and open source systems. These documents focus on the practical aspects of the configuration of these systems and state the concrete values each configuration option should have in order to enhance overall security of real installations. Although CIS refers to these documents as benchmarks they mainly reflect best practices and are not explicitly designed for systems assessment or comparison.

A practical way to characterize the security mechanisms in database systems is proposed in [920]. In this approach database management systems (DBMS) are classified according to a set of security classes ranging from Class 0 to Class 5 (from the worst to the best). Systems are classified in a given class according to the security requirements satisfied. In [50] the authors analyze the security best practices behind the many configuration options available in several well-known DBMS. These security best practices are then generalized and used to define a set of configuration tests that can be used to compare different database installations. An improved set of best practices is then used in [52] to benchmark the security of database servers configurations.

A benchmark that allows database administrators to assess and compare database configurations is presented in [51]. The benchmark provides a trust-based security metric, named minimum untrustworthiness, that expresses the minimum level of distrust the DBA should have in a given configuration regarding its ability to prevent attacks. The use of trust-based metrics as an alternative to security measurement is discussed in [682].

14.5 Resilience Benchmarking

A resilience benchmark should provide generic ways for characterizing a system behavior in the presence of perturbations. If a system is effective and efficient in accommodating or adjusting to perturbations, avoiding failures as much as possible, it is reasonable to consider it as being resilient [33]. This capability can be benchmarked by submitting the system to various types of perturbations and by observing the failures (and their frequency), as well as time and resources dedicated to avoid/recover from them. Still, the perturbations that the system has to face may lead to performance and dependability attributes degradation without leading necessarily to catastrophic system failures. Thus, we need to assess variations of the properties of interest (e.g., performance, availability, integrity) when the system is under varying context conditions, in order to characterize its behavior from a resilience perspective.

Evaluating resilience must consider the system and environment dynamics that are beyond those typically addressed in the evaluation of performance and dependability. While maintaining similar workloads, dependability benchmarks enhanced performance benchmarks by introducing a faultload and dependability metrics, which include performance metrics under faulty conditions. A resilience benchmark must comprise a more wide-ranging set of perturbations, which will certainly include (but will be not limited to) faults. For instance, variations on the workload or in system parameters should be part of those perturbations. New metrics for characterizing resilience are also needed, although some will naturally be based on measures of performance and dependability while facing changes.

In practice, resilience benchmarking includes performance, dependability, and security aspects, and aims at providing generic, repeatable and widely accepted methods for characterizing and quantifying the system (or component) behavior in the presence of faults, and comparing alternative solutions [514]. Although many works have been conducted in the area of performance and dependability benchmarking, it is clear that many key issues must be addressed towards the definition of concrete resilience benchmarks, which, theoretically, should include the following main components:

- *Benchmarking metrics:* the benchmark metrics should allow characterizing and quantifying the system behavior when facing perturbations (i.e., faults, attacks, and operational environment variations). At first sight, resilience benchmarking metrics must characterize performance, dependability and security.
- *Workload:* during the benchmark execution, the system under test must be submitted to a representative set of tasks, which should be as close to real conditions as possible. An important aspect is that a workload cannot be static and must exercise the resilience capabilities of the system, as the real conditions would.
- *Perturbations-load:* a system may be subjected to distinct types of perturbations during its operation, and a benchmark must try to emulate those as realistically as possible. These perturbations may be of three different types: faults, attacks, and perturbations related to system's maintenance.

In the context of the AMBER Coordination Action, funded by the European Union under the Seventh Framework Programme, a set of research needs related to resilience benchmarking have been identified, namely (see details at [127]):

1. Agreed, cost effective, easy to use, fast and representative enough dependability benchmarks for well defined domains.
2. Benchmark frameworks (components and tools) able to be reused to create benchmarks in different benchmarking domains.
3. Inclusion of adequate design methodologies to facilitate benchmark implementation and configuration in future components, systems, and infrastructures.
4. Uniform (standardized) benchmarking process that can be applied by independent organizations to offer certification of the dependability of COTS products (like in the case of standards compliance testing).

These needs raise a set of research challenges that have to be addressed in order to be able to define a (resilience) benchmark, namely (see [127] for details):

1. *Defining benchmark domains* (components, systems, application domains) in order to divide the problem space in adequate/tractable segments.

2. *Defining key benchmark elements* such as measures, workload, faultload, models, to ensure the necessary properties (e.g., representativeness, portability, scalability, repeatability) that allow agreement on benchmark proposals.
3. *Coping with highly complex, adaptable and evolving benchmark targets* (components, systems and services).
4. *Coping with human factors* in the definition and execution of benchmarks.
5. *Assuring proper validation of dependability benchmarks* in order to achieve the necessary agreement to establish benchmarks. This implies the validation of the different benchmark properties.
6. *Assuring reusability of benchmark frameworks* (components and tools) to create benchmarks in different benchmarking domains.
7. *Defining and agreeing on a domain-specific dependability benchmarking process* that can be accepted by the parties concerned (supplier, customer and certifier) and can be adapted to different products in the domain (e.g., in a product line).

14.6 Further Trends in Benchmarking Research

Besides resilience benchmarking we see some further research trends in the area of benchmarking, which we discuss in this section.

14.6.1 Benchmark Engineering

While developing benchmarks, we faced a lack of methodology that describes how to develop good and meaningful benchmarks. Since benchmark development has turned into a complex team effort, there is a need for a development methodology taking the specifics of benchmarks into account. Compared to traditional software, the development process has different goals and challenges. New concepts and processes are needed which address the whole development and life-cycle management of benchmarks. We refer to them including benchmark methodology and measurement techniques with the term *Benchmark Engineering* [793]. First work is already in progress. As example, SPEC is working on development guidelines.

14.6.2 Benchmarking of Large Scale Systems

Large scale, highly distributed systems are increasingly used in mainstream applications. However, for these systems traditional benchmarking approaches fail: how can we benchmark a system with 500,000 nodes? What does a typical workload look like and how does it scale? What should be the distribution of the faultload? etc.

Since it is not feasible to run benchmarks in a realistic environment with thousands of nodes, new methods are needed which allow us to benchmark large scale systems in a realistic way on limited resources. As a consequence, we see a need for research in the area of *simulated benchmarks*.

Similar questions are currently under discussion in several research areas. The authors in [583] discuss requirements for peer-to-peer (P2P) benchmarking and present two exemplary approaches to benchmark such systems. They point out the challenges of developing P2P benchmarks compared to conventional benchmarks.

A very active community can be found in the area of cloud benchmarking. A discussion why traditional benchmarks are not sufficient for evaluating cloud services can be found in [114]. The authors present some initial ideas how a cloud benchmark should be designed including a list of requirements for such a benchmark. In [238] the Yahoo! Cloud Serving Benchmark (YCSB) framework was introduced including a core set of benchmarks. YCSB targets cloud data serving services, allows to create new workloads and is extendible. Another example is the Cloudstone benchmark, which consists of a social-events web application (with PHP and Ruby implementations) and a set of automation tools for load generation and performance measurement [839]. When running the benchmark, the load is generated against the web application, which in turn generates load on the underlying database.

There are still many open questions in the area of P2P and cloud benchmarking. This is the reason, why the SPEC Research Group decided to launch two subcommittees working on these topics.

14.6.3 Power Consumption

In the past, benchmarking focused mainly on computation performance. Since industry and governments are increasingly concerned about the energy use of servers, there is a need to reflect the power consumption in the result of a benchmark. The first standard benchmark providing a metrics, which represents computation performance as well as energy consumption was the SPECpower_ssj2008 benchmark. Nowadays, more and more benchmarks include energy consumption in their result, such as SPEC or TPC benchmarks. Consequently, the SPEC is working on the Server Efficiency Rating Tool (SERT), a tool set to measure and evaluate the energy efficiency of computer servers [889].

A metric for power consumption has to reflect both, traditional performance metrics in relation to the power consumption and not only peak performance is of interest. However, energy consumption *scenarios* are only one example, where traditional benchmark metrics fail or are hard to apply. A major challenge of future benchmark development is the definition of meaningful metrics, which take other aspect than performance and dependability into account (see also Sect. 14.5).

14.7 Conclusion

This chapter presented the state-of-the-art on benchmarking. The work on performance benchmarking has started long ago and has contributed to improve successive generations of systems. Dependability benchmarking efforts both at universities and computer industry sites are quite recent. Security is a newcomer to the benchmarking world and little work has been performed so far.

Although performance benchmarking is a very well established field, further work on dependability benchmarking seems to be necessary in several application areas (e.g., real-time systems, grid computing, parallel systems, etc). Additionally, no dependability benchmark has achieved the status of a real benchmark endorsed by a standardization body. This may be due to several reasons (that need to be studied) but clearly shows that additional work is still needed.

In the area of security benchmarking, a lot of work is clearly needed, as this is a new and quite challenging field for which little work has been developed so far. A key issue is the definition of useful and meaningful security metrics. In fact, the problem of security quantification is a longstanding one. A useful security metric must portray the degree to which security goals are met in a given system, allowing a system administrator to make informed decisions. One of the biggest difficulties in designing such a metric is related to the fact that security is, usually, much more dependent on what is unknown about the system than on what is known about it. In fact, security metrics are hard to define and compute as they involve making isolated estimations about the ability of an unknown individual (e.g., a hacker) to discover and maliciously exploit an unknown system characteristic (e.g., a vulnerability).

To tackle the challenges related to the future implementation of resilience benchmarks, the following research steps are foreseen:

1. Study the metrics that better characterize resilience.
2. Study the definition of dynamic workloads via field studies and analysis of existing workloads.
3. Study the characterization of perturbation loads. This can be based on field studies and on the analysis of already existing faultloads.
4. Define the steps needed for the execution of a resilience benchmark. These steps define the benchmark procedure and should be as generic as possible to allow the portability of the benchmarking approach.
5. Conduct benchmarking campaigns to demonstrate the benchmark and validate its properties.
6. Generalize the resilience benchmarking approach to make possible its application in different domains.
7. Disseminate the benchmarking approach. A key aspect is to identify the best way to foster the adoption by industry and to facilitate the support by a standardization body like TPC and SPEC.

Acknowledgments The work of Marco Vieira and Henrique Madeira was partially funded by the European Commission under project *AMBER - Assessing, Measuring and Benchmarking Resilience*, IST - 216295, funded by the European Union, 2009. The work of Samuel Kounev was partially funded by the German Research Foundation (DFG) under grant No. KO 3445/6-1.

Part V
Testing Techniques

Chapter 15
Resilience Assessment Based on Performance Testing

Alberto Avritzer and Andre B. Bondi

Abstract Performance degradation and/or irregularity are often indicators of system instability. By applying the principle that average performance measures vary little under constant load in a stable system without periodic behavior, we can use performance metrics to anticipate instability within the system. We describe how to use that information to isolate the cause of observed instability and how to structure load tests to identify scenarios in which system instability is likely to occur. We discuss resilience assessment based on performance measurements. Specifically, we present our experience generating load tests and analyzing the performance testing results. We discuss how we have used these results for security, reliability and performance assessment. We discuss the conditions required for system stability and identify some of the causes for system instability, such as security attacks, quality problems, and queuing for system resources. We present a metric that can be used to assess some dimensions of system security, reliability and performance using data obtained from the execution of performance testing. In addition, we present the associated testing activities that are required to collect data for the required modeling and analysis activities, and to help track system security, reliability and performance. We illustrate the presented methodology with empirical results obtained by testing for security, reliability, and performance.

A. Avritzer (✉) · A. B. Bondi
Siemens Corporate Research and Technology, 755 College Road East,
Princeton, NJ 08540, USA
e-mail: alberto.avritzer@siemens.com

A. B. Bondi
e-mail: andre.bondi@siemens.com

K. Wolter et al. (eds.), *Resilience Assessment and Evaluation of Computing Systems*,
DOI: 10.1007/978-3-642-29032-9_15, © Springer-Verlag Berlin Heidelberg 2012

15.1 Introduction

Whether an episode of system instability occurs during execution in production or during laboratory tests, identifying its root cause is very difficult unless expected behavior can be modeled or data have been collected to track the system's behavior and activity during the episode and during the times immediately preceding and following it. It is also difficult to tell whether a modification to the system has improved its stability unless the occurrences of episodes and the conditions and activities leading up to them are carefully tracked.

We also wish to anticipate system instability and take measures to prevent complete failure. This is much easier to do if we have a way of evaluating the current stability of the system, or, perhaps more pessimistically, the instantaneous propensity of the system to fail.

Queuing theory and the theory of stochastic processes tell us that when a system is well behaved and under a constant average load, the performance measures of the system, including resource utilization, average response time, average queue lengths, and memory occupancy will vary within small ranges of values. This is because the rates of change of the probabilities of a system being in a particular state are all zero when the system is in equilibrium. It follows that trends in performance measures over time and large variations over time are indicators of potential instability or even of forthcoming crashes (if observed before the fact) or past crashes (if observed after the fact).

We shall explore potential causes of run-time instability and show how they can be identified by examining the behavior of performance measures. We shall show how performance tests can be structured to reveal potential run-time instability, and illustrate our ideas with some examples.

There are many reasons why a system can become unstable. Among them are:

1. Security intrusion as a result of a security attack. A study of the impact of several different security attacks test cases on system performance metrics was presented in [71]. The test cases used in that research were buffer overflow, stack overflow, SQL injection, denial of service (DOS), and Man-in-the-Middle (MITM) security attacks. It was reported in that research that significant deviations from the stable performance ranges could be detected when the system was under these classes of security attacks.
2. Saturation of one or more hardware resources, such as CPU, I/O, or bandwidth, and/or saturation of software resources such as pools of abstract objects. In the case of the listed hardware resources, the cause is excessive offered load. In the case of software resources, the most simple cause is that the product of the average holding time and the average request rate is close to the size of the resource pool, or even exceeds it.
3. Concurrent programming errors. These can result in deadlocks, livelocks, and/or thread safety and mutual exclusion problems. Concurrent programming errors often lead to Heisenbugs, i.e., bugs that are not always reproducible given the same starting conditions and inputs because they occur non-deterministically.

Concurrent programming errors can result from the incorrect application of synchronization operations, such as semaphore operations, or from the use of scheduling rules that result in cyclic dependency between processes or threads that acquire and release resources that are non-preemptible and non-sharable. These are the necessary conditions for deadlock [404].

4. Memory and object pool leaks are caused by a failure to release the objects in question and return them to their respective free pools. The symptoms of these leaks include the growth of the virtual address space sizes of individual processes and/or threads, shrinking free memory pool and/or object pool sizes, and, in Java-based systems, growing heap sizes. If a leak progresses too far, the system will crash.

5. Memory overflows resulting in unauthorized access and perhaps data corruption, caused, for example, by subscripts being out of range or by bad pointer values.

The foregoing are examples of instability occurring at run or execution time, as opposed to instability of the code base, which can contribute to instability at run time. Specifically,

- Run-time instability manifests itself in performance degradation and/or failures and malfunctions that could be caused by security intrusions, software/hardware defects or contentions for hardware/and or software resources.
- Code base instability sometimes manifests itself in broken builds and compilation failures. It can manifest itself in performance degradation and malfunctions at run time.
- Code base instability may be caused by flaws in the development and change control processes. Although code base instability may be a contributing factor to run-time instability, it is neither a necessary nor a sufficient one. It is not necessary because crashes can occur even if the code base is stable. It is not sufficient because the system can run uneventfully even if the development process was chaotic. Instability of the code base will not be considered further here.

In the remainder of this chapter, the term instability refers to manifestations occurring during execution, i.e., at run time.

The fundamental operational laws of performance modeling [287] can give us guidelines about structuring performance tests to reveal something about stability, and insights on how to interpret performance test results.

In a well behaved system under a constant load:

1. The achieved throughput should equal the offered transaction throughput.
2. The utilizations of the processors, I/O devices, and bandwidth should be proportional to the offered transaction rate. This is known as the Utilization Law.
3. The average response time should increase slowly as a function of the offered load unless the utilization of any resource in the system under test exceeds 90 %.
4. The average values of the following performance measures should be approximately constant during a test run under constant load, except during the rampup and cooldown periods, or when the load is inherently periodic:

   a. The average transaction response times,
   b. All resource utilizations,
   c. The average throughput,
   d. The average memory occupancy,
   e. The average sum total of all address space sizes,
   f. The average address space sizes in all time intervals during the test.

If the load is inherently periodic, the quantities listed in items (a)–(f) should also be correspondingly periodic unless the length of the averaging period exceeds the period of the load. In that case, the average measures should again be approximately constant over the course of the test run.

Any deviation from these rules is a sign of system instability. To recognize the potential for instability, we structure our initial load tests and collect measurements in a way that allows us to verify that the performance measures have the characteristics listed above. If they do not, further investigation is warranted.

In the foregoing, we have focused on how instability can be inferred by deviations from basic operational laws. The results of performance tests that are structured to reveal possible deviations from these laws are illustrated in Sect. 15.4. In Sect. 15.2, we shall present a complementary approach in which the choices of loads are determined based on sets of Markov chains whose states correspond to the operational states of a system. These states are artefacts of both the offered load (whether seen in testing or in production) and of the operational status of the system.

In the Markov state testing approach, for each state of the Markov chain that needs to be tested, we drive the system into that state, and we generate a sustained constant load to measure the system behavior for that load and determine pass/fail conditions. Therefore, the approach presented in Sect. 15.2 identifies the system states that require testing, while the stability conditions presented in this section help determine pass/fail conditions for each test. In addition, the approaches presented in Sect. 15.2, Markov Testing, and Sect. 15.3, Stability Testing are complementary. In the approach presented in Sect. 15.2 a stochastic model is used and a customer affecting metric is derived to represent the fraction of time the system satifies the customer affecting metric. The approaches presented in Sect. 15.3 cover initial tests and performance tests to uncover instability.

Identifying the impact of security intrusions, hardware and software defects, and of system resource contention, when the system is run under constant average load, requires the collection and/or computation of the customer affecting metric. Ideally, we should be able to define a unified metric to cover security, reliability and performance. We describe the application of Markov chain theory to the generation of test cases for the evaluation of security, reliability and performance. We also describe the process that should be used to solve the Markov chain and to compute the customer affecting metric. In the next section we describe in detail the Markov state testing [69, 73] approach and the associated metric of interest.

15.2 Modeling and Analysis Approaches Based on Performance Testing Results

We apply results from performability theory [69] for test case generation and assessment of system reliability by constructing two types of Markov chains:

1. resource failure-based Markov chain of the form $\mathcal{X} = \{X(t), t \geq 0\}$
2. resource usage-based Markov chain of the form $\mathcal{Y} = \{Y(t), t \geq 0\}$

In the resource failure-based Markov chain we capture the number and type of operational/failed components that impact the system's ability to complete its mission. In the resource usage-based Markov chain we capture the system allocated resources for a certain load, for a given state of the failure-based Markov chain. The objective is to test for failures that are most likely to occur and for states that are important in the assessment of the system's ability to satisfy the customer affecting performance metric, for a certain system configuration.

We now present the approach we suggest to use to assess the impact of security attacks, quality issues and resource contention on the defined customer affecting metric.

The customer affecting metric of interest is defined using a performability approach as:

'The fraction of time the system satisfies the defined requirements specifications during an observation period (0, t).'

We define two types of Markov chains:

- a resource failure-based Markov chain, where a state ϕ maps the resources that are unavailable to perform useful work,
- a resource usage-based Markov chain, where a state $s = (s_1, s_2, \ldots, s_M)$ maps the resources that are being used.

The failure-based Markov chain contains failure states and associated events: failures and repairs. The resource usage Markov chain contains the state of resources and associated events: allocation and release. Security intrusions have been shown [71] to impact system performance and therefore can be captured by appropriate definition of the resource usage Markov chain.

The modeling and assessment process defined in [69] to compute the customer affecting metric after the execution of performance tests is:

1. define the customer affecting metric for a given observation period as: the fraction of time the system satisfies the defined requirements specifications during an observation period (0, t).
2. define a resource failure based Markov chain for the system. The system can be in K different system resource configurations. Generate the K Markov resource usage based Markov chains, one Markov chain associated with each resource configuration. System resource configurations could vary because of resource failures and/or security attacks.

3. Using test results, obtain the pass/fail results for each state of the K Markov chains. The pass/fail also conveys the impact of the failure on the customers.
4. Solve the Markov chains and obtain the long term fraction of time each Markov chain spends in each state.
5. Applying uniformisation [273], the customer affecting metric can be computed as a transient metric for the observation interval or as a steady state metric when the observation period becomes very large.

The detailed mathematical derivations is available from [69, 273]. We present a summary of the results below.

We assume that the system under study can be in K different configurations, where each configuration results from combinations of system failures. The failure configurations comprise the set C. The ith configuration is denoted by $c_i \in C$. The cardinalty of C is K. The performance test suite is generated from the most likely states of the resource usage-based Markov chain $\mathcal{Y}$. An important assumption used to simplify testing complexity is that a failing performance test for a certain configuration can be marked as failed for higher loads that use the same configuration. In addition, a configuration $c_i \in C$ that that has a very low probability of occurence, i.e., smaller than a pre-specified cutoff probability, can be marked as failed without testing, thus reducing the complexity of the performance test suite, as the impact of an extremely low probabilility test case on the customer affecting metric is negligible.

$\mathcal{Y}(i)$ represents the resource usage-based Markov chain for configuration c_i and $\alpha_j(i)$ is defined to be:

$$\alpha_j(i) = \begin{cases} 1 \text{ if the performance requirements are} \\ \quad \text{satisfied for configuration } c_i \text{when} \\ \quad \text{in state } s_j \text{ of } \mathcal{Y}(i) \\ 0 \text{ otherwise} \end{cases}$$

$M(i)$ is the total number of states in $\mathcal{Y}(i)$. Let us define $\gamma_j(i)$ to be the long-term fraction of time the system is in state $s_j(i)$ of $\mathcal{Y}(i)$. Then, $\sum_{j=1}^{M(i)} \gamma_j(i)\alpha_j(i)$ is the fraction of time configuration c_i satisfies the target requirements for the system.

The customer affecting metric of interest is defined to be $\Gamma(t)$, the fraction of time the system satisfies the customer affecting requirements specifications during an observation interval $(0, t)$.

The calculation of the customer affecting metric requires an additional assumption: the transition rates between resource usage based Markov chain states are significantly larger than the transition rates between different resource failure based configurations. In addition, we define *reward rates* $r_1, \ldots, r_C$, where reward r_i is the the amount of reward per unit of time accrued, when the system is in configuration c_i. Let us compute r_i as:

$$r_i = \sum_j \gamma_j(i)\alpha_j(i). \tag{15.1}$$

Furthermore, let $\delta_i(t)$ units of time be the time the system is in configuration c_i during the observation period $(0, t)$. Therefore, $\delta_i(t) * r_i$, is the fraction of time during $(0, t)$ that the system is expected to pass the requirements test.

Next we calculate $\Gamma(t)$. We refer the reader to the transient analysis results from Markov reward models (e.g. [273]). Equation (58) of [273] provides an expression for $\Gamma(t)$,

$$\Gamma(t) = \sum_{n=0}^{\infty} \mathcal{C}/\widetilde{=}^{-\Lambda t} \frac{\Lambda t}{n!} \left[\frac{\sum_{j=0}^{n} \mathbf{r} \cdot \mathbf{v}(j)}{n+1} \right], \tag{15.2}$$

where

- $\mathbf{v}(n) = \pi(0)\mathbf{P}_{\mathcal{X}}^{n}$ is the nth state probability vector when $\mathcal{X}$ is uniformized with rate Λ.
- $\pi(0)$ is the initial state probability vector for $\mathcal{X}$ usually equal to $\langle 1, 0, \dots, 0 \rangle$ meaning that the system is fully operational at time 0.

Let Γ be the limit as $t \to \infty$ of fraction of time the system satisfies the customer affecting performance requirements. For $t \to \infty$ we get:

$$\Gamma = \sum_{i=1}^{K} p(\phi_i) \sum_{j=1}^{M(i)} \gamma_j(i)\alpha_j(i), \tag{15.3}$$

where we recall that $p(\phi_i)$ is the probability that the failure-based Markov chain $\mathcal{X}$ is in state ϕ_i.

The security test cases described in [71] have been shown to create increased resource usage and possible system instability. Therefore, the performability approach just described for the assessment of system reliability and performance applies also to the evaluation of a security/performance/reliability related customer affecting metric, when the system is tested using known security attack test cases. The additional work required to assess the impact of security attacks on system performance is related to the developement of security attack test cases, and the extension of the resource usage-based Markov chain and the resource failure based Markoc chain to include configurations related to the security test case and new resource usage based states that are related to the security attacks.

15.3 Test Case Generation Process

15.3.1 Initial Tests

The first step to uncovering possible instability is to subject the system to a well defined mix of transactions at increasing rates. The transaction rate, or a set of rates for different transaction types, must be kept at the same level(s) for a long enough period for the system to settle into and remain in equilibrium. The purpose of testing the system at increasing loads is to verify that the average resource utilizations are linear with respect to the transaction rate, that average response times vary little during the period that the transaction rate is constant, and that the average response time is insensitive to the offered transaction rate unless at least one resource has a utilization exceeding 90 %. We should increase the transaction rate to the point at which the utilization of the bottleneck resource approaches 95 %. Testing at higher levels of utilization risks saturating the system without conveying information about the resource utilization trend. Note: linearity is conditioned on the speed of the resources being constant over time. To verify linearity of utilizations of processors, one must ensure that their clock speeds do not vary with time during performance testing, e.g., to conserve energy.

15.3.2 Performance Tests to Provoke Instability

Comparing Performance on Uniprocessor and Multiprocessor Systems System instability may be caused by a concurrent programming error or by other causes. It is possible that concurrent programming errors that are not apparent during a performance test on a uniprocessor system (or a uniprocessor with hyperthreading turned off) will become apparent on a multiprocessor system (or a uniprocessor system with hyperthreading turned on) or vice versa. This is because running a multithreaded system on one or the other will allow different sequences of intereleaved thread executions to occur. In the authors' experience, there is a concurrent programming problem if the performance is worse or less stable on a multiprocessor system than on a uniprocessor system, other things being equal.

Tests Involving Bursts or Surges of Activity

Periodic Load. Some systems generate constant amounts of work at regular intervals, and then become quiescent. Examples include network management systems and building management and surveillance systems that generate status polling messages at regular intervals.

Quiet and Bursts. Some systems will alternate between quiet intervals and bursts of traffic lasting short or long amounts of time.

For example:

- A network management system must automatically execute procedures in response to traps and alarms, or bursts of traps and alarms.
- An airport conveyor system handling the baggage of arriving and transferring passengers will be subject to a new burst of demand whenever a plane arrives.

Intense Extended Surge of Transactions. An intense extended surge of activity can cause a system to crash. Instead, the system should merely suffer performance degradation during and immediately after the surge and then recover gracefully without further intervention.

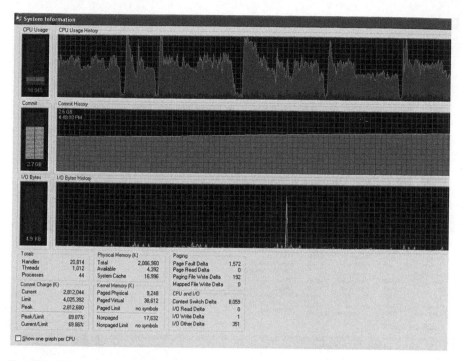

Fig. 15.1 Measurements of an XP-based system showing, from *top* to *bottom*, CPU utilization, committed memory, and I/O byte traffic over time

15.4 Empirical Results

15.4.1 Increasing Memory Occupancy and Periodic Drops in CPU Usage

Figure 15.1 shows measurements of a system under constant load. The processor usage drops off from time to time, and then resumes its normal usage after showing a spike of activity following the dropoff. Memory usage increases at a rate that appears to be nearly constant. The spike in I/O activity may be attributable to logging.

- The drops in CPU usage indicate that a deadlock repeatedly occurred, followed by a release on timeout. An investigation of the code showed that this was indeed the case. The problem was remedied, and troughs went away on a subsequent run.
- The constant increase in memory usage indicates the presence of a memory leak. The cause of the leak was identified and eliminated.

This example illustrates how the analysis of performance measurements was used to identify and eliminate two potential causes of system instability. The repeated deadlocks resulted in intervals in which no useful work was being done. Any trans-

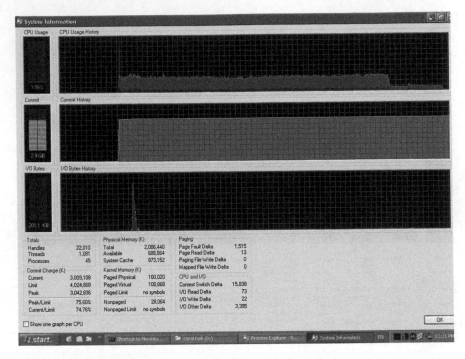

Fig. 15.2 The same XP-based system, after problems were fixed

actions in progress or triggered during those intervals would have had very large response times. In production, the memory leak would have resulted in a system crash or else necessitated the stopping and restarting of the system to prevent it, had it been allowed to continue long enough.

In Fig. 15.2, we see measurements of the same use case after the problems encountered were remedied. The CPU utilization oscillates within a very small range rather than dropping down to zero from time to time as in Fig. 15.1. Also, the number of committed bytes stays fairly constant throughout the run rather than increasing once the initial rampup period is over.

15.4.2 Comparison of a Healthy and an Unhealthy Use Case

In this section, we illustrate how the results of load tests were used to check whether two use cases have suitable performance characteristics.

We first consider Use Case A, which has desirable performance characteristics.

- Figures 15.3 and 15.4 show how the performance measures evolved over time at a load of 300 transactions per second (TPS). The test lasted about 4 min. Observations were collected every fifteen seconds.

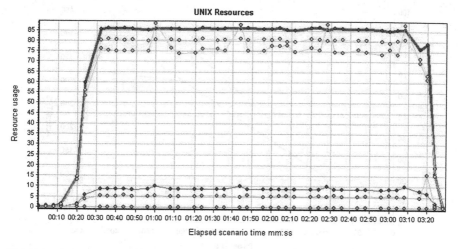

Fig. 15.3 CPU and other utilizations for Use Case A at 300 TPS

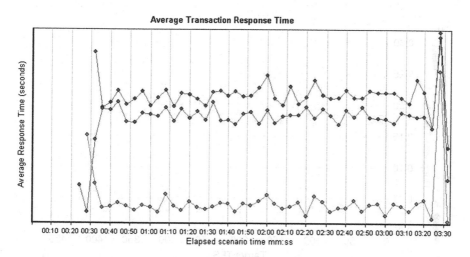

Fig. 15.4 Average response times for Use Case A observed at different load generators at 300 TPS

– Figure 15.3 shows that the CPU utilization (the heavy line) ramped up quickly from zero and leveled off at about 85.
– Figure 15.4 shows the average response times observed at different load generators. These lie within so narrow a range that the performance plotting tool did not print labels on the vertical axis.

• Figure 15.5 is a plot of CPU utilization and actual transaction rate as a function of the offered transaction rate. It shows that the average CPU utilization and the

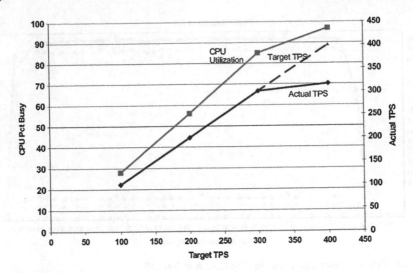

Fig. 15.5 CPU utilization versus load, Use Case A

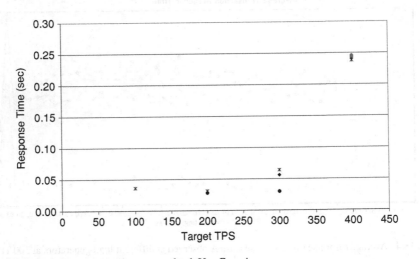

Fig. 15.6 Average response time versus load, Use Case A

actual transaction rate increase linearly to 300 TPS. At 300 TPS, the CPU busy is 85 %. The linearity of the CPU utilization with respect to the target TPS means a target transaction rate of 400 TPS corresponds to a predicted CPU busy in excess of 100.

- Figure 15.6 is a plot of the corresponding average response times. It shows that the average response times are low for target throughputs up to 300 TPS and increase dramatically at 400 TPS. This is consistent with the observations in Fig. 15.4.

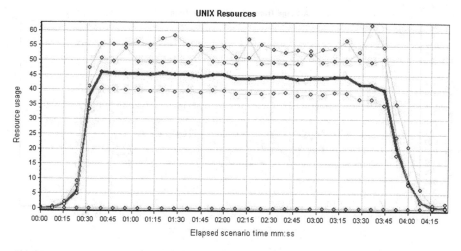

Fig. 15.7   CPU and other utilizations for Use Case B at 2,000 transactions per second

These plots indicate that Use Case A is healthy from a performance standpoint. The CPU utilization is linear with respect to the offered load, while the average response is insensitive to the offered load at transaction rates of 300 per second or less. These results show that the system has good load scalability and indicate the absence of software bottlenecks, concurrency problems, or stability issues, at least in this operating region. We now consider Use Case B, which is unhealthy from a performance standpoint.

- Figure 15.7 shows that the CPU utilization reached a plateau quickly during a test run at 2,000 transactions per second (TPS), but that it diminished slightly during the run well before the offered load was turned off.
- Figure 15.8 shows that the average response times oscillated sharply over a wide range during the same test run, and that the amplitude of the oscillations increased irregularly over time.
- The plot of CPU utilization shows in Fig. 15.9 has an upper bound of about 45 %, and shows that target throughputs in excess of 1,500 per second could not be achieved.
- The plot of average response time increases sharply and linearly with respect to the load when the target transaction rate exceeds 1,500 per second (Fig. 15.10).

Taken together, these curves indicate the presence of a software bottleneck or a concurrency problem. The large oscillations of the average response times suggest that the system would be unstable at loads greater than or equal to 1,500 TPS. Since Use Case B implements a frequently used service, these performance test results have provided us with an early warning of a problem that might have been difficult to isolate by testing the applications themselves.

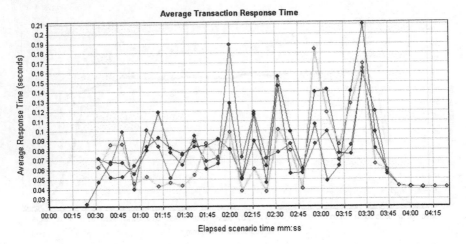

Fig. 15.8 Average response times for Use Case B observed at different load generators at 2,000 transactions per second

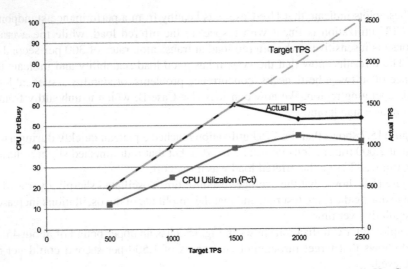

Fig. 15.9 CPU utilization and actual transaction rate versus target transaction rate for Use Case B

15.4.3 Unstable Throughput

When a system is functioning smoothly, the transaction failure rate should be low. In the system we were testing, we increased the number of virtual users at regular time intervals. If the system were functioning smoothly, we would expect the transaction completion rate to increase with the number of virtual users. The graphs of both virtual users and transaction completion rates would have the form of an ascending

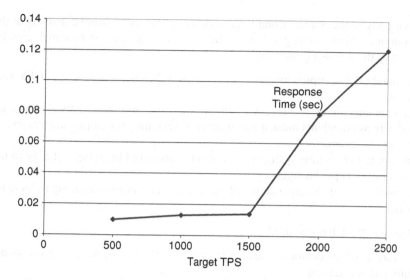

Fig. 15.10   Average response time versus target transaction rate for Use Case B

staircase. In this experiment, the number of virtual users was used as a surrogate for the offered load. Transactions were generated at a set rate by each virtual user. Instead of describing the staircase form, both the transaction completion and failure rates exhibited spikes and troughs, to the extent that the failed and successful transaction rates became the same during an interval near the end of the experiment.

These results indicate system instability with multiple causes. Because the sum of the failed and successful transaction rates does not describe a staircase (since some of the peaks and troughs of both rates occur at the same time), we know that there are intervals in which transaction processing was obstructed altogether, indicating the possibility of deadlocks and timeouts. A detailed examination of the run log and of the data showed the presence of a thread safety problem, in addition to the possibility of a deadlock problem, perhaps caused by the exhaustion of an object pool. Further investigation revealed this to be the case.

15.4.4 Evolution of a System Towards Deadlock

Consider a resource pool that is accessed by threads requesting and relinquishing identical resources through a common FCFS queue. The authors have seen this in more than one computer-controlled system. We resort to a metaphor. The coat rooms in some museums in New York City have this structure. Visitors queueing to leave their coats on hangers and to collect their coats pass through a common queue to reach the attendants who serve them. Deadlock occurs if all hangers are occupied and the first visitor in the queue wishes to leave a coat. This could occur in the middle of a busy day in the winter, when large numbers of visitors might be arriving at the museum in coats or wishing to leave. The onset of deadlock can be anticipated by monitoring the following metrics:

- The number of occupied hangers increases towards it maximum value and hovers near it.
- The queueing times of visitors gradually increase, and then suddenly increase markedly as attendants must spend more time searching for empty hangers and/or for coats to return.
- The number of visitors queueing to collect coats could increase, but only to the number of occupied hangers.
- The utilization of the attendants will increase, thus causing queueing to increase further.

   Once deadlock has occurred,

- The rate at which customers leave the system will be zero, because no coats are being left or returned.
- The utilization of the coat room attendants due to coat distribution and collection will be zero.
- Queue lengths outside the coat room will increase.

In a discrete event simulation of this system, deadlock can be induced by generating a burst of "visitors" who leave their "coats" in the coat room for a long time compared with the times between arrivals. If the size of the simulated burst exceeds the number of hangers, deadlock will occur. Deadlock can be resolved by coat room attendants asking those who wish to collect coats (thus freeing hangers) to come to the head of the queue. It can be prevented by giving priority to those collecting coats, but that might be socially unacceptable and/or architecturally infeasible [138, 139]. Notice that in this instance, the steady increase in the number of occupied hangers would not be caused by a "hanger" leak, because that would correspond to visitors failing to collect their coats. Instead, it is caused by visitors wishing to free a resource (hangers) being delayed by those wishing to acquire one. This in and of itself is a source of system instability, quite apart from the possibility of hangers being exhausted by a high volume of visitors to the museum on a cold day. For this reason, the problem might have gone unnoticed had it not occurred in a laboratory setting. This suggests that performance measurement is not sufficient to reduce the risk of instability. Reviews are also needed to prevent the implementation of design choices that could lead to instability in the first place.

15.4.5 DOS Security Attacks

Figure 15.11 shows the impact of a Denial of Service (DOS) security attack on the memory usage performance signature for a system that is run with constant load. The figure shows both the normal constant load memory usage performance signature,

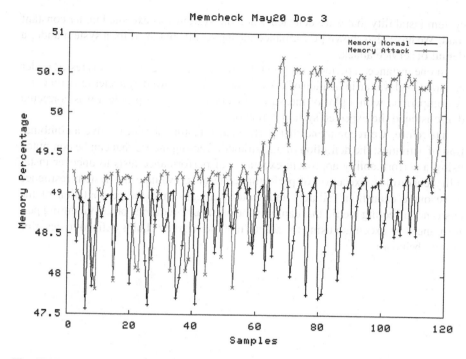

Fig. 15.11 Memory usage performance signature for constant load and DOS security attack

and the DOS attack signature, while the same normal constant load is executed in the background. The DOS attack is initiated mid-way through the test at sample point 60. Similar plots are shown in [71] for stack overflow, buffer overflow, sql injection, and Man in the middle (MITM) security test cases.

15.5 Conclusions

We have presented several approaches for resilience assessment based on performance measurements.

We have shown how a performability approach can be used to define a resilience assessment customer-affecting metric, and we have summarized the modeling and assessment approach to compute the customer-affecting metric using performance measurement results and the solution of the Markov chains that describe system behavior.

We have described the types of tests that we recommend should be included in a performance test suite designed to uncover system instability.

We have illustrated our concepts with several performance testing results showing common instability conditions. Specifically, we have shown several examples of system instability that were detected when the system was executed under constant load. We have also shown performance testing results obtained for a system under a denial of service attack.

In one instance, a periodic drop in CPU usage when the system was tested under constant load was an indication of a system deadlock followed by a release on timeout. A comparison with a plot of measurements taken after the problem was corrected demonstrated the expected system behavior.

Therefore, we recommend that performance testing should involve a combination of activities: the definition of a customer-affecting metric that can be computed using a performability approach; execution of performance tests to uncover instability conditions; execution of performance tests to evaluate the system resilience to resource exaustion, failures, and security attacks; and the computation of the customer-affecting metric that captures the pass/fail conditions observed during performance test execution using the solution of the Markov chains defined to describe system behavior.

Chapter 16
Robustness Testing Techniques and Tools

Zoltán Micskei, Henrique Madeira, Alberto Avritzer, István Majzik, Marco Vieira and Nuno Antunes

Abstract Robustness is an attribute of resilience that measures the behaviour of the system under non-standard conditions. Robustness is defined as the degree to which a system operates correctly in the presence of exceptional inputs or stressful environmental conditions. As triggering robustness faults could in the worst case scenario even crash the system, detecting this type of faults is of utmost importance. This chapter presents the state of the art on robustness testing by summarizing the evolution of basic robustness testing techniques, giving an overview of the specific methods and tools developed for major application domains, and introducing penetration testing, a specialization of robustness testing, which searches for security vulnerabilities. Finally, the use of testing results in resilience modelling and analysis is discussed.

Z. Micskei (✉) · I. Majzik
Budapest Univ. of Technology and Economics,
Budapest, Magyar Tudósok krt. 2, Hungary
e-mail: micskeiz@mit.bme.hu

I. Majzik
e-mail: majzik@mit.bme.hu

H. Madeira · M. Vieira · N. Antunes
DEI/CISUC, University of Coimbra,
3030-290 Coimbra, Portugal
e-mail: henrique@dei.uc.pt

M. Vieira
e-mail: mvieira@dei.uc.pt

N. Antunes
e-mail: nmsa@dei.uc.pt

A. Avritzer
Siemens Corporate Research and Technology,
755 College Road East, Princeton, NJ 08540, USA
e-mail: alberto.avritzer@siemens.com

K. Wolter et al. (eds.), *Resilience Assessment and Evaluation of Computing Systems*,
DOI: 10.1007/978-3-642-29032-9_16, © Springer-Verlag Berlin Heidelberg 2012

16.1 Introduction

Robustness is an attribute of resilience that measures the behaviour of the system under non-standard conditions. Robustness is defined in IEEE Standard 24765:2010 as the degree to which a system operates correctly in the presence of *exceptional inputs* or *stressful environmental conditions* [465]. To further refine the difference between robustness and resilience Avizienis et al. defined robustness as *"dependability with respect to external faults, which characterizes a system reaction to a specific class of faults"* [63].

The goal of *robustness testing* is to activate those faults (typically design or programming faults) or vulnerabilities in the system that result in incorrect operation, i.e., robustness failure, affecting the resilience of the system. Robustness testing mostly concentrates on the internal design faults activated through the system interface. The robustness failures are typically classified according to the CRASH criteria [540]: *Catastrophic* (the whole system crashes or reboots), *Restart* (the application has to be restarted), *Abort* (the application terminates abnormally), *Silent* (invalid operation is performed without error signal), and *Hindering* (incorrect error code is returned–note that returning a proper error code is considered as robust operation). The measure of robustness can be given as the ratio of test cases that exposes robustness faults, or, from the system's point of view, as the number of robustness faults exposed by a given test suite.

Robustness testing can be characterized with the following two components of the tests (stimuli); the *workload* triggers (regular) operation of the system, while the *faultload* contains the exceptional inputs and stressful conditions applied on the system. Depending on how these two loads are balanced, robustness testing can be used for verification and evaluation purposes. Robustness testing can be used as a special kind of conformance testing, where only a faultload is executed against the public interfaces of the system. Overloading can also be considered as a stressful condition, this way stress tests (i.e., submitting only a high amount of workload) are also used to assess the robustness of a system.

Security testing is the process in which the software is verified not for functional purposes but to detect vulnerabilities [634]. A vulnerability is a weakness (an internal fault) that may be exploited by an attacker to cause harm or gain access to the system [860]. *Penetration testing* is a technique to perform security testing and consists of a particular form of robustness testing in which the application execution is analysed when submitted to malicious conditions (i.e., malicious input parameters that try to take advantage of vulnerabilities). Penetration tests are widely used by developers to detect security vulnerabilities in their code [860] and consist of stressing the application from the point of view of an attacker using specific malicious inputs. Penetration testing can be performed manually or automatically. However, automated tools, also referred as web vulnerability scanners, are the typical choice as, comparing to manual tests and inspection, execution time and cost are quite lower. These tools provide an automatic way for searching for vulnerabilities, avoiding the repetitive and tedious task of doing hundreds or even thousands of tests by hand for each vulnerability type.

Robustness benchmarks are agreed-upon and reproducible procedures defined by specifying (1) the workload, (2) the faultload, (3) the standard procedures and rules to execute them, (4) the experimental setup, and (5) the relevant measures that characterize the robustness of the system under benchmarking. As robustness is an attribute of dependability, robustness benchmarks can be considered as a special category of dependability benchmarks.

From the point of view of the lifecycle of systems, on the one hand, robustness testing can be used as an internal step in the development process, complementing the other verification and evaluation activities. On the other hand, robustness tests can be performed (typically by external parties) after the release of the system, to assess its dependability or to compare it to other systems. The detected robustness faults can be handled either by performing corrections in the design and implementation, or by the application of specific wrappers that are able to confine the effects of the faults. This latter solution is particularly relevant when the system contains commercial off-the-shelf (COTS) components, which cannot be modified to correct the faults/robustness weaknesses, thus a wrapper built around the component is the only solution to improve its robustness.

The outline of this chapter is as follows. Section 16.2 discusses the evolution of basic robustness testing techniques. Section 16.3 provides an overview of the specific methods and tools developed for major application domains. Section 16.4 introduces the penetration testing technique and some of the existing tools. Section 16.5 briefly discusses the use of testing results for resilience modelling and analysis. Finally, Sect. 16.6 concludes the chapter.

16.2 Robustness Testing

In the past decades many research projects were devoted to the robustness testing of a specific application or application type. The early methods were mainly based on hardware fault injection, but later the research focus moved to software-implemented techniques. In this section we introduce the main milestones, which can be connected to the introduction of new *testing techniques*.

16.2.1 Injecting Physical Faults

Early work on robustness testing used *fault injection* (FI) tools to induce or simulate the effects of various hardware related faults. Here a clear distinction shall be made between the purposes of general FI and FI for robustness testing. The general technique assesses the ability of a system or component to handle internal hardware or software faults. In a robustness testing framework, FI can be used to assess the ability of a component to handle interaction faults that are triggered by injecting faults into the environment (e.g., interacting components or underlying layers) while keeping the tested component intact. In this way also the robustness of error detection and error handling mechanisms (considered as components to be tested) can be investigated.

FIAT [92] or FTAPE [865] are examples for FI tools that are reported to be used for such robustness testing purposes.

16.2.2 *Using Random Inputs*

One of the first robustness testing techniques was the generation of *random input* for the system. Random inputs are easy to generate, there is a chance that robustness faults are activated by them, and due to the simple acceptance criteria (crash/hung is checked) there is no need to generate reference output.

Fuzz [650] was one of the first tools supporting this technique. It was utilized in three series of experiments to test reliability and robustness of various applications. In 1990, utility programs on seven variants of Unix operating systems were tested. In 1995, the tests were repeated to check whether robustness of these utilities had been improved and support to test X Window applications were added. Lastly, in 2000, Fuzz was used to test 30 GUI applications on Windows NT. Although the method used was really simple, it detected a great deal of robustness errors, namely 40 % of the Unix command line programs and 45 % of the Windows NT programs crashed (terminated abnormally) or hung (stopped responding to input within a reasonable length of time) when called with random input data.

Although random testing is a basic technique, it proves to be useful even for modern COTS software systems. The tests in Fuzz were reapplied to MacOS in a study prepared in 2007 [649] with the following results: 10 command line utilities crashed out of the 135 utilities that were tested (a failure rate of 7 %), 20 crashed and 2 hung out of the 30 GUI programs tested (a failure rate of 73 %). Thus, it turns out that robustness testing using random inputs is still a viable technique as robustness of common software products has not been significantly improved in general in the last fifteen years.

Fuzzing is extensively applied to security related testing, as presented in a recent book [865].

16.2.3 *Using Invalid Inputs*

Basic software robustness testing technique is the systematic calling of the interface functions of the system under test using parameter values that are selected from the boundaries and outside of their allowed domain. E.g., if allowed values of a parameter are positive integers then robustness tests may contain the zero, a negative number and the MAXINT value.

A typical tool supporting this approach was Riddle [377], which used a grammar-based description of the system's input to generate random and invalid tests. According to the grammar definition legally structured inputs were generated that were filled with random values, possible malicious values (like special or non-printable characters) and boundary values (for example numbers like MAXINT + 1). In this way, syntactically correct inputs could also be created (not just totally random streams), and a greater portion of the systems functionality could be accessed by the tests. Results showed that about 10 % of the tests on GNU command line utilities produced unhandled exceptions. The robustness failures observed were mainly memory access violation exceptions, privileged instruction exceptions and illegal instruction exceptions. The typical cause of these exceptions was the improper handling of non-printable characters and (excessively) long input streams.

Another area where invalid inputs can be easily defined and used for robustness testing is low-level OS device drivers. In [643] such drivers were tested by selecting extreme values from the following categories: forbidden value, out of bounds value, invalid pointer assignment, NULL pointer assignment, missing local variable initialization, and missing call of a related function.

16.2.4 Using Type-Specific Tests

The robustness tests can be further refined by using specific invalid inputs for each function in the system's interface. To minimize the amount of manually created test cases a *type-specific approach* was introduced. The basic idea is that valid and invalid inputs are defined for the data types used in the system's interface functions, and the robustness tests are generated by combining the values for the different parameters. The size of the invalid input domain can be further reduced by using inheritance between the types to test.

The Ballista tool [540] used this approach to compare the robustness of 15 POSIX operating systems using a test suite for 233 function calls. The general goal of the research was to implement methods to measure the robustness of the exception handling mechanism of systems. The results could be used to evaluate the dependability of a system and characterize how it responds to the failures of other components. In an experiment performed on the Safe Fast I/O (SFIO) library, the performance drawback of robustness hardening was also measured. The tests showed that the performance penalty of proper data validation and parameter checking was fewer than 2 %. A good summary of the experiences gained using Ballista can be found in [542].

16.2.5 Testing Object-Oriented Systems

The type-specific technique mentioned above can be enhanced in object-oriented (OO) systems with the help of automatically building a parameter graph with the type structure. The parameter graph describes how the specific object types used as parameters in method calls can be generated as results of calling constructors or public methods of other classes. This way the generation of an invalid object (needed to test a given method) can be traced back to the call of another method (possibly having parameters of simpler input types).

The JCrasher tool [249] creates robustness tests for Java programs automatically by analysing which methods could return a type needed for the actual parameters. It examines the type information of the set of Java classes constituting the application and constructs code fragments that will create instances of different types to test the behaviour of public methods with random or invalid data.

In OO applications the testing of exception handling is an important aspect of assessing the robustness of the fault handling and recovery code. Exception flow analysis and testing exception-catch paths is presented in [358].

16.2.6 Applying Mutation Techniques

Code mutation techniques [285] can be also applied to generate robustness tests. Starting from a valid code, e.g., a functional test or an application using the system's interfaces, mutation operators can be applied, which resemble the typical faults causing robustness problems (e.g., omitting calls, interchanging calls, replacing normal values in parameters with invalid values).

Mutation and extension of valid test sequences may also help in state-based systems or components to cover more states and transitions than in case of stateless API testing. In [584], first a set of paths is generated to cover state transitions of the tested component, and normal test cases are applied to traverse these paths and bring the component into specific states. In each state, the available methods are called with invalid inputs to test the robustness in that state. This approach is motivated by the fact that complex components may fail differently in different states.

16.2.7 Model-Based Robustness Testing

The increasingly popular model-driven development paradigm led to the idea of model based testing (using models as formal or semi-formal specification for testing purposes) [159, 290] and also model based automated test generation [26]. Naturally, model based test generation can be tailored to create robustness tests by looking for extreme values and conditions on the basis of the pre- and post-conditions, invariants, and constraints fixed in the design model. Model-based testing is currently a very active field of research; here we mention only a few techniques and tools that are relevant to testing robustness.

The first test generation approaches utilized formal specifications and functional models (B, Z, LOTOS, etc.). Constraint-solving techniques were applied to generate boundary values of input domains as well as the corresponding test cases. In state based formalisms, e.g., in IOLTS [338], path searching and model mutation (on the basis of fault models) were applied in order to find tests for concrete robustness criteria. Timed behaviour was modelled and tested using timed automata [350] or extended interoperability models [628].

In case of communication protocols SDL was the primary modelling language used for generating robustness tests [792]. In another work [741], finite state machine models of communication protocols were extended and faulty protocol data units were generated on the basis of a stress operational profile in a statistical approach to model based robustness testing.

In UML based designs the Object Constraint Language (OCL) was used to specify valid domains, this way providing input information also for robustness testing. Typical examples of UML based test generator tools that support (a subset of) OCL were LTG/UML [902] and ParTeG [932].

Model based configuration and execution of robustness testing is complementary to model based test generation. In [700], a framework was presented that fits to the model based development approach by offering to the tester the model of the tested application (using UML class diagram model elements) and domain-specific extensions that allow the configuration of fault injection and robustness testing experiments. The modifications that are required for robustness testing are implemented automatically (using a Java bytecode manipulation technology) on the basis of the model extensions.

16.2.8 Historical Overview of the Basic Robustness Testing Approaches

A detailed survey of robustness testing techniques was provided in the context of the ReSIST Network of Excellence, in particular in the report summarizing the state of knowledge [772].

To conclude the section on basic approaches and major supporting tools, Table 16.1 presents the historical time line of the techniques and evaluates their current applicability, while Fig. 16.1 presents the relations between the techniques.

Table 16.1 Historical time line of techniques used for robustness testing

Testing technique	Introduced	Evaluation of applicability
Injection of physical faults	Early 1990s	Used in specific domains, e.g., in safety-critical systems
Random inputs	Mid 1990s	Basic technique, still useful for off-the-shelf software
Invalid inputs	Late 1990s	Used as part of type-specific testing
Type-specific testing	Around 2000	Very effective technique, used together with mutation
OO approach	Early 2000s	Extension of type-specific tests to OO languages
Mutation techniques	Early 2000s	Effectively complements type-specific techniques

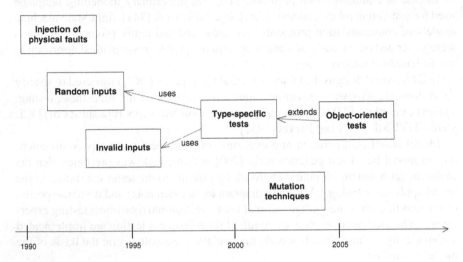

Fig. 16.1 Relations between the basic robustness testing techniques

16.3 Robustness Testing Techniques in Specific Application Domains

In recent years, robustness testing techniques have been successfully used in several application domains. In the following we survey the peculiarities of these techniques.

16.3.1 User Interfaces

State-based testing of graphical user interfaces (GUI) relies on building a graph based model that describes the elements of the interfaces and the connections (e.g., allowed activation sequences) between them. This model is often called an event-flow or event-interaction graph. In a robustness testing scenario the graph describing the normal operation is completed with the connections that are not allowed. The activation of these connections (after reaching the activation of their starting point by a sequence of normal interactions) constitutes the robustness test suite. These robustness tests can be generated automatically based on the possible sequences obtained from the graph model. This technique can be considered as a specific mutation technique that appends an invalid activation after the sequence of valid ones.

The GUITAR framework [638] is a tool set for creating automatic tests for GUIs. It detects the elements of the GUI, e.g., menus, buttons, and constructs an event flow graph. Rapid tests trying to crash the application can be generated from the model after each release of the software. The effectiveness of the method and the tools were demonstrated on the GUI of an open-source office suite.

16.3.2 High Availability Middleware

Robustness is a key factor in middleware systems that are applied to provide high availability (HA) services to applications and to manage the configuration of redundant components: robustness faults in the HA middleware can be activated by poor quality application components, and this way one such component may render the whole application inaccessible. The complexity in testing these middleware implementations comes from the highly state-based nature of these systems: without a proper setup code most of the calls in the public interface result in trivial error messages, this way the robustness of the valid operation cannot be tested.

In [648] the common interface specified by the Service Availability Forum was used to test and compare the robustness of different middleware implementations. A robustness test suite was constructed on the basis of the definition of the common interface using a type-specific approach, i.e., various combinations of valid and invalid inputs were generated automatically. To test stressful environmental conditions an operating system call wrapper was implemented, since the manifestation of many stressful conditions (network errors, lack of disk space etc.) can be simulated by injecting errors into the return values of system calls. Additional robustness tests were generated by mutating tests in the functional test suite (changing the order of calls, applying invalid parameter values when a given state is reached, etc.). These methods exposed distinct robustness faults in the system, demonstrating the importance of combining the different testing techniques.

16.3.3 Real-Time Executives

Microkernels are currently common components in a wide range of applications, ranging from daily-use appliances (e.g., mobile devices) to space-borne vehicles. Robustness testing of microkernels has been addressed using stressful operational conditions, invalid inputs at the public interfaces, and fault injection.

In [56] the MAFALDA tool is used to collect objective failure data and to find ways to improve the error detection capabilities of the Chorus and the Lynx microkernels. A specialized version of that tool, the MAFALDA-RT, is used in [784] to test real-time systems. The tool applies a novel method to cope with the problem of temporal intrusiveness caused by the use of software implemented fault injection (SWIFI) tools. In addition to detecting typical failure modes (e.g., application hang, system hang, exception, etc.), the observation capabilities of MAFALDA-RT are extended to consider temporal properties characterizing both the executive and application layers (e.g., task processing, task synchronization, context-switch, system calls, etc.).

The Xception tool [241] is a SWIFI tool that was used in several experiments of microkernel robustness testing (e.g., [86, 622, 665]). This tool is able to introduce perturbations in the processor to emulate errors at the hardware and software level (in the case of a software fault, the fault is introduced when user applications are being executed). The robustness evaluation consists of assessing the abilities of the microkernel to handle the stressful situation caused by the existence of faults. One interesting work used Xception to emulate hardware transient faults and observe their effects at the user application and at the operating system level [615]. The results demonstrated that errors occurring in one user-mode application can propagate to other user-mode applications through the OS itself.

16.3.4 OLTP and DBMS

Although not as common as other types of systems, Online Transaction Processing (OLTP) systems were also used in past robustness related works. The following two studies are worth noting.

The work presented in [919] presents a dependability benchmark (covering also the attributes of robustness) for OLTP systems. The experiments exposed the OLTP systems under observation to stressful conditions and measured the performance penalty of the Database Management System (DBMS) engine and the rate of occurrence of data corruption on the data tables. The stressful conditions were caused by the injection of operator errors. This work demonstrated that it is possible to assess the dependability and robustness properties of OLTP systems and rank the systems under study according to the results.

The work in [342] presents a method to detect intrusions and malicious data access based on DBMS auditing. Although this work is not exactly inline with the traditional robustness testing described earlier in this section, assessing the ability to prevent or detect intrusion can constitute a measure to the robustness of a system (in this case, the security attributes of that system).

16.3.5 Web Services

Web Services are a widespread technology to implement services accessible over a computer network. Each Web Service has a well-defined interface (usually using a standardized description language). Several methods were proposed to generate robustness tests based on these interfaces for Web Services.

The work described in [897] presents a specification-based robustness testing framework for Web Services. The testing framework includes the analysis of the service specification to ensure that it is complete and consistent. It applies the Covering Scenario Generation algorithm to identify the locations where incompleteness and inconsistency exist. The testing framework also includes the robustness testing of the Web Service by generating positive test cases (that should be successful) as well as negative test cases (that should not be successful).

WebSob [626] is a tool that automates the generation and execution of test cases for Web Services. This tool executes Web Service requests using the provider's Web Service Description Language (WSDL) specification. This tool was applied to freely available implementation of Web Services and revealed several robustness problems. The tool does not require the knowledge of the implementation of the Web Services under test.

The work presented in [916] defines a benchmark to assess the robustness of Web Services. It uses invalid data in the method invocations to discover both programming and design errors. The parameter values used in the method invocations are modified (i.e., corrupted) based on the data types and the semantics of the method parameters. The Web Services are classified according to the type and number of failures observed during the tests.

The WS-FIT tool [604] performs a dependability analysis of Web Services. For the sake of robustness testing it performs network fault injection by capturing, modifying, and retransmitting SOAP messages. This technique allows for easy corruption of RPC method invocation within SOAP messages and can emulate the following errors: corruption of input and output data, omission or duplication of messages, delay of messages.

16.4 Penetration Testing

Penetration testing, a specialization of robustness testing, consists of the analysis of the program execution in the presence of malicious inputs (based on a database of know malicious values for specific applications or generated based on predefined rules), searching for potential vulnerabilities [860]. Hackers are nowadays moving their focus to this kind of vulnerabilities and explore applications' inputs with specially tampered values trying to find weaknesses. These vulnerabilities cannot be mitigated by traditional security mechanisms such as firewalls and intrusion detection system, thus highlighting the importance of detecting these vulnerabilities before deployment. Also, the exposition of web applications makes them particularly prone to attacks that try to exploit code vulnerabilities. In this section we overview existing techniques, focusing particularly in web applications and code vulnerabilities, as penetration testing is particularly important in these scenarios. Penetration testing is, in fact, the most common approach to detect vulnerabilities in web applications and web applications are the most frequent context in which penetration testing is used.

16.4.1 "Black-Box" Penetration Testing

In "black-box" penetration testing the tester does not know the internals of the application and it uses fuzzing techniques over the applications requests [860]. The tester needs no knowledge of the implementation details and tests the inputs of the application from the user's point of view by applying malicious inputs. The number of tests can reach hundreds or even thousands for each vulnerability type. The vulnerability detection is based essentially on the analysis of the application output. The human tester or the tool analyses the contents of this output including values or errors returned and exceptions raised. The vulnerabilities are detected when certain patterns found in the response are caused by the attacks launched. Many black-box penetration testing techniques were proposed in the past. We introduce a few in the next paragraphs due to the relevant innovations they introduced.

WAVES [450] is a black-box technique for testing web applications for SQL-Injection vulnerabilities. The technique is based on a reverse engineering process that identifies the data entry points of a Web application and attacks them with malicious patterns. An algorithm is proposed to allow "deep injection" and to eliminate false negatives. During the attack phase, the application's responses to the attacks are monitored and machine-learning techniques are used to improve the attack methodology.

SecuBat [504] is an open-source penetration testing tool that uses a black-box approach to crawl and scan web sites for the presence of exploitable SQL injection and cross-site scripting (XSS) vulnerabilities. SecuBat does not rely on a database of known bugs. Instead, it tries to exploit the distinctive properties of application-level vulnerabilities. To increase the confidence in the correctness of the results, the tool also attempts to automatically generate proof-of-concept exploits in certain cases.

A black-box taint-inference technique for the detection of injection attacks is proposed in [817]. The technique does not require any intrusive source-code or binary instrumentation of the application to be protected; instead, it intercepts the inputs and outputs of the application. Then, the technique infers tainted data in the intercepted SQL statements, and then employs syntax and taint-aware policies to detect unintended use of tainted data.

In [631] an automated penetration testing tool is presented that can find reflected and stored XSS vulnerabilities in web applications. The proposed technique improves the effectiveness of penetration testing by leveraging input from real users as a starting point for its testing activity. The technique follows an entire user's session using recorded real user inputs to generate test cases to launch fuzzing attacks. This way, the technique increases the code coverage by exploring pages that are not reachable for other tools. The experiments show that the approach is able to test more thoroughly the web applications and identify more bugs than a number of open-source and commercial tools.

A vulnerability scanner for web services that performs better than the commercial ones currently available is presented in [48]. The work focuses on the detection of SQL Injection vulnerabilities, one of the most common and most critical types of vulnerabilities in web environments. The proposed approach is based on a large set of attacks and includes an enhanced response analysis (i.e., vulnerability detection) algorithm. Experimental evaluation shows that the proposed approach performs much better than well-known commercial tools, achieving very high detection coverage while maintaining the false positives rate quite low.

16.4.2 *"Gray-Box" Penetration Testing*

The main limitation of black-box approaches is that vulnerability detection is limited by the output of the tested application. Gray-box approaches consist of complementing black-box testing with white-box techniques to overcome such limitation.

Dynamic program analysis is based on the analysis of the behaviour of the software while executing it [860]. The idea is that by analysing the internal behaviour of the code in the presence of realistic inputs it is possible to identify bugs and vulnerabilities. Obviously, the effectiveness of dynamic analysis depends strongly on the input values (similarly to black-box testing), but it takes advantage of the observation of the source code. For improving the effectiveness of dynamic program analysis, the program must be executed with sufficient test inputs. Code coverage analysers help guaranteeing an adequate coverage of the source code [409, 664].

Source code or bytecode instrumentation is a technique that can be used to develop runtime anomaly detection tools. In [578] it is proposed an anomaly detection approach to secure web services against SQL and XPath Injection attacks. The web service is instrumented in such way that all the SQL/XPath commands used are intercepted before being issued to the data source. The approach consists of two phases. First, in the learning phase, the approach learns the regular patterns of the queries being issued. Then, at runtime, the commands are compared with the patterns learned previously in order to detect and abort potentially harmful requests. The problem of this technique is that it does not include the generation of the requests to use during the learning phase and so it requires the user to exercise the web service in the learning phase.

In [47] the runtime anomaly detection approach from [578] was enhanced with an automated workload and attackload generation technique in order to be able to detect SQL and XPath Injection vulnerabilities in web services. This way, after the instrumentation of the web service a workload is generated using information about the domains of the parameters of the web service operations. Learning takes place while executing the workload to exercise the web service. Afterwards, an attackload is generated and used to attack the web service. Vulnerabilities are detected by comparing the incoming commands during attacks with the valid set of commands previously learned.

"Acunetix AcuSensor Technology" [19] is a technique introduced by Acunetix that combines black-box scanning with feedback obtained during the test execution. This feedback is provided by sensors previous placed, using code instrumentation, inside the source code or bytecode. Using this technique it is possible to find more vulnerabilities, to indicate in the code exactly where they are, and to report less false positives. This technology is available for web applications, specifically .NET and PHP web applications. In case of .NET this technology can be injected in bytecode.

Two techniques that combine static and dynamic analysis have been proposed to perform automated test generation to find SQL Injection vulnerabilities. SQLUnit-Gen, presented in [824], is a tool that combines static analysis with unit testing to detect SQL injection vulnerabilities. The tool uses a third-party test case generator and then modifies the test cases to introduce SQL injection attacks. In practice, these attacks are obtained by using static analysis to trace the flow of user input values to the point of query generation. Sania, presented in [544], is a testing framework to detect SQL Injection vulnerabilities in web applications during development and debugging phases. Sania intercepts the SQL queries between a web application and a database and constructs parse trees of these queries. Terminal leafs of parse trees typically represent vulnerable spots. The technique then generates attacks according to the syntax and semantics of these potentially vulnerable spots. Finally, Sania compares the parse trees of the original SQL query with the ones resulting after an attack to assess the safety of these spots. The differences between the parse trees are considered vulnerabilities, originating a warning.

Runtime anomaly detection tools are used as attack detection systems to protect the applications at runtime. However, a detected attack is typically the exploitation of an existing vulnerability, therefore showing that at least a vulnerability exists. Examples include [578] and AMNESIA (Analysis and Monitoring for NEutralizing SQL-Injection Attacks) [406]. AMNESIA combines static analysis and runtime monitoring to detect and avoid SQL injection attacks. Static analysis is used to analyse the source code of a given web application building a model of the legitimate queries that such application can generate. At runtime, AMNESIA monitors all dynamically generated queries and checks them for compliance with the statically generated model. When a query that violates the model is detected it is classified as an attack and is prevented from accessing the database. The problem is that the model built during the static code analysis may be incomplete and unrealistic because it lacks a dynamic view of the runtime behaviour of the application.

While other works focused on identifying vulnerabilities related to the use of external inputs without sanitizations, the work presented in [84] introduces an approach that combines static and dynamic analysis techniques to analyse the correctness of sanitization processes in web applications. First, a technique based on static analysis models the modifications that the inputs suffer along the code paths. This approach uses a conservative model of string operations, which might lead to false positives.

Then, a second technique based on dynamic analysis works bottom-up from the sinks and reconstructs the code used by the application to modify the inputs. The code is then executed, using a large set of malicious input values to identify exploitable flaws in the sanitization process.

16.4.2.1 Examples of Penetration Testing Tools

Penetration testing tools provide an automatic way to search for vulnerabilities avoiding the repetitive and tedious task of doing hundreds or even thousands of tests by hand for each vulnerability type. The most common automated security testing tools used in web applications are generally referred to as web security scanners (or web vulnerability scanners). Web security scanners are often regarded as an easy way to test applications against vulnerabilities. These scanners have a predefined set of tests cases that are adapted to the application to be tested. In practice, the user only needs to configure the scanner and let it test the application. Once the test is completed the scanner reports existing vulnerabilities (if any detected). Most of these scanners are commercial tools, but there are also some free application scanners often with limited use, since they lack most of the functionalities of their commercial counterparts.

Two very popular free security scanners that support web services testing are Foundstone WSDigger [351] and WSFuzzer [712]. WSDigger is a free open source tool developed by Foundstone that executes automated penetration testing in web services. Only one version of this software was released up to now (in December 2005). The tool contains sample attack plug-ins for SQL Injection, cross-site scripting (XSS), and XPath Injection, but it was released as open-source to encourage users to develop and share their own plug-ins and its test files are simple to edit to add new test cases. WSFuzzer is a free open source program that mainly targets HTTP based SOAP services. This tool was created based on real-world manual SOAP penetration testing work, automating it. Nevertheless, the tool is not meant to replace a solid manual human analysis. One problem of this tool is that its configuration is very complex. The main problem of both WSDigger and WSFuzzer is that, in fact, they do not detect vulnerabilities: they attack the web service under testing and log the responses leaving to the user the task of examining those logs and identify the vulnerabilities. This requires the user to be an "expert" in security and to spend a huge amount of time to examine all the results.

As for commercial scanners, three brands currently lead the market: HP WebInspect [434], IBM Rational AppScan [459] and Acunetix Web Vulnerability Scanner [755].

HP WebInspect is a tool that *"performs web application security testing and assessment for today's complex web applications, built on emerging Web 2.0 technologies. HP WebInspect delivers fast scanning capabilities, broad security assessment coverage and accurate web application security scanning results"* [433]. This tool includes pioneering assessment technology, including simultaneous crawl and audit (SCA) and concurrent application scanning. It is a broad application that can be applied for penetration testing in web-based applications.

IBM Rational AppScan "*is a leading suite of automated Web application security and compliance assessment tools that scan for common application vulnerabilities*" [459]. This tool is suitable for users ranging from non-security experts to advanced users that can develop extensions for customized scanning environments. IBM Rational AppScan can be used for penetration testing in web applications, including web services.

Acunetix Web Vulnerability Scanner "*is an automated web application security testing tool that audits a web applications by checking for exploitable hacking vulnerabilities*" [755]. Acunetix WVS can be used to execute penetration testing in web applications or web services and is quite simple to use and configure. The tool includes numerous innovative features, for instance the "AcuSensor Technology" [19].

16.5 Resilience Modelling and Analysis Using Testing Results

The Chapter on performance testing contains an overview of the application of modelling and analysis based in performance testing results. The described approach can be generalized to robustness testing with proper metrics and model state definitions.

The approach consists of defining a resilience related metric that can be derived from the system security, reliability or performance requirements as follows:

The fraction of time the system satisfies the defined resilience requirements specifications during an observation period (0,t)

The steps required to implement the approach presented in [69] are:

1. A resilience-based state definition needs to be devised. The resilience-based state definition maps system resources to the events the system is designed to be resilient to. For example, resilience related events could be faults, security intrusions, or any other system activity that needs to be modelled.
2. Resilience modelling using the approach introduced in [69, 72] requires the definition of Markov chains to contain the states and associated events. For example, a failure-based Markov chain captures failures and repair events. Detailed descriptions of the approaches with examples are presented in [68, 71].
3. The utilization of testing results pass/fail conditions to assess the resilience metric requires the association of each resilience test with a Markov chain state. All states associated with a test case result pass condition are used in the resilience metric computation. In addition, the solution of the Markov chain enables the association of the resilience metric with a notion of system reliability.

As topics for future research, we foresee a probabilistic quantification of robustness related initiation and completion events (e.g., failure/repair) in some of the relevant non-functional requirement domains, such as performance, reliability and security.

16.6 Conclusion

This chapter presented an overview on robustness testing techniques, providing examples of applications to several domains. In particular, we have introduced penetration testing, where black-box and gray-box tests are used for detecting security vulnerabilities. Finally, we have also introduced a quantitative approach for the evaluation of a robustness metric by using robustness testing results. The approach is based on the definition of a system model in terms of robustness, the definition of test cases that are related to the model states, and assessing pass/fail for each test case executed in the robustness testing phase.

The robustness testing approaches presented in this chapter can be used to define a systematic process that includes robustness metric definition, modelling of system robustness, robustness test cases generation, automated tools for robustness testing, and the assessment of the system robustness metric by using the pass/fail robustness test case results. The current state of the art in robustness testing emphasizes the need for additional studies on the identification of the most useful robustness models, and the associated probabilistic quantification of the robustness states that are visited by failures and security penetration events.

Acknowledgments The work presented in this chapter was partially funded by the European Commission under project AMBER-Assessing, Measuring and Benchmarking Resilience, IST-216295, funded by the European Union, 2009.

Part VI
Case Studies

Chapter 17
Case Study: Mobile Networks

Samir Bellahsene, Leïla Kloul, Philipp Reinecke
and Katinka Wolter

Abstract In order to be resilient, a network must possess means to ensure connectivity even in the presence of disturbances. This chapter will study two different approaches to increase resilience of mobile networks, in the context of the handover procedure. Seamless handovers between base stations is a prerequisite for service continuity in mobile networks. The handover process consists in handing off a call to a new base station when the mobile user moves to its corresponding cell while the call is in progress. It imposes frequency synchronicity requirements which imply strict bounds on the tolerable frequency deviations of base-station clocks. The preferred protocol for frequency synchronisation in packet-switched backhaul networks is the Precision Time Protocol (PTP). In the first part of this chapter, the suitability of backhaul networks for accurate frequency synchronisation using PTP is investigated. Two solutions for improving accuracy are derived. While the first is applicable to networks of any topology, but may require costly reconfiguration, the second is limited to specific setups, but can be applied without changing the network. The second part of this chapter is dedicated to the performance analysis of a Markov-based prediction model. Mobility prediction constitute an important

S. Bellahsene · L. Kloul (✉)
PRiSM, Université de Versailles,
45 Avenue des Etats Unis,
78000 Versailles, France
e-mail: kle@prism.uvsq.fr

S. Bellahsene
e-mail: sabe@prism.uvsq.fr

P. Reinecke · K. Wolter
Institute of Computer Science,
Free University Berlin,
Takustr. 9, 14195 Berlin, Germany
e-mail: philipp.reinecke@fu-berlin.de

K. Wolter
e-mail: katinka.wolter@fu-berlin.de

K. Wolter et al. (eds.), *Resilience Assessment and Evaluation of Computing Systems*,
DOI: 10.1007/978-3-642-29032-9_17, © Springer-Verlag Berlin Heidelberg 2012

solution to enable seamless handovers in cellular networks. The mobility trace is the main information used to perform mobility prediction. However, using solely this information makes the prediction process difficult when the mobile user is new in the network, that is, when its mobility trace is poor. The efficiency of the prediction model relies on both the ability of the model to predict successfully the next move of a mobile user and its ability to perform such a prediction in a short delay. In order to assess the Markov-based prediction model, data sets of a real cellular network in a major US urban area are used.

17.1 Introduction

Service continuity is one of the main quality of service requirements in mobile networks. However the continuity of user sessions is not always guaranteed as the changes of radio channel, namely handovers, during mobile users movements between the network cells, imposes short session disconnections. Thus, in the case of applications such as multimedia applications where a session discontinuity cannot be transparent to the users, the continuity of a service like VoIP is not guaranteed unless an efficient handover procedure is implemented (see Chap. 2 for more details).

The handover procedure consists in handing off a call to a new cell when the mobile user crosses the current-cell boundaries and moves to an adjacent cell while the call is in progress. Such a procedure is initiated if the RSRP (Reference Signal Received Power) of the adjacent cell is greater than the RSRP of the current cell by a parameter value called the *hysteresis*. Moreover, this condition should last longer than a time threshold called the Time-To-Trigger (TTT). Once the handover procedure is initiated, if the new cell does not have enough channels to support the handoff, or if the session disconnection between the mobile user and the old base station lasts longer than a critical time while the connection with the new base station is not established yet, the call is dropped.

Thus, in order to avoid communication failures during the handover procedure, most solutions in the literature rely on the optimisation of the handover parameters, which consists mainly in finding a tradeoff between the values of the TTT and the hysteresis thresholds [482, 566, 829]. Some of these solutions were introduced in the last releases of current mobile network standards [11, 12]. The other proposed solutions focus on either speeding up the handover phases, or enhancing the existing communication protocols between the network entities [166, 660, 669].

In this chapter, we present two distinct approaches that can be taken into account by the mobile networks standardisation bodies. The objective of the first approach is to enhance the communication protocols by elaborating solutions for an accurate time synchronisation between the base stations of the current 3GPP and future LTE mobile networks. Precise time synchronisation, especially frequency synchronisation between base stations is an important factor for seamless handovers. To that end, the frequency on the air interface of wide area mobile base stations must not deviate by more than $50\,ppb^1$ (parts per billion) relative to the nominal frequency. In existing networks, base stations can derive the frequency from the bit clock of the mobile backhaul network. As backhaul network operators migrate towards packet switching backhaul networks, this is no longer possible, since packet switching networks are inherently asynchronous. In these networks, Timing-over-Packet (ToP) according to the Precision Time Protocol (PTP, IEEE 1588) is employed for frequency synchronisation. A PTP master clock located at the network controller site sends Sync messages of only a few bytes to the PTP slave clocks in the base stations. The frequency accuracy obtainable with PTP deteriorates with increasing Packet Delay Variation (PDV) of the Sync messages which again is caused by queueing in the switches that can happen despite the high priority of the PTP packets. Packet delay variation does not measure the deviation of the frequency of the air interface as caused by an inaccurate clock in the base station. Instead, PDV measures the difference in transmission time of the timing packets from the network master clock to the slave clock in the base station.[2]

As synchronisation packets have to share the backhaul network with other traffic, characteristics of the backhaul network and the background traffic may cause packet delay variation to become too large to guarantee frequency synchronisation of the base-station clocks, resulting in failed handover of calls between neighbouring base-stations. Hence, resilience of time synchronisation in the backhaul network to disturbances from background traffic becomes an important issue in modern mobile networks.

Detailed network impairment models of transmission links and network nodes can provide more insight in timing behaviour of complex networks than the common black-box testing. In the first part of this chapter, we illustrate this by identifying backhaul networks and load scenarios in which PTP accuracy is insufficient, and, consequently, resilience of the handover procedure cannot be guaranteed. Based on the insights gained from the detailed simulation analysis a modified version of PTP is presented. It considerably reduces the PDV of the Sync messages in tree-structured networks as they are typically deployed for mobile backhauling. Since the detailed simulation models require runtimes of several days, sometimes even weeks, the phase-type approximation technique from Chap. 5 has been applied. Assuming identical switches is not too far from reality and enables us to simulate long paths between the master clock and the base station in a short time at high accuracy.

The objective of the second approach presented in this chapter is to speed up the handover phases by predicting the next moves of mobile users. This approach has been developed for a specific and advanced mobile network architecture called the Two-nodes IP Network [99]. This architecture is comparable to the emerging fourth generation networks like the LTE and mobile WiMAX. As the coverage density in mobile networks is often important, in particular in urban areas, the time needed by any mobile terminal to scan all its neighbouring cells and classify those offering the best RSRP can be long enough to become a critical issue for service continuity.

---

[1] To illustrate: a frequency of $2\,GHz = 2 \times 10^9/s$ may vary in the range $[2 \times 10^9 \pm 100]$ per second.

[2] The main difference between master and slave is the quality of the oscillator and hence the clocks' frequency stability.

For example, in LTE [6, 9, 10, 563], a 6 ms measurement gap allows measuring up to 3 neighbouring cells. The periodicity of this measurement gap is generally 40 or 120 ms. Thus, if the neighbourhood of the cell serving a mobile terminal consists of 8 cells, at least 98 ms will be required to scan all the cells in the neighbourhood. In addition, in a critical situation, the handover in LTE can introduce an interruption time, which can last more than 100 ms [6]. Consequently, for multimedia applications that require service interruption time lower than 150 ms, the number of cells to scan, before a handover is initiated, may play a decisive role in service continuity.

Mobility prediction, if well performed, may constitute a solution to limit the number of cells to scan. The objective is to predict the next cell(s) to be visited by the mobile user. If a unique cell is predicted, no RSRP measurement is needed, otherwise only the predicted cells will be scanned and the one offering the best quality of signal is selected. Thus, mobility prediction allows also the network to anticipate the preparation of the handover in the predicted cell, enabling seamless handovers and thus limited call dropping rates. The efficiency of such an approach relies not only on its ability to predict the next cell to be visited by the mobile user, but also by its ability to perform such a prediction in a short delay giving thus the network enough time to prepare the handover to the new cell before the mobile user is disconnected.

Besides the mobility history of the user, the prediction approach presented in the second part of this chapter takes into account an important characteristic of mobile networks which is the ping pong handover phenomenon between neighbouring cells. This phenomenon is related to the fixed values of signal threshold and the propagation conditions when managing handovers [840]. The variation of propagation conditions introduces fluctuations in the network coverage which leads to ping pong handovers. Using simulation, we assess the efficiency of our prediction algorithm by assessing its ability to predict successfully the next cell to be visited by a mobile user. We use data sets from a real cellular network in a major US urban area [756].

Structure of the chapter: The first part of the chapter illustrates the performance analysis of future frequency synchronisation techniques, that is the precision time protocol in sufficient detail as to identify otherwise hidden problems. We first present the backhaul networks models. Then we investigate the optimisation of the delay variation in tree-structured networks. We finally present the performance analysis of the investigated approaches. The second part of this chapter is dedicated to performance analysis of a Markov-based mobility prediction model and the ability assessment of a mobile network implementing such an approach in anticipating the mobile user resource needs in a near future. After defining the context of the analysis, that is a two-nodes IP network architecture, the mobility prediction model is presented. Using data sets from a real cellular network, the performance results of the model are discussed.

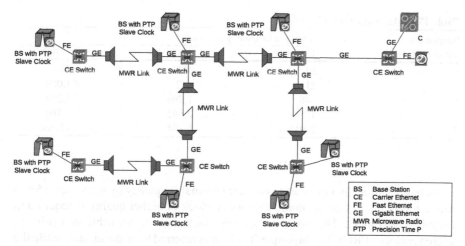

Fig. 17.1 Network using synchronisation of base stations with ToP using PTP. The master clock (in the upper *right corner*) sends PTP Sync messages to the slaves. Note that in this application domain, network traffic is usually shown as flowing from *right* to *left*

Table 17.1 Frequency accuracy requirements on the air interface of FDD base stations

BS class	Minimum accuracy (ppb)
wide area BS	±50
Medium range BS	±100
Local area BS	±100
Home BS	±250

17.2 Frequency Synchronisation in Mobile Backhaul Networks

One technical prerequisite for service-continuity in mobile networks is seamless handover of calls from one base station (BS) to the next. Handover requires precise frequency synchronisation (also referred to as *syntonisation*),[3] as radio access networks (RAN) apply frequency division duplex (FDD) in compliance with [14] for separating the downlink from the uplink carriers. As standardized by 3GPP in [14], the frequency on the air interface of wide area GSM, WCDMA, and LTE base stations must not deviate by more than 50 ppb[4] from the nominal frequency. See Table 17.1 below for the frequency requirements of all BS classes. Since part of the frequency accuracy is lost between the local oscillator of the BS and the air interface, in practice, the clock signal generated by the local oscillator has to be as accurate as some 15 ppb.

---

[3] Note that base stations of an FDD-RAN typically have no need for phase synchronisation or time-of-day synchronisation.

[4] Parts per billion, defined as 10^{-9}.

Table 17.2 Maximum tolerable PDV

Integration window size (h)	Required frequency accuracy		
	15 ppb (μs)	30 ppb (μs)	75 ppb (μs)
1	54	108	270
4	216	432	1,080
12	648	1,296	3,240
24	1,296	2,592	6,480
48	2,592	5,184	12,960

As base-station clocks depart from their nominal frequency by more than 15 ppb after a few hours of free-running operation, a clock of higher quality is required to discipline the clocks. The frequency of this clock must be traceable to a primary reference clock (PRC). The high-quality clock is referred to as the master, while the local oscillators in the base stations are called slaves. The master clock is typically deployed at the centre of the mobile network, while the slave clocks are in the base stations of the radio access network. This scenario is depicted in Fig. 17.1.

Frequency is related to the transmission time of packets across the network. It can only be determined by integrating transmitted packets over a time interval. The longer the integration the more robust the syntonization becomes against low-frequency packet delay variations.

If the transmission time of packets was constant, the frequency would not vary and packet delay variation would be zero. Since the frequency variation must be bounded this translates to a bound on the packet delay variation (PDV) which means that the difference in speed between fast and slow packets is of interest. In practice, only a given portion of the fastest packets is considered at all. Since both the PTP slave clock algorithm as well as the oscillator in the BS are vendor-discretionary the packet delay variation (PDV) boundaries given in Table 17.2 have to be regarded as guide values rather than precise limits.

However, in the evaluation of our models we consider a limit of 216 μs as total packet delay variation between master and slave to be a reasonable target.

In current mobile networks employing TDM-based transport technology, the frequency of the master clock can be conveyed to the base stations via the bit clock of the backhaul network. With the migration of backhaul networks to inherently asynchronous packet-switching technologies like Carrier Ethernet (CE), however, the bit clock of the backhaul network can no longer convey the frequency of the master clock to the slave clocks. In such networks, timing-over-packet (ToP) using the precision time protocol (PTP) according to IEEE standard 1588–2008 [444] is the preferred syntonisation method.

Let us illustrate the operation of PTP using the scenario shown in Fig. 17.1: A PTP master clock syntonises PTP slave clocks in a single-ended way. The network has one PTP master clock on the right-hand side and several PTP slave clocks at the end of the paths through this tree-structured graph, with packets travelling from right to left, and top to bottom. The leaves of the graph represent the PTP slave nodes.

At constant time intervals, the PTP master clock sends Sync messages to the slaves. The slaves use the interarrival time of consecutive Sync messages to synchronise their local clock frequency to that of the master clock. Network connections in this scenario can be either optical fiber Gigabit Ethernet (GE) links connecting carrier-ethernet (CE) switches, or microwave radio (MWR) links with radio antennas on either end. The last link to the mobile base station (BS) usually is a fast ethernet (FE) connection. Some PTP slaves are only a few links away from the master clock, but others can be at a distance of up to 20 links.

As long as the packet transfer delay (PTD) of the Sync messages over the backhaul network stays constant, the PTP slave clocks can discipline their local oscillator to the frequency of the PTP master clock. Variation in the PTD measurements, however, adversely impacts the operation of ToP and this the accuracy achievable by the BS clocks. This is formally described by the packet delay variation (PDV),[5] which is defined for each packet i as

$$PDV_i := PTD_i - \min\{PTD_j | j \in \mathbf{N}\}$$

The upper PDV bound tolerable by a PTP implementation depends on the PTP implementation and the accuracy of the local clocks. Although these are typically kept secret by vendors, a total PDV between master and slave of 216 μs is a reasonable target.[6] PTP implements a mechanism for reducing PDV, based on the transmission delay of packets: The master and slave clocks timestamp each packet upon transmission and receipt. Based on the difference between both timestamps, the slave clock can identify packets that have experienced high delay, and are therefore likely to increase PDV. The slave clock then uses only the fastest packets, with typical thresholds being at the 1 % quantile or below. Therefore, the target is to keep the 1 % quantile of PTD below 216 μs.

The strict requirements discussed in the previous paragraphs pose two important problems when engineering mobile backhaul networks for use with ToP. First, PDV obviously depends on the topology of the network and on the delays accumulated within the switches. These delays in turn are affected by the background load and by the internal structure of the switch. This raises the question whether the backhaul network will be able to support syntonisation with the required accuracy. Secondly, in cases where the network will not be able to provide PTP PDV at or below the upper limit of 216 μs, the network may be redesigned or more master clocks may be added. However, these solutions are quite expensive, and thus the question arises whether more cost-efficient methods can be found.

In this section we illustrate how to apply modelling and simulation techniques in tackling both questions. We base the example on the work in [939], where the focus was on tree-structured networks, which are typical for backhaul-networks in

---

[5] Note that there exist different terminologies and definitions for the variation in transmission delays (sometimes also referred to as jitter), e.g. the instantaneous PDV, as defined by [284].

[6] We refer the reader to [939] for details and only mention here that this upper bound corresponds to a maximum deviation of 15 ppb and an integration window size of 4 h.

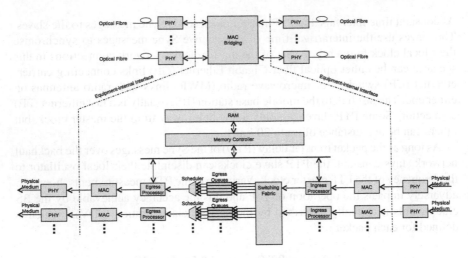

Fig. 17.2 Functional model of a single-stage Carrier Ethernet switch

mobile communication. From the point of view of PTP, these networks can be considered as sequences of links and switches, where the length of the sequences varies. Our approach proceeds as follows: first, we construct a simulation model for the link/switch sequence between the master clock and the slave clock. Using discrete-event simulation, we use this model to gain a deeper understanding of the dependence of PDV on the network. Secondly, we derive a closed-form expression for PDV in a network of a special structure. This expression implies a simple method for minimising PDV. In a third step, we use our simulation model to evaluate the improvement.

17.2.1 Precise Modelling of Backhaul Networks

The need for simulation models is dictated by the limited coverage of measurement studies on practical networking equipment. Even though there exist guidelines for conducting such studies [475], cost and time constraints render exhaustive tests infeasible. Discrete-event simulation allows evaluation in a much more efficient way. However, the high accuracy requirements of PTP demand very precise models, because the delay variation experienced by the fastest packets when passing through a node must be quantifiable with microsecond precision under a wide range of load and other conditions. As these requirements are far beyond those of typical applications, sufficiently accurate models for networking equipment are not available in state-of-the-art network simulators. In particular, current models do not include many of the internal structures that have an effect on the PDV.

Our first step in [939] was thus to develop the required models based on detailed structural analyses of the networking equipment. Consider the functional structure of a typical single-stage Carrier Ethernet switch with a MAC bridging device as its central component, as shown in Fig. 17.2. Virtually all MAC bridging devices implement a transmit FIFO buffer between the egress scheduler and the transmit MAC. Since this transmit FIFO buffer is behind the priority queueing and scheduling block (which applies strict priority queueing, SPQ) it completely ignores any packet priority. It can be modelled as a rate-matching buffer with the low and the high thresholds being vendor-discretionary and sometimes configurable. The arbitrary delay element represents the PDV attributable to equipment-internal packet processing and forwarding including packet storage and physical interfaces. MAC bridging device manufacturers usually guarantee the PDV caused by device internal packet processing and forwarding to be less than $1\,\mu s$. With the PDV of a Gigabit Ethernet interface being less than $0.2\,\mu s$ [463], and leaving some margin, the PDV of the arbitrary delay element may be assumed to be less than $2\,\mu s$. Based on this analysis, we model the switch as shown in Fig. 17.3. Note that a state-of-the-art delay model of a packet switch would neither include the transmit FIFO buffer nor the arbitrary delay element. Therefore, it would fail to represent PDV characteristics at the accuracy required for evaluating ToP using PTP. This can be shown by simulating one link at different background load levels, using both the detailed model and a model of a switch provided in a typical library of simulation models for network equipment. Figure 17.4 shows the 1 % quantile of the packet-transfer delay for growing background load. Note that the delay equals almost zero at all load levels if we use the default model, while there is a sharp delay increase at 100 % load when we use the detailed model. This delay step at 100 % background load occurs when the Transmit FIFO runs into overload and throttles the egress scheduler in order to reduce the load. This effect is well-known from experiments, but, as we observe, is not reproduced by standard CE switch models.

In [939] we constructed chains of CE switch models as models of the network paths between master and slave clocks. We simulated the CE chain with a variable number of links and using different types of background traffic. With such highly-detailed models one typically encounters scalability issues, as simulation times increase quickly when the complexity of the scenario increases. In our case we could solve these issues by applying the hybrid approach described in Chap. 6: we simulated the delay behaviour of a single switch and modelled the resulting PDT distribution by a Phase-type (PH) distribution (cf. Chap. 5). This allowed us to simulate the effect of the switch on the PDV by drawing random delays from the approximating PH distribution, instead of simulating individual background packets. Simulating longer chains then just required chaining these delay models together.

The developed simulation model provides sufficient insight to allow us to propose solutions to the problem of excessive PDV. One solution consists in avoiding the delay step. This can be achieved by implementing a leaky-bucket egress shaper in the switch. The leaky-bucket egress shaper reduces the data rate of the packet stream entering the Transmit FIFO buffer, but in contrast to the Transmit FIFO buffer it honours priority of the PTP packets. If the egress shaper is configured such that the input to the Transmit FIFO buffer has a lower date rate than the output, the FIFO buffer does not encounter overload situations, and thus no throttling of the scheduler

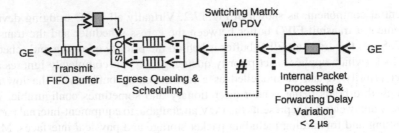

Fig. 17.3 Delay model of a CE switch

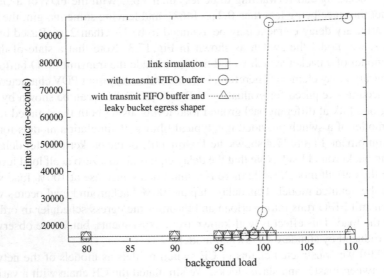

Fig. 17.4 PDV 1 % over a GE-link between two single-stage CE switches

becomes necessary. Figure 17.4 confirms this effect, as the delay step vanishes when the egress shaper is active.

17.2.2 Minimising Delay Variation in Tree-Structured Networks

The leaky-bucket egress shaper may require costly network reconfiguration and may not be sufficient to reduce PDV to acceptable levels. Therefore, in [939] we devised a second method for minimising PDV that works without changing the network, simply by adjusting PTP. Note, however, that this method as described here is limited to special network topologies, which is not the case for the leaky-bucket egress shaper.

The development of the method started with an analysis of the way high-priority packets traverse a tree-structured network, with the aim of investigating possible waiting times. The lower diagram in Fig. 17.5 shows the architecture of a tree-structured

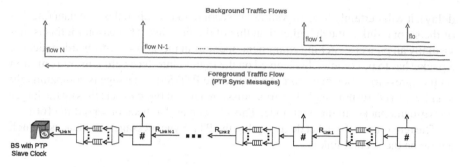

Fig. 17.5 Delay model for tree-structured backhaul network (PTP path in Fig. 17.1). Note that the CE switch models do not contain the transmit FIFO buffer, as we assume that its effects have been eliminated by implementing a Leaky-Bucket Egress Shaper [939]

network while the upper part shows the packet flows that exist in such a network. All traffic is generated at the PTP master and radio controller while the background traffic might leave the route of the PTP packet at any one of the switches on the way to the PTP slave. This means in particular, that no background traffic enters this network at an intermediate switch. Therefore, no new background packets will appear just in front of an arbitrary PTP packet.

Based on this analysis, we derived a closed-form expression for the PDV in such a network. In order to do so, we had to take into account the best case and the worst case of the packet transfer delay (PTD). Without loss of generality we considered a fixed but arbitrary PTP packet. The probability P_r that the packet experiences a remainder-of-packet delay depends on the background traffic load and is

$$P_r = \left(\sum_{n=1}^{N} R_{bt} \right) \Big/ R_{l1},$$

with R_{bt} being the data rate of the nth background traffic flow (as shown in Fig. 17.5), and R_{l1} being the data rate of the first link. If the packet message has to wait, the remainder-of-packet delay t_{r1} is uniformly distributed between zero and the transmission time of a complete background packet:

$$0 \leq t_{r1} \leq \frac{S_{bp}}{R_{l1}},$$

where S_{bp} is the size of the background packet and R_{l1} is the data rate of the first link. Since the background and the PTP Sync traffic flows are aligned on the first link, the arrival times of background and PTP Sync packets are no longer statistically independent on the succeeding links, where the packets arrive one after the other. If a 94 byte PTP Sync message has to wait in the first node until the background packet (e.g. of size 1522 bytes) is completely transmitted over the first link (remainder-of-packet delay), it will certainly have to wait in the second node again unless the data rate R_{l2} of the second link is much higher than that of the first link. The reason for this is that virtually all packet switching equipment operates in the store-and-forward mode, i.e. a packet has to be completely received on the ingress interface before its transmission on the egress interface may start. And since the PTP Sync message is comparatively small with a correspondingly low transmission delay, it typically catches up the larger background packet in the next node. This effect has also been observed in [166].

The remainder-of-packet delay t_{r2} that the PTP Sync message experiences in such a case on the second link is

$$t_{r2} = (S_{bp}/R_{l2}) - (S_{\text{PTP Sync}}/R_{l1}) \tag{17.1}$$

with $S_{\text{PTP Sync}}$ being the size of the PTP Sync message. This kind of remainder-of-packet delay further accumulates over all links down to the PTP slave clock in the BS.

Packet delay variation is reduced if the PTP Sync messages can be prevented from catching up the background packets, i.e. t_{r2} has to be zero. Using (17.1) for the calculation of t_{r2} one obtains

$$t_{r2} = (S_{bp}/R_{l2}) - (S_{\text{PTP Sync}}/R_{l1}) = 0 \tag{17.2}$$
$$S_{\text{PTP Sync}}/R_{l1} = S_{bp}/R_{l2} \tag{17.3}$$
$$S_{\text{PTP Sync}} = S_{bp} * R_{l1}/R_{l2} \tag{17.4}$$

This means that the optimal size of the PTP Sync messages depends on the size of the background packets and the data rate of the ingoing and outgoing link to a switch. In consequence, the PTP Sync messages have to be enlarged. The optimal size depends on the PDV the ToP implementation can accept, the background traffic load and packet size distribution, and the number of links and their data rates. Figure 17.6 demonstrates the impact of the PTP Sync message size in various scenarios. To illustrate the above reasoning we formally analyse a special case, where we first assume all links have the same data rate R_l.

Let $u = S_{bp}/R_l$ and $p = S_{\text{PTP Sync}}/R_l$ then we can define the maximum packet transfer delay $PTD_{\max}$ and the minimum packet transfer delay $PTD_{\min}$ of a PTP Sync message across N links as

$$PTD_{\max} = u + p + (N-1) * \max(0, (u-p))$$
$$+ (N-1) * p = p + N * u$$
$$PTD_{\min} = n * p \tag{17.5}$$

In the worst case, the PTP Sync packet has to wait for transmission of a full background packet at the first link, then adding its own transmission time. At all subsequent links the PTP Sync message needs to wait for the remaining transmission time of the (usually larger) background packet. Finally, the transmission time of the PTP Sync message across all links needs to be added. In the best case, the PTP

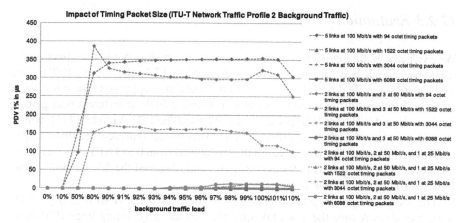

Fig. 17.6 Impact of timing packet size on the 1 % quantile of the packet-delay distribution

Sync message never needs to wait for remaining transmission delay of a background packet.

We can distinguish two cases:

$$1)\ u >= p,$$
$$2)\ u < p.$$

Only the first case is of practical interest and therefore we can omit the maximum in the definition of PTD_{max} in what follows. Peak-to-peak packet delay variation $p2pPDV$ is defined as the difference between PTD_{max} and PTD_{min}. Let us assume that $u = i * p$, where $i \geq 1$ in order to respect case 1) and i can be any real-valued number. Then

$$p2pPDV = PTD_{max} - PTD_{min} = p(1 + (i - 1) * N).$$

The value of i that minimises the delay variation and obeys all restrictions is $i = 1$. For the optimal value of i the PTP Sync packet has the same size as the (largest) background packet and then the peak-to-peak packet delay variation reduces to

$$p2pPDVopt = p.$$

For the second case ($u < p$) the delay variation is always p. The more general case with links of variable speed requires heavy notation and is omitted here. The reasoning is analogous.

17.2.3 Evaluation

We evaluated both approaches using our simulation model. While the effect of the leaky-bucket egress shaper is straightforward (see Fig. 17.4), the optimal PTP packet size requires a more detailed discussion of the evaluation approach. We simulated a chain of 5 links at different link speeds and observed PDV at different background load levels and different PTP packet sizes. We used the ITU-2 model [475], which describes a mix of packets of different sizes. This model represents realistic traffic conditions in a network. We increased the traffic load from 0 to 110 %. We increased the size of PTP packets from 94 to 6088 bytes and observed the 1 % quantile of the PDV for each combination of the parameters.

The results are presented in Fig. 17.6. It is immediately obvious that larger PTP Sync messages reduce the 1 % PDV quantile considerably. Using large PTP Sync packets the packet delay variation across 5 links stays below 200 µs and hence the PDV requirements for ToP are fulfilled. The same cannot be said for the default packet size of 94 bytes, where the 1 % quantile of PDV is much larger.

17.3 Service Continuity in a Two-Node IP Network

The objective of the second approach we present in this chapter is to speed up the handover phases, by anticipating the future needs of mobile users in terms of channels. At each newly entered cell, the approach consists in predicting the next cell(s) the more likely to be visited by the mobile user. If a unique cell is predicted, no RSRP measurement is needed and the handover procedure can start. However, if more than one cell is predicted, the mobile terminal scans the predicted cells and the one offering the best quality of signal is selected.

The prediction model has been developed for the two-nodes IP network architecture depicted in Fig. 17.7 [99, 185]. In such an architecture, the mobility functions are ensured by an Enhanced Access Gateway (EGW) and the Base Station (BS), these nodes being interconnected using an all-IP network. Although very attractive, meeting common mobile networks requirements using such an architecture may not always be straightforward. The use of only two nodes (EGW, BS) implies that the mobility anchor point, which is assumed to be in the access gateway, may be very far from the base stations. Thus, in order to guarantee the continuity of a service like VoIP in such an architecture, an efficient handover procedure is required. The efficiency of this procedure relies on its ability to handoff an ongoing call to the new host cell in a delay sufficiently short to guarantee an uninterrupted service.

Due to the fixed values of signal threshold and various conditions of signal propagation, mobile users can switch several times and randomly between two neighbouring cells [566]. This phenomenon, which is known as the *ping pong handover* phenomenon, can occur even with small position changes of the mobile user. This is a major problem faced by the mobility management procedure. Because of signal

Fig. 17.7 Two nodes-based
IP architecture

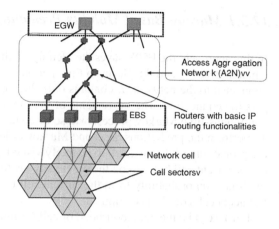

shadowing [644], the signal attenuation is important and can last for a while. If this attenuation makes the strength of the received signal by a mobile terminal lower than the fixed threshold, the handover will be triggered even if the mobile user remains geographically in the same cell. Thus even if the behaviour of the mobile user is very regular, one can still observe a certain factor of randomness in its behaviour, due to the ping pong handover. This randomness factor is more important in dense urban areas because the degree of cells neighbouring and the density of buildings are very important.

Currently, several types of mobility prediction models exist in the literature [99, 188]. Some rely on the history of the user's mobility patterns only, and thus are sensitive to the changes of the user behaviour. Others are based on both the history of the mobility patterns and formal models. For several of these procedures, a high degree of randomness in the user's behaviour may have an impact on their efficiency.

As the mobility prediction approaches are much more needed in urban areas, any mobility prediction model based on the mobility trace should take into account the randomness factor due to ping pong handover. In [98], a mobility prediction approach which tries to efficiently correlate the mobility data of the user while taking into account the ping pong handover phenomenon between neighbouring cells, has been proposed for the two-nodes network architecture in Fig. 17.7. This approach, which has been designed to be implemented at the enhanced gateway level of the two-node IP architecture, is mainly based on the history of the user's mobility patterns and Markovian models. In the following, we discuss and show how such a mobility prediction approach can be used to guarantee a certain service continuity for such an architecture.

17.3.1 Markov-Based Mobility Prediction Approach

In this approach, the EGW saves the identity of all cells crossed by a mobile user, in a history trace L and this during a time period T. Initially, when a mobile user gets connected to the network, L contains only the identity of the cell where the mobile gets the connection. The movements of the mobile user are then modelled using a continuous-time Markov process whose discrete states are the cells of the network. To perform the predictions, a second Markov order-based predictors are used. Thus the probability of the next cell to be likely visited by a mobile user depends not only on its current cell but also on the previously visited cell. Then for each mobile user, the transition probability from the current cell C_i to each cell in its neighbourhood (adjacent cells of C_i) is computed.

Let $\Gamma(C_i)$ be the neighbourhood of cell C_i and let (M, N, r) be the tuple associated with each mobile user such that:

- for each pair of cells C_j and C_k in $\Gamma(C_i)$, $M(C_k C_i, C_j)$ is the number of transitions of the mobile user from cell C_i to cell C_j in the past, knowing that each time such a transition occurred the mobile user was previously in cell C_k.
- $N(C_k, C_i)$ is the number of transitions of the user from cell C_k to cell C_i.
- $r(C_i)$ denotes the average residence time of a mobile user in cell C_i.

The memory allowed for each mobile user is limited to a fixed size of L. Limiting this memory to time period T is compensated by the global knowledge of the tuple (M, N, r). Let $L = C_1 C_2 C_3 ... C_{i-1} C_i$ be the mobility history trace of a mobile user and let $X = C_{i-1} C_i$ be the sequence, in L, of the previously visited cell and the current cell of this user. Assuming that $Y = C_i C_{i+1}$ is the sequence of the current cell and the future cell to be visited, the estimated transition probability P_{e_1} is given by:

$$P_{e_1} = P(X_{i+1} = Y / X_i = X) = \frac{M(X, C_{i+1})}{N(C_{i-1}, C_i)}$$

Unfortunately, if the previous cell C_{i-1} appears for the first time in the mobility trace of the user, $M(X, C_{i+1})$ is equal to zero for all neighbouring cells of C_i. In [840], when the second-order Markov based predictor fails, Song et al. propose to fall back on the first order, that is, only sequences formed by two consecutive visited cells $(C_i C_{i+1})$ are used. In this case, the transition probability given by the first-order Markov chain is:

$$P_{e_2} = P(X_{i+1} = Y / X_i = X) = \frac{N(C_i, C_{i+1})}{Z(C_i)}$$

where $Z(C_i)$ is the number of times cell C_i appears in the mobility history of the mobile user. However, their algorithm remains inefficient whenever the current cell C_i appears for the first time in the mobility trace of a mobile user.

In order to overcome this problem, in [98] an additional information is used: the visit frequency to each cell C_i from its neighbouring cells. For that, let

$H(C_i) = \sum_{j=1}^{K} Z(C_j)$ where K is the total number of adjacent cells to C_i. In this case, the estimated probability value of the next move to $C_{i+1} \in \Gamma(C_i)$ is given by:

$$P_{e_3} = \frac{Z(C_{i+1})}{H(C_i)}$$

The occurrence of the ping pong phenomenon is detected in the mobility trace of a mobile user thanks to cell sequences of the form $C_{i-1}C_iC_{i+1}$ where $C_{i-1} = C_{i+1}$. This phenomenon is taken into account by considering a randomness factor denoted by α. This factor is set to 1 each time a sequence $C_{i-1}C_iC_{i+1}$ where $C_{i-1} = C_{i+1}$ occurs, that is the current handover is considered 100 % as a ping pong handover. In all the other cases, $0 < \alpha < 1$ and its optimal value is set, during the simulation, to the value that maximises the prediction accuracy.

The overall prediction process using the approach is very simple. During each step of the prediction, the approach algorithm checks if the mobility trace provides enough information about the mobile user behaviour and the probability to move towards an adjacent cell takes the general form: $P_e = P \times \alpha$, where P depends on the contents of the trace. If the trace provides enough information about the previously visited cell, the second order Markov based-predictor is used and $P = P_{e_1}$. Otherwise, the first order Markov based-predictor is used and $P = P_{e_2}$. However, as the first order cannot be used if $N(C_i, C_{i+1}) = 0$ the algorithm uses the ultimate information that can be found in the mobility trace, that is, the visit frequency to a cell. In this case, $P = P_{e_3}$.

If the Markov-based mobility prediction approach (MMPA) fails in predicting a unique cell, a multicast is performed by the EGW; a group of cells, the more likely to be visited by the mobile user in the near future, is selected, and all users packets are sent to the base stations of these cells. Then, instead of scanning all the cells in its neighbourhood, the mobile user makes the necessary measurements to select among only the group of predicted cells the one offering the best quality of signal. The complete algorithm is described in [98].

17.3.2 Applying the Mobility Prediction Approach

The efficiency of a mobility prediction model relies on both the ability of the model to predict successfully the next move of a mobile user and its ability to perform such a prediction in a short delay. The accuracy of the predictions made by the model can be assessed using the following equation:

$$Accuracy = \frac{Number\ of\ successful\ predictions}{Total\ number\ of\ predictions}$$

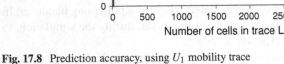

Fig. 17.8 Prediction accuracy, using U_1 mobility trace

17.3.2.1 Experimental Data Sets

The assessment of the prediction model is performed using data sets collected from ten user terminals in Houston, a major US urban area [756]. These data sets are measured on Wi-Fi enabled GSM cellular phones and contain logs of the network availability, the signal levels, and the context information. Using these data sets, the mobility trace for each participant is extracted. Each mobility trace represents the sequence of all visited cells during a month, which is the time period of the data collection. The size of the traces varies between 1629 cells (shortest) and 7286 cells (the longest).

As the data sets provide, for each log period, the list of visible cells for a mobile user, this list is used to build for each cell its neighbourhood. The number of cells in the covered area differs from a participant to another. The smallest network contains 151 cells and the biggest one contains 551 cells. We define the mean degree of neighbouring in each trace as the mean cells number in the neighbourhood, in each network, from which the mobility trace is built.

17.3.2.2 Numerical Results

Both Figs. 17.8 and 17.9 show the accuracy level reached by the prediction model (MMPA) when applied on the experimental data set of U_1, the first mobile participant. The results in Fig. 17.8 show that even if changes occur in the behaviour of the mobile user, the model succeeds in making good predictions. Examples of the impact of the changes in the user's behaviour can be observed mainly when $|L| \approx 200$, $|L| \approx 650$ and $|L| \approx 1200$. Around these points, we can see slight degradations in the performance of the prediction model, degradations which are quickly corrected. This can be explained by the fact that the mobile user visits new cells, not met before in the mobility trace, or by the appearance of new sequences of cells in the trace.

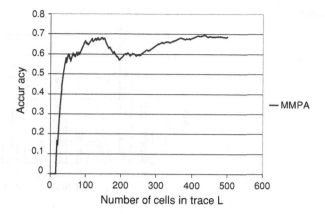

Fig. 17.9 Prediction accuracy, using U_1 mobility trace

Figure 17.9 provides a more detailed representation of the accuracy level of the predictions when the mobility trace of mobile user U_1 is limited to the 500 first cells ($0 \leq |L| \leq 500$). When the user is newly connected ($|L| \leq 50$) the prediction accuracy level increases rapidly to reach 60 %. This shows that the prediction model learns rapidly the behaviour of the mobile user. This learning capability is due to the use of Markov-based predictors and the visit frequency. Because of new cell sequences in the trace, at $|L| = 160$ the prediction accuracy starts to decrease until reaching its lowest level at $|L| = 200$. The system starts to get better predictions again only when the size of the trace becomes more significant.

In Fig. 17.10, the performances of the prediction model (MMPA) are compared to two other prediction models, the global prediction algorithm (GPA) [99], and the second-order Markov based algorithm with fall back (O(2) fallback) [840]. We consider the data sets of the ten mobile participants and for each mobile user, the average gain of using MMPA instead of GPA and O(2) fallback approaches, is computed in terms of prediction accuracy. The value of the time period T used for each mobile user is optimised after experimentations. The results obtained show that, for all data sets, the MMPA outperforms the other prediction models. As stated before, the aim of MMPA is to perform better predictions when the user's behaviour is characterised by a high degree of randomness like when the user is newly connected to the network.

Figure 17.11 shows the importance of the role that the mobility prediction plays to guarantee session continuity in mobile networks. In these experiments, we consider the mobility trace of mobile user U_4 for which the mean number of neighbouring cells is 32 cells. In this case, if no mobility prediction approach is used, the mobile terminal has to scan 32 cells in order to make signal quality measurements. Consequently, if t_s is the time needed to scan one neighbouring cell, the total time to scan all the cells is $V_s = 32 \times t_s$. In contrary, if a list of 5 cells is predicted (Fig. 17.11), the accuracy level of the prediction is 0.9 and the time needed to scan these cells is $V'_s = 5 \times t_s$, that is 84.4 % less time than if all cells have to be scanned. Clearly if less cells are

Fig. 17.10 Average gain

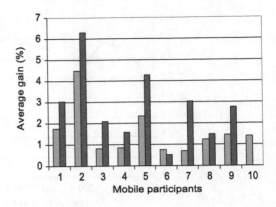

Fig. 17.11 Prediction accuracy versus scanning time

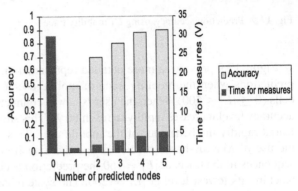

Fig. 17.12 Optimal value of α

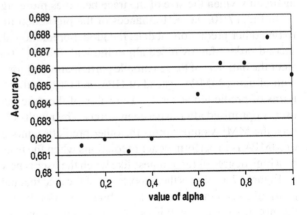

predicted, the corresponding accuracy level is lower, but the time needed to make measurements is also lower. Moreover, Fig. 17.11 shows that it is always possible to have a satisfying trade-off between the accuracy level and the measurement time.

Note that in the case of mobile participant U_4, the mobility trace does not allow reaching a significant accuracy level when predicting a single cell; the prediction accuracy is about 49 %. This is due to the fact that the mobile user does not have regular movements and the cell sequences in its history trace show that it follows different paths.

In all the experiments, the optimal value of α is set to 0.9. This value matches the best prediction accuracy level possible for the mobility traces used. Figure 17.12 shows the prediction accuracy level at the end of the mobility trace of mobile user U_1 for different values of α. In particular, it shows that the prediction accuracy is better when $0.70 < \alpha \le 0.90$. Thus, to get benefit from the ping pong handover in the prediction procedure the value of α is set to 0.90.

17.4 Conclusion

In this chapter, two major aspects of the handover procedure have been investigated using modelling. The first aspect is the time synchronisation between base stations and its impact on seamless handovers. We have presented the application of modelling, simulation, and analytical methods to solving the problem of frequency synchronisation in mobile backhaul networks. We have seen that, although this appears to be a standard problem, standard models are not sufficiently accurate to reflect important properties of real systems. Consequently, a highly-detailed modelling approach had to be taken. This approach gave us the required insight into the complex internal details of networking equipment that affect the accuracy of frequency synchronisation. Based on this knowledge, we could derive two solutions for improving accuracy. While the first is applicable to networks of any topology, but may require costly reconfiguration, the second is limited to specific setups, but can be applied without changing the network. We showed how such methods can be evaluated using our highly-detailed models, and we demonstrated that both methods are effective in reducing packet-delay variation. In practice, the choice between these methods would not only be determined by the efficacy of either, but also by their costs, an aspect we did not consider in formal terms here.

The second aspect related to the handover procedure is the ability of the network to anticipate the resource needs of a mobile user in a near future. The presented mobility prediction approach is based on the history trace of the mobile user and a Markovian model. Besides the regular movements, this approach takes into account random movement conditions, in particular those related to the ping pong phenomenon in urban areas. Using real data sets, we have assessed the efficiency of the prediction model by assessing its ability to predict successfully the next cell to be visited by a mobile user. We have showed, through an example, the impact of mobility prediction in reducing the scanning time performed by a cell.

Such a mobility prediction approach can be implemented at the EGW level [98]. Moreover, it can be combined with another approach such as the cells sectorization-based approach, which has been designed to be used within the base stations [99]. However, as the Markov-based model uses the history traces which contain the mobile users data, for security and confidentiality reasons, it will be more appropriate to implement such an approach within the mobile terminals.

Acknowledgments Samir Bellahsene and Leïla Kloul are supported by the European Celtic project HOMESNET [8]. Katinka Wolter and Philipp Reinecke are supported by the German Research Council under grant Wo 899/2-1. They would like to thank highstreet technologies GmbH and Alfons Mittermaier for the provided insights into frequency synchronisation techniques.

Chapter 18
Case Study on Critical Infrastructures: Assessment of Electric Power Systems

Silvano Chiaradonna, Felicita Di Giandomenico and Paolo Lollini

Abstract Critical Infrastructures (CI) are increasingly responsible for vital services our society relies on; therefore, assessing their resilience is of utmost importance for improving trustworthiness on their services. Given the many challenges and open issues involved, a number of initiatives have been ongoing in the last decade, researching methods and developing tools for resilience assessment of critical infrastructures. Moving from the major challenges posed by CI from the point of view of resilience assessment and assessment needs, this chapter overviews a modelling framework for the analysis of interdependencies in Electric Power Systems (EPS), adopting a state-based stochastic approach. First, it is shown how the selected approach deals with the interdependencies, complexity, heterogeneity and scalability dictated by the infrastructures involved in framework implementation are then discussed, and some illustrative examples of different typologies of analysis are provided on selected EPS scenarios.

18.1 Introduction

In chapter on "Assessing Dependability and Resilience in Critical Infrastructures: Challenges and Opportunities" in this book, the emergent and extensive sector of Critical Infrastructures (CI) employing cyber control subsystems has been discussed, as a motivating application area where assessment and evaluation of resilience is a

S. Chiaradonna · F. Di Giandomenico
ISTI Department, Italian National Research Council, via Moruzzi 1, 56124 Pisa, Italy
e-mail: chiaradonna@isti.cnr.it

F. Di Giandomenico
e-mail: digiandomenico@isti.cnr.it

P. Lollini(✉)
University of Firenze, Viale Morgagni 65, 50134 Florence, Italy
e-mail: lollini@unifi.it

K. Wolter et al. (eds.), *Resilience Assessment and Evaluation of Computing Systems*,
DOI: 10.1007/978-3-642-29032-9_18, © Springer-Verlag Berlin Heidelberg 2012

ACRONYMS
AC	Alternating current
CI	Critical infrastructure
DC	Direct current
EI	Electric infrastructure
EPS	Electric power system
ITCS	Information-technology based control system
LCS	Local control system
$\mathcal{RS}$	Reconfiguration strategy
RTS	Regional tele-control system
SAN	Stochastic activity networks

primary concern, given their central role in providing vital services our everyday life relies on. There, general considerations on major challenging issues presented by CI, on requirements for resilience assessment of CI as well as an overview of currently ongoing studies have been sketched out. In this chapter, the emphasis is on providing a case study showing a concrete approach to resilience analysis in the target field. To this purpose, we have selected the sector of Electric Power Systems (EPS).

EPS have long been recognised as being critical infrastructures of all countries. They constitute representative examples of hybrid complex systems, being composed of the Electrical Infrastructure (EI) characterised by physical electrical parameters and by the Information-Technology based Control System (ITCS). Due to their strong interconnection, interdependencies are therefore a major challenge in EPS, since they become a formidable vehicle through which a failure in a subsystem propagates to the others, possibly resulting in cascading or escalating failure [778]. A number of blackouts both in Europe and in US during the years 2000s and the consequent damages experienced have raised research on Electric Power Systems protection to a hot topic, triggering initiatives at both national and international levels. An overview of such initiatives is included in chapter on "Assessing Dependability and Resilience in Critical Infrastructures: Challenges and Opportunities" in this book. Here, we focus on the presentation of the modelling framework for the analysis of interdependencies in EPS, that was one of the major efforts of the EU project CRUTIAL [245]. It resorts to a model-based approach, where the structure and correct/incorrect behaviour of major system components, as well as their interactions, are represented at a certain abstraction level. To keep the framework general and theoretically applicable to any EPS configuration, a modular development is adopted. Simulation is then applied to assess metrics well representative of the quality of service perceived by end users (blackout size indicators). This framework, whose incremental developments are documented in a few publications [200–203], has been recognised as a novel contribution to the analysis of interdependencies in EPS, since it integrates the electrical infrastructure and the cyber control via explicit modelling of the main entities of the two subsystems and of the interactions between them.

The rest of this chapter is organised as follows. Section 18.2 introduces the logical structure of EPS, the simplifying assumptions made in the study and exemplifies effects of failures due to reciprocal dependencies. Discussion of the features required

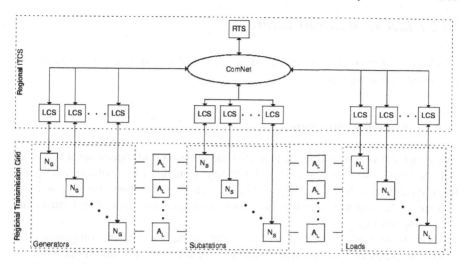

Fig. 18.1 Logical structure of a regional transmission grid, with the associated information control system. Reproduced from [201] ©2009 Springer

from an evaluation framework to satisfactorily deal with EPS assessment is conducted in Sect. 18.3, where an overview of formalisms for EPS evaluation is also included. The implementation of the proposed evaluation framework is addressed in Sect. 18.4. There, the sub-models representing the components of the EPS logical structure are first synthetically described, and then details for just one sub-model are provided. The sub-model was chosen as a representative one to show how peculiarities of EPS components have been implemented. Some illustrative evaluation examples of indicators of the blackout size experienced under selected failure scenarios are presented in Sect. 18.5, to concretely demonstrate the applicability and utility of the evaluation framework in assessing resilience of this critical infrastructure. Finally, our conclusions are drawn in Sect. 18.6.

18.2 Logical Structure of EPS

In this section we discuss the modelling of EPS, presenting how we have represented its main characteristics and which are the simplifying assumptions. This modelling framework has been already introduced in [203]; however, it is recalled here to provide a complete context overview of the system under analysis.

The target of our analysis is a regional EPS system, whose logical structure is illustrated in Fig. 18.1.

The modelling of each infrastructure is first preceded by the description of its logical structure, where the main logical components are identified. The section concludes with a brief description of possible interdependencies between ITCS and EI.

18.2.1 Electrical Infrastructure

The logical structure and the model of the electrical infrastructure are presented in the following two subsections.

18.2.1.1 EI Logical Structure

The high level structure of EI includes the transmission grid (operating at very high voltage levels), the distribution grid (operating at medium/low voltage levels), the high, medium and low voltage generation plants, and the high, medium and low voltage loads. The transmission and distribution grids are divided in regions and areas, respectively.

In the lower part of Fig. 18.1 we have depicted the main logical components that constitute the electric infrastructure, and that we took into account in our model of the regional transmission power grid: generators (N_G), loads (N_L), substations (N_S) and power lines (A_L). From a topological point of view, the power transmission (as well as the distribution) grid can be considered like a network, or a graph, in which the nodes of the graph are the generators, substations and loads, while the arcs are the power lines. The topology T of the grid is typically a meshed graph for the transmission grid, and a partially meshed, radial or ring graph for the distribution grid.

Power plants can include one or more *generators*. The energy produced by the generators is then adapted by transformers, to be conveyed with minimal dispersion to the different types of end-user customers (*loads*), through the power grid. The *power lines* are components that physically connect the substations with the power plants and the end-user customers, and the *substations* are structured components in which the electric power is transformed and split over several power lines. In the substations there are transformers and several kinds of connection components (bus-bars, protections and breakers). In particular, each substation is logically divided into different sections, which are characterised by certain voltage levels and are connected to each other through transformers. Each section consists of a single or double bus-bar.

Voltage, frequency, current, angle, active and reactive power are some of the main (not independent) physical parameters associated with the electric equipment constituting EI (generators, substations, power lines and loads); their specific values are of primary importance in determining the current status of the overall EI. In fact, they affect the behaviour of the electric equipment they are referring to (e.g., in terms of availability and reliability of the electric equipment), thus also influencing the evolution of the overall power grid.

18.2.1.2 Modelling EI

Let us consider that the region is composed by n_A power lines and n_N nodes, divided in n_G generators, n_L loads and n_S substations, with $n_N = n_G + n_L + n_S$. The topology of the network is described by the $n_A \times n_N$ *adjacency matrix* $A = [a_{li}]$, where:

$$a_{li} = \begin{cases} 1 & \text{if line } l \text{ exits node } i, \\ -1 & \text{if line } l \text{ enters node } i, \\ 0 & \text{otherwise.} \end{cases}$$

The power associated with each node i is P_i, which is positive for the generators, negative for the loads and zero for the substations. The maximum power that a generator i can supply is P_i^{max} and the maximum power flow that a transmission line l can carry is F_l^{max}. A line is overloaded if the power flow exceeds F_l^{max}. The susceptance of each line l is b_l.

Some simplifying assumptions have been made to represent the power flow through the transmission grid, following the same approach used in [42, 190, 301, 785]. Therefore, the state and the evolution of the transmission grid are described by the active power flow F on the lines and the active power P of the nodes (generators, loads or substations) at steady-state, i.e., when they have reached an equilibrium condition. P and F satisfy the linear equations for a direct current (DC) load flow approximation of the alternating current (AC) system:

$$P = B \cdot \Theta \tag{18.1}$$

$$F = D \cdot A \cdot \Theta \tag{18.2}$$

with

$$\sum_{i=0}^{n_N-1} P_i = 0, \tag{18.3}$$

where:

- $B = A^T \cdot D \cdot A$ is the $n_N \times n_N$ *susceptance matrix*, and A^T is the transpose of A,
- $D = diag(b_0, b_1, \ldots, b_{n_A-1})$ is the $n_A \times n_A$ diagonal matrix with the l-th diagonal entry representing the susceptance b_l,
- $\Theta = (\theta_0, \theta_1, \ldots, \theta_{n_N-1})^T$ is the *node voltage angle vector*,
- $P = (P_0, P_1, \ldots, P_{n_N-1})^T$ is the *node power vector*,
- $F = (F_0, F_1, \ldots, F_{n_A-1})^T$ is the *line power flow vector*.

As shown in [751], DC can be considered a good approximation of active power flows in the network, although not every network is suitable for DC power flow calculations, especially when power flow controlling devices (like phase shifting transformers (PST)) are involved [908]. In these cases, a full AC power flow including voltage support, reactive power management and transmission losses must be considered, which requires iterative solution of a set of non-linear algebraic equations [707]. In order to reduce calculation time and to simplify the developed model, we considered the simplified power flow problem, thus concentrating our interest on the interdependencies problem. However, our model can be extended to fully account for the AC system aspects.

An autoevolution function $\mathcal{AS}()$ is considered to represent the automatic evolution of EI when an event modifying the grid topology occurs. In this case, EI tries to find a new electrical equilibrium for the new grid topology, by changing the values of the power flow through the lines but leaving the generated and consumed power unchanged (i.e., only through redirection of current flows). The new power flow F through the lines depends on the power injected on the nodes of the grid, on the electrical characteristics of the power lines (e.g., the susceptance) and on the grid topology. In particular, the output values of $\mathcal{AS}()$ are derived by solving the linear power flow equation system (18.1–18.3) for fixed values of P. The other activities typically performed by electric operators in case of failures, such as modulation of production (affecting the input power) and partial/total disconnection of some loads (affecting the output power), have been considered and modelled in the information control part of the system (see Sect. 18.2.2.2).

18.2.2 Information-Technology Based Control System

The logical structure and the model of the ITCS infrastructure are presented in the following two subsections.

18.2.2.1 ITCS Logical Structure

In the upper part of Fig. 18.1 we have depicted a possible logical structure of a regional ITCS, i.e., the part of the information control system controlling and operating on a region of the transmission grid. The components LCS (Local Control System) and RTS (Regional Tele-control System) differ in their criticality and in the locality of their decisions, and they can exchange grid status information and control data over a (public or private) network ($ComNet$ component). LCS guarantees the correct operation of a node (generator, substation or load) and reconfigures the node in case of breakdown of some apparatus. It includes the data acquisition and control equipment (sensors and actuators). RTS monitors its assigned region in order to diagnose faults in the power lines. In case of breakdowns, it chooses the most suitable corrective actions to restore the functionality of the grid. Since RTS is not directly connected to the substations, the corrective actions are put in operation through the pertinent LCS.

18.2.2.2 Modelling ITCS

The control operations performed by ITCS on EI are not represented in detail, but a simplified model is considered where only the effects on the transmission grid of mitigation methods to cope with EI malfunctions, namely generation redispatch, load shedding or grid reconfigurations, are accounted for. In particular, the ITCS actions are abstracted at two levels on the basis of the locality of the EI state considered by ITCS to decide on proper reactions to disruptions (the same approach as adopted in [785]). Each level is characterised by an activation condition (that specifies the events that enable the ITCS reaction), a reaction delay (representing the overall computation and application time needed by ITCS to apply a reconfiguration) and a reconfiguration strategy ($\mathcal{RS}$), based on generation redispatch (i.e., varying the generated power) and/or load shedding (i.e., varying the load demand). The reconfiguration strategy $\mathcal{RS}$ defines how the configuration of EI changes when ITCS reacts to an event that has compromised the electrical equilibrium.[1] For each level, a different reconfiguration function is considered:

- $\mathcal{RS}_1()$. The reconfiguration function $\mathcal{RS}_1()$ represents the effect of the ITCS reaction on the complete transmission grid when only the state local to the affected EI components is considered. Given the limited information necessary to issue its output, $\mathcal{RS}_1()$ is deemed to be local and fast in providing its reaction. $\mathcal{RS}_1()$ is performed by LCS components, and it can be triggered by RTS (as actuation of a global reconfiguration) or can be directly triggered by LCS when it locally detects a lack of (electrical) equilibrium.
- $\mathcal{RS}_2()$. The reconfiguration function $\mathcal{RS}_2()$ represents the effect of the ITCS reaction on the complete transmission grid when the state global to all the EI system under the control of ITCS is considered. Therefore, differently from $\mathcal{RS}_1()$, $\mathcal{RS}_2()$ is deemed to be global and slower in providing its reaction. $\mathcal{RS}_2()$ is performed by RTS.

The activation condition, the reaction delay and the definition of the functions $\mathcal{RS}_1()$ and $\mathcal{RS}_2()$ depend on the policies and algorithms adopted by the specific tele-operation system. It is out of the scope of this chapter to discuss in detail definitions of reconfiguration policies. We adopt definitions for $\mathcal{RS}_1()$ and $\mathcal{RS}_2()$ functions inspired by [42, 190, 301, 785], where redispatch is formulated as an optimisation problem minimising the amount of load shed subject to the system constraints, which is reasonable for the purpose of our study. Of course, providing different specifications would imply different implementations but without changing how the method incorporates and uses them.

The output values of $\mathcal{RS}_1()$ and $\mathcal{RS}_2()$ for defining the new P and F vectors are derived considering that for a given power demand, the power flow equations do not have a unique solution. The adopted definition for the function $\mathcal{RS}_1()$ is given by the solution (values for P and F) of the power flow Eqs. (18.1–18.3) while minimising a

---

[1] Events that impact on the electrical equilibrium are typically an EI component's failure or the insertion of a new/repaired EI component.

simple cost function, which indicates the cost incurred in having loads not satisfied and having the generators producing more power. The output values of $\mathcal{RS}_2()$ for P and F are derived by solving an optimisation problem to minimise the change in generation or load shedding with respect to the initial configuration, considering more sophisticated system constraints as described in [204, 785].

18.2.3 Interdependencies

Dependencies of the state of EI on failures of ITCS mainly pertain to the topology T and the values of the physical parameters associated with each electric equipment, depending on the logical components affected by the failures and obviously on the type of the failures. For example, consequences of a failure of the component LCS associated with a component N_S, N_G or N_L (see also Fig. 18.1) can be:

- Omission failure of LCS, fail silent LCS: no (reconfiguration) actions are performed on N_G, N_S, N_L or A_L.
- Time to failure of LCS: the above (reconfiguration) actions on N_G, N_S, N_L or A_L are performed after a certain delay (or before the instant of time they are required).
- Value failure of LCS: incorrect closing or opening of the power lines A_L directly connected to the failed component is performed.

The failure of the component RTS corresponds to an erroneous (request of) reconfiguration of the state of EI (namely an unneeded or a missing reconfiguration) affecting one or more components of the controlled region. The effect of the failure of RTS on a component N is the same as the failure of the component LCS associated with that component N.

On dependencies generated from the EI side, a malfunction of EI components (resulting in a partial or total blackout) can bring the ITCS infrastructure in a weakened state [577]. In general, parts of the ITCS control can no longer implement their functions due to constraints originated from failures of EI, e.g., shortage of electricity supply for ITCS parts when UPS power backup units are missing. For example, a blackout can switch off a number of nodes of the communication network, thus overloading the nodes that remain online and reducing the performance of the network. A blackout can also increase the expected time to react to failures of EI components (that is, evaluation and application of RTS), or degrade the optimality of the reconfiguration actions because not all the local control equipment involved in the reconfiguration is reachable. Additionally, a blackout can increase the expected time to repair a failed EI component (for example, the communication network $ComNet$). In our evaluation model, dependencies from EI to ITCS are accounted for when defining the failure and repair rates of the ITCS components, such as $ComNet$, RTS and LCS, and the repair rate of the EI components, which are considered dependent upon the blackout size. The modularity of the developed modelling framework allow to include in the analysis other aspects of interdependencies between EI and ITCS, when explicitly identified.

18.3 EPS Modelling and Evaluation Framework: Requirements and Available Formalisms

In this section we first describe the main characteristics that a modelling and evaluation framework should satisfy for the analysis of EPS. Then we provide an overview of the formalisms introduced in the literature to model and evaluate critical infrastructures, specifically focusing on the electric power systems. Finally we shortly discuss how well such formalisms actually fulfil the identified set of requirements, which has led us to adopt the Stochastic Activity Networks (SAN) formalism [805] for our analyses.

18.3.1 Framework's Requirements

To represent and model the behaviour of EI and ITCS and their interactions, the modelling and evaluation framework should possess a number of features encompassing the following aspects : i) modelling power, i.e., the basic modelling mechanisms required to build the EPS model; ii) modelling efficiency, i.e. the advanced modelling mechanisms required to build the EPS model more efficiently; and iii) solution power, i.e., the ability to provide efficient solution methods adequate for the EPS modelling complexity and for the assessment of the specific measures of interest. With respect to the structural and behavioural aspects of EPS systems, major requirements on a suitable modelling framework include:

R1 The system has a natural hierarchical structure, as shown in the examples of logical schemes of Fig. 18.1. Therefore, the modelling framework should support hierarchical composition of different sub-models. The model for the overall EPS could be facilitated considering replication of (anonymous and not anonymous[2]) sub-models, and the replicated and composed models should share part of the state (common state).

R2 The state of EI is completely described through the physical parameters associated with each electric equipment (voltage, current, etc.) and through the topology (T) of the grid: the first set of parameters defines the current status of each EI component, while the topology defines how such components are connected together to form the overall EI. Therefore, it is crucial that the modelling framework should support the representation of a hybrid-state composed by a discrete part (the topology) and a continuous one (the electric parameters).

R3 The time to failures of the components N_S, N_G, N_L and A_L depends also on the value of the electric parameters associated with the components. This means that the framework should support time and probability distributions, as well as conditions enabling the time consuming events (e.g., for the activation of a local protection) that can depend both on the discrete and on the continuous state.

---

[2] Not anonymous replicas can be identified by an index.

R4 We need to consider the reconfiguration actions triggered by the ITCS components, e.g., by LCS and RTS. Moreover, the automatic evolution (autoevolution)
of the electric parameters in case of instability events, e.g. in correspondence of
a power line failure, should also be considered. Therefore, the framework should
support the call to the functions implementing the reconfiguration algorithms,
as well as the autoevolution algorithm.

R5 To manage complexity at solution level, ability to perform separate evaluation
of different sub-models and combination of the obtained results should be supported.

R6 Risk analysis of EPS based on a stochastic approach requires the definition
of measures of performability, which is a unified measure proposed to deal
simultaneously with performance and dependability. To this purpose, a reward
structure should be set-up by associating proper costs/benefits to generators/loads
and interruption of service supply.

18.3.2 Formalisms and Approaches for EPS Modelling and Simulation

Several approaches to EPS modelling and simulation have been adopted in the literature, each having different levels of detail, modelling power, user-friendliness,
and computational efficiency. The works in [115, 376] provide a general understanding of common methods for critical infrastructure analysis, including visualisation
and data presentation techniques, while a specific survey focused on modelling and
simulation can be found in [777]. A specific discussion on the usage of graphical
formalisms for critical infrastructures analysis has been presented in [137]. In the
following we shortly discuss some of the typical formalisms that have been used for
modelling critical infrastructures in general and electric power systems in particular,
grouping them in few macro categories.

18.3.2.1 Graph-Based Techniques

In graph-based techniques, the physical topology and configuration of the infrastructure is mapped to some kind of a graph, which can then be analysed to reveal useful
information about the system. To perform assessments with respect to faults or external attacks, critical infrastructures are often modelled as networks, and then nodes
are progressively removed to evaluate the possible cascading effects on the system.
These kinds of analyses are used to compare infrastructure designs and topologies,
for example showing the maximum number of random attacks that a certain topology
may handle before becoming disconnected (e.g., see [281]).

Network flows based approaches are typically used to model resource requirements and utilisation among different infrastructures. In such paradigm, interdepen-

dent infrastructures are viewed as networks, with movement of commodities (i.e., material, electric power, etc.) corresponding to flows, and with services corresponding to a desired level of these flows. Approaches based on network flow are easily modelled using supply-demand graph; in such kind of graphs, nodes are seen either as supply, transhipment, or demand nodes, while arcs represent links through which commodities flow from producers (supply) to consumers (demand) nodes. Supply-demand graphs have been used for example in [582] to identify the telecommunication components which are more vulnerable to power components failures within a certain critical infrastructure design.

18.3.2.2 Petri Nets

Petri nets (PN) and their extensions are modelling formalisms that are widely used in dependability analysis. Many variants of Petri nets formalisms exist, which may have different properties and modelling power, such as Generalised Stochastic Petri Nets (GSPN), Stochastic Activity Networks (SAN), Stochastic Well-formed Nets (SWN), hybrid Petri nets [262] and Fluid Stochastic Petri Nets (FSPN) [444]. Although they have a simple graphical representation, they provide a great modelling power and are therefore well suited for the modelling of complex systems like critical infrastructures in general, and electric power systems in particular.

In [195] a GSPN model is developed to evaluate the impact of a potential intrusion due to a cyber attack on the Supervisory Control and Data Acquisition (SCADA) system, which is in charge of controlling and monitoring the electric power system. There are two submodels in the Petri net model: a firewall model and a password model, which are instantiated based on the configuration of the internal SCADA network and its possible access points. A combined modelling approach for the evaluation of the interdependencies between the Electric Power infrastructure and its SCADA system has been developed in [95], where the quantification is achieved through the integration of two models. The first is a SAN model, which concentrates on the structure of the power grid and its physical quantities; the second is a Stochastic Well-formed Net (SWN) model, which concentrates on the algorithms of the control system and on the behaviour of the attacker. The scenario modelled in this work considers a situation in which a load shedding activity is needed to re-establish the nominal working conditions upon an electrical failure, but the control system is not working properly due to a Denial of Service (DoS) attack. In [609] Petri nets have been employed in the evaluation of pricing issues related to congestion in de-regulated power market systems. In [522] a new general methodology, based on FSPN, is proposed for the modelling and reliability evaluation of small isolated power systems (which include wind turbines, photovoltaics, and diesel generators).

18.3.2.3 General Simulation Environments

A simulator is a tool that tries to mimic the behaviour of a system. A large collection of simulation environments exists for critical infrastructures analysis, which can essentially be categorised in single domain and multiple domain simulators. The electric power infrastructure, together with telecommunication networks and transportations, has been the focus of development of domain specific simulators, featuring many simulation tools having different granularity [472].

An ad-hoc simulator for the evaluation of dependability and performability measures in electrical power systems has been presented in [785]. The stochastic model is composed by separated and simple submodels of the dynamics of the EI and ITCS subsystems. The impact on the dependability and performability of the cascading or escalating failures has been analysed by providing explicit modelling of the interdependencies between the main subsystems.

18.3.2.4 Agent Based Modelling and Simulation

The agent based paradigm follows a bottom–up approach to manage system complexity. The simulator is built as a population of interacting, intelligent agents. An agent is "an autonomous system (software and/or hardware) that is situated in an environment (possibly containing other agents) and acts on it in order to pursue its own goals, and it is often able to learn from previous experiences" [777]; each agent is an individual entity with location, capabilities, and memory. Using such approach, a simulator is developed, where an agent may model physical components of infrastructures, decision policies or, possibly, the external environment [716].

As other modelling techniques, the agent based paradigm can be applied at different levels of detail, which are sometimes referred to as micro- and macro-agent based simulation [179]. The micro-agent based approach uses a bottom up approach modelling every single component of an infrastructure, putting them together to simulate the whole infrastructure(s). The macro-agent based simulation represents a whole infrastructure with a single agent, hiding the implementation details from the other agents. Using such approach, it is also possible to apply the federated simulation approach, leaving the physical, detailed simulation of each infrastructure to some specific sector tool controlled by an associated agent, and expose only a predefined interface to other agents.

18.3.3 Discussion on Requirements

In this section we discuss how the framework's requirements identified in Sect. 18.3.1 are actually fulfilled by the available formalisms and analysis approaches that have been proposed in the literature, discussing them at macro categories level.

Graphs are a natural way to represent relations between elements, so graph-based techniques are in general appropriate for defining the hierarchical structure of the system (R1) and the interdependencies that exist between infrastructures. For the same reason, cascading failures may be represented as well. On the contrary, they do not support the call to external functions (R4), and are often tailored to a specific measure or analysis type, thus not allowing the definition of different measures (R6).

Formalisms belonging to Petri net category are usually well suited to support time and probability distributions, conditions enabling the time consuming events (R3), as well as the definition of performance and performability measures (R6). The extent to which they are able to represent the hierarchical structure of the system (R1) and the hybrid-states (R2) highly depends on each individual Petri net formalism but, with some exceptions, the scalability of the model is often a limiting factor and they are usually tailored to model discrete state systems. The invocation of external functions (R4) is usually precluded, with the exception of few Petri net formalisms, as well as the separate evaluation of different sub-models and the combination of the obtained results (R5).

Simulation packages may easily represent non-discrete system states (R2), time and probability distributions, as well as conditions enabling the time consuming events (R3). Invocation of external functions (R4) is possible, although it often requires a significant effort to be achieved. Simulation environments may allow evaluation of very complex measures (R6), but they are usually able to evaluate a limited pre-defined set of them.

Thanks to the macro-agent and federation approach, external functions as well as tools can be usually integrated quite easily in agent-based frameworks (R4), as well as specific evaluation approaches to mitigate the complexity at solution level (R5). Agent-based simulation frameworks have the same limitations that arise in other simulation frameworks for what concerns the available measures that can be evaluated (R6).

Although there is no formalism category that, as a whole, is capable to fulfil all the identified requirements, some specific formalisms belonging to specific categories feature more advanced capabilities that can be used to feasibly model complex systems like EPS. An example belonging to the Petri net class is the SAN formalism, which provides the modeller with some primitives that can be profitably exploited to fulfil the identified requirements, thus overcoming the limitations of most of the other Petri net based models.

The SAN formalism [805] is a stochastic extension of Petri nets based on four primitives: places, activities, input gates, and output gates. Places and activities represent the state and the actions of the modelled system, respectively. Special places, called "extended places", allow the representation of the primitive data types of the programming language C++, like int, float, double, including structures and arrays of primitive data types. Input gates control the enabling of activities and define the marking changes that will occur when an activity completes (fires). Output gates define the marking changes when an activity completes. The attributes of the SAN primitives are defined by using sequences of C++ statements.

The SAN formalism supports the hierarchical composition of different sub-models (R1) thanks to the Join and Rep compositional operators [803], also allowing the replication of not anonymous submodels (see Sect. 18.4.2.1—paragraph "The indexing of the replicas"—for more details on this feature). The representation of both discrete and continuous states (R2) is another SAN feature: the SAN formalism supports continuous valued tokens, thanks to a special primitive called "extended place" that allows token of complex data types to be included in the model. The SAN formalism also supports time and probability distributions, as well as marking-dependent enabling conditions (R3—see Sect. 18.4.2 for some specific examples). Moreover, SAN allows the modeller to include C++ code inside input and output gates, as well as custom functions using C++ header files and libraries implementing, for example, the required reconfiguration and autoevolution algorithms (R4). Also, the definition and evaluation of both dependability and performance-oriented metrics (R6) is fully met by resorting to the Performance Variable (PV) reward model [804], which can be used to represent either dependability or performability measures. Finally, SAN does not offer any specific feature to support the separate evaluation of different sub-models and the combination of the obtained results (R5).

18.4 Overview of the Framework Implementation

Following the discussion at the previous section, we have selected the SAN formalism and the modelling and solution tool Möbius [242] to implement our framework.

The software tool Möbius supports multiple high-level modelling formalisms (including the SAN) and multiple solution techniques. It also allows the construction of composed models from previously defined models, supporting a hierarchical approach to modelling based on state-sharing. In this approach the submodels are linked together through sharing of state variables of each submodel. A state variable captures some portion of the state of a model, and is represented differently by different formalisms. In a SAN model, a place represents a state variable. Thus, it is possible to compose SAN models by holding one or more state variable in common. Möbius features the Replicate/Join composed model formalism [803]. A Join is a state-sharing composition node used to compose two or more submodels. A Replicate is a special case of the Join node used to construct a model consisting of a number of identical copies of a submodel.

The goal of the models is to assess metrics well representative of the quality of service perceived by end users, such as indicators of blackout size (for example, the percentage of power demand that is not met, the cost of a blackout, the number of loads involved in a blackout, etc.).

In Sect. 18.4.1 we will briefly introduce all the atomic models composing the overall EPS model, while in Sect. 18.4.2 we will detail the implementation of the model representing a generic node with connected transformers, showing how we have concretely realised some of the general framework's characteristics detailed in

Sect. 18.3.1. The whole set of implemented models can be found in [204] and it is not reported here for the sake of brevity.

18.4.1 The Composed EPS model

When modelling the considered EPS, we have followed a modular and compositional approach. The following atomic models (the leaves in Fig. 18.2) have been identified as building blocks to generate the overall EPS model:

- PL_SAN, which represents the generic power line with the connected transformers.
- PR1_SAN and PR2_SAN, which represent the generic protections and the breakers connected to the two extremities of the power line.
- N_SAN and LCS_SAN, which represent, respectively, a node of the grid (a generator, a load or a substation) and the associated LCS (see Fig. 18.1).
- AUTOEV_SAN, which represents the automatic evolution (autoevolution) of EI when an event modifying its state occurs.
- RS_SAN, which represents the computation and application of the local reconfiguration strategy $RS_1()$, and the computation of the regional reconfiguration action $RS_2()$ (its application is modelled in RTS_SAN).
- RTS_SAN and COMNET_SAN, which represent, respectively, the Regional Telecontrol System RTS, where the regional reconfiguration strategy $RS_2()$ is applied, and the public or private networks ($ComNet$ of Fig. 18.1).

In Fig. 18.2, it is shown how these atomic models are composed and replicated to obtain the composed model representing the EPS region.

The model AL represents a power line with the associated protections and it corresponds to the A_L logical component of Fig. 18.1. This model is then replicated to obtain all the necessary non anonymous A_L components of the grid. The model N_LCS is obtained by composing the atomic models N_SAN and LCS_SAN. Then the model is replicated to obtain all the necessary non anonymous N_G, N_S and N_L components of the grid, with the associated LCS. The model Auto_Control is obtained by composing the atomic models AUTOEV_SAN and RS_SAN, so it represents both the autoevolution function and the reconfiguration strategy locally applied by the LCS components. The overall EPSREG model is finally obtained through composition of the different models and it represents the EPS under study.

The different (atomic or non atomic) submodels interact with each other through sharing of some common places that represent the parameters or part of the states of the EPS.

These models populate our modelling framework as template models, which are used to represent a large variety of specific scenarios in the EPS sector. The overall model results from the composition and replication of such template models. For example, the power line model including the protections (AL) is built by composing (through the Join operator) the template model for the power line (PL_SAN) with those for the protections (PR1_SAN and PR2_SAN), and it is replicated (through

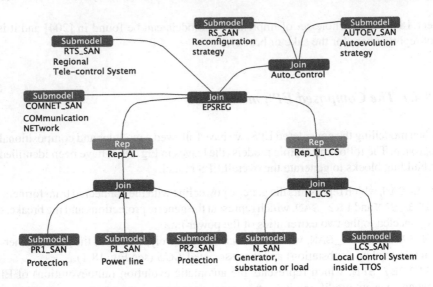

Fig. 18.2 Composed model for an EPS region

the Rep operator) to obtain the model representing the whole set of power lines with
protections composing the grid (Rep_AL). In the following we provide some more
details on the system's aspects captured by the N_SAN model, representing a node
of the grid.

18.4.2 The Atomic Model N_SAN

The atomic model N_SAN, shown in Fig. 18.3, represents the generic node (generator,
substation or load). For the sake of conciseness, we concentrate here on three main
aspects of this model: i) the indexing of the replicas, ii) the failure modes, and iii) the
failures propagation to the connected lines. The boxes in Fig. 18.3 group the basic
modelling elements related to these three aspects, and are discussed in the following
subsections.

18.4.2.1 The Indexing of the Replicas

As discussed in [203], the modelling activity can be facilitated considering replication of non anonymous submodels, i.e., distinguishing each replicated submodel
by an associated index. In this section we will detail the part of the N_SAN
model implementing this specific aspect, as shown in box i) of Fig. 18.3. As
detailed in Sect. 18.4.1, the N_SAN model is anonymously replicated n_N times
(one replica for each node in the system) using the Rep compositional operator

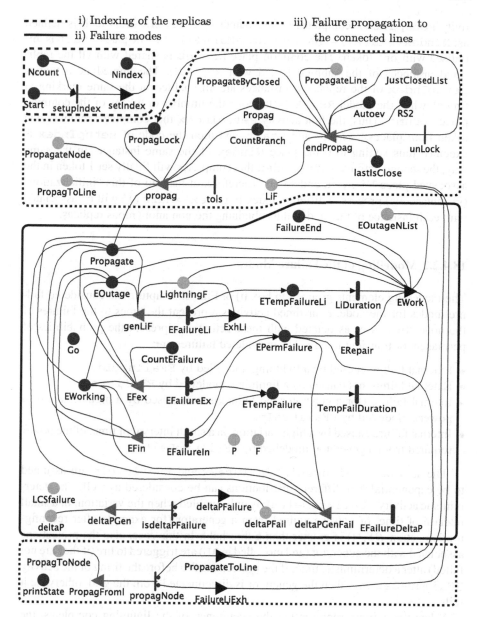

Fig. 18.3 The model N_SAN for a generic node. Reproduced from [202]. ©2011 Elsevier

[803]. In order to distinguish between the different replicas, we need to associate an index with each node, which is represented by the number of tokens in the place Nindex. This place is shared with atomic model LCS_SAN, but it is local to composed model N_LCS. Place Nindex is set when the immediate ac-

tivity `setupIndex` completes by the output gate `setIndex`, which is defined as: `Nindex->Mark()=(nN-Ncount->Mark())-1`. The place `Start` is initialised with one token. The common place `Ncount` is shared with all the replicated instances of the submodel (and with `RS_SAN`). The immediate activities `setupIndex` of the replicated instances are all enabled in the same marking at time 0, when the model `RS_SAN` sets to n_N the number of tokens of the common place `Ncount`. Thus, the first activity `setupIndex` that completes removes one token from places `Ncount` and `Start`, and then the code of `setupIndex` is executed thus setting to 0 the place `Nindex` of the same instance. In the same way, the second activity `setupIndex` that completes will finally set 1 token in the associated place `Nindex`, and so on. Therefore, at the end of this "initialization" (instantaneous) process, a different index (`Nindex->Mark()`) will be associated with each instance of the model, thus obtaining the non anonymous replicas.

18.4.2.2 Modelling of the Failure Modes

The modelling elements grouped in box ii) of Fig. 18.3 capture the considered failure modes for the node. Four timed activities represent the times to the failures of the node. Two cases associated with these activities represent the probabilities of permanent or transient failure. The considered failures are:

- External failure caused by a lightning, modelled by `EFailureLi`.
- External failure not caused by a lightning, modelled by `EFailureEx`.
- Internal failure (for a generator the failure is not caused by a high variation of power), modelled by `EFailureIn`.
- Internal failure caused by a high variation, in a small interval of time, of the power required from a generator, modelled by `EFailureDeltaP`.

The activities `EFailureLi`, `EFailureEx` and `EFailureIn` are assumed to be exponential, but different distributions can be considered as well. The deterministic activity `EFailureDeltaP` is only enabled when the variation, in a small interval of time, of power required from a generator (for example after reconfiguration action $RS_1()$) is greater than a threshold. In this case, also the protections associated with the generator (and modelled in `AL`) are triggered to fire (if they are not failed) after a deterministic time. If the protections fire before the firing of the activity `EFailureDeltaP`, then the generator is disconnected from the grid, otherwise it fails.

When the activity representing the occurrence of the lightning completes, the input gate `GenLiF` sets the place `LightningF` with a value randomly generated from an uniform distribution, which represents the power of the lightning.

The duration of a transient failure is represented by the timed activities `LiDuration` (deterministic for the lightning) and `TempFailDuration` (exponential for the other failures). The repair time of a permanent failure is represented by the deterministic activity `ERepair`.

In the current model we have represented an on/off behaviour of the node, which becomes completely unavailable and disconnected from the grid in case of failure. It is worthwhile mentioning that this is not a methodology limitation and the same model could be extended considering different degraded operational modes, each one corresponding to a given degree of failure (e.g., disconnecting only a part of the lines linked to the failed node). The same extensions could be done for power lines models, provided that the different failure modes actually have an impact on the metrics of interest, and that the propagation rules towards the connected lines have been established.

18.4.2.3 Modelling the Failures Propagation to the Connected Power Lines

The modelling elements grouped in box iii) of Fig. 18.3 capture the failures propagation from the node to the connected power line and vice versa. This propagation is instantaneous; it starts from the failed node (or line) and stops:

1. when the propagation reaches an open breaker, or
2. when a protection fires (thus opening a breaker), or
3. when the propagation reaches a failed line or node, or
4. when the propagation reaches a node already touched by the propagation process.

The effect of the propagation of a lightning or a failure is to isolate the failed component from the rest of the grid. If the propagation reaches a generator or a load (because the protections did not fire), these components are considered failed.

The propagation to the lines connected to the modelled node is represented by the enabling and completion of immediate activities and by places common to different submodels. In particular, when a failure occurs in the node, the immediate activity tols is enabled. It locks the propagation (by subtracting one token to the place PropagLock) and, for each line l linked to the node, it sets the entry l of the extended place PropagToLine with the failure propagation info. Place PropagToLine is an array representing the failure propagation to the power lines linked to the node. In the case of lightning propagation, entries of place PropagToLine are also set with the power of the lightning divided by the number of lines connected to the node (power of lightning is uniformly distributed to all lines connected to node).

One immediate activity PropagNode enabled when a failure (e.g., a lightning) propagates from the model PR1_SAN or PR2_SAN associated to a power line. Two cases on the activities represent the probability that the propagation causes a failure of the node (if it is not already failed). In the case of failure of the node, the propagation ends, otherwise the propagation moves to the model PR1_SAN or PR2_SAN, represented by the extended place PropagateToLine.

The failures propagation modelled in box iii) of Fig. 18.3 represents the immediate impact of the failure on the connected grid elements. Then, based on the resulting EI configuration, the application of the EI autoevolution or LCS reconfiguration functions will determine the new EI stable state. Afterwards, a failure propagation could occur over time, for example, if in the resulting stable state some power lines are overloaded and no appropriate reconfiguration actions are performed by RTS in due time, thus leading to a cascading effect (e.g., in the case of the 2003 Italian blackout [491]).

18.5 Illustrative Examples of EPS Evaluation

In this section we present some examples of analysis of a few representative scenarios. The conducted analyses targeted the following objectives to be assessed:

1. the impact of failures of one or more nodes and of the ITCS infrastructure (through the DoS affecting the communications network $ComNet$) on the measures of interest, and
2. the criticality level of generators, substations or loads with respect to the measures of interest.

The analysed case study and the results of the numerical evaluations are discussed in the following subsections.

18.5.1 Analysed Power Grid, Measures of Interest and Failure Scenarios

The reference power grid under analysis is the IEEE Reliability Test System–1996 (RTS-96),[3] typically used in bulk power system reliability evaluation studies [468]. RTS-96 is a multi-area reliability test system obtained by modifying, updating and linking various single IEEE RTS-79 areas [467]. For sake of simplicity, we analyse a single area of RTS-96 (referred as RTS96A in the following). Figure 18.4 shows the topology of RTS96A and the power associated with each node and line at the time zero. The label "P_i / P_i^{max}" associated with the generators (circles) represents the initial (active) power P_i and the maximum power that the generator i can supply P_i^{max}. The label "D_i" associated with the loads (squares) represents the power demand of the load i. The label "$F_l / F_l^{max} (b_l)$" associated with each line l represents the initial power flow F_l through the line, the maximum power flow F_l^{max} that a transmission line can carry and the susceptance b_l of the line. A negative F_l value means that the current is flowing in the opposite direction of the corresponding arrow.

The measures of interest considered in this study are user-oriented indicators of the blackout size. They are:

1. $P_{UD}(0, l)$, defined as the mean of the percentage of power demand $UD(0, l)$ that is not met in the interval $[0, l]$, and
2. discrete probability distribution function PDF of $UD(0, l)$, defined as the probability that $UD(0, l)$ is equal to 0 or it is in the interval $(a, a + 10]\%, a = 0, 10, \ldots, 90$.

---

[3] http://www.ee.washington.edu/research/pstca

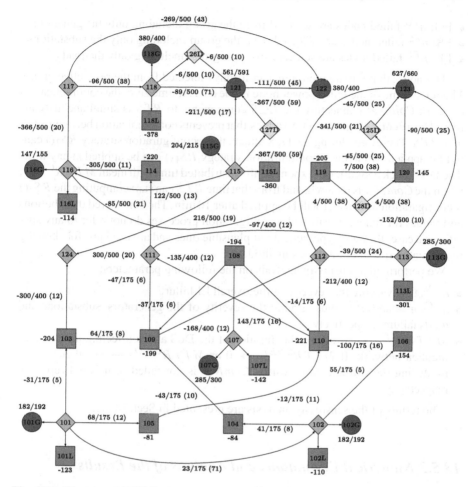

Fig. 18.4 Diagram of RTS96A (generators are *circles*, loads are *squares* and substations are *rhombi*). For the sake of clarity, only the integer part of the original values associated with generators, power lines and loads are shown (in MW). Reproduced from [202]. ©2011 Elsevier

In the considered scenario, the EI state is in electrical equilibrium and it is initially set as depicted in Fig. 18.4, where, for each generator i, all the ratios P_i / P_i^{max} are equal to a fixed value 0.95, called the power grid stress level. Moreover, power demand is constant in time. At time zero, we suppose that one or more nodes are simultaneously affected by a permanent disruption (e.g., due to a tree fall or a terrorist attack), thus becoming unavailable. Pessimistic assumption is made that all power lines linked to failed nodes are disconnected and the failed node is isolated from the grid. The number of simultaneously failed nodes are alternatively selected from different groups of nodes. Therefore, four cases are considered:

- FN: n^{FN} failed nodes are selected from the group including all the types of nodes (generators, substations or loads),

- FG: n^{FG} failed nodes are selected from the group including only the generators,
- FS: n^{FS} failed nodes are selected from the group including only the substations,
- FL: n^{FL} failed nodes are selected from the group including only the loads.

The nodes that fail are randomly (uniformly) selected from the respective group. The repair time of the failed power node is one day. At time zero, the communication network $ComNet$ connecting the LCS components to RTS is simultaneously affected by a denial of service (DoS) attack that prevents communication between LCS and RTS. Therefore, during the DoS attack, the reconfiguration strategy $RS_2()$ cannot be applied, while the reconfiguration strategy $RS_1()$ can be applied at any time. The DoS attack ends after an exponentially distributed time with mean $MTTR^{CNET}$, when the $ComNet$ is repaired, and from that time RTS can start computing the $RS_2()$ reconfiguration action that will be applied after 10 min. The considered distributions and values for failure, repair and reconfiguration processes do not refer to any specific real case; they are hypothetical but plausible ones and are used just for showing the potentialities of our analysis method.

We performed a sensitivity analysis on the following parameters:

- n^{FN}, thus varying the severity of the overall EI failure.
- n^{FG}, n^{FS} and n^{FL}, thus varying the severity of the generators, substations and loads failure, respectively.
- $MTTR^{CNET}$, thus varying the duration of the DoS attack affecting the communication network. If $MTTR^{CNET} \to 0$ or $MTTR^{CNET} \to \infty$, then we are modelling the extreme cases where $ComNet$ is not failed or it is not repaired, respectively.

The results of the sensitivity analysis are presented in Sect. 18.5.2.

18.5.2 Numerical Evaluations and Analysis of the Results

In this section we present some of the results that we obtained through the solution of the overall model for the RTS96A in the scenario previously described. A transient analysis has been performed, using the simulator provided by the Möbius tool [242]. For each numerical result, 20,000 simulation runs (batches) were executed using a confidence level equal to 0.95. The confidence intervals obtained for the result are less than 10 % wide and are shown in the plots, although they are so small that they are similar to points. The parameters values used as default values are $l = 1$ day (for the considered interval of time), $MTTR^{CNET} = 3h$ (for the duration of the DoS attack exponentially distributed) and $n^{FN} = n^{FG} = n^{FS} = n^{FL} = 1$ (for the number of nodes affected by failure).

Figures 18.5 and 18.6 show the impact on $P_{UD}(0, l)$ of failures of nodes and $ComNet$ at varying the number of simultaneous failed nodes and the expected duration of the DoS attack $MTTR^{CNET}$, for different types of failed nodes. In particular, Fig. 18.5a–d, focus on the assessment of $P_{UD}(0, l)$ for cases FN, FG, FS and FL, respectively. The lines shown in the legend are sorted by decreasing values of $P_{UD}(0, l)$.

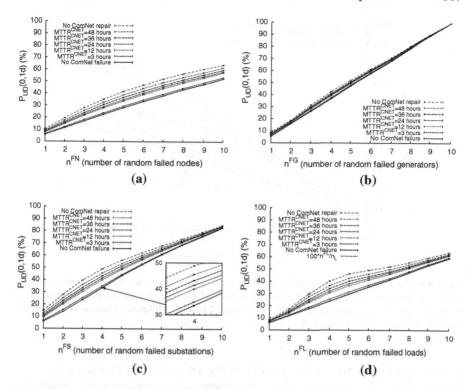

Fig. 18.5 Percentage of the expected power demand $P_{UD}(0, l)$ that is not met in the interval $[0, l]$, with $l = 1$ day, as a function of a different number of random simultaneous failed nodes, for different values of $MTTR^{CNET}$ and for different types of failed nodes: cases FN (**a**), FG (**b**), FS (**c**) and FL (**d**)

In Fig. 18.5d the formula $100n^{FL}/n_L$ is also plotted, representing the expected loss of power demand in the interval $[0, l]$ because of the failure of only n^{FL} loads (and no other failure), under the assumption that power lines linked to failed loads are not disconnected. This plot is a lower bound for the curves shown in Fig. 18.5d. In fact, it overlaps with the curve for which no failure of $ComNet$ occurs, then showing that, when reconfiguration strategy $RS_2()$ can be applied by RTS, the failures of loads, even though disconnecting lines linked to them, do not impact on the expected power demand of the other loads, for the considered topology and setting.

In Fig. 18.5, as expected, $P_{UD}(0, l)$ increases considering higher number of failed nodes n^X, being $X = FN, FG, FS, FL$. Fixing the value for n^X, $P_{UD}(0, l)$ gets worse in the case in which the DoS attack lasts longer. However, the impact of $MTTR^{CNET}$ on $P_{UD}(0, l)$ tends to decrease at low and high values of n^X (that is, at the extremes of the curves). In fact, when n^X is low, the impact of the failed nodes on $P_{UD}(0, l)$ is so small that the effect of the reconfiguration strategy $RS_2()$ applied by RTS is negligible. Similarly, when n^X is high, the impact of the failed nodes on $P_{UD}(0, l)$ is so big that no reconfiguration exists to reduce the loss of power demand,

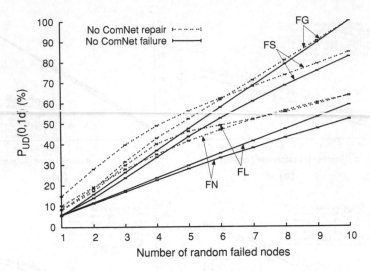

Fig. 18.6 Percentage of the expected power demand $P_{UD}(0, l)$ that is not met in the interval $[0, l]$, with $l = 1$ day, as a function of a different number of random simultaneous failed nodes, for different types of failed nodes (cases FN, FG, FSand FL) and for different values of $MTTR^{CNET}$ (corresponding to no $ComNet$ failure and no $ComNet$ repair)

as shown in Fig. 18.5b, for the case FG. In Fig. 18.5b, $P_{UD}(0, l)$ reaches the value of 100 % for $n^{FG} = 10$, since all the 10 generators of the grid are failed. Finally, we note that for values of $MTTR^{CNET}$ varying from 0 to 3 h, the values of $P_{UD}(0, l)$ vary less than 2 %.

Figure 18.6 focuses on the comparison of the assessment of $P_{UD}(0, l)$ for different types of failed nodes, corresponding to the cases FN, FG, FS and FL, when only the two extreme values are considered for $MTTR^{CNET}$ (i.e., no $ComNet$ failure and no $ComNet$ repair), being the other values omitted for simplicity. This figure shows how the level of criticality of each type of node with respect to the measure of interest $P_{UD}(0, l)$ can depend on the number of failed nodes and on the delay to apply the reconfiguration strategy $RS_2()$ by RTS. When the reconfiguration strategy $RS_2()$ can be applied by RTS (no $ComNet$ failure), or when it cannot be applied only for a small duration of time (e.g., less than 3 h, for the topology and setting used in our case study), the criticality of the types of nodes is shown by the lines labelled "No $ComNet$ failure" and their rank is not affected by the number of failed nodes, but by the type of failed nodes. In this case, it is interesting to note that failures involving nodes randomly selected from the overall set of nodes (FN, which includes generators, substations and loads) are less critical than failures that only involve loads, substations, or generators. On the other extreme, when the reconfiguration strategy $RS_2()$ cannot be applied and the number of nodes is less than 6, the failures of substations are more critical than the failures of generators.

In both Figs. 18.5 and 18.6 we have provided mean values for the percentage of unsatisfied power demand $UD(0, l)$ in an interval $[0, l]$. Figure 18.7 shows the dis-

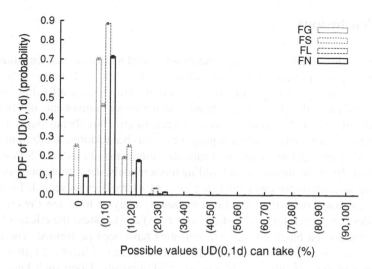

Fig. 18.7 Probability (PDF) that $UD(0, 1d)$ is equal to 0 or it is in the interval $(a, a + 10]\%$, $a = 0, 10, \ldots, 90$, for $MTTR^{CNET} = 3$ h and for one random failed node selected from different types of nodes (cases FN, FG, FS and FL)

crete probability distribution function PDF of $UD(0, l)$, for $MTTR^{CNET} = 3$ h, when only one randomly selected node is affected by failure and for different types of failed nodes (cases FN, FG, FS and FL). The values on the x-axis represent different levels of criticality of the failures of nodes with respect to the measure of interest $UD(0, l)$. Each level of criticality corresponds to 0 or to a range $(a, a + 10]\%$, $a = 0, 10, \ldots, 90$. The most critical failures are those leading to higher values of $UD(0, l)$. The PDF values of $UD(0, l)$ on the y-axis multiplied by 100 represent the percentage of failures of nodes occurred for each level of criticality. Thus, Figure 18.7 shows the percentage of failures of nodes associated with each level of criticality, depending on the types of nodes affected by the failures. In the case that the failed node is randomly selected from the set of generators, substations or loads (case FN), the most critical failures are the 1.4 % of the overall occurred failures (corresponding to the (20, 30]% range). When the random failed node is a generator (case FG), or a substation (case FS), or a load (case FL), then the most critical failures are the 0.4 % for the generators, the 3 % for the substations and the 0.5 % for the loads, respectively. Thus, loss of power demand is mainly caused by failures of substations. Finally, we note that, for all types of failed nodes, the highest percentage of nodes failures is associated with the criticality level for which $UD(0, l) \in (0, 10]\%$. Also, the mean $P_{UD}(0, l)$ varies in the range $(0, 10]\%$, from 5.4 for case FG to 6.7 for FS.

18.6 Conclusions

This chapter described a modelling framework suited to analyse and quantitatively assess the impact of interdependencies between the electric grid and the cyber level control in electric power systems, in presence of malfunctions affecting both infrastructures. First, the abstract specification of the major components under analysis in both EI and ITCS infrastructures, in terms of structural/behavioural aspects, failure patterns and related mutual dependencies, has been addressed. Requirements imposed to assessment methods and available formalisms to model EPS have been pointed out. Then, the definition of building block models, their implementation and their composition to represent a whole region of EPS have been presented. To provide concrete estimates of resilience-related measures, analyses to assess the criticality of EI nodes (i.e., generators, substations and loads) and to assess the effects of a few failure scenarios on blackouts related indicators have been performed. The results of these analyses allow to understand the relative impact of involved failure/repair processes and the criticality of the electric grid elements. Upon such knowledge, indications can be derived on appropriate actions to take towards making the design and development of EPS system more resilient.

Tackling interdependencies is a very challenging and still in progress research area. This chapter goes in the direction to explore the field and to provide a step forward. Given the high inner complexity, the study has been performed under some limiting conditions. Enhancements of this study are foreseen in several directions. Extending the view to the nation and cross-nations wide vision as well as enriching the EPS analysis in several directions are among current and planned future work. Exploring whether and to which extent the proposed modelling framework would adapt to analyse interdependencies in other critical infrastructures (e.g., in oil/water production and transportation systems and in the telecommunication sector) is another stimulating research option.

Acknowledgments This work has been partially supported by the European Community through the IST Projects CRUTIAL [245](Contract n. 027513) and by the Italian Ministry for Education, University, and Research (MIUR) in the framework of the Project of National Research Interest (PRIN) "DOTS-LCCI: Dependable Off-The-Shelf based middleware systems for Large-scale Complex Critical Infrastructures".

Chapter 19
Providing Dependability and Performance in the Cloud: Case Studies

Nikolaus Huber, Fabian Brosig, Nicholas Dingle,
Kaustubh Joshi and Samuel Kounev

Abstract Cloud Computing promises a variety of opportunities but also brings up several challenges. The three case studies presented in the following are examples on how challenges in the field of capacity management, dependability, and scalability can be addressed and how opportunities of Cloud Computing can be leveraged to, e.g., maintain performance requirements or to increase dependability.

19.1 Introduction

As discussed in Chap. 4, Cloud Computing has several challenges and opportunities. The increased flexibility and the shared resources cause challenges like security or performance issues, to mention only some examples. However, the increasing

N. Huber (✉) · F. Brosig · S. Kounev
Karlsruhe Institute of Technology,
76131 Karlsruhe, Germany
e-mail: nikolaus.huber@kit.edu

F. Brosig
e-mail: fabian.brosig@kit.edu

S. Kounev
e-mail: kounev@kit.edu

N. Dingle
School of Mathematics,
University of Manchester,
Manchester M13 9PL, UK
e-mail: nicholas.dingle@manchester.ac.uk

K. Joshi
AT & T Labs Research,
Florham Park, NJ, USA
e-mail: kaustubh@research.att.com

K. Wolter et al. (eds.), *Resilience Assessment and Evaluation of Computing Systems*,
DOI: 10.1007/978-3-642-29032-9_19, © Springer-Verlag Berlin Heidelberg 2012

flexibility provides also opportunities like higher availability and fault tolerance, resilience to attacks, or improved resource efficiency.

In this chapter we present three case studies as examples on how the previously mentioned challenges can be addressed and how the opportunities can be used to add value to systems running in Cloud Computing environments. The first two case studies are approaches on managing performance and dependability in Cloud Computing environments. The third case study is a scalability study of two different tools for performance analysis in Cloud Computing environments. For related work and state-of-the-art on approaches for resilience assessment and managing dependability and performance, the reader is referred to Chap. 4.

In Sect. 19.2, we demonstrate how prediction techniques based on performance models can be used to maintain the service-level agreements (SLAs) while using available resources efficiently. The approach uses the Palladio Component Model [96] and its simulator to predict service response times and resource utilizations. Section 19.3 presents an architecture and algorithm on balancing the trade-off between performance and dependability. It uses performance and availability models to react to changes in the underlying infrastructure which are results of failures or upgrades. The hierarchical optimization algorithm extends queuing models to balance the needs of availability and performance. Several scenarios show the applicability of this approach even in a cloud scenario with different data centers. Finally, we present a case study on the computational and communication scalability of a Cloud Computing environment by transferring two HPC applications to a Cloud Computing environment (Amazon EC2). Both tools calculate the full distributions of response times in Continuous Time Markov Chains (CTMCs) but require a different amount of interprocessor communication, and hence scale differently in Cloud Computing environments.

19.2 Elastic Capacity Management

In this section we present results of our case study on self-adaptive resource management in virtualized environments [451]. To avoid violations of service-level agreements (SLAs) or inefficient resource usage, capacity management has to be adopted continuously during system operation. For example, in Cloud Computing scenarios resources allocated to services need to be increased or decreased to reflect changes in application workloads. This is an approach on elastic capacity management based on online architecture-level performance models [556]. The goal is to maintain performance and efficient resource usage during run-time. In our evaluation we use the new SPECjEnterprise2010 benchmark.[1]

---

[1] SPECjEnterprise2010 is a trademark of the Standard Performance Evaluation Corp. (SPEC). The SPECjEnterprise2010 results or findings in this publication have not been reviewed or accepted by SPEC, therefore no comparison nor performance inference can be made against any published

19.2.1 Self-Adaptive Resource Management

Our self-adaptive resource management follows the control loop model [196] which consists of four phases: *collect, analyse, decide* and *act*. For the *collect* phase, we assume that changes of the application workload are either announced by the customers (e.g., for an upcoming sales promotion) or by techniques like workload forecasting [147]. We then use the Palladio Component Model [96] and its performance prediction techniques to *analyze* the impact of these changes and to *decide* which actions to take. In this case study, the *act* phase covers the reconfiguration operations adding/removing application server cluster nodes and increasing/decreasing the number of virtual CPUs of a cluster node's virtual machine.

19.2.1.1 Resource Allocation Algorithm

The following algorithm is executed if SLAs are violated or resources are used inefficiently. The goal is to find a new system configuration which again maintains performance and resource efficiency. The algorithm is specified in generic terms, such that it can be applied to different types of resources and resource allocation operations. In short, the algorithm works on a set of services, resource types and SLAs. The SLAs specify, e.g., the requested average response time for a service at a given arrival rate. Each time there is a change of a specified SLA (e.g., a new client workload is scheduled for execution or a change in the workload intensity of an existing workload is forecast), we use our architecture-level performance models to predict the effect of this change on all SLAs. The algorithm can be divided into two phases: PUSH phase and PULL phase. If an SLA violation is detected, the PUSH phase of our algorithm is executed which allocates additional resources (PUSH additional resources into the system) until all client SLAs are satisfied. After the PUSH phase finishes, the PULL phase is executed to optimize the resource efficiency. If no SLAs are violated, the PULL phase starts directly to reduce the amount of used resources (PULL them out of the system).

PUSH:

Basically, while there exists a client response time SLA that is violated, in this phase the algorithm increases the amount of allocated resources for all resource types used by the service which are overutilized. Increasing the number of allocated resources works as follows: If a there is an instance of an overutilized resource type (e.g., a VM) which has some processing resources available that are not allocated yet,

----

(Footnote 1 continued)
SPEC result. The official web site for SPECjEnterprise2010 is located at http://www.spec.org/jEnterprise2010.

additional resources are allocated (e.g., virtual CPUs). Otherwise, a new instance of this resource type is added (e.g., a new VM is started).

PULL:

The PULL phase aims to optimize the resource efficiency by releasing resources that are not utilized efficiently under the current client workloads. In our algorithm, inefficient usage means the delta of maximum utilization and current utilization of a resource type is greater than a predefined constant, e.g., 20 %. While there is a resource type assigned to service of the currently considered workload which is used inefficiently, the amount of resources allocated to this service will be decreased, i.e., for a resource type instance the capacity (e.g., virtual CPUs) is reduced. If the client SLAs are predicted to be violated after this change, the change is reversed.

19.2.2 Evaluation

In this section we briefly explain the SPECjEnterprise2010 benchmark and the experimental environment we used to evaluate our approach. Finally, we present the experimental results.

19.2.2.1 SPECjEnterprise2010 Benchmark

We selected the SPECjEnterprise2010 benchmark application as a basis for our case study because it models a representative, state-of-the-art enterprise system. SPECjEnterprise2010 is a benchmark developed by SPEC's Java subcommittee to measure the end-to-end performance and scalability of Java EE-based application servers. The benchmark workload is generated by an application that is modeled after an automobile manufacturer. As business scenarios, the application comprises customer relationship management (CRM), manufacturing and supply chain management (SCM).

The benchmark driver executes five benchmark operations. A dealer may *browse* through the catalog of cars, *purchase* cars or *manage* his dealership inventory, i.e., sell cars or cancel orders. A manufacturer may place *work orders* for manufacturing vehicles, either triggered per WebService or RMI call. In our experiments these benchmark operations function as the different services. To control the request arrival rate of each service individually, we had to slightly modify the benchmark driver. We split up the two driver domains and three manufacturing domains into five different domains, each invoking its own service. The resulting five independent services are called *Purchase*, *Manage*, *Browse*, *CreateVehicleEJB* and *CreateVehicleWS*.

19.2.2.2 Architecture-Level Performance Model

To make decisions in our control loop, we use a PCM model [96] as architecture level performance model to predict the service response times and resource utilizations of the SPECjEnterprise2010 application for a specific load. The PCM model is semi-automatically extracted from a running benchmark application instance. As extraction method, we use the method in [150]. However, for this case study we extracted the entire benchmark application, i.e., including supplier domain, dealer domain, web tier and the asynchronous communication between the three domains. For reasons of brevity, the reader is referred to [451] for a detailed description of the PCM model instance.

19.2.2.3 Experimental Setup

As hardware environment for the experiments, we use six blade servers from a cluster environment. Each server is equipped with two Intel Xeon E5430 4-core CPUs running at 2.66 GHz and 32 GB of main memory. The machines are connected by a 1 GBit LAN. On top of each machine, we run Citrix XenServer 5.5 as the virtualization layer. Inside the XenServer's VMs, we run the benchmark components. Each component runs in its own VM, initially equipped with two virtual CPUs (VCPUs). As operating system, these VMs execute CentOS 5.3. As Java EE application server, we use the Oracle Weblogic Server (WLS) 10.3.3. The load balancer is haproxy 1.4.8 using round-robin as load balancing strategy. The database is an Oracle 11g database server instance deployed on a VM with eight VCPUs on a separate node on Windows Server 2008. The SPECjEnterprise2010 benchmark application is deployed in a cluster of WLS nodes. For the evaluation, we considered reconfiguration options concerning the WLS cluster and the VCPUs the VMs are equipped with: WLS nodes are added to or removed from the WLS cluster, VCPUs are added to or removed from a WLS node's VM. These reconfigurations are applicable at run-time, i.e., can be applied while the benchmark application is running.

19.2.2.4 Results

In the following section we present experimental results of our approach. First, we demonstrate how the approach behaves when the system workload increases. Next, we give an example how this approach can be used for elastic capacity management and show its benefits.

Workload Growth:

In this scenario, we evaluate our approach when increasing the workload of all services deployed in our environment. We increase the load in two steps from 2x to 4x

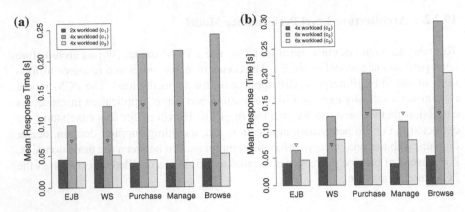

Fig. 19.1 The response times when changing workload from (**a**) 2x to 4x and (**b**) 4x to 6x, respectively (SLAs denoted by $\triangledown$). The three bars depict the response times for all five services before the load increase, after the load increase, and after system reconfiguration

and 4x to 6x (see Fig. 19.1). The standard workload (1x) is defined as request arrival rate (requests/second) for each service: (CreateVehicleEJB, 15), (CreateVehicleWS, 15), (Purchase, 12.5), (Manage, 12.5) and (Browse, 25). Our starting point is that five services are running on one node with three VCPUs (c_1) with 2x the standard workload and the following SLAs (CreateVehicleEJB, 30, 74 ms), (CreateVehicleWS, 30, 74 ms), (Purchase, 25, 130 ms), (Manage, 25, 130 ms), (Browse, 50, 130 ms) which are initially satisfied. Now, we increase the workload to 4x the standard load. For this new workload, the reallocation algorithm detects a violation of the SLAs and recommends to reallocate the system resources using two nodes, one with four VCPUs and one using three VCPUs (c_2). Applying this configuration to our benchmark, the SLAs are satisfied. For the measurement results see Fig. 19.1a.

In the second step, we increase the workload to 6x the standard load and do not change the SLAs. Again, this leads to a violation of the SLAs in our simulation results. Therefore, we apply our algorithm, finding a new suitable configuration with three nodes, two with four VCPUs and one with three VCPUs (c_3). The experiment results are depicted Fig. 19.1b. However, the results show that after reallocation the SLA of the Browse service is still slightly violated. This is not due to inaccuracy of our model, but rather due to scalability problems of the database machine, which is not powerful enough to handle the new workload while satisfying the original SLAs. Hence, we are confident that given a more powerful database, the SLAs would be satisfied. The way this problem would be addressed in practice would be to either scale the database or renegotiate the SLAs. As both solutions can be handled with our online performance prediction mechanism, we plan to extend our approach with this solution in the future.

Fig. 19.2 Assigned capacity and servers for a workload distribution over seven days

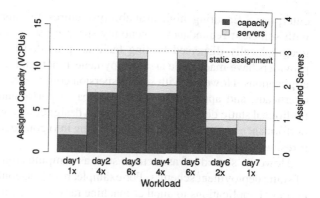

Resource Usage and Efficiency:

After evaluating the functionality of our approach, this section discusses its benefits. Imagine a workload distribution over seven days like the one depicted in Fig. 19.2. In a static scenario, one would assign three dedicated servers to guarantee the SLAs for the peak load. However, with our approach one can dynamically assign the system resources. In the static scenario, one would use three servers for seven days, whereas our approach needs only $1 + 2 + 3 + 2 + 3 + 1 + 1 = 13$ server days. Hence, in such a scenario, only 62% of the resources of the static assignment are needed and thereby almost 40% of the resources available can be saved.

19.2.3 Conclusions

This case study on self-adaptive resource management demonstrates that architecture-level performance models in combination with resource allocation algorithms can be applied to react to changes during runtime. It shows that it is possible to achieve elastic capacity management while satisfying specified SLAs. Exemplary, we showed how the system reacts on changes in the workload and how such an approach can save up to 40% of the resources. Also important to note is that this case study demonstrates that architecture-level performance models can be used effectively at runtime to support self-adaptiveness.

19.3 Case Study: Balancing Performance and Dependability Tradeoffs

Availability and responsiveness are crucial, although often conflicting, requirements for the multitier applications that implement critical business functionality for many enterprises. Ensuring high availability requires the applications to be deployed with sufficient redundancy, potentially spanning several data centers. Today, large geographically dispersed hosting facilities provided by leading cloud computing providers have made wide area deployments practical for even low to medium scale applications. However, with such dispersion come consistency and synchronization overheads, and applications must often pay a performance penalty as a result. In traditional static deployments, application designers often tune such availability and performance tradeoffs manually after taking into consideration application architectures, workloads, and requirements.

However, shared infrastructures such as compute clouds necessitate a rethinking of static deployment schemes. For example, resource contention might require relocation of applications to another machine rack, cluster, or even another data center. Additionally, current trends in system and data center design emphasize the use of large numbers of machines running cheap, less reliable commodity components that can fail often. For example, Google reported an average of 1000 node failures/yr in their typical 1800 node cluster for a cluster MTBF of 8.76 h [274]. At the same time, skilled manpower is quickly becoming the most expensive resource, thus encouraging data center operators to batch repairs and replacement, thus increasing MTTR in the process. In fact, portable "data-center in a box" designs (e.g., [407]) that contain tightly packed individual components that are completely non-serviceable, i.e., with an infinite MTTR, are emerging.

These trends imply that applications will run in increasingly dynamic environments in which parts of the infrastructure are in a failed state and static solutions to availability and performance tradeoffs will no longer suffice. However, dynamic solutions that redeploy multitier applications are challenging because they must not only balance availability against performance, but they must also factor in resource allocation between competing applications. Poor placement of a critical resource such as a database server may cause it to be a bottleneck for the whole application and as a result, the hosts where other tiers of the application are placed may become underutilized.

In this study, we show how online performance and availability models can be used to address these challenges and drive dynamic multitier application redeployment in the event of failures so as to minimize performance degradation while maintaining availability constraints. We build an online controller based on the models that regenerates the affected software components across clusters or data centers in the event of infrastructure failures, and reconfigures the entire system to run optimally on the remaining resources. Using simulation and fault injection studies, we show that this approach can provide high availability with far fewer resources than traditional approaches.

19.3.1 Performance and Availability Models

We consider a consolidated data-center environment in which a set of multitier applications A are deployed on a set of physical hosts H located in a number of data centers. The hosts are organized into racks, clusters, and data centers in a resource hierarchy $(R, \leq_R)$, where R is the set of "resource groups" (i.e., machine, rack, cluster, data-center) and $\leq_R$ specifies a direct hosting relationship between the groups, e.g., Host1 $\leq_R$ Rack1 $\leq_R$ DataCenter1. The hosting relation $\leq_R^*$ is the transitive closure of $\leq_R$, e.g., Host1 $\leq_R^*$ DataCenter1 indicates that Host1 is directly or indirectly hosted in DataCenter1. Figure 20.3a shows an example resource hierarchy with 20 machines distributed across four racks in two data centers. Two resource groups are said to be at the same "level" $rl \in RL$ if they are of the same type, e.g., Rack1 and Rack2. The example in the figure has three levels.

Hosts are interconnected by a data center network and the network latency between hosts depends on how close they are to one another in the hierarchy, i.e., hosts placed in the same rack have a lower network latency between them than hosts across different racks, which have a lower latency than hosts in different data centers. We denote by $L(rl)$ the maximum latency between two hosts separated at resource level rl. Finally, we denote the mean time between failures for each resource group r by $MTBF_r$. In general, MTBF increases with increasing resource level, i.e., MTBF for hosts is smaller than MTBF for a rack, which is smaller still than the MTBF for an entire data center.

Each application a consists of a set N_a of component types (e.g., web server, database), each of which contains several replicated components to avoid single points of failure. Each application a may support multiple transaction types T_a. For example, the RUBiS [184] auction site benchmark used in our testbed has transactions that correspond to login, profile, browsing, searching, buying, and selling. Each transaction can initiate a sequence of function calls between application components. The application's workload w_a is given by a vector of request rates w_a^t for its transactions. Each application-component replica executes in its own virtual machine (VM) [90] on a physical host anywhere in the resource hierarchy that it can share with other VMs. Each VM is allocated a share of the host's CPU capacity that is enforced by Xen's credit-based scheduler.

Availability Models: We consider an application to be available when at least one replica of each component is running on an operational machine, and define availability as the fraction of time the application is available. A replication level of at least two for each of the application's component types is necessary to avoid single points of failure, but not always sufficient. If all replicas of the same type are contained within a single resource tier, e.g., a rack, then a failure of that tier causes application failure. Therefore, we allow each application to specify its desired availability and use information about the system's recovery policy codified by the MTTR to calculate the application's minimum desired "mean time between failures", or $MTBF_a$ as:
$$MTBF_a \geq \frac{Availability_a \cdot MTTR}{1 - Availability_a}.$$

We can now calculate the application's actual MTBF for a given placement of its components across the resource hierarchy. Assume that each resource group r fails independently according to a Poisson failure process with rate $\lambda_r = 1/MTBF_r$ and each failure disables all the application components the group contains. If the replication level of any application component type drops to zero as a result of a resource failure, then the application fails. For each of application a's component types n_a, let $r^{max}(n_a)$ be the highest level resource group such that all replicas of n_a are contained in that resource group. e.g., if an application's database had two replicas hosted in DataCenter1:Rack1:Host1 and DataCenter1:Rack2:Host3, then $r^{max}(db_a) = $ DataCenter1. Only failures at resource levels $r^{max}(n_a)$ or higher will cause an application failure by causing the replication level of the component type n_a to fall to zero.

Under these assumptions, the overall failure arrival process is also Poisson with rate $\sum_{r \in R} \lambda_r$. A failure event affects resource group r with probability $\lambda_r / \sum_{r \in R} \lambda_r$, and causes application a to fail if r is such that there is at least one component type n_a with a value of $r^{max}(n_a)$ that is lower than r. i.e., $r^{max}(n_a) \leq_R^* r$ (Condition 1). Since this condition only filter resource groups, the application failure process is also Poisson with a rate equal to the sum of λ_r over resource groups for which condition 1 is true. $r^{max}(n_a)$ depends only the exact system configuration, so the application failure process has a constant rate until the system is reconfigured by the controller. Thus, the MTBF for application a in a system configuration c is given by:

$$\text{MTBF}_a(c) = \left(\sum_{\substack{\forall r \in R \text{ s.t. } \exists n_a \in N_a \\ \text{s.t. } r^{max}(n_a) \leq_R^* r}} MTBF_r^{-1} \right)^{-1} \tag{19.1}$$

This equation assumes that no additional failures occur in the time window between the first failure and the time the controller finishes reconfiguring the system. While this is not strictly true, it is a reasonable assumption because the reconfiguration actions (VM instantiation, migration, CPU capacity changes) are very short compared to typical resource MTBF values.

Performance Models: To quantify the performance of alternative system configurations, we construct application models using the layered queuing network modeling formalism [943] to predict the response times of application transactions and the corresponding resource utilization demands for each replica for a given workload and system configuration (i.e., the CPU capacity assigned to each application VM). Each application component is modeled as a FCFS queue, while hardware resources (e.g., CPU and disk) are modeled as processor sharing (PS) queues. Interactions between tiers triggered by a transaction are modeled as synchronous calls in the queuing network, and our models also account for the resource sharing overhead imposed by Xen. The parameters for models (e.g., per-transaction service time at each queue) are measured in an offline measurement phase, where each application is instrumented using system call interception. Then, delays between incoming and

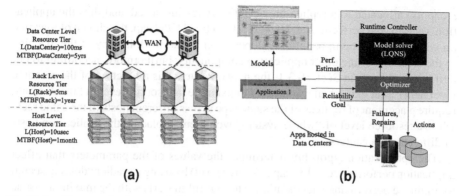

Fig. 19.3 a Resource levels example, **b** approach overview

outgoing messages are measured per transaction. Details of the LQN models and their validation can be found in [498].

We compute the application's mean response time in a new configuration as the sum of the response time $RT(a, t)$ of each transaction t weighted by the fraction $\gamma(a, t)$ of the transaction in the application's workload mix. The response time degradation in a potential new configuration is simply the difference between the predicted mean response time in the new configuration and the mean response time in the original configuration before the failure.

19.3.2 Online Optimization Algorithm

Our approach, as shown in Fig. 19.3b, is realized by a runtime controller that monitors the system and which, when a failure or recovery occurs, reconfigures all applications. To do so, it uses standard virtual machine techniques—it can either migrate each application component's VM to another host, or change the CPU share allocated to the VM on its current host. The controller chooses actions that minimize the mean performance degradation (across all applications) as a result of the failure while still maintaining the desired level of replication and application MTBF (Eq. 19.1). It has to balance several factors in doing so. Maximizing performance dictates that application components be placed close to one another to minimize the impact of network latency, but packing components too closely (e.g., on the same machine) may actually degrade performance by forcing VMs to use less CPU resources than they require. Additionally, applications requiring high levels of reliability will have to be distributed across higher resource levels to prevent single failures from impacting multiple replicas.

The optimization is carried out over the large space of all possible system configurations $c \in C$, each of which specifies: (a) the assignment of each replica n_k to a physical host $c.\text{host}(n_k)$, and (b) the CPU share cap $c.\text{cap}(n_k)$. The CPU cap

c.cap(n_k) allocated to a replica impacts it's processing speed, and thus the application's end-to-end performance. However, for a fixed CPU cap, the choice of machine on which to host a component only depends on the network latency of the machine to the locations of the other application components. Furthermore, according to our definition of $L(rl)$ in Sect. 19.3.1, the network latency is a function of the resource level rather than individual resources. For example, a resource level of "rack" would require placement of replicas of the same type across different hosts in the same rack, while a resource level of "whole system" would entail placing the replicas on hosts in different data centers.

The optimization algorithm determines the values of the parameters that affect application performance (CPU cap, resource level) by using a gradient descent search to minimize performance degradation. The algorithm starts with the maximum value of CPU cap (i.e., 1.0) for every replica and the lowest permissible resource level for each tier such that the application's MTBF given by Eq. 19.1 is higher than the application's desired MTBF. Once an initial value for the parameters is chosen for every replica, the algorithm attempts to fit the replicas into the available resources using a bin-packing algorithm that respects each replica's choice of resource level and uses the CPU cap as the "volume" of the replica. If a fit cannot be found, the algorithm executes an additional iteration of the gradient descent to either choose to lower the value of the CPU cap of a single replica to reduce CPU requirements, or to increase the resource levels of a single application tier to increase the flexibility the bin-packer has when distributing replicas. The option (which application, which tier, and whether to reduce the CPU cap or increase the resource level) that results in the least amount of performance degradation is chosen. The LQNS queuing model described above is used to estimate the performance degradation in the new configuration. The bin-packing is attempted again and the process repeats until a successful fit in the available resources is found. More details of the optimization algorithm can be found in.

Upon finding a successful fit, the optimizer calculates the difference between the original configuration and new configuration for each replica, and returns the set of actions (migrate VM, adjust CPU cap, re-instantiate VM) needed to affect the change. The durations of these actions are relatively short compared to typical MTBF values, and range from a few milliseconds to a few minutes at most. Furthermore, they can be performed without causing VM downtime [224].

19.3.3 Simulation Based Evaluation

In this section, we present simulation results using a simulator written in the Java based SSJ framework [581]. The target application for the experiments is the RUBiS online auction benchmark. We created the LQNS model using offline measurements from [498] and execute the model using transaction workload rates representing user behavior according to the "browsing mix" defined by the RUBiS test client generator.

We compare our approach (Opt) with two reference strategies: a) the Static strategy that relies on design redundancy to tolerate failures, and b) the "least loaded" (LL) strategy that reinstantiates each failed replica (VM) in the order of decreasing CPU utilization on the least loaded host within the same level of the resource hierarchy. The utilization of the target host is then updated to take into account the reinstantiated VM before choosing a host for the next failed VM. Once the VMs have been reassigned, the controller reallocates the CPU capacities to the VMs on each host proportional to their measured CPU utilization with a lower bound of 10% CPU. When a host is recovered/replaced, LL migrates the original VMs running on the host before it failed from their current locations back to the host.

We simulate the three strategies in a cloud setup consisting of two data centers with three clusters each, 3 racks in each cluster, and 4 machines in each rack. The communication delays between two machines in a rack are D, while those between machines in different racks, clusters, and data centers are $1.5D$, $2D$, and $2.5D$, respectively. The setup hosts 6 instances of RUBiS: 3 gold instances each with a weight of 5 (Eq. 19.1) and 3 silver instances with a weight of 1 (Eq. 19.1). For all instances, each of the three tiers is replicated twice. The gold instances offer higher availability and require tiers to be replicated at-least across separate clusters, while the silver application tiers can be replicated across racks in the same cluster. The workload is set to 30 and 60 requests/sec for the gold and silver instances, respectively. Each VM is initially allocated 80% of one physical CPU.

For each strategy, we run fault injection experiments in which failures are simulated at different levels of the hierarchy (i.e., data center, cluster, rack, host) using a Poisson process with different failure and repair rates. Specifically, if the MTBF and MTTR on the host-level are M_f and M_r, then at the rack, cluster, and data center levels they are $4M_f$ and $4M_r$, $16M_f$ and $16M_r$, and $160M_f$ and $160M_r$, respectively. To make the results applicable for systems with different MTBFs and MTTRs, we report all times normalized to the host-level MTBF M_f, which was set to 1.0, For repair, we vary the per host relative MTTR from 0.1 to 1, indicating that repair takes from 10 to 100% of the MTBF. Each simulation runs for a normalized time period of 10 (i.e., 10 failures per run on the average), and we repeat each experiment 10 times. For each experiment, we calculate both the availability of the system and the performance degradation.

Figure 19.4a shows the unavailability of the system as a function of the relative MTTR. Both the Opt and LL strategies achieve 100% availability, while the unavailability of the Static strategy increases significantly with the relative MTTR. Since both LL and Opt regenerate VMs as soon as a failure occurs, this result is expected. In practice, both strategies may not achieve 100% availability for two reasons. First, the controllers require time to make a reconfiguration decision after a failure event and second, instantiation of new VMs is not instantaneous. During both intervals, the system may be vulnerable to additional failures. Fortunately, both windows are very short compared to typical MTBF values in practice.

Figure 19.4b shows the mean performance degradation D of the applications computed over the period that they are available vs. the MTTR. The results show that in some cases LL does not perform much better than Static. This is because if a set of

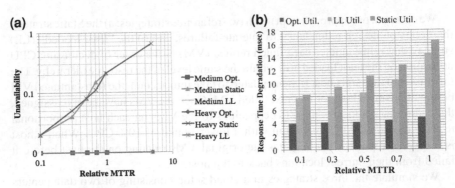

Fig. 19.4 Simulation results. **a** Unavailability, **b** performance degradation

hosts (a whole rack, cluster, or data center) fails and the failed hosts contain the VMs of the silver applications, it may be better to do nothing (i.e., Static) than reallocating those VMs to machines that are running Gold instances (which have a higher impact on the weighted mean response time) and slowing them down. LL also cannot determine which components are bottlenecks and often makes decisions based on small differences in host CPU utilizations (since all of them are high). It can end up co-locating a regenerated VM with a bottleneck resource, thereby greatly degrading the response time. On the other hand, the Opt strategy can avoid these bottlenecks using its queuing model, and performs significantly better than both Static and LL by exhibiting little performance degradation even at high relative MTTR values.

19.3.4 Fault Injection Based Evaluation

Next, we experimentally evaluate the Opt, Static, and LL strategies using fault injection experiments on a system subjected to actual failures and a realistic workload. Our testbed contains 10 machines divided into two racks of 5 each. Each host has an Intel Pentium 4 1.80 GHz processor, 1 GB RAM, and a 100 Mb Ethernet interface and runs Linux kernel 2.6.18 guest OS VMs on an open-source Xen version 3.2.0 hypervisor. The controller is run on a separate server.

The hosted applications are two instances of the 3-tier servlet version of RUBiS running on an Apache 2.0.54 webserver, a Tomcat 5.0.28 application server, and MySQL 3.23.58 database. Each tier has two replicas, and each replica runs in its own VM for a total of 12 VMs. Replicas for the same tier are constrained to run in different racks. In the initial configuration, the VMs hosting the Tomcat and MySQL replicas are allocated 80% of a physical CPU capacity and the VMs hosting Apache replicas are allocated 40% of a physical CPU capacity. For the Static strategy, the placement and the capacity allocation of the VMs remains the same throughout

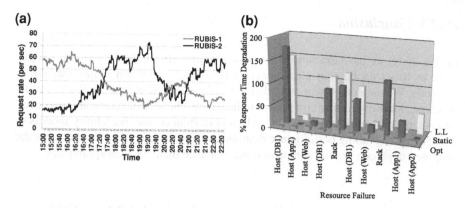

Fig. 19.5 Cloud simulation results. **a** Workload, **b** response time degradation (%) after failure and reconfiguration

the experiment, while the Opt and LL strategies adjust the location and capacity allocation of the VMs based on the workload at the time of the failure.

The applications are subjected to the workloads shown in Fig. 19.5a. These traces are produced by a user emulator that simulates actual users using a semi-Markov model. Each state of semi-Markov model corresponds to a single transaction. Transitions between states $s' \rightarrow s''$ encode the probability that a user issues the transaction destination transaction s'' after visiting the source page s'. The user is assumed to spend a normally distributed amount of random "think time" between every consecutive transaction. The number of concurrent users at any given time varies, and we obtain these variations from actual user traces from publicly available web-site logs [57, 293].

Both individual host and rack failures (i.e., a correlated failure of all machines in a rack due to a common cause such as power supply or switch/router) are injected. In each run, a single failure is injected at a random time instant and the mean response time of the applications before and after the injection and reconfiguration is measured to calculate the performance degradation. Each strategy is subjected to failures at the same time instant and workload and the mean performance degradation across all transactions is reported. The results for one of the RUBiS instances are shown in Fig. 19.5b. Across all failures, the average performance degradation for the Static and LL strategies is 46 and 47 %, respectively, while it is only 9.5 % for the Opt. controller. The gap between Static and LL is small because the initial configuration has no single point of failure, and the relatively light workload allows the Static approach to operate after a single failure without requiring VM regeneration. The large differences between Opt. and LL demonstrate the benefits of taking performance bottlenecks into account during reconfiguration. The Opt. controller has very low degradation even when entire racks fail.

19.3.5 Conclusion

In this study, we have examined how online controllers can be constructed to optimize multitier application placements by balancing performance and availability tradeoffs. We use component redundancy to tolerate single machine failures, virtual machine cloning to restore component redundancy whenever machine failures occur, and smart component placement based on performance and availability models to minimize the resulting performance degradation. Experimental results show that the proposed approach provides improved performance than classical approaches.

19.4 Computational and Communication Scalability of EC2

Stochastic models of real-life computer and communication systems allow engineers to analyse the correctness and performance of such systems at design time. This allows for problems to be detected and choices to be investigated much more quickly and cheaply than if such investigations are delayed until after the system has been implemented. Markov chains are one of the most commonly-encountered modelling formalisms, but to capture even the most essential behaviour of a real-life system may require a Markov chain with many millions of states. The analysis of such chains will require the combined compute power and memory capacity of a number of computers in parallel; for example, see [106, 148, 164, 297, 399, 411, 699, 892]. Typical quantities of interest are long-run or *steady-state* probability distributions and the distributions of response times between specified initial and goal states.

To exploit the power of these implementations, the user is typically required to possess a dedicated computational cluster or network of workstations. Such hardware is, however, expensive to buy and to run, requires sufficient space with associated power and cooling to house it, and staff to maintain it. With the ever-present pressure on academic research budgets, it is conceivable that individual research groups will struggle to continue to acquire such resources for themselves. Cloud computing holds the promise of dramatically reducing these overheads.

A key concern in using existing performance analysis tools in the cloud is how well those tools themselves perform in this environment, as their performance in this shared environment could well differ from their performance on dedicated hardware. Cloud computing offers the ability to make use of large numbers of processors far more cheaply than we could ourselves own, but if our tools cannot efficiently use these extra resources then they will need either to be modified or to be replaced with tools that can.

In this section we study the scalability of two of our previously-presented performance analysis tools: a Laplace transform-based response time analyser [295] and the HYpergraph-based Distributed Response-time Analyser (HYDRA) [296, 297]. Both tools calculate the full distributions of response times in Continuous Time

Markov Chains (CTMCs), which can then be used to reason about a wide range of performance requirements in formal models of systems.

We compare the two tools' scalabilities in a cloud computing environment (Amazon EC2) and a variety of traditional environments in the context of a case study analysis of a CTMC model. We expect that the Laplace transform-based tool will scale well in all environments because of the minimal amount of interprocessor communication that it requires, but that HYDRA may suffer in environments with limited network bandwidth despite the fact that it employs a data partitioning scheme that minimises interprocessor communication.

19.4.1 Performance Analysis Tools

We will study the scalability of two previously-presented performance analysis tools: a Laplace transform inverter [295] and the HYpergraph-based Distributed Response-time Analyser (HYDRA) [296, 297]. Although the core computation carried out by both tools is repeated sparse matrix–vector multiplication, the way in which they parallelise the problem is different and consequently they place very different communication loads on the network.

19.4.1.1 Laplace Transform Inverter

The distributed Laplace transform inverter is written in C++ and uses the Message Passing Interface (MPI) [397] standard, so it is portable to a wide variety of parallel computers and workstation clusters. It features a master-slave architecture that ensures a good load balance and very high utilisation of slave processors. In addition, there is no inter-slave communication and the amount of master-slave communication is low. We therefore expect this tool to exhibit good scalability.

19.4.1.2 HYDRA

HYDRA is also implemented in C++ and uses MPI. The key opportunity for parallelism in HYDRA is in the repeated sparse matrix–vector multiplications that form the core of the implemented algorithm. To perform these operations efficiently in parallel it is necessary to map the non-zero matrix elements onto processors such that the computational load is balanced and communication between processors is minimised. To achieve this, we use hypergraph partitioning to assign matrix rows and corresponding vector elements to processors [182]. Our previous work has observed that this gives HYDRA good scalability on both parallel computers with fast interconnection networks and also on networks of workstations connected via switched Ethernet [296, 297] .

Table 19.1 Summary of the four architectures on which the Laplace transform inversion tool was executed

Name	Type	CPU (GHz)	RAM	Network
PC (2004)	Workstation	Intel Pentium 4 2.0	512 MB	100 Mbps Ethernet
PC (2010)	Workstation	Intel Core2 Duo 3.0	4GB	1 Gbps Ethernet
Camelot	Cluster	Opteron dual-core 2.2	8GB	2.5 Gbps Infiniband
Amazon	EC2 Small instance	c. Opteron 1.0–1.2	1.7GB	Unknown

Table 19.2 Average run-times in seconds (T), speed-ups (S_p) and efficiencies (E_p) for p-processor response time density calculations in a 537,768 state model using the Laplace transform inversion tool

	PC (2004)			PC (2010)			Camelot			Amazon		
p	T	S_p	E_p	T	S_p	E_p	T	S_p	E_p	T	S_p	E_p
1	5,096.0	1.0	1.0	1,190.5	1.00	1.00	4,181.3	1.00	1.00	2,835.9	1.00	1.00
2	2,582.6	1.97	0.99	592.4	2.00	1.00	2,149.1	1.95	0.97	1,522.4	1.86	0.93
4	1,298.4	3.92	0.98	301.4	3.95	0.99	1,083.1	3.86	0.97	776.2	3.65	0.91
8	675.8	7.54	0.94	150.9	7.89	0.99	587.6	7.12	0.89	422.7	6.71	0.83
16	398.4	12.79	0.80	78.0	15.26	0.95	350.3	11.94	0.75	218.8	12.96	0.81

19.4.2 Amazon Elastic Compute Cloud

Amazon Elastic Compute Cloud (Amazon EC2) is a service that allows users both to purchase computing resources on-demand and also to reserve them to guarantee availability in the future. Central to EC2 are Amazon Machine Images (AMIs), which are instantiations of the Linux or Windows operating system that are brought into being by the user and run as virtual machines. Amazon have a range of standard AMIs, based on Windows and various versions of Linux, that come pre-installed with commonly-used packages as well as providing tools to enable users to build their own AMIs containing exactly the applications and packages that they require.

Both of our tools described in the previous section require MPI and, although none of the standard Amazon AMIs include this, there is a user-produced AMI that does [835, 836]. This AMI costs $0.085 per instance per hour,[2] and is only available in the US-N. Virginia region of EC2. Similarly, although Amazon has recently released a dedicated Cluster Compute image (Cluster Compute Quadruple Extra Large, or cc1.4xlarge) with access to 10 Gbps Ethernet interconnection, this image does not come with MPI installed as standard. By following the publicly-available instructions of [835], however, we were able to create our own custom Cluster Compute image that included MPI. It costs $1.60 per instances per hour to run and is also only available in the US-N. Virginia region.

---

[2] See http://aws.amazon.com/ec2/pricing/ for a full list of rates.

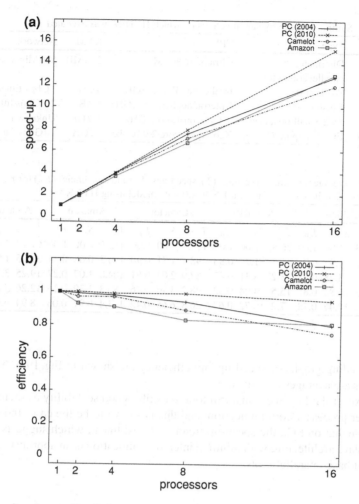

Fig. 19.6 a Speed-up and **b** efficiency graphs for p-processor response time density calculations in a 537, 768 state model using the Laplace transform inversion tool

19.4.3 Results

19.4.3.1 Laplace Transform Inverter

These results are presented for four architectures and are reproduced from [294] to provide a basis for comparison with the new results in the next section. Table 19.1 summarises the processor speeds, main memory and network bandwidths of the four architectures. Table 19.2 shows the run-times, speed-ups and efficiencies for the calculation of response time densities on p processors for a 537,768 state model.

Table 19.3 Summary of the five architectures on which HYDRA was executed

Name	Type	CPU	RAM	Network
AP3000	Distributed-memory parallel computer	UltraSparc 300 MHz	256 MB	520 Mbps mesh
PC (2010)	Workstation	Intel Core2 Duo 3.0 GHz	4 GB	1 Gbps Ethernet
Camelot	Cluster	Opteron dual-core 2.2 GHz	8 GB	2.5 Gbps Infiniband
Amazon	EC2 Small Instance	c. Opteron 1.0–1.2 GHz	1.7 GB	Unknown
Amazon CC	EC2 Compute Cluster	Xeon quad-core 2.93 GHz	23 GB	10 Gbps Ethernet

Table 19.4 Average run-times in seconds (T), speed-ups (S_p) and efficiencies (E_p) for p-processor response time density calculations in a 1,639,440 state model using HYDRA

	AP3000			PC (2010)			Camelot			Amazon			Amazon CC		
p	T	S_p	E_p	T	S_p	E_p	T	S_p	E_p	T	S_p	E_p	T	S_p	E_p
1	1,243.3	1.00	1.00	76.8	1.00	1.00	178.1	1.00	1.00	112.5	1.00	1.00	61.77	1.00	1.00
2	630.5	1.97	0.99	43.5	1.76	0.88	98.7	1.81	0.90	166.2	0.68	0.34	31.05	1.99	0.99
4	328.2	3.78	0.95	23.2	3.31	0.83	87.9	2.03	0.51	104.8	1.07	0.27	19.25	3.21	0.80
8	182.3	6.82	0.85	15.5	4.94	0.62	48.2	3.70	0.46	86.3	1.30	0.16	12.26	5.04	0.63
16	99.7	12.47	0.78	7.2	10.72	0.67	26.8	6.65	0.42	123.4	0.91	0.06	8.94	6.91	0.43

Corresponding graphs of speed-up and efficiency are shown in Fig. 19.6. Note that run-times are averages of 5 runs.

We expect the Laplace transform tool to exhibit good scalability as there is very little inter-processor communication, and this is shown to be the case. Indeed, it is noticeable that on EC2 the speed-up trend is almost linear, which suggests that the master-slave architecture with minimal intercommunication is an appropriate design for cloud-based parallel tools.

19.4.3.2 HYDRA

These results are presented for five architectures, four of which are reproduced from [294]. Table 19.3 summarises the processor speeds, main memory and network bandwidths of the five architectures. As parallel sparse matrix–vector multiplication potentially requires a great deal of data to be exchanged at each iteration of the solution, we also investigate HYDRA's scalability when executed on Amazon's Compute Cluster instances. In an effort to ensure that we the effect of the network is included in our results, we used two instances for all values of $p > 1$, with at least one process assigned to each instance.

Table 19.4 shows the run-times, speed-ups and efficiencies for the calculation of response time densities on p processors for a 1,639,440 state model. Corresponding graphs of speed-up and efficiency are shown in Fig. 19.7. Once again, these run-times were averaged over 5 runs. Although the use of hypergraph partitioning minimises

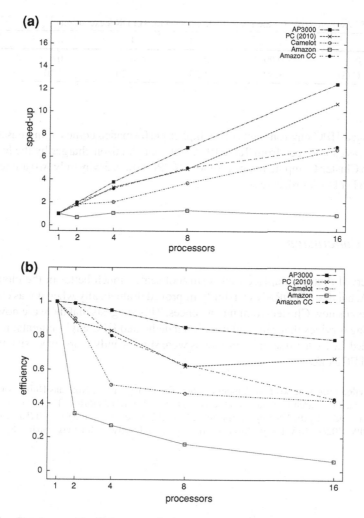

Fig. 19.7 a Speed-up and **b** efficiency graphs for p-processor response time density calculations in a 1,639,440 state model using HYDRA

the amount of data that must be sent, we observe that the speed-ups achieved are accordingly lower than for the Laplace transformer inverter. We also observe that the scalability of HYDRA on the standard Amazon EC2 instances is the worst of all five architectures. Although we expected the speed-up and efficiency to be lower than on the dedicated hardware platforms, it is still very surprising to see just how badly HYDRA fares in the cloud.

The interconnection of the Cluster Compute instances clearly provides far higher bandwidth than the network connecting standard EC2 instances, and as a result HYDRA's scalability on these AMIs is much more in line with that experienced

Table 19.5 Average costs (to the nearest whole cent) for HYDRA execution

	1	2	4	8	16
Amazon ($)	0.09	0.17	0.34	0.68	1.36
Amazon CC ($)	1.60	3.20	3.20	3.20	3.20

in dedicated HPC environments. This higher performance comes at increased cost, however, as can be seen from Table 19.5. Note that Amazon charges by the hour and that each Cluster Compute image provides 8 CPUs (hence why only two are required to run 16 HYDRA processes).

19.4.4 Conclusion

We observed that the Laplace transform tool scaled much better in the cloud than HYDRA, but that HYDRA's scalability improved dramatically when it was executed on Amazon's new Cluster Compute instances. This suggests that there are now cloud computing services that can rival traditional dedicated HPC environments; it should be recalled, however, that these instances were significantly more expensive than the standard EC2 ones.

Acknowledgments The work of Samuel Kounev, Fabian Brosig and Nikolaus Huber was funded by the German Research Foundation (DFG) under grant No. KO 3445/6-1. The work of Nicholas Dingle was funded by the UK Engineering and Physical Sciences Research Council (EPSRC) under grant EP/I006702/1 "Novel Asynchronous Algorithms and Software for Large Sparse Systems".

Part VII
Conclusions and Outlook

Chapter 20
Future of Resilience Assessment: The AMBER Research Roadmap

Andrea Bondavalli, Henrique Madeira and Paolo Lollini

Abstract This chapter provides a condensed description of a roadmap for research in technologies for assessment, measurement and benchmarking (AMB) of the resilience of information, computer and communication systems. The research roadmap is the result of the EU-funded AMBER Coordination Action, integrating the consortium experience in the field with the insights resulting from discussions and interviews with a variety of stakeholders about motivating scenarios, drivers and priorities. A set of motivating scenarios help understand the current needs and challenges in resilience assessment. These scenarios present viewpoints of industrial players, end users, system operators and regulators. The research roadmap then provides a detailed list of research needs and challenges grouped in three categories: (i) scientific and technological foundations, (ii) measurement and assessment, and (iii) benchmarking. The foundations make the case for two types of research advances, which we could label as 'back to basics' and 'holistic'. The measurement and assessment category identifies a number of topics of acute interest and that are particularly challenging. Resilience benchmarking aims at providing generic, repeatable and widely accepted methods for characterising and quantifying the system (or component) behaviour in the presence of faults, and comparing the resilience of alternative solutions. In addition to the above research issues, we also identified the challenges we see in education as well as standardization.

A. Bondavalli (✉) · P. Lollini
University of Firenze,
Viale Morgagni 65, I-50134 Firenze, Italy
e-mail: bondavalli@unifi.it

P. Lollini
e-mail: lollini@unifi.it

H. Madeira
DEI/CISUC, University of Coimbra,
3030-290 Coimbra, Portugal
e-mail: henrique@dei.uc.pt

K. Wolter et al. (eds.), *Resilience Assessment and Evaluation of Computing Systems*,   415
DOI: 10.1007/978-3-642-29032-9_20, © Springer-Verlag Berlin Heidelberg 2012

20.1 Introduction

Assessment, measurement and benchmarking of resilience (of computer systems) are related concepts, but each of them reflects a different form of characterizing computer resilience and demands specific methods and tools. These three terms may mean slightly different things to different communities, but the scope of the AMBER Coordination Action was easily defined as covering activities that involve quantitative descriptions of the qualities of dependability and resilience. Within this scope, we will use "assessment" to mean obtaining any kind of quantitative statement about these qualities, especially supporting decision-making; use "measurement" in its usual meaning in science and engineering, of mapping empirical observations to numbers in a rigorous manner, through rigorous procedures and calibrated instruments; and distinguish "benchmarking" activities as those in which the main purpose is ranking systems using a simple and standardized method, even at the cost of it being somewhat simplistic [514]. We have generally excluded from consideration, or described only as the background for the research on which AMBER aims to foster debate and coordination, many activities that may be called "measurement" in a broad sense of the word, like formal proofs and checklist-oriented methods.

The very word "resilience" is not uniformly defined (or accepted) in what can be called the "resilience research community". "Resilience" tends to be a synonym for fault tolerance in a broad sense, but it is also used to convey ideas of dependability measures, sought for less narrowly defined and less static scenarios than assumed in currently established methods [38]. AMBER addresses the communities of researchers and users dealing with assessment of qualities related to these two broad concepts. Therefore, in this chapter we will use "resilience" as a generalisation of terms like dependability, security, fault tolerance, and so on, to encompass all attributes of the quality of "working well in a changing world that includes failures, errors and attacks". The changes can be planned or predictable, such as software upgrades and configuration changes, or rather unforeseen such as failures [771]. The main objective of AMBER is the synthesis of a research roadmap on assessing, measuring, and benchmarking resilience (abbreviated in the following as AMB resilience), to be understood as a list of research directions that seem worth pursuing now, with associated priorities.

The rest of this chapter is structured as follows: Sect. 20.2 proposes a set of ad-hoc scenarios (in the form of "stories") to exemplify the current needs and challenges in the fields of future Internet, embedded systems, enterprise computing, supervision of IT infrastructure, safety certification and regulation. In Sect. 20.3 we consider the drivers that can influence the research needs on assessing, measuring and benchmarking resilience, as well as the effective transfer of resilience assessment best practices to European industry and the adoption of standards for resilience assessment and benchmarking. The research roadmap is then provided in Sect. 20.4, integrating the consortium experience in the field with the insights resulting from a long discussion of scenarios, drivers and inputs from stakeholders and experts. Four main areas have been identified: (i) scientific and technological foundations, (ii) measurement and assessment, (iii) benchmarking, and (iv) education, training, standardization and take up. For each of them we identify recognised needs, challenges to be overcome to satisfy these needs, and objectives for specific actions to be performed in a short or medium term period. Moreover, we provide a synoptic diagram of the connections between needs, challenges and (short and medium term) objectives. Finally, Sect. 20.5 presents a series of alternate viewpoints about priorities within the research roadmap. Each viewpoint has been written by one of the partners in consultation with one or more industrial experts in a specific sector, to give examples of priorities that specific stakeholders would identify within the roadmap. The parts of this section are accordingly titled: "embedded systems", "transportation", "certification authorities and assessors", "future internet technological platforms", "service architectures, platforms and infrastructures" and "enterprise security". By showing samples of viewpoints within each sector, this last section allows a reader to appreciate how the roadmap could be tailored to specific points of view, offering a particular perspective for reading the roadmap itself. The original version of the Roadmap presented expanded descriptions and discussions of a subset of the needs, challenges, objectives and actions from Sect. 20.4, to clarify their meaning and justification and to provide a more concrete view of the underlying issues. These could not be reported here for the sake of space but can be accessible in [37].

20.2 Motivating Scenarios

In this section we present a set of example scenarios to illustrate some of the challenges that need to be addressed in resilience assessment, measurement and benchmarking. The scenarios belong to the fields of Future Internet, embedded systems, enterprise computing, supervision of IT infrastructure, and safety certification and regulation.

20.2.1 Future Internet

The two scenarios presented in this section show how the Future Internet could, and possibly will, shape the lives of all Europeans in or around 2020. They concern the green urban transport (Sect. 20.2.1.1) and the information on the move (Sect. 20.2.1.2).

20.2.1.1 Green Urban Transport

Scenario.[1] It is 2020. As problems caused by vehicular traffic increase, several cities are considering new mechanisms to control the emissions of pollutants. A proposed method is based on monitoring drivers' behaviour. Drivers are allowed to use personal or public transport, but this depends on the level of traffic congestion at the time; each driver has a personal carbon credit. Drivers can save their own carbon credit by using public transport or reducing the use of their personal vehicles; users pay more or less for travelling, depending on their carbon credit.

Several service providers propose their own ICT implementations of this mechanism. A large European capital city decides to introduce this service in the city centre, since the pollution levels and people's use of personal vehicles have become excessive. The Traffic and Mobility Office of the city thus needs to choose a service provider; the choice must be based on cost as well as trustworthiness qualities, like reliability, resilience to attacks and operational failures, ability to guarantee quality of service and to protect user data to ensure privacy. The Traffic and Mobility Office gives the technical tasks of evaluating the different service providers, identifying the best candidate, and defining a suitable combination of network infrastructure and wireless connections, to an external company. This company has access to usage and failure data concerning the various service providers, from other cities where the service is already in use.

Will this company be able to assess the level of trustworthiness of each service provider, so as to correctly select the best one for the needs of this specific city?

Open challenges. This scenario highlights several challenges in the measurement and assessment of trustworthiness of future Internet-based applications. First of all we have a problem of requirements: what are the main attributes of the service, to be measured for assessing a Green Urban Transport application? Which aspects of security and trustworthiness are measurable and quantifiable, and which metrics are appropriate, for this application? If experimental evaluation (of subsystems or of small-size pilot implementations of the complete service) is sought, are there any reference fault loads and attack loads that are known to be appropriate for assessing resilience to attacks and operational failures? There is also the problem of compositionality of measurements: after estimating low-level metrics referring to subsystems (e.g., performance/reliability/security of the wireless network, of the wired infrastructure network and of vehicles' on-board subsystems) how can these measures be aggregated to extract meaningful information about the trustworthiness of the overall system, that is, of the service? Even with access to usage and failure data about the competing service providers in other cities (in which the service is already used), how can the assessments obtained from these data, representing trustworthiness attributes of each provider's service in a different set of cities, be extrapolated to the new application environment (the specific city), or even meaningfully compared?

---

[1] This scenario is partly inspired by *Application scenarios and functional requirements for wireless sensor and actuator networks in Future Internet* (*F. Forest*) presented at the 10th LETI Annual Review (24–25 June 2008).

20.2.1.2 Network 2020: Information on the Move

Scenario.[2] This scenario is about seamless mobility: Matti moves effortlessly between his home, car and office interacting with family, friends and work colleagues as he goes, and always with his whole world of personal information at his fingertips. The scenario illustrates how in the Future Internet our personal information, content and services will be available to us anywhere, at any time. Our everyday environments will be context-aware: systems and devices will be able to sense how, where and why information is being accessed and respond accordingly. The Internet will be our personal global network. This new world of seamless applications, services and content requires a new network infrastructure, in which advanced features such as semantics and trustworthiness must be built in. Matti's world is more seamless than anything we have today. There is no lost connectivity, no waiting for logons, no poor quality content, no systems that don't talk to each other. What's more, all environments (Matti's home, office, car) are context-aware: systems and devices are able to sense how, where and why information and content are being accessed and respond accordingly.

Open challenges. From the point of view of measurement, assessment and benchmarking of trustworthiness of future Internet-based applications, this scenario opens several challenges. First of all, trust and security are paramount in this scenario: they will be key enablers in realising the potential of the new online world, and measurement, assessment and benchmarking of trust and security will be necessary in order to be able to assure the quality of new future services as perceived by their users (also allowing a fair comparison of alternative services and avoiding mistrust). Secondly, developments such as cloud computing, social networks, and service mash-ups require new approaches to regulation of privacy at a worldwide scale. This should be complemented by research into privacy enhancing technologies (where data protective features and services are built in from the ground up) as well as into their assessment techniques that provide public characterization of the level of built-in privacy and security, together with accountability. To counter these strategies, and to create a level playing field, regulators and public policymakers should strive for open standards (including standard benchmarking of trustworthiness), open interfaces (including hooks for measuring and assessment), and federated architectures (reliable and secure interoperable platforms). Adherence to these principles should be a mandatory requirement for developers of public Internet services. Thirdly, resilience AMB technologies and large-scale testbeds are required to identify (and then reduce) the major technology-related roadblocks that may be in the way to the realization of this scenario: vulnerable architectures, lack of adaptability (also in quality of content and energy consumption), non-scalable connectivity and accessibility, performability bottlenecks, lack of resilience to attacks and operational failures.

---

[2] It is a shortened version of the scenario "Network 2020: Information on the Move" taken from Future Internet 2020, Visions of an industry expert group, which is an industry expert group report. This document was found on the website http://www.future-internet.eu, the European Future Internet Portal, a project initiative which hosts the Europe-wide debate on the Future of the Internet.

20.2.2 Embedded Sub-Systems in Automobiles

Scenario. Large automotive manufacturers rely on third parties for many aspects of their products—that is, they are 'Original Equipment Manufacturers' (OEM). As an example, assume that an automotive manufacturer has decided to offer an infotainment server as an optional feature for its next generation of cars. Such an infotainment server has interfaces to the car's backbone network to connect user panels, the wireless communication unit, and the GPS receiver. This poses high risks, and the OEM must make absolutely sure that an infotainment system supplied by a second company does not endanger the operation of the car.

In more detail, the infotainment server is required to implement a set of "sandboxing"[3] techniques that provide confined execution environments for downloaded applications. This must prevent hidden malware to access or manipulate restricted data and should ensure that software design faults (software bugs) in a downloaded application do not cause interruptions of the infotainment or system programs. It also continuously monitors the application for failures and must take appropriate action for benign as well as severe failures.

Assume now that the OEM has two offers from infotainment systems suppliers. How do the OEM's test engineers decide between offerings from different infotainment system suppliers?

Open challenges. The OEM test engineers must measure both security and dependability properties since malicious attacks as well as accidental failures must be anticipated. However, the OEM test engineers have almost no standard ways of testing the two offerings. This is true even if we assume that both suppliers offer easily accessible and usable test infotainment systems. There are no widely accepted attack or fault loads, nor methods to derive such attack and fault loads. In addition, the products delivered by the suppliers do not have standard measurement hooks, fault/attack injection hooks or measurement data formats. Finally, there are no well-understood and reusable grading criteria for ranking of contending systems.

The knowledge gaps apply to theoretical underpinnings, modelling and emulation of software faults and security attacks, injection and data collection techniques, as well as definition of useful measures.

20.2.3 Information Security Management in a Financial Enterprise

Scenario. A Chief Information Security Officer (CISO) of a large investment firm carries large responsibilities and must balance pressures and objectives from many angles. In this particular case, the CISO must decide about the Information Rights

---

[3] *Sandboxing* means providing a strictly-controlled set of resources for a guest program to run in, such as a limited space on disk and memory. Many features, such as reading from input devices or accessing the host system, are restricted.

Management (IRM) solution that is appropriate for the company. IRM solutions are variations of digital rights management solutions, associating metadata to information such that management software can regulate access. The CISO's first objective is to determine whether such a system is worth the investment, and the question therefore is: how does the CISO assess the value of the IRM solution to the company?

To address this question, the CISO uses as starting point a variety of company security policies and government regulations to determine the requirements for an IRM solution. These pertain to often contradictory requirements, such as the need to protect data from unauthorised access, allow for auditing of purposely retained data, and the need to avoid employees accessing or receiving information that would endanger the lawfulness of a financial interaction. In addition, the CISO must take into consideration the employees sustained productivity as well as the behavioural patterns of employees with respect to the use of technological security solutions. Finally, a cost-benefit and risk analysis would need to be carried out.

Open challenges. The CISO is far less concerned with technology matters than the engineers in the embedded sub-systems in automobiles scenario, but the lack of well-established reusable tools, techniques and methodology is as much (if not more) a pain point as in the technology oriented scenario. To carry out assessment of socio-technological solutions such as the deployment of IRM, the CISO would want a set of well-established tools and techniques that allow for integration of many concerns. The CISO needs input parameters about security offered by the IRM solution, thus requiring well-established security benchmarks. The CISO also needs analysis methods and techniques to associate IT decisions with the financial and business implications for the company. Finally, the CISO would like a knowledge base or other reusable way of identifying and taking into account human factors. The knowledge gaps apply to dealing with the multi-facetted nature of socio-technical systems in an integrated fashion.

20.2.4 High-Level Education for IT Administration

Scenario. We assume a fresh computer science graduate whose first job is in IT administration. The company he works for manages the IT for its customers and the young IT administrator is made responsible for the daily operation of the office applications of the customer enterprise. His tasks are relatively focussed, but become more challenging over time, and include elements such as configuration management, application monitoring, service level agreement management and eventually software purchasing. The core in all jobs of this young administrator is assessment, both of the existing applications and of possible newly acquired applications. The question this scenario asks is: how well did the computer engineering degree prepare the young graduate for the job?

The young graduate is fortunate enough to work for a company that is at the forefront of management technologies. A considerable tool suite is available to help the IT administrator with tasks such as network, server and service monitoring and for more advanced analysis such as root cause analysis of failures. The R&D division of the company develops advanced software tools for IT management, for instance using model-based approaches. These model-based approaches include UML design and domain-specific deployment models that include resilience properties in their abstractions. In addition, the various divisions for which the young administrator manages applications pose service level agreements (SLAs) that they want the IT systems to fulfil.

Open challenges. As we mentioned, in this scenario we are especially interested in the question whether the young computer engineering graduate received the right training for his job. In particular, without proper fundamental training in assessment methods and techniques the IT administrator is in danger of misusing the existing set of elaborated management software tools (such as for monitoring, data processing and root cause analysis). Of increasing importance is the ability of an IT administrator to work with model-based abstractions, including UML in the design of application integration solutions, domain-specific abstractions for deployment and assessment, and model-based prediction to determine if SLAs can be met and operation can be optimised with respect to SLAs. The increasing virtualisation-based dynamism through service-oriented software and service provision solutions (including cloud computing) create further challenges in dealing with partners through contracts that include service level agreements.

20.2.5 Safety Certification and Regulation

Scenario. An assessor (working in an independent assessment body or in a regulatory agency) has to recommend or approve the use of a certain safety system for a dangerous plant in a regulated industrial sector. The proposed system is a combination of off-the-shelf platforms with industry-specific and plant-specific application software, developed with the help of off-the-shelf development tools. Accidents in this plant may have very serious consequences, so the assessment must give assurance of very low probability of accident, and very high confidence in this assessment is required before the plant can be allowed to operate with the proposed safety system. Following accepted practice, the assessor relies on aspects such as 'proven in use' evidence, quality of the development process and resilience of the architecture. But, how confidently can the assessor assess system safety using such evidence?

The assessor uses the common set of techniques and tools in his work, but this does not prevent him from being uncertain about the quality of the assessment. In fact, the value of "proven in use" evidence is difficult to assess: it is hard to know whether, for instance, the new safety system uses the same set of features of the off-the-shelf platform as the one from which statistics were derived, and how many failures may have gone unreported in the past. Another potential problem is the reliance on process quality evidence: even though the assessor accepts the practical constraints that dictate such reliance, he can feel particularly uneasy justifying the relation between process quality and system safety. A third significant difficulty is in deciding how much "credit" to give for resilience features in the architecture: while the architecture may include useful redundancy, statistically reliable information about its effectiveness is limited, both because it is intrinsically hard to obtain and because the vendors are often reluctant to provide it.

Open challenges. The difficulties that the assessor needs to overcome indicate some hard open problems. More statistical evidence would be available if widely accepted and uniform ways for collecting data such as amount of use, profile of use and failure data were available, and if vendors were willing to make such data available. Even given abundant statistical data, the assessor needs methods and techniques for extrapolating resilience measures from such data. The issues include for instance: estimating the coverage of error detection and reporting facilities; characterising how process quality relates to resilience of a product; improving the modelling and empirical knowledge of how system architecture affects resilience, so as to characterise and, where possible, reduce the uncertainty of predictions based on probabilistic models. Last, we can notice two other open challenges: (i) the need to increase awareness of the sources of uncertainty in the evaluation of computing systems and (ii) the lack of uniform and standard ways to collect, filter and report results obtained in experimental evaluations. These methods should be based on sound statistical and metrological science.

The knowledge gaps are in the quality and appropriateness of the data available for assessing the resilience of systems, and the mathematical methods needed for sound inference given the inevitable limits of the data that can feasibly be collected.

20.3 Drivers

In this section we discuss the main drivers for expanding research activities on assessing, measuring and benchmarking resilience. Two important ever-present drivers are **complexity**, which often sprouts difficulties in project development, and **the pace of change**, which requires companies to rapidly understand the benefits and drawbacks of any new technology or development process, while maintaining legacy systems and obsolete technologies. The focus here is on the drivers that have recently emerged and that are expected to continue motivating the development of resilience-related techniques in the foreseeable future. The description is organized into the following four groups:

- Information technologies are global and pervasive;
- There is a greater awareness for the environment;
- Socio-economic factors affect the resilience market;
- Technological innovation creates new issues.

20.3.1 Information Technologies are Global and Pervasive

New ways of distributing services. Recent initiatives promote new technologies and new forms of distributing services through the Internet. Economic factors boost the market for everything as a service, including software, aiming at reducing acquisition and maintenance costs. For an organization to move to this service model there is a strong need for assessing how good a given provider is, and how service failures affect the return on investment. Using software through the Internet, as pushed by cloud computing initiatives, will also open up opportunities for competition among different providers of cloud applications. This creates the need to benchmark, measure and forecast the service provided through these new means, as that decision-making requires factual data on resilience, availability, integrity, etc.

Emerging cyber-threats. Most enterprise systems cannot be considered secure unless they are dependable, and vice versa. For this reason, organizations are interested in evaluating the resilience of their infrastructures to attacks. There are numerous emerging cyber-threats, such as the increasing number of malware objects on the Internet and the growing concerns with botnets (groups of infected computers that are controlled by attackers), which may be used for data theft and other malicious intents. Here, AMB technologies can be used as instruments for deciding whether to migrate an organization to a newer version of a given software product, for evaluating whether an infrastructure requires improvements, etc.

20.3.2 There is a Greater Awareness for the Environment

New technologies for a greener world. Technology is the means for guiding and monitoring green policies, with governments and businesses wishing to analyze energy usage and carbon footprints, to control traffic congestion, and so on. On the end-user's side, environmental and economical concerns motivate an increasing number of professionals to carry out their activities using computer systems at home—telecommuting. Those end-users, and the organizations they work for, are interested in benchmarking different Internet access providers (wired and wireless) as well as the resilience of online collaboration tools. The success of most green initiatives depends on the resilience of the underlying technologies. For this reason, governments, organizations and end-users would benefit from trustworthy techniques for resilience assessment and measurement.

Green computing. There are several ongoing initiatives to make efficient use of computer resources, to reduce the emission of greenhouse gases and to improve the environment. The energy consumption of computer systems may be reduced by centralizing processing power in a server and using thin clients as terminals. This provides also the means for reducing storage requirements, by storing only a single copy of each file and allowing multiple users to access it. The large-scale usage of this type of architecture and associated techniques (e.g., virtualisation) is unprecedented.

Resilience assessment and measurement are fundamental to guarantee the success of these initiatives, as it is necessary to optimize the balance between resilience and environmental impact of computer infrastructures.

20.3.3 Socio-Economic Factors Affect the Resilience Market

Regulatory demands. Society is becoming increasingly dependent on large-scale ICT systems, as well as advanced embedded systems, which in the event of major service failures may cause not only significant economic loss, but also severe accidents or loss of vital government and public services. This dependence has increased the efforts to regulate many issues related to resilience and dependability, as regulators seek to protect the public. Data retention regulations are motivated by cyber-crime and terrorism; the Sarbanes-Oxley (SOX) act requires corporations to maintain and retain correct financial records, related to assessment through the need for monitoring, logging, auditing and analysis; the upcoming ISO 26262 standard for the automotive industry recommends fault injection as a means to assess the effectiveness of safety-related functions; the Basel 2 agreement has created a need for banks to apply quantitative forecast to "operational risk" (which includes risks from ICT failure). These are examples of the trend to regulate resilience and its assessment, which compels companies to adopt the necessary AMB technologies.

Human factors. It is well known that the dependability of complex IT systems relies to a large extent on human operators and their ability to handle failures and other critical events. Experience shows that outages of systems that have been designed to be highly resilient (e.g., telephone systems and large file servers) are often caused by operator mistakes. While human reliability analysis has a long history, there is a striking lack of adequate techniques for assessing and modelling users and operators in complex roles for a large range of IT-based systems.

20.3.4 Technological Innovation Creates New Issues

Component-based and off-the-shelf products. Computer systems and systems-of-systems are often built using off-the-shelf products, and software is increasingly designed by decomposing a system into subcomponents that can be purchased from different suppliers. This way of developing systems is now widespread, including its use for building critical systems and infrastructures. Consequently, it is necessary to create and adapt AMB techniques to compare different suppliers regarding the resilience of their products. Since system integrators have less control over the development process, it is increasingly important to evaluate the offerings of multiple vendors of a given component.

Hardware and software reliability. Hardware failure modes are likely to change significantly with new field-programmable devices and new integrated circuit technologies, as these are increasingly susceptible to soft errors (data corruption), device aging, and variations in manufacturing processes. This will force chip manufacturers to add more fault tolerance to their circuits, thereby changing the way hardware failures are manifested at the system level. Regarding software, a few trends of interest are the increasing use of programming frameworks (that implement generic functionality and handle flow-control), new development processes (such as agile and model-driven development), and automatic code generation, as well as the use of thread-level parallelism in multi-core programming. These advances are likely to change the rate and nature of software and hardware faults, calling for new fault models and new techniques for understanding how to mitigate their effects.

20.4 AMBER Research Roadmap: Abridged View

The AMBER roadmap for research on assessing, measuring and benchmarking (AMB) resilience is reported here in a condensed version. The detailed description of the roadmap can be found in [37]. This roadmap was the final synthesis of inputs coming from the work described in the previous sections, on the analysis of the state of the art reported in [36], of other past research roadmaps, specifically AMSD [39], GRID [396], and ReSIST [771]; on the feedback received during the organized AMBER panels and workshops, as well as on the received responses to the AMBER questionnaire.

Four main areas for investigation have been identified:

- **Research area 1—Scientific and technological foundations**: addresses the foundations that underlie the other areas of research discussed in this roadmap. Among the foundational issues, we identify for example the need for sound metrology-based assessment principles, the handling of complex models and multi-faceted arguments, and the inclusion of human behaviour.
- **Research area 2—Measurement and assessment**: deals with the challenges and research directions related to measurement and assessment activities as typically used to characterize a system alone, as opposed to ranking different systems. Measurement and assessment require sound and well-defined methods, although these need not be standardized.
- **Research area 3—Benchmarking**: targets benchmarking activities, which can be seen as the evolution of current resilience assessment techniques into more standardized approaches. Resilience benchmarks offer generic, repeatable and widely accepted methods for characterising the system behaviour in the presence of faults, and allow the comparison of the resilience of alternative solutions.
- **Research area 4—Education, training, standardization and take up**: discusses the educational, training and standardization issues related to resilience AMB. Some of these issues identify actions that can be performed or supported by the research community, while some others are related to more general policy actions.

For each of these research areas there are:

- *Needs* (linked to the various drivers identified), which are felt by stakeholders and research activities would aim to satisfy (although full satisfaction may be more an ideal state than a feasible objective).
- *Challenges*, the most probable difficulties and obstacles to be overcome, in view of the context, the present state, the objectives and the nature of the problem to be solved.
- *Objectives*, which identify either tangible results to be achieved or research directions to be followed. For each objective we specify:

  - The **short** (0–3 years) or **medium** (3–8 years) **term** in which the specified results should be achieved or progress in the research directions should be made;

- *Actions*, i.e., specific activities that should be carried out to achieve the result or to pursue the research direction.

The approach used to present the AMBER research roadmap for AMB resilience consists of identifying in first place the needs and related challenges for each research areas. This is presented in a table format to allow a very condensed view. Then, a second table presents, for each area, the research objectives and the actions related to attaining such objectives. A synoptic diagram shows the links among research needs, challenges and objectives for each area, providing a snapshot of what is proposed to each of the four research areas.

We found this approach richer than the classic list of research topics that is normally used to present research roadmaps. Unfortunately, presenting the research needs, challenges, objectives and actions for each research area identified in the research roadmap occupies a large number of pages. For that reason, the detailed description of the AMBER research roadmap for AMB resilience is accessible in [37], and the rest of this section just provides a condensed view with the topmost research priorities.

The rest of this section presents a short list of topics, from those described in [37], that are seen by the AMBER consortium as having the highest priority for a possible research programme. This list was selected by consensus, through rounds of debate within the consortium and advisors from industry. Each partner contributed its knowledge of specific stakeholders and of the links between required scientific advances. An agreed constraint was to keep the final list short: there are six top priorities from the research areas of "scientific and technological foundations", "measurement and assessment", "benchmarking"; plus two concerning "education, training, standardization and take up".

We have not ranked the priorities within this short list. We have instead identified, for each item in the list, the more general goals, in terms of desired changes in the landscape of application of assessment, measurement and benchmarking (AMB) of resilience, which it aims to satisfy. Ranking within our list, for example for the purpose of a funded research programme or to set the roadmap of a research group or community, would depend on a necessary political choice between these more general goals. These **general goals for the research roadmap on AMB resilience** are the following:

1. Extension and combination of AMB methods to ensure that resilience assessment integrates security issues together with accidental faults, design faults with physical faults, human behaviour with machine behaviour, even in very complex systems.

This integration is necessary now, and becoming more essential, with the increasingly complex and integrated systems that characterise "future Internet" scenarios and generally the Information Society.

2. Addressing the fundamental difficulties in quantitative assessment of high consequence and low probability events: predictive value of past experience, combination of diverse evidence, uncertainties about the models and assumptions used.

This goal is crucial for society: these difficulties affect the assessment of systems and infrastructures with great societal value but also great potential risk. Advances that expand the range of applicability of AMB in these critical areas would also have beneficial fall-out for the less critical applications.

3. Empirical validation of the practices already developed in measurement, modelling and benchmarking, so that industry has a basis for steering its own investment regarding these techniques.

Adoption of new techniques throughout industry requires companies to invest in adapting and implementing the techniques. But this requires sufficient empirical evidence of how effective each technique is, and within which constraints.

4. Making the current practices of measurement, assessing and benchmarking more rigorous, e.g., via better use of the established principles of metrology, better matching of the choice of metrics to the measurement needs, and widely.

A more rigorous approach in applying AMB techniques would significantly increase the benefits they offer and reduce the risks from inappropriate application.

5. Building AMB techniques or tools for specific systems and application areas where these are currently inadequate.

In some applications, for instance in Future Internet scenarios with their characteristics of dynamicity, large scale, heterogeneity, developing AMB techniques and tools poses new research challenges, beyond the application of known principles and solution.

6. Gaining widespread acceptance of AMB practices and results.

Promoting AMB practices in the form of, for example, state-of-the-art reports, cookbooks, and success stories, and disseminate the AMB results to the parties that can benefit from them would contribute to the achievement of this goal.

Table 20.1 shows the topmost research topics for each area and the related general goals. The research topics (column in the middle) can be seen in detailed in [37], including their relationship with research challenges, objectives and related actions.

Table 20.1 Topmost priorities for each area and related general research goals

Investigation areas	Topmost research priority topics	Related general goals
Scientific and technological foundations	Validated methods for extrapolating measurements to predictions of system behaviour despite differences between the system in operation and its environment and the system/environment where the measurements were taken	2, 4, 5
	Investigation of principles for successful integration of resilience assessment, measurement and benchmarking technologies into different phases of the life cycle of IT systems, including methods for evaluating technical efficiency and economic impact	1, 4, 5
	Improved "argumentation" processes, which correctly formulate, communicate and verify complex arguments combining "hard" evidence (measurement, mathematical models) and "soft" evidence (judgement), with proper treatment of epistemic uncertainty, levels of confidence, and "unknown unknowns"	1, 2, 4
Measurement and assessment	Development of efficient on-line mechanisms to monitor the environment conditions of the system and to dynamically evaluate and assess its resilience	1, 5
	Practical, trustworthy and widely applicable tools for measurement and assessment in large-scale dynamic systems, adaptable and evolving infrastructures, and other domains where these are lacking	1, 5
Benchmarking	Validated reference faultloads (i.e., sets of faults that are representative of specific domains) and corresponding injection tools (that allow easy implementation and portability of the faultloads) to be used in the development of resilience benchmarks	3, 5
Education, training, standardization and take up	Dissemination of research results, including benchmark prototypes, showing that resilience benchmarks are technically achievable and cost effective	6
	Promotion of proper and fair resilience assessment practices for specific classes of systems/services	6

20.5 Tailoring the Roadmap to Specific Roles and Industrial Domains: Specific Examples

This section proposes a set of viewpoints about priorities as seen from the perspective of specific industrial domains or professional roles, thus providing alternative ways of reading the roadmap. We considered the following perspectives:

- Embedded Systems;
- Transportation;
- Certification Authorities and Assessors;
- Future Internet technological platforms;
- Service Architectures, Platforms and Infrastructures; and
- Enterprise Security.

Within each industrial domain and each role there will be a variety of actors with different interests and opinions. From this range of different positions, this section documents a sample of opinions of senior experts. The discussion provided for each "perspective" in the above list is the outcome of rounds of interaction between AMBER members and AMBER Advisory Board members working in these different domains, aiming to identify the items of the roadmap that may get priority in the respective domain. In addition to the topmost research priorities presented in Sect. 20.4, the interested reader should consider the detailed description of the research roadmap presented in [37] in order to follow the discussion of roadmap in each domain.

20.5.1 Embedded Systems

This section describes the research priorities for the embedded systems domain, specifically addressing the perspective of a systems integrator. The role of a systems integrator is to bring together components manufactured by several companies into a complete product, such as an airplane, a power plant or a satellite. One important concern in this domain is to be able to predict the robustness of a design as early as possible in the development. This calls for techniques to extrapolate measurements from previous designs to new ones, taking advantage of the field data usually collected by mature companies. To facilitate adoption, those techniques must be easily integrated with existing methodologies, such as model-driven development. Therefore, two important research problems are:

- Validated methods for extrapolating measurements to predictions of system behaviour despite differences between the system in operation and its environment and the system/environment where the measurements were taken.
- Reducing the cost of resilience assessment and measurement by developing methods and tools that are easy to integrate into existing development methods and tool chains.

It is fundamental for an integrator to understand the resilience of components and subsystems that are purchased from suppliers. Simple metrics such as the mean time between failures are insufficient, and suppliers are gradually expected to provide more information on the failure modes and failure rates of their components. This can be achieved, at least in part, if suppliers perform resilience benchmarking and make the results available. A relevant step in this direction is:

- Development of concrete examples of resilience benchmarks (possibly as an evolution/standardization of benchmark prototypes).

Manufacturers of safety-critical systems must provide arguments sustaining the safety of their products. In some industries these arguments are documented and compiled into "safety cases", which enable all stakeholders to examine the available evidence. It is important for manufacturers to have high confidence in the argumentation. Otherwise, they are forced to incorporate more redundancy in their designs, in order to remain on the safe side. Thus, an important issue is:

- Improved "argumentation" processes, which correctly formulate, communicate and verify complex arguments combining "hard" evidence (measurement, mathematical models) and "soft" evidence (judgement), with proper treatment of epistemic uncertainty, levels of confidence, and "unknown unknowns".

When a systems integrator acquires a hardware module from a supplier and installs its own software on it, there is a need to assess the final configuration. The supplier can ease this assessment by equipping the hardware modules with:

- Standard set of monitoring features and "hooks" to facilitate AMB and the development of related tools.

Lastly, companies learn from accumulated experience how to improve the resilience of their systems. Field failure data is fundamental to enable this process. However, it is often difficult to obtain and log the necessary data, particularly when using commercial off-the-shelf components. The embedded systems industry would therefore benefit from ways to understand the circumstances under which components fail. To this end, progress should be made in:

- Development of efficient on-line mechanisms to monitor the environment conditions of the system and to dynamically evaluate and assess its resilience.

20.5.2 Transportation

The commercial transport industry is developing ever more complex Intelligent Transport Systems (ITS) to enhance the range of features provided to drivers, fleet managers and other stakeholders. The present focus is on enabling vehicle connectivity to numerous other systems and infrastructures, including traffic management centres, local authorities, weather stations, toll offices, other vehicles and the Internet.

The growing complexity of systems in this domain needs to be met with appropriate (and possibly new) methods and tools for assessing, measuring and benchmarking resilience. One important challenge lies in reducing the cost and time required to perform certification, verification and validation activities. To this end, the most relevant foundational issues are:

- Investigation of principles for successful integration of resilience assessment, measurement and benchmarking technologies into different phases of the life cycle of IT systems, including methods for evaluating technical efficiency and economic impact.
- Adoption of sound measurement practices in the transportation domain.
- Standard set of monitoring features and "hooks" to facilitate AMB and the development of related tools.

Transportation systems are becoming more dynamic and capable of adapting to changes in the operation environment. There is a lack of cost-efficient resilience assessment techniques for such systems, especially techniques that are able to cope with the unpredictability of the environment and the customized configuration of each vehicle. For these reasons, there are two main issues in measurement and assessment that should be addressed:

- Development of efficient on-line mechanisms to monitor the environment conditions of the system and to dynamically evaluate and assess its resilience.
- Development of experimental techniques for resilience assessment in ITS.

In the long term, transportation systems are expected to provide greater flexibility, allowing for instance third-party applications to be downloaded onto onboard units. Such runtime changes to the system's configuration would benefit from ways to benchmark the impact of software changes on the overall resilience. Two relevant steps in this direction are:

- Cost effective, easy to use, and fast enough resilience benchmark prototypes for the transportation domain.
- Validated reference faultloads (i.e., sets of faults that are representative of specific domains) and corresponding injection tools (that allow easy implementation and portability of the faultloads) to be used in the development of resilience benchmarks.

Lastly, the transportation industry faces very rapid changes in processes and technologies, leading managers and engineers to invest in self-learning. Such initiatives require teaching material to be made available. Thus, an important step would be:

- Definition of specific syllabus and course material for resilience assessment.

20.5.3 Certification Authorities and Assessors

The main goal of an assessor (or of a certification authority) is to check if the dependability requirements (e.g., in terms of safety, security, resilience) of a specific system are satisfied or not. This is a very crucial work that usually concerns safety-critical systems where a failure can lead to catastrophic consequences, so the assessor should rely on a set of techniques and tools which allows him to trustfully rely on the quality of the assessment process and on the outcomes produced. In other words, there is a general need of

- practical, trustworthy and widely applicable tools for measurement and assessment in large-scale dynamic systems, adaptable and evolving infrastructures, and other domains where these are lacking.

Another important issue is that, in the assessment process, the system's behaviour is usually predicted using data collected in the past, sometimes related to different environments or even to similar systems. To take correct decisions, corrective factors should be known to extrapolate future system's behaviour from slightly different scenarios, or at least indications should be available about the uncertainty of predictions. In other words, there is the need to

- validate methods for extrapolating measurements to predictions of system behaviour despite differences between the system in operation and its environment and the system/environment where the measurements were taken.

As detailed in the scenario concerning safety certification and regulation, there are several open challenges that still need to be managed and that currently prevent an assessor from being certain about the quality of the assessment, so both *short term* and *medium term* research activities are considered important. With respect to the AMBER roadmap, the focus will be on the "measurement and assessment" and "scientific and technological foundations" areas rather than on "benchmarking", since the objective is not comparing different systems or different products to select the most dependable one, but just to assess the dependability properties of a critical system to allow the claim they are above the required threshold.

The main *medium term* objective is the

- Development of (domain-specific) compositional framework for a holistic assessment process.

The complexity of current critical system, in terms of heterogeneity, evolvability, largeness, dynamicity, inhibits the application of well-proven traditional methods "as they are", but requires the development of an assessment framework where the synergies between different evaluation techniques and tools are exploited to provide realistic assessments. A related challenge is the

- provision of domain-specific rules to compose/integrate different methods and tools for resilience assessment.

This challenge concerns the expressive power of the formalisms (for efficient modelling) as well as the complexity that the supporting solution tools can handle. The elaboration of proper rules to divide the problem and then compose/integrate the results of the different methods and tools used to solve the sub-problems is a possible encouraging approach to attack this challenge.

To achieve this medium term objective, particular effort should be put on:

- The identification of the base types of measurement and modelling techniques and tools (applicable in a given application domain) and the possible interactions among them to provide realistic assessments;
- The assessment of the impact of the approximations introduced in modelling on the resilience-related predictions sought.

Moreover, a number of *short term* (thus preliminary) actions and objectives should be pursued:

- Identification of simple and universally accepted resilience-related metrics.
- Development of efficient methods coping with model and size complexity.

Besides these research activities, there are other issues seen of primary importance in education as well as standardization. Among them:

- The need of comprehensive state-of-the-art reports (including research gaps, limitations and success stories) on resilience assessment techniques.
- The availability of a cookbook (or cookbooks for different domains) on resilience assessment and benchmarking.
- The short term objective of defining specific syllabus and course material for resilience assessment.

20.5.4 Future Internet Technological Platforms

The most important theme in the Future Internet domain is currently related to the proliferation of *digital identities* and their *trustworthiness*.

The services that will constitute the basis of the Future Internet should offer simple access methods, and the problem of guaranteeing security in the usage of these services should be always considered.

Future Internet will have characteristics of high dynamicity, with the necessity of maintaining a high level of resilience. Future Internet systems will be characterized by needs of adaptability to different environment conditions. Future Internet systems will also be evolving systems. In this scenario, enhancing the capabilities of on-line monitoring systems will thus be of uttermost importance, as it would be difficult to obtain off-line trustable (measurement) results of resilience.

We identified the following six points of the AMBER final research roadmap as the topmost research priorities in this domain.

- Development of efficient on-line mechanisms to monitor the environment conditions of the system and to dynamically evaluate and assess its resilience.
- Standard set of monitoring features and "hooks" to facilitate AMB and the development of related tools (at least for computer systems in specific domains).
- Coping with highly complex, adaptable and evolving benchmark targets (components, systems and services).
- Designing and developing test beds for emerging applications, e.g., cloud computing and collaborative services.

- Reusable benchmark components and tools to facilitate the development of benchmarks in different benchmarking domains.
- Understanding the value of assessing and benchmarking resilience in the business domain and in the user/customer community.

20.5.5 Service Architectures, Platforms and Infrastructures

Service architectures, platforms and infrastructures become increasingly vulnerable as they become more complex, thus they need to be made resilient to attacks and operational failures. Minimizing the risk associated with loss of IT services or user data needs precise *metrics* and *models* for evaluating risk, augmenting *tools* with resilience evaluation capabilities in the design phase, and rigorous resilience *measurement* and *real-time assessment* techniques in the operational phase. These needs are reflected in the following objectives and actions selected by the domain experts.

Various *resilience metrics* are used in an increasing extent in IT systems. However, only a few metrics can be used all over the lifetime of an IT system in a consistent way, due to the lack of proper definitions and a related metrological foundation. There is a significant ambiguity in the definitions used and in the specification of the validity of measures and benchmark results, thus reducing the portability and reusability of results used in quantitative evaluation. In case of security, adaptation of metrics to evolving threats is largely unsolved. Accordingly, the first priority is the following:

- Elaboration of easy-to-use, practically measurable (domain-specific) resilience metrics (including metrics for security) and establishing common "operational" definitions for them, in domains where these are lacking.

Resilience *predictive capability* is useful for system developers, designers, architects, and IT practitioners who assemble the hardware and software into even more complex systems, and the clients who purchase and operate the systems. Methods to evaluate resilience metrics prior to build, at the various levels of build, are especially useful, permitting design change before design commitment. In addition to failures, it would be valuable to assess the resilience to planned change activities such as release and version updates, hardware or software configuration changes, etc. The ability to quantify resilience along the dimensions of the application environment so as to more accurately predict for a specific use case of the application is highly desirable:

- Elaboration of extrapolation methods and tools to generalize observations (measures).

The correspondence between IT based *risks* and business (or more generically environmental) processes is of high concern. The qualification and quantification of risks must heavily dependent upon their implications on business processes using the IT systems under analysis. Resilience measures need to focus on the risks with the highest impacts on the supported processes. In the area of IT, because of inadequate or non-existent models and tools, assessing business risks is not always done or done with insufficient rigour. If there were trustworthy and accurate models for evaluating risk associated with loss if IT service, it is expected that they would be extensively used by researchers (to demonstrate risk reduction capabilities), designers, consultants and clients (expectation is that clients would invest in resilience if risks were clearly understood). The development of models for risk analysis has to be extended with the elaboration of corresponding methods and tools. The related research priorities are the following:

- Understanding the economics and overall impact of resilience assessment on the lifecycle of IT systems.
- Development of business and economic models for risk analysis.

Real-time assessment of operational system resilience is a critical need. Even when a system has been properly configured for resilience, changes can take place which reduce or even eliminate protection measures. If there is no notification of the degraded state, outages may occur that the system was assumed to be guarded against. Advanced analytics to evaluate system state and to identify states where resilience is exposed have great value. Initially, real-time assessment should provide for issue notification and recommendations on corrective actions. Over time, automation of the required responses can be expected. In large-scale dynamic systems new, efficient model-driven online assessment methods are needed. Accordingly, the following research is of high priority:

- Development of efficient online mechanisms to monitor the environment conditions of the system and to dynamically evaluate and assess its resilience.

Aimed more at the system design stage than the operational phase, the development of methods and tools for resilience assessment would be beneficiary. These would be used by system and application architects, that is, those who design and build complex IT systems at various levels. Augmenting design tools with resilience evaluation capabilities is an excellent objective as is resilience assessment of design models:

- Development of methods for resilience assessment through automated analysis.

The advances in resilience assessment and measurement shall face (and serve) the paradigm shift observable in this domain nowadays: details of computing resources are abstracted from the users who no longer need control over the technology infrastructure. Virtualisation and other technologies enable convenient, on-demand access to a shared pool of configurable computing resources as outsourced services. In this context, assessment and measurement of resilience is closely related to Service Level Agreement/Quality of Service management and charging of customers. Accordingly, the used techniques have to be precise, rigorous, and at the same time widely accepted and agreed among service providers and customers.

20.5.6 Enterprise Security

CISOs need to make tools to make security investment decisions. Currently, they base their decision making on trends in the industry, general intuition of good practice, etc; they also need to be salesmen to sell their proposed decisions to management. The tools CISOs would want fall in two camps: (i) tools that allow objective decision-making and (ii) ways to share knowledge (an interesting aspect not covered by AMBER much). The hope of a CISO is not that much in expert systems that act as oracle and divine answers, but in tools that provide objective suggestions, communicate the important aspects to consider, etc. A CISO is very much aware of the importance in understanding human tendencies, often economically driven, possibly influenced by training, and often driven by personal tendencies such as risk averseness. Although a CISO understands that assessment is a key ingredient in making decisions in more objective fashion, assessment per se is not a goal they pursue: it's the decision making that matters, not the assessment.

Priorities:

- Integration of considerations related to human behaviour in the assessments of resilience of computer systems as affected by the behaviour of their users, system managers, and adversaries.
- Practical, trustworthy and widely applicable tools for measurement and assessment in large-scale dynamic systems, adaptable and evolving infrastructures, and other domains where these are lacking.
- Finding whether representative types of faults exist through field studies and analysis. (In this case, faults should be read as 'attacks'. The fact that it is field studies and analysis is irrelevant, any means are okay.)

20.6 Summary and Conclusion

This chapter presents a roadmap for research in technologies for assessment, measurement and benchmarking (AMB) of the resilience of information, computer and communication systems. It is the result of the EU-funded AMBER Coordination Action, integrating the consortium experience in the field with the insights resulting from discussions and interviews with a variety of stakeholders about motivating scenarios, drivers and priorities.

The chapter starts by describing a set of motivating scenarios that help understand the current needs and challenges in resilience assessment. These scenarios present viewpoints of industrial players, end users, system operators and regulators. Based on these scenarios, we identify opportunities and challenges that we believe will act as drivers for investment in improved resilience AMB technologies. In particular, we believe that the establishment of standardized and sound assessment technologies and benchmarks will be a catalyst for the acceptance of AMB solutions. If done well, it will lead to improved competition by providing easy to communicate measurable objectives for manufacturers, system integrators and users alike. In addition, the increasing demands by regulators as well as the continuing technological progress in software and hardware will create challenges to be addressed by research in the field.

The research roadmap (described in a condensed form in Sect. 20.4 and fully detailed in [37]) first provides a detailed list of research needs and challenges grouped in four categories: (i) scientific and technological foundations, (ii) measurement and assessment, (iii) benchmarking, and (iv) education, training, standardization and take up.

The **foundations** make the case for two types of research advances, which we could label as 'back to basics' and 'holistic'. The desire to go 'back to basics' refers to the creation of a standardized set of sound but simple techniques and tools for assessment, based on, for instance, insights from metrology (the science of measurement). The 'holistic' view refers to the identification that the context (human, socio-economic, political) in which computer systems operate should be considered and assessment should take a holistic view, thus requiring the study of human factors, business impact and the integration of tools and/or arguments, as well as identifying practical limits for the applicability of each class of methods as a function of the environment. A two-prong strategy is therefore needed, on the one hand to keep advancing our assessment methods and techniques to deal with increasingly complex system deployments, and address hard theoretical problems, on the other to work towards standardized basic tools that through widespread use can dramatically change the way resilience is viewed and perceived.

The **measurement and assessment** category identifies a number of topics of acute interest and that are particularly challenging (in addition to topics already discussed above). In particular, this concerns on-line assessment for run-time system adaptation and optimisation (often referred to as self-adaptive systems), the quantitative assessment of security (that is, the ability of a system to withstand attacks and malicious interference), and the analysis of collected data in a structured and powerful a way.

Resilience **benchmarking** aims at providing generic, repeatable and widely accepted methods for characterising and quantifying the system (or component) behaviour in the presence of faults, and comparing the resilience of alternative solutions. A problem is that, typically, benchmarks may not be designed well enough and are prone to being 'gamed', in which case the benchmark may have unintended negative consequences. The research challenges identified in the benchmarking section are therefore about constructing benchmarks that are robust while easy to use. Questions that need to be addressed are how to subdivide application domains, how to create acceptance through standardization, how to include measurement and fault injection hooks into systems, etc.

We also identified the challenges we see in **education** as well as **standardization**. It has become apparent to us that to fulfil some of the potential of broadly applied resilience assessment, both these aspects need to be addressed. We already touched on the importance of standardized basic assessment techniques and standardized benchmarks. In addition, advances in education of assessment techniques are critical for computer system engineers to appreciate the power of quantitative assessment as well as the pitfalls of poorly conducted assessment.

Acknowledgments The authors acknowledge the support given by the European Commission to the AMBER Coordination Action [38].

References

1. Amazon Web Services Discussion Forums, http://developer.amazonwebservices. com/connect/thread.jspa?threadID=21401&tstart=15. Accessed June 2009
2. Cloud Security Alliance, http://www.cloudsecurityalliance.org. Accessed Jan 2011
3. Cloutage—tracking cloud incidents, security, and outages, http://cloutage.org/. Accessed June 2011
4. IEEE Standard for Local and Metropolitan Area Networks, Part 16: Air Interface for Fixed and Mobile Broadband Wireless Access Systems. Technical report 802.16e, Dec 2005
5. IDC Enterprise Panel, N = 244, Aug 2008
6. Universal Mobile Telecommunications System (UMTS) and LTE; Radio Resource Control (RRC); Feasibility study for E-UTRA and UTRAN. (3GPP TS 25.912, v9.0.0, release 9), Oct 2009
7. NTT Communications, Cloud or Fog? The business realities of cloud computing for UK enterprises (2009)
8. HOMESNET: Home Base Station, An emerging network paradigm. CELTIC European project (2009–2011), http://www.celtic-initiative.org/Projects/Celtic-projects/Call6/ HOMESNET/homesnet-default.asp
9. Evolved Universal Terrestrial Radio Access (E-UTRA) and Evolved Universal Terrestrial Radio Access Network (E-UTRAN); Overall description, Stage 2 (3GPP TS 36.300, release 9, v9.1.0), Oct 2010
10. LTE and Evolved Universal Terrestrial Radio Access (E-UTRA); Radio Resource Control (RRC); protocol specification. (3GPP TS 36.331, release 9, v9.4.0), Sept 2010
11. Universal Mobile Telecommunications System UMTS; Radio Resource Control (RRC); protocol specification. (3GPP TS 36.331, release 9, v9.1.0), Feb 2010
12. Summary of the Amazon EC2 and Amazon RDS Service Disruption in the US East Region, http://aws.amazon.com/message/65648/. Accessed April 2011
13. FEMA Reference manual to mitigate potential terrorist attacks against buildings [electronic resource]: providing protection to people and buildings (Federal Emergency Management Agency, Washington, 2003)
14. 3GPP. TS 25.104, V9.5.0 (2010-09) Base station (BS) radio transmission and reception (FDD) (2010), http://www.3gpp.org/ftp/specs/html-info/25104.htm
15. R.J. Abbott, Resourceful systems for fault tolerance, reliability, and safety. ACM Comput. Surv. 22(1), 35–68 (1990)
16. R. Abdalla, Q. Cheng, V. Tao, J. Li, Network-centric approach for modeling infrastructure interdependency. Photogramm. Eng. Remote Sens. 27(6), 681–690 (2007)
17. T. Abdelzaher, K. Shin, N. Bhatti, Performance guarantees for web server end-systems: a control-theoretical approach. IEEE Trans. Parallel Distrib. Syst. 13(1), 80–96 (2002)

A. Avritzer et al., *Resilience Assessment and Evaluation of Computing Systems*,   441
DOI: 10.1007/978-3-642-29032-9, © Springer-Verlag Berlin Heidelberg 2012

18. B. Abrahao, V. Almeida, J. Almeida. Self-adaptive SLA-driven capacity management for internet services, in *International Workshop on Distributed Systems: Operations and Management* (2006)

19. Acunetix. AcuSensor Technology (2011), http://www.acunetix.com/vulnerability-scanner/acusensor.htm

20. B. Addis, D. Ardagna, B. Panicucci, L. Zhang, Autonomic management of cloud service centers with availability guarantees, in *IEEE International Conference on Cloud Computing* (2010), pp. 220–227

21. V. Adve, R. Bagrodia, J. Browne, E. Deelman, A. Dube, E. Houstis, J. Rice, R. Sakellariou, D. Sundaram-Stukel, P. Teller, M. Vernon, Poems: end-to-end performance design of large parallel adaptive computational systems. IEEE Trans. Softw. Eng. **26**(11), 1027–1048 (2000)

22. M.K. Agarwal, M. Gupta, V. Mann, N. Sachindran, N. Anerousis, L.B. Mummert, Problem determination in enterprise middleware systems using change point correlation of time series data, in *IEEEIFIP Network Operations and Management Symposium NOMS*, Vancouver, Canada, April 2006, pp. 471–482

23. S. Agarwala, F. Alegre, K. Schwan, J. Mehalingham, E2eprof: automated end-to-end performance management for enterprise systems, in *Proceedings of the 37th Annual IEEE/IFIP International Conference on Dependable Systems and Networks*, June 2007, pp. 749–758

24. P. Agrawal, Fault tolerance in multiprocessor systems without dedicated redundancy. IEEE Trans. Comput. **37**, 358–362 (1988)

25. M.K. Aguilera, J.C. Mogul, J.L. Wiener, P. Reynolds, A. Muthitacharoen. Performance debugging for distributed system of black boxes, in *Proceedings of the Nineteenth ACM Symposium on Operating Systems Principles*, Bolton Landing, NY, Oct 2003, pp. 74–89

26. B. Aichernig, W. Krenn, H. Eriksson, J. Vinter, State of the Art Survey–Part A: Model-based Test Case Generation (2008), http://www.mogentes.eu/

27. J. Aidemark, J. Vinter, P. Folkesson, J. Karlsson, GOOFI: Generic object-oriented fault injection tool, in *Proceedings of the 2001 International Conference on Dependable Systems and Networks (DSN 2001)*, July 2001, pp. 83–88

28. M. Albano, S. Chessa, R. Di Pietro, Information assurance in critical infrastructures via wireless sensor networks, in *Information Assurance and Security, 2008. ISIAS '08. Fourth International Conference on*, Sept 2008, pp. 305–310

29. A. Albinet, J. Arlat, J.-C. Fabre, Characterization of the impact of faulty drivers on the robustness of the linux kernel, in *Proceedings of the 2004 International Conference on Dependable Systems and Networks (DSN 2004)*, IEEE Computer Society, June–July 2004, pp. 867–876

30. J. Almasizadeh, M.A. Azgomi, Intrusion process modeling for security quantification, in: *International Conference on Availability, Reliability and Security* (IEEE Computer Society, Los Alamitos, 2009), pp. 114–121

31. J. Almeida, V. Almeida, D. Ardagna, I. Cunha, C. Francalanci, M. Trubian, Joint admission control and resource allocation in virtualized servers. J. Parallel Distrib. Comput. **70**(4), 344–362 (2010)

32. R. Almeida, N. Mendes, H. Madeira, Sharing experimental and field data: the amber raw data repository experience, in *30th IEEE International Conference on Distributed Computing Systems Workshops—ICDCSW*, Genova (2010)

33. R. Almeida, M. Vieira, Benchmarking the resilience of self-adaptive software systems: perspectives and challenges, in *6th International Symposium on Software Engineering for Adaptive and Self-Managing Systems, SEAMS* 2011, Honolulu, Hawaii, May 2011

34. J. Alonso, L. Silva, A. Andrzejak, P. Silva, J. Torres, High-available grid services through the use of virtualized clustering, in *Grid Computing, 2007 8th IEEE/ACM International Conference on*, Sept 2007, pp. 34–41

35. G.A. Alvarez, F. Cristian, Centralized failure injection for distributed, fault-tolerant protocol testing, in *Proceedings of the 17th IEEE International Conference on Distributed Computing Systems (ICDCS'7)* (1997), pp. 78–85

36. AMBER Consortium. D2.2—state of the art (final version). Technical report, AMBER Consortium, July 2009
37. AMBER Consortium. D3.2—final research roadmap. Technical report, AMBER Consortium, Dec 2009
38. AMBER—Assessing, Measuring and BEnchmarking Resilience (Coordinated Action ICT-FP7-216295), http://www.amber-project.eu/
39. AMSD Consortium. D1.1—amsd: A dependability roadmap for the Information Society in Europe, part 3—towards a dependability roadmap. Technical report, AMSD Consortium (2003)
40. H. Ando, R. Kan, Y. Tosaka, K. Takahisa, K. Hatanaka, Validation of hardware error recovery mechanisms for the SPARC64 V microprocessor, in *Proceedings of the 38th Annual IEEE/IFIP International Conference on Dependable Systems and Networks (DSN)*, IEEE Computer Society, Anchorage, Alaska, June 2008, pp. 62–69
41. A. Andrzejak, L. Silva, Deterministic models of software aging and optimal rejuvenation schedules, in *10th IFIP/IEEE International Symposium on Integrated Network Management, 2007. IM '07* (2007), pp. 159–168
42. M. Anghel, K.A. Werley, A.E. Motter, Stochastic model for power grid dynamics, in *40th Hawaii International Conference on System Sciences (CD-ROM)*, Waikoloa, Big Island, January 2007 IEEE, pp. 113–122 (10 pages)
43. L. Angrisani, S. D'Antonio, M. Esposito, M. Vadursi, Techniques for available bandwidth measurement in ip networks: a performance comparison. Int. J. Comput. Telecommun. Netw. **50**, 332–349 (2006)
44. L. Angrisani, A. Pescape, G. Ventre, M. Vadursi, Performance measurement of IEEE 802.11b-based networks affected by narrowband interference through cross-layer measurements. IET Commun. **2**(1), 82–91 (2008)
45. M. Anji, Y. Jiaxi, G. Zhizhong, Electric power grid structural vulnerability assessment, in *Power Engineering Society General Meeting* (IEEE, 2006), p. 6
46. L. Antoni, R. Leveugle, B. Fehér, Using run-time reconfiguration for fault injection applications, in *Proceedings of the IEEE Instrumentation and Measurement Technology Conference*, vol 3, May 2001, pp. 1773–1777
47. N. Antunes, N. Laranjeiro, M. Vieira, H. Madeira, Effective detection of SQL/XPath injection vulnerabilities in web services, in *Proceedings of the 2009 IEEE International Conference on Services Computing (SCC '09)* (2009), pp. 260–267
48. N. Antunes, M. Vieira, Detecting SQL injection vulnerabilities in web services, in *Fourth Latin-American Symposium on Dependable Computing (LADC '09)* (2009), pp. 17–24
49. P. Apparao, R. Iyer, X. Zhang, D. Newell, T. Adelmeyer, Characterization & analysis of a server consolidation benchmark, in *Proceedings of 4th International ACM Conference on Virtual Execution Environments (VEE '08)* (2008), pp. 21–30
50. A. Araujo Neto, M. Vieira, Towards assessing the security of dbms configurations, in *Proceedings of the 2008 IEEE/IFIP International Conference on Dependable Systems and Networks (DSN 2008)* (2008)
51. A. Araujo Neto, M. Vieira, A trust-based benchmark for dbms configurations, in *Proceedings of the 15th IEEE Pacific Rim International Symposium on Dependable Computing (PRDC 2009)* (2009)
52. A. Araujo Neto, M. Vieira, H. Madeira, An appraisal to assess the security of database configurations, in *Proceedings of the Second International Conference on Dependability (DSN 2009)* (2009)
53. J. Arlat, M. Aguera, L. Amat, Y. Crouzet, J.-C. Fabre, J.-C. Laprie, E. Martins, D. Powell, Fault injection for dependability validation: a methodology and some applications. IEEE Trans. Softw. Eng. **16**(2), 166–182 (1990)
54. J. Arlat, A. Costes, Y. Crouzet, J.-C. Laprie, D. Powell, Fault injection and dependability evaluation of fault-tolerant systems. IEEE Trans. Comput. **42**(8), 913–923 (1993)

55. J. Arlat, Y. Crouzet, J. Karlsson, P. Folkesson, E. Fuchs, G.H. Leber, Comparison of physical and software-implemented fault injection techniques. IEEE Trans. Comput. **52**(9), 1115–1133 (2003)
56. J. Arlat, J.-C. Fabre, M. Rodríguez, F. Salles, Dependability of COTS microkernel-based systems. IEEE Trans. Comput. **51**(2), 138–163 (2002)
57. M. Arlitt, T. Jin, Workload characterization of the 1998 world cup web site, in HP technical report, HPL-99-35 (1999)
58. M. Armbrust, A. Fox, R. Griffith, A.D. Joseph, R. Katz, A. Konwinski, G. Lee, D. Patterson, A. Rabkin, I. Stoica, M. Zaharia, A view of cloud computing. ACM Commun. **53**(4), 50–58 (2010)
59. S. Asmussen, O. Nerman, M. Olsson, Fitting phase-type distribution via the em algorithm. Scand. J. Stat. **23**, 419–441 (1996)
60. M.H. Assat, S.R. Das, E.M. Petriul, L. Jin, C. Jin, D. Biswas, V. Groza, M. Sahinoglu, Hardware and software co-design in space compaction of digital circuits, in *IEEE Instrumentation and Measurement Technology Conference*, vol. 2 (2004), pp. 1503–1508
61. A. Avizienis, Fault-tolerance and fault-intolerance: Complementary approaches to reliable computing, in *International Conference on Reliable Software* (ACM, Los Angeles, 1975), pp. 458–464
62. A. Avizienis, Design diversity and the immune system paradigm: Cornerstones for information system survivability, in *Third Information Survivability Workshop (ISW-2000)*, Boston, MA (2000)
63. A. Avizienis, J.-C. Laprie, B. Randell, C.E. Landwehr, Basic concepts and taxonomy of dependable and secure computing. IEEE Trans. Depend. Secure Comput. **1**(1), 11–33 (2004)
64. D. Avresky, J. Arlat, J.-C. Laprie, Y. Crouzet, Fault injection for formal testing of fault tolerance. IEEE Trans. Reliab. **45**(3), 443–455 (1996)
65. A. Avritzer, A. Bondi, M. Grottke, K.S. Trivedi, E.J. Weyuker, Performance assurance via software rejuvenation: monitoring, statistics and algorithms, in *International Conference on Dependable Systems and Networks, 2006 (DSN 2006)*, June 2006, pp. 435–444
66. A. Avritzer, A. Bondi, E.J. Weyuker, Ensuring stable performance for systems that degrade, in *5th International Workshop on Software and Performance, 2005. Proceedings (WOSP '05)* (ACM, New York, 2005), pp. 43–51
67. A. Avritzer, R.G. Cole, E.J. Weyuker, Using performance signatures and software rejuvenation for worm mitigation in tactical manets, in *6th International Workshop on Software and Performance, 2007. Proceedings (WOSP '07)* (ACM, New York, 2007), pp. 172–180
68. A. Avritzer, R.G. Cole, E.J. Weyuker, Methods and opportunities for rejuvenation in aging distributed software systems. J. Syst. Softw. **83**(9), 1568–1578 (2010)
69. A. Avritzer, E. de Souzae Silva, R. Leão, E. Weyuker, Automated generation of test cases using a performability model. IET Softw. **5**(2), 113–119 (2011)
70. A. Avritzer, B. Larson, Load testing software using deterministic state testing, in *ISSTA'93* (1993), pp. 82–88
71. A. Avritzer, R. Tanikella, K. James, R.G. Cole, E.J. Weyuker, Monitoring for security intrusion using performance signatures, in *1st joint WOSP/SIPEW International Conference on Performance Engineering, 2010. Proceedings (WOSP/SIPEW '10)* (ACM, New York, 2010), pp. 93–104
72. A. Avritzer, E.J. Weyuker, Generating test suites for software load testing, in *ISSTA '94: Proceedings of the 1994 ACM SIGSOFT international symposium on Software Testing and Analysis* (ACM, New York, 1994), pp. 44–57
73. A. Avritzer, E.J. Weyuker, The automatic generation of load test suites and the assessment of the resulting software. IEEE Trans. Softw. Eng. **21**(9), 705–716 (1995)
74. A. Avritzer, E.J. Weyuker, Monitoring smoothly degrading systems for increased dependability. Empire Softw. Eng. **2**(1), 59–77 (1997)

75. G.A.G.B. Schroeder, The computer failure data repository (CFDR), in *Workshop on Reliability Analysis of System Failure Data—RAF'07* (MSR Cambridge, Cambridge, 2007)

76. J. Bachmann, M. Riedl, J. Schuster, M. Siegle, An efficient symbolic elimination algorithm for the stochastic process algebra tool caspa, in *Proceedings of the 35th Conference on Current Trends in Theory and Practice of Computer Science (SOFSEM '09)* (Springer, Berlin, 2009), pp. 485–496

77. P. Bahl, R. Chandra, A.G. Greenberg, S. Kandula, D.A. Maltz, M. Zhang, Towards highly reliable enterprise network services via inference of multi-level dependencies, in *Proceedings of the 2007 Conference on Applications, Technologies, Architectures, and Protocols for Computer Communications*, Kyoto, August 2007, pp. 13–24

78. C. Baier, J.-P. Katoen, H. Hermanns, Approximate symbolic model checking of continuous-time markov chains, in *Proceedings of the 10th International Conference on Concurrency Theory (CONCUR '99)* (Springer, London, 1999), pp. 146–161

79. R. Bakhshi, L. Cloth, W. Fokkink, B. Haverkort, Mean-field analysis for the evaluation of gossip protocols, in *Proceedings 6th International Conference on the Quantitative Evaluation of Systems (QEST'09)*, IEEE Computer Society (2009), pp. 247–256

80. R. Bakhshi, L. Cloth, W. Fokkink, B. Haverkort, Mean-field framework for performance evaluation of push–pull gossip protocols. Perform. Eval. **68**(2), 157–179 (2011)

81. G. Balbo, in *Introduction to Stochastic Petri nets*, ed. by B. Brinksma, H. Hermanns, J.-P. Katoen (Springer, Berlin, 2001)

82. G. Balbo, S. Bruell, M. Sereno, Embedded processes in generalized stochastic petri nets, in *Proceedings 9th International Workshop on Petri Nets and Performance Models* (2001), pp. 71–80

83. S. Balsamo, A. Di Marco, P. Inverardi, M. Simeoni, Model-based performance prediction in software development: a survey. IEEE Trans. Softw. Eng. **30**(5), 295–310 (2004)

84. D. Balzarotti, M. Cova, V. Felmetsger, N. Jovanovic, E. Kirda, C. Kruegel, G. Vigna, Saner: composing static and dynamic analysis to validate sanitization in web applications, in *IEEE Symposium on Security and Privacy (SP 2008)* (2008), pp. 387–401

85. Y. Bao, X. Sun, K.S. Trivedi, Adaptive software rejuvenation: degradation model and rejuvenation scheme, in *2003 International Conference on Dependable Systems and Networks, 2003. Proceedings*, June 2003, pp. 241–248

86. Y. Bao, X. Sun, K.S. Trivedi, A workload-based analysis of software aging, and rejuvenation. IEEE Trans. Reliab. **54**(3), 541–548 (2005)

87. R. Barbosa, N. Silva, J. Durães, H. Madeira, Verification and validation of (Real Time) COTS products using fault injection techniques, in *Proceedings of the Sixth International IEEE Conference on Commercial-off-the-Shelf (COTS)-Based Software Systems* (2007), pp. 233–242

88. P. Barger, J.-M. Thiriet, M. Robert, Dependability analysis of a distributed control or measurement architecture, in *Instrumentation and Measurement Technology Conference, 2003. IMTC '03. Proceedings of the 20th IEEE*, vol. 1, May 2003, pp. 473–477

89. P. Barham, A. Donnelly, R. Isaacs, R. Mortier, Using magpie for request extraction and workload modelling, in *Proceedings of the 6th conference on Symposium on Operating Systems Design & Implementation—vol. 6*, San Francisco, Dec 2004, pp. 259–272

90. P. Barham, B. Dragovic, K. Fraser, S. Hand, T. Harris, A. Ho, R. Neugebauer, I. Pratt, A. Wareld, Xen and the art of virtualization, in *Proceedings of 19th Symposium on Operating System Principles* (2003), pp. 164–177

91. L. Barroso, U. Hölzle, The case for energy-proportional computing. IEEE Comp. **40** (2007)

92. J.H. Barton, E.W. Czeck, Z. Segall, D.P. Siewiorek, Fault injection experiments using FIAT. IEEE Trans. Comput. **39**(4), 575–582 (1990)

93. F. Baskett, M. Chandy, R. Muntz, F. Palacios, Open, closed and mixed networks of queues with different classes of customers. J. ACM **22**, 248–260 (1975)

94. F. Bause, P. Buchholz, P. Kemper, A toolbox for functional and quantitative analysis of deds, in *Computer Performance Evaluation*. Lecture Notes in Computer Science, vol. 1469, ed. by R. Puigjaner, N. Savino, B. Serra (Springer, Berlin, 1998), pp. 356–359

95. M. Beccuti, G. Franceschinis, S. Donatelli, S. Chiaradonna, F. Di Giandomenico, P. Lollini, G. Dondossola, F. Garrone, Quantification of dependencies in electrical and information infrastructures: the crutial approach, in *Fourth International Conference on Critical Infrastructures, 2009. CRIS 2009* (2009), pp. 1–8

96. S. Becker, H. Koziolek, R. Reussner, The palladio component model for model-driven performance prediction. J. Syst. Softw. **82**, 3–22 (2009)

97. S. Bellahsene, Algorithmique et Evaluation de performances pour les réseaux de télécommunication. Ph.D. thesis, Université de Versailles (2012)

98. S. Bellahsene, L. Kloul, A new markov-based mobility prediction algorithm for mobile networks, in *Proceedings of the 7th European Performance Engineering Workshop (EPEW 2010)* (Springer, Bertinoro, 2010)

99. S. Bellahsene, L. Kloul, D. Barth, A hierarchical prediction model for two nodes-based IP mobile networks, in *Proceedings of the 12th ACM International Conference on Modeling, Analysis and Simulation of Wireless and Mobile Systems (MSWiM '09)*, Tenerife, Canary Islands, Oct 2009, pp. 173–180

100. M. Bena, J.-Y. Le Boudec, A class of mean field interaction models for computer and communication systems. Perform. Eval. **65**(11–12), 823–838 (2008)

101. F. Benevenuto, C. Fernandes, M. Santos, V. Almeida, J. Almeida, G. Janakiraman, J. Santos, Performance models for virtualized applications, in *Frontiers of High Performance Computing and Networking—ISPA 2006 Workshops*. Lecture Notes in Computer Science, vol. 4331 (Springer, Heidelberg, 2006), pp. 427–439

102 M.N. Bennani, D. Menascé, Resource allocation for autonomic data centers using analytic performance models, in *Proceedings of the Second International Conference on Automatic Computing* (2005)

103. A. Benoit, Méthodes et algorithmes pour l'évaluation des performances des systèmes informatiques à grands espaces d'états. Ph.D thesis (2003)

104. A. Benoit, L. Brenner, P. Fernandes, B. Plateau, W. Stewart, The peps software tool, in *Computer Performance Evaluation. Modelling Techniques and Tools*. Lecture Notes in Computer Science, vol. 2794, ed. by P. Kemper, W. Sanders (Springer, Berlin, 2003), pp. 98–115

105. A. Benoit, B. Plateau, W. Stewart, Memory-efficient algorithms with applications to the modelling of parallel systems. Future Gener. Comput. Syst. **22**, 838–847 (2006)

106. M. Benzi, M. Tuma, A parallel solver for large-scale Markov chains. Appl. Numer. Math. **41**, 135–153 (2002)

107. C. Beounes, M. Aguera, J. Arlat, S. Bachmann, C. Bourdeau, J.-E. Doucet, K. Kanoun, J.-C. Laprie, S. Metge, J. Moreira de Souza, D. Powell, P. Spiesser, Surf-2: a program for dependability evaluation of complex hardware and software systems, in *The Twenty-Third International Symposium on Fault-Tolerant Computing, FTCS-23. Digest of Papers*, June 1993, pp. 668–673

108. L. Berardinelli, V. Cortellessa, A. Di Marco, Performance modeling and analysis of context-aware mobile software systems, in *FASE* (2010), pp. 353–367

109. The Berkeley/Stanford recovery-oriented computing (ROC) project (2008), http://roc.cs.berkeley.edu. Accessed 26 Nov 2011

110. S. Bernardi, J. Merseguer, D. Petriu, A dependability profile within marte. Softw. Syst. Model. **10**, 313–336 (2009)

111. L. Bernstein, Innovative technologies for preventing network outages. AT&T Tech. J. **72**(4), 9–10 (1993)

112. A. Bessani, P. Sousa, M. Correia, N.F. Neves, P. Verissimo, The CRUTIAL way of critical infrastructure protection. IEEE Secur. Priv. **6**(6), 44–51 (2008)

113. S. Bhatia, A. Kumar, M.E. Fiuczynski, L.L. Peterson, Lightweight, high-resolution monitoring for troubleshooting production systems, in *Proceedings of the 8th USENIX Conference on Operating Systems Design and Implementation*, San Diego, CA (2008), pp. 103–116

114. C. Binnig, D. Kossmann, T. Kraska, S. Loesing, How is the weather tomorrow? Towards a benchmark for the cloud, in *Proceedings of the 2nd International Workshop on Testing Database Systems, DBTest '09* (ACM, New York, 2009), pp. 9:1–9:6

115. R. Bloomfield, L. Buzna, P. Popov, K. Salako, D. Wright, Stochastic modelling of the effects of interdependency between critical infrastructures, in *4th International Workshop on Critical Information Infrastructures Security (CRITIS 2009)*. Lecture Notes in Computer Science, vol. 6027, ed. by E. Rome, R. Bloomfield (Springer, Berlin, 2010), pp. 201–212

116. R. Bloomfield, N. Chozos, K. Salako, Current capabilities, requirements and a proposed strategy for interdependency analysis in the UK, in *4th International Workshop on Critical Information Infrastructures Security (CRITIS 2009)*. Lecture Notes in Computer Science, vol. 6027, ed. by E. Rome, R. Bloomfield (Springer, Berlin, 2010), pp. 188–200

117. A. Bobbio, S. Garg, M. Gribaudo, A. Horvath, M. Sereno, M. Telek, Modeling software systems with rejuvenation, restoration and checkpointing through fluid stochastic petri nets, in *Proceedings of the 8th International Workshop on Petri Nets and Performance Models* (1999), pp. 82–91

118. A. Bobbio, M. Gribaudo, M. Telek, Analysis of large scale interacting systems by mean field method, in *Proceedings of the 5th Conference on the Quantitative Evaluation of Systems (QEST'08)*, IEEE Computer Society (2008), pp. 215–224

119. A. Bobbio, A. Puliafito, M. Telek, K.S. Trivedi, Recent developments in non-Markovian stochastic petri nets. *J. Circuits Syst. Comput.* 119–158 (1998)

120. A. Bobbio, M. Sereno, C. Anglano, Fine grained software degradation models for optimal rejuvenation policies. Perform. Eval. **46**, 45–62 (2001)

121. A. Bobbio, K. Trivedi, An aggregation technique for the transient analysis of stiff markov chains. IEEE Trans. Comput. **C-35**(9), 803–814 (1986)

122. N. Bobroff, A. Kochut, K. Beaty, Dynamic placement of virtual machines for managing SLA violations, in *Integrated Network Management, 2007. IM '07* (2007), pp. 119–128

123. P. Bodik, M. Goldszmidt, A. Fox, D.B. Woodard, H. Andersen, Fingerprinting the datacenter: automated classification of performance crises, in *Proceedings of the 5th European Conference on Computer systems*, Paris, France (2010), pp. 111–124

124. L. Bodrog, A. Horváth, M. Telek, Moment characterization of matrix exponential and markovian arrival processes. Ann. OR **160**(1), 51–68 (2008)

125. G. Bolch, S. Greiner, H. de Meer, K.S. Trivedi, *Queueing Networks and Markov Chains: Modeling and Performance Evaluation with Computer Science Applications* (Wiley-Interscience, New York, 1998)

126. E. Bompard, R. Napoli, F. Xue, Analysis of structural vulnerabilities in power transmission grids. Int. J. Crit. Infrastruct. Prot. **2**(1–2), 5–12 (2009)

127. A. Bondavalli, Research roadmap—deliverable D3.2, amber—assessing, measuring and benchmarking resilience, Ist—216295 funded by the European Union, 2009. Technical report, AMBER consortium (2009), http://amber-dbserver.dei.uc.pt:81/roadmap

128. A. Bondavalli, A. Ceccarelli, L. Falai, M. Vadursi, Foundations of measurement theory applied to the evaluation of dependability attributes, in *DSN '07: Proceedings of the 37th Annual IEEE/IFIP International Conference on Dependable Systems and Networks* (IEEE Computer Society, Washington, 2007), pp. 522–533

129. A. Bondavalli, A. Ceccarelli, L. Falai, M. Vadursi, A new approach and a related tool for dependability measurements on distributed systems. IEEE Trans. Instrum. Meas. **59**(4), 820–831 (2010)

130. A. Bondavalli, S. Chiaradonna, D. Cotroneo, L. Romano, Effective fault treatment for improving the dependability of cots and legacy-based applications. IEEE Trans. Depend. Secure Comput. **1**, 223–237 (2004)

131. A. Bondavalli, S. Chiaradonna, F. Di Giandomenico, F. Grandoni, Discriminating fault rate and persistency to improve fault treatment, in *IEEE FTCS International Symposium on Fault-Tolerant Computing* (1997), pp. 354–362

132. A. Bondavalli, S. Chiaradonna, F. Di Giandomenico, F. Grandoni, Threshold-based mechanisms to discriminate transient from intermittent faults. IEEE Trans. Comput. **49**, 230–245 (2000)

133. A. Bondavalli, S. Chiaradonna, F. Di Giandomenico, I. Mura, Dependability modeling and evaluation of multiple-phased systems using deem. IEEE Trans. Reliab. **53**(4), 509–522 (2004)

134. A. Bondavalli, M. Dal Cin, D. Latella, I. Majzik, A. Pataricza, G. Savoia, Dependability analysis in the early phases of uml based system design. J. Comput. Syst. Sci. Eng. **16**(5), 265–275 (2001)

135. A. Bondavalli, O. Hamouda, M. Kaâniche, P. Lollini, I. Majzik, H.-P. Schwefel, The hidenets holistic approach for the analysis of large critical mobile systems. IEEE Trans. Mob. Comput. **10**(6), 783–796 (2011)

136. A. Bondavalli, P. Lollini, L. Montecchi, Qos perceived by users of ubiquitous umts: compositional models and thorough analysis. J. Softw. **4**(7) (2009)

137. A. Bondavalli, P. Lollini, L. Montecchi, Graphical formalisms for modeling critical infrastructures, in *Critical Infrastructure Security: Assessment, Prevention, Detection, Response*, ed. by F. Flammini (WIT Press, London, 2011, to appear)

138. A.B. Bondi, Characteristics of scalability and their impact on performance, in *The 2nd International Workshop on Software and Performance*, Sept 2000, pp. 195–203

139. A.B. Bondi, V.Y. Jin, A performance model of a design for a minimally replicated distributed database for database-driven telecommunications services. Distrib. Parallel Databases (4), 295–318 (1996)

140. L. Bononi, M. Di Felice, A. Molinaro, S. Pizzi, A cross-layer architecture for effective channel assignment with load-balancing in multi-radio multi-path wireless mesh networks, in *7th ACM International Symposium on Mobility Management and Wireless Access (Mobiwac09)* (2009)

141. S. Bornot, J. Sifakis, On the composition of hybrid systems, in *Proceedings of the First International Workshop on Hybrid Systems: Computation and Control* (Springer, London, 1998), pp. 49–63

142. L. Bortolussi, A. Policriti, Stochastic concurrent constraint programming and differential equations, in *QAPL'07, 5th Workshop on Quantitative Aspects of Programming Languages*. Electronic Notes in Theoretical Computer Science, **190**(3), 27–42 (2007)

143. S. Bose, P. Mishra, P. Sethuraman, R. Taheri, Performance evaluation and benchmarking, in *Performance Evaluation and Benchmarking*, chapter Benchmarking Database Performance in a Virtual Environment, ed. by R. Nambiar, M. Poess (Springer, Berlin, 2009), pp. 167–182

144. R. Boucherie, A characterisation of independence for competing markov chains with applications to stochastic petri nets. IEEE Trans. Softw. Eng. **20**(7), 536–544 (1994)

145. H. Boudali, P. Crouzen, M. Stoelinga, Dynamic fault tree analysis using input/output interactive markov chains, in *Proceedings of the 37th Annual IEEE/IFIP International Conference on Dependable Systems and Networks*, DSN '07 (IEEE Computer Society, Washington), pp. 708–717

146. J.-Y. Le Boudec, D. McDonald, J. Mundinger, A generic mean field convergence result for systems of interacting objects, in *Proceedings of the 4th Conference on the Quantitative Evaluation of Systems (QEST'07)* (IEEE Computer Society, 2007), pp. 3–18

147. G. Box, G. Jenkins, *Time Series Analysis: Forecasting and Control* (Prentice Hall PTR, Upper Saddle River, 1994)

148. J. Bradley, N. Dingle, W. Knottenbelt, H. Wilson, Hypergraph-based parallel computation of passage time densities in large semi-Markov models. Linear Algebra Appl. **386**, 311–334 (2004)

149. S. Brocklehurst, B. Littlewood, New ways to get accurate reliability measures. IEEE Softw. **9**, 34–42 (1992)

150. F. Brosig, S. Kounev, K. Krogmann, Automated extraction of palladio component models from running enterprise java applications, in *Proceedings of ROSSA 2009*, ACM, Oct 2009

151. A. Brown, L. Chung, W. Kakes, C. Ling, D.A. Patterson. Dependability benchmarking of human-assisted recovery processes, in *Proceedings of the 2004 International Conference on Dependable Systems and Networks (DSN 2004)* (2004)

152. A. Brown, J. Hellerstein, M. Hogstrom, T. Lau, S. Lightstone, P. Shum, M.P. Yost, Benchmarking autonomic capabilities: promises and pitfalls, in *Proceedings of the 1st International Conference on Autonomic Computing (ICAC 2004)* (2004)

153. A. Brown, G. Kar, A. Keller, An active approach to characterizing dynamic dependencies for problem determination in a distributed environment, in *IFIP/IEEE International Symposium on Integrated Network Management*, Seattle, WA, May 2001, pp. 377–390

154. A. Brown, C.L. Chung, D.A. Patterson, Including the human factor in dependability benchmarks, in *DSN 2002 Workshop on Dependability Benchmarking* (2002)

155. A. Brown, D.A. Patterson, Towards availability benchmarks: a case study of software raid systems, in *Proceedings of the 2000 USENIX Annual Technical Conference* (2000)

156. A. Brown, D.A. Patterson, To err is human, in *First Workshop on Evaluating and Architecting System Dependability (EASY)* (2001)

157. E. Brown, J. Place, A. van de Liefvoort, Generating matrix exponential random variates. Simulation **70**, 224–230 (1998)

158. R.G. Brown, P.Y.C. Hwang, *Introduction to Random Signals and Applied Kalman Filtering* (Wiley, New York, 1997)

159. M. Broy, B. Jonsson, J.-P. Katoen, M. Leucker, A. Pretschner, *Model-Based Testing of Reactive Systems: Advanced Lectures*. Lecture Notes in Computer Science (Springer, New York, 2005)

160. R. Bryant, Graph-based algorithms for boolean function manipulation. IEEE Trans. Comput. **C-35**(8), 677–691 (1986)

161. K. Buchacker, M. Dal Cin, H.-J. Hoxer, R. Karch, V. Sieh, O. Tschache, Reproducible dependability benchmarking experiments based on unambiguous benchmark setup descriptions, in *Proceedings of the IEEE/IFIP 2003 International Conference on Dependable Systems and Networks (DSN 2003)* (2003)

162. P. Buchholz, Compositional analysis of a markovian process algebra, in *Proceedings of the 2nd Process Algebra and Performance Modelling Workshop*, ed. by U. Herzog, M. Rettelbach (1994)

163. P. Buchholz, Hierarchical structuring of superposed GSPNs. IEEE Trans. Softw. Eng. **25**(2), 166–181 (1999)

164. P. Buchholz, M. Fischer, P. Kemper, Distributed steady state analysis using Kronecker algebra, in *Proceedings of the 3rd International Conference on the Numerical Solution of Markov Chains (NSMC'99)*, Zaragoza, Spain, Sept 1999, pp. 76–95

165. P. Buchholz, P. Kemper, Numerical analysis of stochastic marked graphs, in *Proceedings of International Workshop on Petri Nets and Performance Models* (IEEE-Computer Society Press, Durham, 1995), pp. 32–41

166. D. Bui, A. Dupas, M. Le Pallec, Packet delay variation management for a better ieee1588v2 performance, in *International IEEE Symposium on Precision Clock Synchronization for Measurement, Control and Communication*, Brescia, Oct 2009, pp. 1–6

167. R. Buyya, R. Ranjan, R.N. Calheiros, Modeling and simulation of scalable cloud computing environments and the CloudSim toolkit: challenges and opportunities, in *Proceedings of the 7th High Performance Computing and Simulation (HPCS 2009) Conference* (2009)

168. J.P. Buzen, A.W. Shum, MASF—multivariate adaptive statistical filtering, in *International Computer Measurement Group Conference*, Nashville, TN, Dec 1995, pp. 1–10

169. C.M.T. Calafate, P. Manzoni, A multi-platform programming interface for protocol development. *Euromicro Conference on Parallel, Distributed, and Network-Based Processing* (2003), p. 243

170. J. Campos, S. Donatelli, M. Silva, Structured solution of stochastic dssp systems, in *Proceedings of International Workshop on Petri Nets and Performance Models* (IEEE-Computer Society Press, St Malo, 1997), pp. 91–100

171. G. Candea, E. Kiciman, S. Zhang, P. Keyani, A. Fox, Jagr: an autonomous self-recovering application server, in *Autonomic Computing Workshop*, 25 June 2003, pp. 168–177

172. L. Cardelli, On process rate semantics. J. Theor. Comput. Sci. **391**(3), 190–215 (2008)

173. R. Carmo, L. de Carvalho, E. de Souza e Silva, M. Diniz, R. Muntz, Tangram-ii: a performability modeling environment tool, in *Computer Performance Evaluation Modelling Techniques and Tools*. Lecture Notes in Computer Science, vol. 1245 (Springer, Berlin, 1997), pp. 6–18

174. S. Carpenter, B. Walker, J.M. Anderies, N. Abel, From metaphor to measurement: resilience of what to what? Ecosystems **4**, 765–781 (2001)

175. J. Carreira, H. Madeira, J.G. Silva, Xception: a technique for the experimental evaluation of dependability in modern computers. IEEE Trans. Softw. Eng. **24**(2), 125–136 (1998)

176. G. Casale, R.R. Muntz, G. Serazzi, Special issue on tools for computer performance modeling and reliability analysis. SIGMETRICS Perform. Eval. Rev. **36**, 2–3 (2009)

177. E. Casalicchio, E. Galli, Metrics for quantifying interdependencies, in *IFIP International Federation for Information Processing*, vol. 290, edited by M. Papa, S. Shenoi (Springer, Boston, 2009), pp. 215–227

178. E. Casalicchio, E. Galli, S. Tucci, Federated agent-based modeling and simulation approach to study interdependencies in IT critical infrastructures, in *11th IEEE International Symposium on Distributed Simulation and Real-Time Applications*, Oct 2007, pp. 182–189

179. E. Casalicchio, E. Galli, S. Tucci, Macro and micro agent-based modelling and simulation of critical infrastructures, in *Complexity in Engineering (COMPENG '10)*, Rome, Italy, Feb 2010, pp. 79–81

180. K. Cassidy, K. Gross, A. Malekpour, Advanced pattern recognition for detection of complex software aging phenomena in online transaction processing servers, in *Proceedings of the International Conference on Dependable Systems and Networks, 2002*, DSN 2002, pp. 478–482

181. V. Castelli, R.E. Harper, P. Heidelberger, S.W. Hunter, K.S. Trivedi, K. Vaidyanathan, W.P. Zeggert, Proactive management of software aging. IBM J. Res. Dev. **45**(2), 311–332 (2001)

182. U. Catalyürek, C. Aykanat, Hypergraph-partitioning-based decomposition for parallel sparse-matrix vector multiplication. IEEE Trans. Parallel. Distrib.Syst. **10**(7), 673–693 (1999)

183. The computer failure data repository (CFDR) (2008)

184. E. Cecchet, A. Chanda, S. Elnikety, J. Marguerite, W. Zwaenepoel, Performance comparison of middleware architectures for generating dynamic web content, in *Proceedings of the ACM/IFIP/USENIX 4th International Middleware Conference*, Rio de Janeiro, Brazil (2008)

185. Celtic, TWINBOARD: two-nodes IP network for better and optimized aggregation, routing and delivery, European Celtic project (2007–2009), http://www.celtic-initiative.org/Projects/TWINBOARD/default.as

186. Center for Internet Security, The CIS Security Metrics v 1.0.0, May 2009

187. R. Chandra, R.M. Lefever, M. Cukier, W.H. Sanders, Loki: a state-driven fault injector for distributed systems, in *2000 International Conference on Dependable Systems and Networks (DSN 2000) (Formerly FTCS-30 and DCCA-8) (DSN 2000)*, New York, NY, USA, June (IEEE Computer Society, 2000), pp. 237–242

188. R. Chellappa, A. Jennigs, N. Shenoy, A review on current work in mobility prediction for wireless networks, in *Proceedings of the 3rd Asian International Mobile Computing Conference* (Kasetsart University, 2004)

189. D. Chen, S. Dharmaraja, D. Chen, L. Li, K. Trivedi, R. Some, A. Nikora, Reliability and availability analysis for the jpl remote exploration and experimentation system, in *Conference on Dependable Systems and Networks, 2002. DSN 2002. Proceedings of the International*, pp. 337–342

190. J. Chen, J.S. Thorp, I. Dobson, Cascading dynamics and mitigation assessment in power system disturbances via a hidden failure model. Int. J. Electr. Power Energy Syst. **27**(4), 318–326 (2005)

191. L. Chen, A. Avizienis, On the implementation of N-version programming for software fault tolerance during program execution, in *1st International Computer Software and Applications Conference, (COMPSAC 77)* (New York, 1977), pp. 149–155

192. M. Chen, E. Kiciman, E. Fratkin, E. Brewer, A. Fox, Pinpoint: problem determination in large, dynamic, internet services, in *Proceedings of the 2002 International Conference on Dependable Systems and Networks*, Bethesda, MD, June 2002, pp. 595–604

193. M. Chen, A. Zheng, J. Lloyd, M. Jordan, E. Brewer, Failure diagnosis using decision trees, in *Proceedings of the First International Conference on Autonomic Computing*, New York, May 2004, pp. 36–43

194. Y. Chen, A. Das, W. Qin, A. Sivasubramaniam, Q. Wang, N. Gautam, Managing server energy and operational costs in hosting centers, in *SIGMETRICS International Conference* (2005)

195. L. Chen-Ching, T. Chee-Wooi, M. Govindarasu, Cybersecurity of SCADA systems: vulnerability assessment and mitigation, in *IEEE/PES Power Systems Conference and Exposition (PSCE '09)*, March 2009, pp. 1–3

196. B.H.C. Cheng, R. de Lemos, H. Giese, P. Inverardi, J. Magee, Software engineering for self-adaptive systems: a research roadmap, in *Software Engineering for Self-Adaptive Systems* (2009)

197. K.-T. Cheng, S.-Y. Huang, W.-J. Dai, Fault emulation: a new methodology for fault grading. IEEE Trans. CAD Integr. Circuits Syst. **18**(10), 1487–1495 (1999)

198. L. Cherkasova, K.M. Ozonat, N. Mi, J. Symons, E. Smirni, Anomaly? Application change? or Workload change? Towards automated detection of application performance anomaly and change, in *2008 IEEE International Conference on Dependable Systems and Networks with FTCS and DCC DSN*, Anchorage, Alaska, June 2008, pp. 452–461

199. Artemis-ju-100022 chess—composition with guarantees for high-integrity embedded software components assembly, http://www.chess-project.org

200. S. Chiaradonna, F. Di Giandomenico, P. Lollini, Evaluation of critical infrastructures: challenges and viable approaches, in *Architecting Dependable Systems V*. Lecture Notes in Computer Science, vol. 5135, ed. by R. De Lemos, F. Di Giandomenico, C. Gacek, H. Muccini, M. Vieira (Springer, Heidelberg, 2008), pp. 52–77

201. S. Chiaradonna, F. Di Giandomenico, P. Lollini, Interdependency analysis in electric power systems, in *3rd International Workshop on Critical Information Infrastructures Security (CRITIS 2008)*. Lecture Notes in Computer Science, vol. 5508, ed. by R. Setola, S. Geretshuber (Springer, Berlin, 2009), pp. 60–71

202. S. Chiaradonna, F. Di Giandomenico, P. Lollini, Definition, implementation and application of a model-based framework for the analysis of interdependencies in electric power systems. Int. J. Crit. Infrastruct. **4**(1), 24–40 (2011)

203. S. Chiaradonna, P. Lollini, F. Di Giandomenico, On a modeling framework for the analysis of interdependencies in electric power systems, in *IEEE/IFIP 37th International Conference on Dependable Systems and Networks (DSN 2007)*, June 2007, Edinburgh, UK, pp. 185–195

204. S. Chiaradonna, P. Lollini, F. Di Giandomenico, Modelling framework of an instance of the electric power system: functional description and implementation. Technical report RCL071202—version 4, University of Florence, Dip. Sistemi Informatica, RCL Group,

March 2010, http://dcl.isti.cnr.it/Documentation/Papers/Techreports.htm

205. R. Chillarege, Orthogonal defect classification, Section 9, *Handbook of Software Reliability Engineering* (IEEE Computer Society Press/McGraw-Hill, New York, 1995)

206. R. Chillarege, I.S. Bhandari, J.K. Chaar, M.J. Halliday, D. Moebus, B. Ray, M. Wong, Orthogonal defect classification—a concept for in-process measurement. IEEE Trans. Softw. Eng. **18**(11), 943–956 (1992)

207. R. Chillarege, N. Bowen, Understanding large system failures-a fault injection experiment, in *Fault-Tolerant Computing, 1989. FTCS-19. Digest of Papers. Nineteenth International Symposium on*, June 1989, pp. 356–363

208. G. Chiola, G. Franceschinis, R. Gaeta, M. Ribaudo, Greatspn 1.7: graphical editor and analyzer for timed and stochastic petri nets. Perform. Eval. **24**, 47–68 (1995)

209. G.S. Choi, R.K. Iyer, FOCUS: an experimental environment for fault sensitivity analysis. IEEE Trans. Comput. **41**(12), 1515–1526 (1992)

210. H. Choi, V.G. Kulkarni, K.S. Trivedi, Markov regenerative stochastic petri nets. Perform. Eval. **20**(1–3), 337–357 (1994)

211. M. Choraś, A. Flizikowski, R. Kozik, W. Holubowicz, Decision aid tool and ontology-based reasoning for critical infrastructure vulnerabilities and threats analysis, in *Proceedings of the 4th International Conference on Critical Information Infrastructures Security*, CRITIS'09 (Springer, Berlin, 2010), pp. 98–110

212. H. Christiansson, E. Luiijf, Creating a European scada security testbed, in *Critical Infrastructure Protection*. IFIP International Federation for Information Processing, vol. 253, ed. by E. Goetz, S. Shenoi (Springer, Boston, 2007), pp. 237–247

213. J. Christmansson, R. Chillarege, Generation of an error set that emulates software faults, in *26th IEEE Fault Tolerant Computing Symposium, FTCS-26*, Sendai, Japan, June 1996, pp. 304–313

214. J. Christmansson, R. Chillarege, Generation of an error set that emulates software faults based on field data, in *Proceedings of the Twenty-Sixth International Symposium on Fault-Tolerant Computing*, Washington, 25–27, June 1996, pp. 304–313

215. C. Ciardo, M. Tilgner, On the use of kronecker operators for the solution of generalized stochastic petri nets. Technical report 96-35. Institute for Computer Applications in Science and Engineering, Hampton, VA, May 1996

216. G. Ciardo, R. German, C. Lindemann, A characterization of the stochastic process underlying a stochastic petri net. IEEE Trans. Softw. Eng. **20**(7), 506–515 (1994)

217. G. Ciardo, G. Luettgen, R. Siminiceanu, Saturation: an efficient strategy for symbolic state-space generation, in *Proceedings of TACAS'01*, Genova, Italy. Lecture Notes in Computer Science, vol. 2031 (Springer, 2001), pp. 328–342

218. G. Ciardo, A. Miner, Smart: simulation and markovian analyzer for reliability and timing, in *Proceedings of IEEE International Computer Performance and Dependability Symposium, 1996*, , Sept 1996, p. 60

219. G. Ciardo, A. Miner, A data structure for the efficient Kronecker solution of GSPNs, in *PNPM'99*, ed. by P. Buchholz, M. Silva (IEEE Computer Society, New York, 1999), pp. 22–31

220. G. Ciardo, A.S. Miner, Efficient reachability set generation and storage using decision diagrams, in *Application and Theory of Petri Nets 1999 (Proceedings of the 20th International Conference on Applications and Theory of Petri Nets)* (Springer, 1999), pp. 6–25

221. G. Ciardo, D. Nicol, K. Trivedi, Discrete-event simulation of fluid stochastic petri nets. IEEE Trans. Softw. Eng. **25**(2), 207–217 (1999)

222. I. Cidon, I. Gopal, Paris: an approach to integrated high-speed private networks. Int. J. Digit. Analog Cabled Syst. **1**, 77–86 (1988)

223. P. Civera, L. Macchiarulo, M. Rebaudengo, M.S. Reorda, M. Violante, New techniques for efficiently assessing reliability of SOCs. Microelectron. J. **34**(1), 53–61 (2003)

224. C. Clark, K. Fraser, S. Hand, J.G. Hansen, E. Jul, C. Limpach, I. Pratt, A. Warfield, Live migration of virtual machines, in *Proceedings of the USENIX Symposium on Networked System Design and Implementation* (2005)

225. G. Clark, T. Courtney, D. Daly, D. Deavours, S. Derisavi, J.M. Doyle, W.H. Sanders, P. Webster, The möbius modeling tool, in *9th International Workshop on Petri Nets and Performance Models*, 2001, pp. 241–250

226. G. Clark, S. Gilmore, J. Hillston, N. Thomas, Experiences with the pepa performance modelling tools, in *UKPEW'98, Proceedings of the 14th UK Performance Engineering Workshop*, Edinburgh, July 1998

227. E. Clarke, M. Fujita, P. McGeer, K. McMillan, J. Yang, X. Zhao, Multi-terminal binary decision diagrams: an efficient data structure for matrix representation, in *IWLS: International Workshop on Logic Synthesis*, Tahoe City, May 1993

228. L. Cloth, B.R. Haverkort, Model checking for survivability, in *2nd International Conference on the Quantitative Evaluation of Systems* (IEEE Computer Society, 2005), pp. 145–154

229. A. Coccoli, P. Urban, A. Bondavalli, A. Schiper, Performance analysis of a consensus algorithm combining stochastic activity networks and measurements, in *IEEE DSN— International Conference on Dependable Systems and Networks (IPDS Track)* (IEEE Computer Society Press, Washington, 2002), pp. 551–560

230. I. Cohen, J.S. Chase, M. Goldszmidt, T. Kelly, J. Symons, Correlating instrumentation data to system states: a building block for automated diagnosis and control, in *Proceedings of the 6th Conference on Symposium on Operating Systems Design & Implementation*, vol. 6, San Francisco, CA, Dec 2004, pp. 231–244

231. I. Cohen, S. Zhang, M. Goldszmidt, J. Symons, T. Kelly, A. Fox, Capturing, indexing, clustering, and retrieving system history, in *Proceedings of the Twentieth ACM Symposium on Operating Systems Principles*, Brighton, UK, Oct 2005, pp. 105–118

232. Commission of the European Communities. Information Technology Security Evaluation Manual (ITSEM) (1993)

233. Common Criteria Common criteria for information technology security evaluation: user guide (1999)

234. C. Constantinescu, Trends and challenges in VLSI circuit reliability. IEEE Micro **23**, 14–19 (2003)

235. C. Constantinescu, Neutron SER characterization of microprocessors, in *Proceedings 2005 International Conference on Dependable Systems and Networks (DSN 2005)* (IEEE Computer Society, Yokohama, 2005), pp. 754–759

236. C. Constantinescu, Neutron ser characterization of microprocessors, in *Proceedings of the 2005 International Conference on Dependable Systems and Networks (DSN 2005)* (2005)

237. M. Conti, G. Maselli, G. Turi, S. Giordano, Cross-layering in mobile ad hoc network design. Computer **37**(2), 48–51 (2004)

238. B.F. Cooper, A. Silberstein, E. Tam, R. Ramakrishnan, R. Sears, Benchmarking cloud serving systems with ycsb, in *Proceedings of the 1st ACM Symposium on Cloud computing (SoCC '10)* (2010)

239. V. Cortellessa, R. Mirandola, Prima-uml: a performance validation incremental methodology on early uml diagrams. Sci. Comput. Program. **44**(1), 101–129 (2002)

240. D. Costa, R. Barbosa, R. Maia, F. Moreira, *Dependability Benchmarking for Computer Systems*, chapter DeBERT—Dependability Benchmarking for Embedded Real-time Off-the-Shelf Components for Space Applications (Wiley, New York, 2008), pp. 255–283

241. D. Costa, H. Madeira, J. Carreira, J.G. Silva, Xception: a software implemented fault injection tool, in *Fault Injection Techniques and Tools for Embedded Systems Reliability Evaluation*, ed. by A. Benso, P. Prinetto (Springer, New York, 2004), pp. 125–139

242. T. Courtney, S. Gaonkar, K. Keefe, E.W.D. Rozier, W.H. Sanders. Möbius 2.3: an extensible tool for dependability, security, and performance evaluation of large and complex system models, in *39th Annual IEEE/IFIP International Conference on Dependable Systems and*

Networks (DSN 2009), Estoril, Lisbon, Portugal, June 29–July 2 2009, pp. 353–358

243. P. Courtois, *Decomposability, Queueing and Computer System Applications*. ACM Monograph Series (1977)

244. G.F. Cretu-Ciocarlie, M. Budiu, M. Goldszmidt, Hunting for problems with Artemis, in *USENIX Workshop on Analysis of System Logs*, San Diego, CA, Dec 2008

245. CRUTIAL, European Project CRUTIAL—Critical utility infrastructural resilience, http://crutial.erse-web.it

246. Crutial—critical utility infrastructural resilience (project ist-fp6-027513), http://crutial.cesiricerca.it (2006)

247. CRUTIAL Consortium, The crutial modelling framework (final version). Technical report, http://crutial.rse-web.it/Dissemination/DELIVERABLES-OF-THE-PROJECT, March 2009

248. CRUTIAL Consortium, On eps-ict interdependencies in the testbeds. Technical report, http://crutial.rse-web.it/Dissemination/DELIVERABLES-OF-THE-PROJECT, March 2009

249. C. Csallner, Y. Smaragdakis, JCrasher: an automatic robustness tester for Java. Softw. Pract. Exp. **34**, 1025–1050 (2004)

250. B. Cully, G. Lefebvre, D. Meyer, M. Feeley, N. Hutchinson, A. Warfield, Remus: high availability via asynchronous virtual machine replication, in *Proceedings of the USENIX Symposium on Networked System Design and Implementation* (2008), pp. 161–174

251. A. Cumani, On the canonical representation of homogeneous Markov processes modelling failure-time distributions. Microelectron. Reliab. **22**, 583–602 (1982)

252. I. Cunha, J. Almeida, V. Almeida, M. Santos, Self-adaptive capacity management for multi-tier virtualized environments, in *IM'07: Proceedings of the 10th IFIP/IEEE International Symposium on Integrated Network Management*, Munich, Germany (2007), pp. 129–138

253. K. Czarnecki, S. Helsen, Classification of model transformation approaches, in *Proceedings of OOPSLA'03*, Anaheim, CA, USA (2003)

254. Software reliability dataset (2008)

255. A. Daidone, F. Di Giandomenico, A. Bondavalli, S. Chiaradonna, Hidden Markov models as a support for diagnosis: formalization of the problem and synthesis of the solution, in *Proceedings of the 25th IEEE Symposium on Reliable Distributed Systems*, Leeds, UK, Oct 2006, pp. 245–256

256. D. Daly, D.D. Deavours, J.M. Doyle, A.J. Stillman, P.G. Webster, Möbius: an extensible tool for performance and dependability modeling, in *11th International Conference, TOOLS 2000*. Lecture Notes in Computer Science, pp. 332–336 (2000)

257. P.R. D'Argenio, H. Hermanns, J.-P. Katoen, R. Klaren, Modest: a modelling language for stochastic timed systems, in *Process Algebra and Probabilistic Methods. Performance Modelling and Verification*. Lecture Notes in Computer Science, vol. 2165, ed. by L. de Alfaro, S. Gilmore (Springer, Berlin, 2001), pp. 87–104

258. S.R. Das, S. Mukherjee, E.M. Petriu, M.H. Assaf, M. Sahinoglu, W.-B. Jone, An improved fault simulation approach based on verilog with application to iscas benchmark circuits, in *Instrumentation and Measurement Technology Conference IMTC 2006)*, April 2006, pp. 1902–1907

259. A. Dasilva, J.-F. Martínez, L. López, A.B. García, L. Redondo, Exhaustif: a fault injection tool for distributed heterogeneous embedded systems, in *Proceedings of the 2007 Euro American conference on Telematics and Information Systems, EATIS 2007*, Faro, Portugal, May 14–17, 2007, ed. by R.P.C. do Nascimento, A. Berqia, P. Serendero, E. Carrillo (2007), p. 17

260. A. David, S. Larry, The least variable phase-type distribution is erlang. Stoch Models **3**, 467–473 (1987)

261. R. David, H. Alla. On hybrid petri nets. Discret. Event Dyn. Syst. **11**, 9–40 (2001)

262. R. David, H. Alla, *Discrete, Continuous, and Hybrid Petri Nets*, 2nd edn. (Springer, Berlin, 2010)

263. M.H.A. Davis, Piecewise-deterministic markov processes: a general class of non-diffusion stochastic models. J. R. Stat. Soc. Ser. B (Methodol.) **46**(3), 353–388 (1984)

264. S. Dawson, F. Jahanian, T. Mitton, A software fault injection tool on real-time mach, in *IEEE Real-Time Systems Symposium* (1995), pp. 130–140

265. S. Dawson, F. Jahanian, T. Mitton, Experiments on six commercial TCP implementations using a software fault injection tool. Softw. Pract. Exp. **27**(12):1385–1410 (1997)

266. S. Dawson, F. Jahanian, T. Mitton, T.-L. Tung, Testing of fault-tolerant and real-time distributed systems via protocol fault injection, in *Proceedings of the Twenty-Sixth International Symposium on Fault-Tolerant Computing*, Washington, June 25–27 (1996), IEEE, pp. 404–414

267. T. Dayar, Iterative methods based on splittings for stochastic automata networks. Eur. J. Oper. Res. **110**, 166–186 (1998)

268. Dbench Project, project funded by the European community under the information society technology programme (1998–2002), http://www.dbench.org/

269. D. de Andres, J.C. Ruiz, D. Gil, P.J. Gil, Run-time reconfiguration for emulating transient faults in VLSI systems, in *Proceedings 2006 International Conference on Dependable Systems and Networks DSN 2006, Dependable Computing and Communications Symposium (DCCS)*, Philadelphia, Pennsylvania, USA (IEEE Computer Society, New York, 2006), pp. 291–300

270. D. de Andrés, J.C. Ruiz, D. Gil, P.J. Gil, Fault emulation for dependability evaluation of VLSI systems. IEEE Trans. Very Large Scale Integr. (VLSI) Syst. **16**(4), 422–431 (2008)

271. H. de Meer, K.S. Trivedi, Guarded repair of dependable systems. Theor. Comp. Sci. **128**, 179–210 (1994)

272. E. de Souza e Silva, D.R. Figueiredo, R.M. Leão, The TANGRAMII integrated modeling environment for computer systems and networks. SIGMETRICS Perform. Eval. Rev. **36**(4), 64–69 (2009)

273. E. de Souza e Silva, H. Gail, Transient solutions for markov chains. *Computational Probability* (Kluwer, Dordrecht, 2000), pp. 43–79

274. J. Dean, Software engineering advice from building large-scale distributed systems. Stanford CS295 class lecture (2007), http://research.google.com/people/jeff/stanford-295-talk.pdf

275. J. Dean, S. Ghemawat, Mapreduce: simplified data processing on large clusters. Commun. ACM **51**, 107–113 (2008)

276. D. Deavours, G. Clark, T. Courtney, D. Daly, S. Derisavi, J. Doyle, W. Sanders, P. Webster, The moebius framework and its implementation. IEEE Trans. Softw. Eng. **28**(10), 956–969 (2002)

277. D. Deavours, W. Sanders, "On-the-fly" solution techniques for stochastic petri nets and extensions. IEEE Trans. Softw. Eng. **24**(10), 889–902 (1998)

278. D.D. Deavours, W.H. Sanders, An efficient disk-based tool for solving large markov models. Perform. Eval. **33**(1), 67–84 (1998)

279. J. Dejun, G. Pierre, C.-H. Chi, EC2 performance analysis for resource provisioning of service-oriented applications, in *Proceedings 3rd Workshop on Non-Functional Properties and SLA Management in Service-Oriented Computing*, Nov 2009

280. A.H. Dekker, Simulating network robustness for critical infrastructure networks, in *Proceedings of the Twenty-Eighth Australasian conference on Computer Science*, vol. 38, ACSC '05, Darlinghurst, Australia (Australian Computer Society, 2005), pp. 59–67

281. A.H. Dekker, B. Colbert, Scale-free networks and robustness of critical infra-structure networks, in *Proceedings of the 7th Asia-Pacific Conference on Complex Systems (Complex 2004)*, Cairns, Australia, Dec 2010

282. S. Delamare, A.A. Diallo, C. Chaudet, High-level modelling of critical infrastructures' interdependencies. Int. J. Critic. Infrastruct. **5**(1/2), 100–119 (2009)

283. T.A. DeLong, B.W. Johnson, J.A. Profeta III, A fault injection technique for VHDL behavioral-level models. IEEE Des. Test. Comput. **13**(4), 24–33 (1996)

284. C. Demichelis, P. Chimento, IP Packet Delay Variation Metric for IP Performance Metrics (IPPM). RFC 3393 (Proposed Standard), Nov 2002

285. R. DeMillo, D. Guindi, W. McCracken, A. Offutt, K. King, An extended overview of the Mothra software testing environment, in *Proceedings of the Second Workshop on Software Testing, Verification, and Analysis* (1988), pp. 142–151

286. A. Dempster, N. Laird, D. Rubin, Maximum likelihood from incomplete data via the em algorithm. J. Roy. Stat. Soc. Ser. B (Methodol.) **39**, 1–38 (1997)

287. P. Denning, J.P. Buzen, The operational analysis of queueing network models. ACM Comput. Surv. **10**(3), 225–261 (1978)

288. Department of Defense, Trusted computer system evaluation criteria (1985)

289. S. Derisavi, A symbolic algorithm for optimal markov chain lumping, in *TACAS 2007*, ed. by O. Grumberg, M. Huth (Springer, Heidelberg, 2007), pp. 139–154

290. A.C. Dias Neto, R. Subramanyan, M. Vieira, G.H. Travassos, A survey on model-based testing approaches: a systematic review, in *Proceedings of the 1st ACM International Workshop on Empirical Assessment of Software Engineering Languages and Technologies, WEASELTech '07* (2007), pp. 31–36

291. Database Language SQL (1992)

292. E.W. Dijkstra, Hierarchical ordering of sequential processes. Acta Inform. **1**, 115–138 (1971)

293. J. Dilley, Web server workload charaterization, in HP Technical report, HPL-96-160 (1996)

294. N. Dingle, An empirical study of the scalability of performance analysis tools in the cloud, in *Proceedings of the 26th UK Performance Engineering Workshop (UKPEW'10)*, Warwick, July 2010, pp. 9–16

295. N. Dingle, P. Harrison, W. Knottenbelt, Response time densities in generalised stochastic petri net models, in *Proceedings of the 3rd International Workshop on Software and Performance (WOSP'02)*, Rome, July 24th–26th 2002, pp. 46–54

296. N. Dingle, P. Harrison, W. Knottenbelt, HYDRA: hypergraph-based distributed response-time analyser, in *Proceedings of the International Conference on Parallel and Distributed Processing Techniques and Applications (PDPTA'03)*, Las Vegas, June 23rd–26th 2003, pp. 215–219

297. N. Dingle, P. Harrison, W. Knottenbelt, Uniformization and hypergraph partitioning for the distributed computation of response time densities in very large Markov models. J. Parallel. Distr. Com. **64**(8), 908–920 (2004)

298. S. Distefano, A. Puliafito, Dependability evaluation with dynamic reliability block diagrams and dynamic fault trees. IEEE Trans. Depend. Secure Comput. **6**(1), 4–17 (2008)

299. A. Dixit, R. Heald, A. Wood, Trends from ten years of soft error experimentation, in *5th IEEE Workshop on Silicon Errors in Logic—Systems Effects (SELSE-5)*, March 2009

300. I. Dobson, B. Carreras, V. Lynch, D.E. Newman, Complex systems analysis of series of blackouts: cascading failure, critical points, and self-organization. CHAOS **17**(2) (2007)

301. I. Dobson, B.A. Carreras, V. Lynch, D.E. Newman, An initial model for complex dynamics in electric power system blackouts, in *34th IEEE Hawaii International Conference on System Sciences (CD-ROM)*, Maui, Hawaii, Jan 2001, 9 pp

302. J. Dobson, B. Randell, Building reliable secure systems out of unreliable insecure components, in *IEEE Conference on Security and Privacy* (IEEE, Oakland, 1986), pp. 187–193

303. T. Dohi, K. Goseva-Popstojanova, K.S. Trivedi, Analysis of software cost models with rejuvenation, in *Fifth IEEE International Symposium on High Assurance Systems Engineering, 2000. HASE 2000* (2000), pp. 25–34

304. T. Dohi, K. Goseva-Popstojanova, K.S. Trivedi, Statistical non-parametric algorithms to estimate the optimal software rejuvenation schedule, in *2000 Pacific Rim International Symposium on Dependable Computing, 2000: Proceedings* (2000), pp. 77–84

305. T. Dohi, K. Goseva-Popstojanova, K.S. Trivedi, Estimating software rejuvenation schedules in high-assurance systems. Comput. J. **44**(6), 473–485 (2001)

306. S. Donatelli, Superposed generalised stochastic petri nets: definition and efficient solution, in *Proceedings of 15th International Conference on Application and Theory of Petri Nets*, ed. by M. Silva (1994)

307. S. Donatelli, P. Kemper, Integrating synchronization with priority into a kronecker representation. Perform. Eval. **44**(1–4), 73–96 (2001)
308. Drupal (2009), http://www.drupal.org/
309. X. Du, Y. Qi, D. Hou, Y. Chen, X. Zhong, Modeling and performance analysis of software rejuvenation policies for multiple degradation systems, in *33rd Annual IEEE International Computer Software and Applications Conference, 2009. COMPSAC '09*, vol. 1, July 2009, pp. 240–245
310. S. Duan, S. Babu, Guided problem diagnosis through active learning, in *Proceedings of the 2008 International Conference on Autonomic Computing*, Chicago, IL, June 2008, pp. 45–54
311. R.B. Duffey, The quantification of resilience: learning environments and managing risk, in *3rd Symposium on Resilience Engineering* (Antibes, France, 2008), pp. 75–81
312. J. Durães, H. Madeira, Characterization of operating systems behavior in the presence of faulty drivers through software fault emulation, in *9th Pacific Rim International Symposium on Dependable Computing (PRDC 2002)* (IEEE Computer Society, New York, 2002), pp. 201–209
313. J. Durães, H. Madeira, Definition of software fault emulation operators: a field data study, in *Proceedings 2003 International Conference on Dependable Systems and Networks (DSN 2003)* (IEEE Computer Society, San Francisco, 2003), pp. 105–114
314. J. Durães, H. Madeira, Generic faultloads based on software faults for dependability benchmarking, in *Proceedings of the IEEE/IFIP 2004 International Conference on Dependable Systems and Networks (DSN 2004)* (2004)
315. J. Durães, H. Madeira, Emulation of software faults: a field data study and a practical approach. IEEE Trans. Softw. Eng. **32**(11), 849–867 (2006)
316. J. Durães, M. Vieira, H. Madeira, Dependability benchmarking of web-servers, in *Proceedings of the 23rd International Conference on Computer Safety, Reliability and Security (SAFECOMP 2004)* (2004)
317. E-UTRA, Evolved Universal Terrestrial Radio Access (E-UTRA) and Evolved Universal Terrestrial Radio Access Network (E-UTRAN); Overall description, Stage 2, release 9, version 9.1.0, Oct 2010
318. K. Echtle, M. Leu, The EFA fault injector for fault-tolerant distributed system testing, in *Proceedings of the IEEE Workshop on Fault-Tolerant Parallel and Distributed Systems* (IEEE Computer Society Press, Amherst, 1992), pp. 28–35
319. K. Echtle, M. Leu, Test of fault tolerant distributed systems by fault injection, in *Proceedings of IEEE Workshop on Fault-Tolerant Parallel and Distributed Systems*, June 1994, pp. 244–251
320. EEC Directive 90/C81/01, Emission test cycles for the certification of light duty vehicles in Europe, EEC emission cycles (1999), http://www.dieselnet.com/standards/cycles
321. A. Ejlali, S.G. Miremadi, Error propagation analysis using FPGA-based SEU-fault injection. Microelectron. Reliab. **48**(2), 319–328 (2008)
322. Y. el-Khamra, H. Kim, S. Jha, M. Parashar, Exploring the performance fluctuations of HPC workloads on clouds, in *2nd IEEE International Conference on Cloud Computing Technology and Science (CloudCom)* (2010)
323. I. Elia, J. Fonseca, M. Vieira, Comparing SQL injection detection tools using attack injection: an experimental study, in *21st International Symposium on Software Reliability Engineering, ISSRE-2010* (IEEE, San Jose, 2010)
324. R. Elling, I. Pramanick, J. Mauro, W. Bryson, D. Tang, *Analytical RAS Benchmarks, Dependability Benchmarking for Computer Systems* (Wiley, Chichester, 2008)
325. Embedded Microprocessor Benchmark Consortium, Eembc homepage (2011), http://www.eembc.org/home.php
326. EMC, Automating root cause analysis: EMC Ionix codebook correlation technology vs. rule-based analysis. Technical report h5964, EMC, Nov 2009
327. R. Enders, T. Filkorn, D. Taubner, Generating BDDs for symbolic model checking in CCS, in *Computer Aided Verification*, vol. 6 (Springer, Berlin, 1993), pp. 155–164

328. B.C. Ezell, Infrastructure vulnerability assessment model (I-VAM). Risk Anal. **27**(3), 571–583 (2007)
329. L. Falai, Observing, monitoring and evaluating distributed systems. Ph.D. thesis, University of Florence (2008)
330. L. Falai, A. Bondavalli, F. Di Giandomenico, Quantitative evaluation of distributed algorithms using the neko framework: the nekostat extension, in *Dependable Computing*. Lecture Notes in Computer Science, vol. 3747, ed. by C.A. Maziero, J.G. Silva, A.M.S. Andrade, F.M. de Assis Silva (Springer, Berlin, 2005), pp. 35–51
331. N. Falliere, L.O. Murchu, E. Chien, W32.sutxnet Dossier, http://www.symantec.com/content/en/us/enterprise/media/security_response/whitepapers/w32_stuxnet_dossier.pdf. Accessed 7 Oct 2010
332. E. Farr, R. Harper, L. Spainhower, J. Xenidis, A case for high availability in a virtualized environment (HAVEN), in *International Conference on Availability, Reliability and Security* (2008)
333. N. Feamster, H. Balakrishnan, Detecting BGP configuration faults with static analysis, in *Proceedings of the 2nd Conference on Symposium on Networked Systems Design & Implementation*, vol. 2, Boston, MA, May 2005, pp. 43–56
334. N. Fenton, Software measurement: a necessary scientific basis. IEEE Trans. Softw. Eng. **20**, 199–206 (1994)
335. N. Fenton, S.L. Pfleeger, *Software Metrics—A Rigorous and Practical Approach*, 2nd edn. (PWS Publishing, Boston, 1997)
336. P. Fernandes, Méthodes numériques pour la solution de systèmes Markoviens àà grand espace d'états. Ph.D. thesis (1998)
337. P. Fernandes, B. Plateau, W. Stewart, Efficient descriptor-vector multiplications in stochastic automata networks. JACM **3**(45), 381–414 (1998)
338. J.-C. Fernandez, L. Mounier, C. Pachon, A model-based approach for robustness testing, in *Testing of Communicating Systems*, (2005), pp. 333–348
339. F. Flammini (ed.), *Critical Infrastructure Security: Assessment, Prevention, Detection, Response* (WIT Press Royal, in press, 2011)
340. F. Flammini, V. Vittorini, N. Mazzocca, C. Pragliola, *A study on Multiformalism Modeling of Critical Infrastructures* (Springer, Berlin, 2009), pp. 336–343
341. J. Fonseca, M. Vieira, Mapping software faults with web security vulnerabilities, in *IEEE/ IFIP International Conference on Dependable Systems and Networks*, June 2008
342. J. Fonseca, M. Vieira, H. Madeira, Online detection of malicious data access using DBMS auditing, in *Proceedings of the 2008 ACM Symposium on Applied Computing, SAC '08*, pp. 1013–1020 (2008)
343. J. Fonseca, M. Vieira, H. Madeira, Vulnerability & attack injection for web applications, in *International Conference on Dependable Systems and Networks with FTCS and DCC— DSN 2009* (IEEE, Lisbon, 2009)
344. J. Fonseca, M. Vieira, H. Madeira, Vulnerability & attack injection for Web applications, in *The 39th Annual IEEE/IFIP International Conference on Dependable Systems and Networks, DSN 2009*, Lisbon, Portugal (2009)
345. J. Fonseca, M. Vieira, H. Madeira, The web attacker perspective—a field study, in *21st International Symposium on Software Reliability Engineering, ISSRE-2010* (IEEE, San Jose, 2010)
346. R. Fonseca, G. Porter, R. Katz, S. Shenker, I. Stoica, X-Trace: a pervasive network tracing framework, in *Proceedings of the Fourth USENIX Symposium on Networked Systems Design and Implementation (NSDI 2007)*, Cambridge, MA, April 2007, pp. 271–284
347. C. Forgy, Rete: a fast algorithm for the many pattern/many object pattern match problem. Artif. Intell. **19**(4), 17–37 (1982)
348. M. Fossi, E. Johnson, D. Turner, T. Mack, J. Blackbird, D. McKinney, M.K. Low, T. Adams, M.P. Laucht, J. Gough, Symantec report on the underground economy. Symantec Security report, Symantec (2008)

349. I.T. Foster, Globus toolkit version 4: software for service-oriented systems, in *Proceedings of the 2005 IFIP International Conference on Network and Parallel Computing* (2005), pp. 2–13
350. H. Fouchal, A. Rollet, A. Tarhini, Robustness testing of composed real-time systems. J. Comp. Methods Sci. Eng. **10**, 135–148 (2010)
351. Foundstone, Inc. Foundstone WSDigger (2011), http://www.foundstone.com/us/resources/proddesc/wsdigger.htm
352. J.-M. Fourneau, L. Kloul, F. Valois, Performance modelling of hierarchical cellular networks using pepa. Perform. Eval. **50**(2–3), 83–99 (2002)
353. K. Fowler, Dependability [reliability]. Instrum. Meas. Mag. IEEE **8**(4), 55–58 (2005)
354. G. Franceschinis, M. Gribaudo, M. Iacono, N. Mazzocca, V. Vittorini, DrawNET++: model objects to support performance analysis and simulation of systems, in *Computer Performance Evaluation: Modelling Techniques and Tools*. Lecture Notes in Computer Science, vol. 2324, ed. by T. Field, P. Harrison, J. Bradley, U. Harder (Springer, Berlin, 2002), pp. 55–60
355. M. Frank, P. Wolfe, An algorithm for quadratic programming. Naval Res. Logist. Q. **3**(1–2), 95–110 (1956)
356. F.C. Freiling, Introduction to security metrics, in *Dependability Metrics*, ed. by I. Eusgeld, F.C. Freiling, R. Reussner (Springer, Berlin, 2008), pp. 129–132
357. R. Fricks, C. Hirel, S. Wells, K. Trivedi, The development of an integrated modeling environment, in *Proceedings of the World Congress on Systems Simulation (WCSS '97)*, Singapore, Sept 1997, pp. 471–476
358. C. Fu, A. Milanova, B.G. Ryder, D.G. Wonnacott, Robustness testing of Java server applications. IEEE Trans. Softw. Eng. **31**, 292–311 (2005)
359. G. Fuchs, R. German, UML2 activity diagram based programming of wireless sensor networks, in *ICSE Workshop on Software Engineering for Sensor Network Applications, SESENA '10* (ACM, New York, 2010), pp. 8–13
360. M. Fujita, P. McGeer, J.-Y. Yang, Multi-terminal binary decision diagrams: an efficient data structure for matrix representation. Formal Methods Syst. Des. **10**(2/3), 149–169 (1997)
361. M. Fukushima, A modified frank-wolfe algorithm for solving the traffic assignment problem. Transp. Res. B Methodol. **18**(2), 169–177 (1984)
362. S. Gaisbauer, J. Kirschnick, N. Edwards, J. Rolia, VATS: virtualized-aware automated test service, in *QEST'08, Fifth International Conference on the Quantitative Evaluation of Systems* (IEEE, St Malo, 2008), pp. 93–102
363. A. Gandhi, M. Harchol-Balter, R. Das, C. Lefurgy, Optimal power allocation in server farms, in *International Joint Conference on Measurement and Modeling of Computer Systems* (2009)
364. Ganglia, Ganglia monitoring system (2007), http://ganglia.inf
365. D.A. Garbin, J.F. Shortle, Measuring resilience in network-based infrastructures, in *Critical Thinking: Moving from Infrastructure Protection to Infrastructure Resilience*, CIP Program Discussion Paper Series (George Mason University, 2007), pp. 73–86
366. S. Garg, Y. Huang, C. Kintala, K.S. Trivedi, Time and load based software rejuvenation: policy, evaluation and optimality, in *Proceedings of the 1st Fault-Tolerant Symposium* (1995), pp. 22–25
367. S. Garg, A. Puliafito, M. Telek, K.S. Trivedi, Analysis of software rejuvenation using markov regenerative stochastic petri nets, in *Proceedings of the Sixth International Symposium on Software Reliability Engineering*, Oct 1995, pp. 180–187
368. S. Garg, A. Puliafito, M. Telek, K.S. Trivedi, Analysis of preventive maintenance in transactions based software systems. IEEE Trans. Comput. **47**(1), 96–107 (1998)
369. S. Garg, A. van Moorsel, K. Vaidyanathan, K.S. Trivedi, A methodology for detection and estimation of software aging, in *Proceedings of the the Ninth International Symposium on Software Reliability Engineering*, Nov 1998, pp. 283–292

370. N. Gast, B. Gaujal, J.-Y. Le Boudec, Mean field for markov decision processes: from discrete to continuous optimization. Technical report, arXiv:1004.2342v2 (2010)

371. D. Gay, P. Levis, R. von Behren, M. Welsh, E. Brewer, D. Culler, The nesC language: a holistic approach to networked embedded systems. SIGPLAN Not **38**(5), 1–11 (2003)

372. E. Gelenbe, Product-form queueing networks with negative and positive customers. J. Appl. Probab. **28**(3), 656–663 (1991)

373. A. Georges, L. Eeckhout, Performance metrics for consolidated servers, in *Proceedings of HPCVirt 2010* (Association for Computing Machinery (ACM), 2010)

374. R. German, C. Kelling, A. Zimmermann, G. Hommel, Timenet-a toolkit for evaluating non-markovian stochastic petri nets, in *Proceedings of the Sixth International Workshop on Petri Nets and Performance Models*, Oct 1995, pp. 210–211

375. S. Ghemawat, H. Gobioff, S.-T. Leung, The google file system, in *Proceedings of the 19th ACM Symposium on Operating Systems Principles* (2003)

376. A.A. Ghorbani, E. Bagheri, The state of the art in critical infrastructure protection: a framework for convergence. Int. J. Crit. Infrastruct. **4**, 244–251 (2008)

377. A.K. Ghosh, M. Schmid, An approach to testing COTS software for robustness to operating system exceptions and errors, in *Proceedings of the 10th International Symposium on Software Reliability Engineering, ISSRE '99* (1999), pp. 166–174

378. R. Ghosh, F. Longo, V. Naik, K. Trivedi, Quantifying resiliency of iaas cloud, in *29th IEEE Symposium on Reliable Distributed Systems, 2010*, Oct 31–Nov 3, 2010, pp. 343–347

379. K. Gilly, S. Alcaraz, C. Juiz, R. Puigjaner, Analysis of burstiness monitoring and detection in an adaptive web system. Comput. Netw. **53**, 668–679 (2009)

380. K. Gilly, C. Juiz, N. Thomas, R. Puigjaner, Adaptive admission control algorithm in a qos-aware web system. Inf. Sci. (in press, 2011)

381. S. Gilmore, J. Hillston, M. Ribaudo, An efficient algorithm for aggregating pepa models. IEEE Trans. Softw. Eng. **27**(5), 449–464 (2001)

382. N. Glombitza, D. Pfisterer, S. Fischer, Using state machines for a model driven development of web service-based sensor network applications, in *ICSE Workshop on Software Engineering for Sensor Network Applications* (2010)

383. *Critical thinking: Moving from Infrastructure Protection to Infrastructure Resilience*. CIP Program Discussion Paper Series, George Mason University School of Law, Feb 2007

384. J. Goldberg, SIFT: A provable fault-tolerant computer for aircraft flight control, in *IFIP Congress 1980* (International Federation for Information Processing Tokyo, 1980), pp. 151–156

385. L. Gönczy, S. Chiaradonna, F.D. Giandomenico, A. Pataricza, A. Bondavalli, T. Bartha, Dependability evaluation of web service-based processes, in *3rd European Performance Engineering Workshop (EPEW2006)*, Budapest, Hungary, June 21–22, 2006. Lecture Notes in Computer Science, vol. 4054, ed. by A. Horváth, M. Telek (Springer, Berlin, 2006), pp. 166–180

386. W. Gordon, G. Newell, Closed queueing systems with exponential servers. Oper. Res. **15**(2), 254–265 (1967)

387. K.K. Goswami, R.K. Iyer, L.T. Young, DEPEND: a simulation-based environment for system level dependability analysis. IEEE Trans. Comput. **46**(1), 60–74 (1997)

388. M. Grabowski, A. Premnath, J. Merrik, J. Harrald, K. Roberts, Leading indicators of safety in virtual organizations. Saf. Sci. **45**, 1013–1043 (2007)

389. M. Grabowski, Z. You, H. Song, H. Wang, J. Merrick, Sailing on Friday: developing the link between safety culture and performance in safety-critical systems. IEEE Trans. Syst. Man Cybern. A Syst. Hum. **40**(2), 263–284 (2010)

390. J. Gray, Why do computers stop and what can be done about it? In *5th Symposium on Reliability in Distributed Software and Database Systems (SRDSDS-5)* (IEEE Computer Society Press, Los Angeles, 1986), pp. 3–12

391. J. Gray (ed.), *The Benchmark Handbook for Database and Transaction Systems*, 2nd edn. (Morgan Kaufmann, San Mateo, 1993)

392. J. Gray, D.P. Siewiorek, High-availability computer systems. IEEE Comput. **9**, 39–48 (1991)
393. M. Gribaudo, A. Remke, Hybrid petri nets with general one-shot transitions for dependability evaluation of fluid critical infrastructures, in *Proceedings of the 2010 IEEE 12th International Symposium on High-Assurance Systems Engineering, HASE '10* (IEEE Computer Society, Washington, 2010), pp. 84–93
394. M. Gribaudo, M. Sereno, Simulation of fluid stochastic petri nets, in *Proceedings of the 8th International Symposium on Modeling, Analysis and Simulation of Computer and Telecommunication Systems, MASCOTS '00* (IEEE Computer Society, Washington, 2000), pp. 231–239
395. M. Gribaudo, M. Telek, Fluid models in performance analysis, in *Proceedings of the 7th International Conference on Formal Methods for Performance Evaluation, SFM'07* (Springer, Berlin, 2007), pp. 271–317
396. GRID Consortium, D11—ICT vulnerabilities of power systems: a roadmap for future research. Technical report, GRID Consortium, Dec 2007
397. W. Gropp, E. Lusk, A. Skjellum, *Using MPI: Portable Parallel Programming with the Message Passing Interface* (MIT Press, Cambridge, 1994)
398. R. Grossman, The case for cloud computing. IT Prof. **11**(2), 23–27 (2009)
399. M. Grottke, V. Apte, K. Trivedi, S. Woolet, Response time distributions in networks of queues, in *Queueing Networks*. International Series in Operations Research and Management Science, vol. 154, ed. by R. Boucherie, N. Dijk (Springer, New York, 2011), pp. 587–641
400. M. Grottke, R. Matias, K.S. Trivedi, The fundamentals of software aging, in *Software Reliability Engineering Workshops, 2008. IEEE International Conference on ISSRE Wksp 2008*, Nov 2008, pp. 1–6
401. M. Grottke, K.S. Trivedi, Fighting bugs: remove, retry, replicate, and rejuvenate. Computer **40**(2), 107–109 (2007)
402. U. Gunnejlo, J. Karlsson, J. Torin, Evaluation of error detection schemes using fault injection by heavy-ion radiation, in *International Symposium on Fault-Tolerant Computing (FTCS '89)* (IEEE Computer Society Press, Washington, 1989), pp. 340–347
403. J. Güthoff, V. Sieh, Combining software-implemented and simulation-based fault injection into a single fault injection method, in *The Twenty-Fifth International Symposium on Fault-Tolerant Computing (FTCS '95)*. (IEEE Computer Society Press, Los Alamitos, 1995), pp. 196–206
404. A.N. Habermann, *Introduction to Operation Systems Design* (Science Research Associates, Chicago, 1976)
405. A. Hale, T. Heijer, Is resilience really necessary? The case of railways, in *Resilience Engineering. Concepts and Precepts*, ed. by E. Hollnagel, D.D. Woods, N. Leveson (Ashgate, Aldershot, UK, 2006), pp. 125–148
406. W.G.J. Halfond, A. Orso, AMNESIA: analysis and monitoring for NEutralizing SQL-injection attacks, in *Proceedings of the 20th IEEE/ACM International Conference on Automated Software Engineering, ASE '05* (2005), pp. 174–183
407. J.R. Hamilton, An architecture for modular data centers, in *Proceedings of the Conference on Innovative Data System Research* (2007), pp. 306–313
408. S. Han, K.G. Shin, H.A. Rosenberg, DOCTOR: an integrated software fault injection environment for distributed real-time systems, in *International Computer Performance and Dependability Symposium* (1995), pp. 204–213
409. R. Hansen, SQL Injection cheat sheet (2006) http://ha.ckers.org/sqlinjection/
410. R. Hansen, XSS (Cross Site Scripting) Cheat Sheet (2009)
411. P. Harrison, W. Knottenbelt, Passage time distributions in large Markov chains, in *Proceedings of ACM SIGMETRICS 2002*, Marina Del Rey, CA, June 2002, pp. 77–85
412. P.G. Harrison, Turning back time in markovian process algebra. Theor. Comput. Sci. **290**(3):1947–1986 (2003)

413. P.G. Harrison, Compositional reversed markov processes, with applications to g-networks. Perform. Eval. **57**(3), 379–408 (2004)

414. P.G. Harrison, B. Strulo, Spades—a process algebra for discrete event simulation. J. Logic Comput. **10**(1), 3–42 (2000)

415. H. Hashempour, L. Schiano, F. Lombardi, Evaluation, analysis, and enhancement of error resilience for reliable compression of vlsi test data. *IEEE Trans. Instrum. Meas.*, **54**(5) 1761–1769 (2005)

416. M. Hauswirth, A. Diwan, P. Sweeney, M. Hin, Vertical profiling: understanding the behavior of object-oriented applications, in *ACM Conference on Object-Oriented Programming, Systems, Languages, and Applications*, Vancouver, BC, Canada (2004), pp. 251–269

417. B.R. Haverkort, H. Hermanns, J.-P. Katoen, On the use of model checking techniques for quantitative dependability evaluation, in *Proceedings of the 19th IEEE Symposium on Reliable Distributed Systems, SRDS 2000* (IEEE Computer Society Press, Los Alamitos, 2000), pp. 228–237

418. B.R. Haverkort, I.G. Niemegeers, Performability modelling tools and techniques. Perform. Eval. **25**(1), 17–40 (1996)

419. R. Hayden, Scalable performance analysis of massively parallel stochastic systems. Ph.D. thesis, Department of Computing, Imperial College London (2011)

420. R. Hayden, J.T. Bradley, A fluid analysis framework for a Markovian process algebra. J. Theor. Comput. Sci. **411**(22–24) 2260–2297, (2010). doi:10.1016/j.tcs.2010.02.001

421. R. Hayden, A. Stefanek, J.T. Bradley, Fluid computation of passage time distributions in large markov models. Technical report, Department of Computing, Imperial College London, Nov 2010 http://pubs.doc.ic.ac.uk/fluid-passage-time/ (under review)

422. Q.-M. He, H. Zhang, Spectral polynomial algorithms for computing bi-diagonal representations for phase type distributions and matrix-exponential distributions. Stoch. Models **22**, 289–317 (2006)

423. X. He, W. Wei, X. Gui, The software rejuvenation model with pre-start technology, in *2008 International Symposiums on Information Processing (ISIP)*, May 2008, pp. 723–727

424. P.E. Heegaard, K.S. Trivedi, Network survivability modeling. Comput. Netw. **53**(8), 1215–1234 (2009)

425. W. Henderson, P. Taylor, Embedded processes in stochastic petri nets. IEEE Trans. Softw. Eng. **17**(2), 108–116 (1991)

426. J.N. Herder, H. Bos, B. Gras, P. Homburg, A.S. Tanenbaum, Fault isolation for device drivers, in *Proceedings of the 39th Annual IEEE/IFIP International Conference on Dependable Systems and Networks (DSN 2009)*, June–July 2009, pp. 33–42

427. H. Hermanns, M. Kwiatkowska, G. Norman, D. Parker, M. Siegle (2003) On the use of MTBDDs for performability analysis and verification of stochastic systems. J. Logic Algebraic Progr. **56**(1–2), 23–67

428. H. Hermanns, J. Meyer-Kayser, M. Siegle, Multi terminal binary decision diagrams to represent and analyse continuous time markov chains, in *3rd International Workshop on the Numerical Solution of Markov Chains*, ed. by B. Plateau, W. Stewart, M. Silva (Prensas Universitarias de Zaragoza, 1999), pp. 188–207

429. H. Hermanns, M. Rettelbach, Syntax, semantics, equivalences, and axioms for MTIPP, in *Proceedings of the 2nd Workshop on Process Algebras and Performance Modelling (PAPM '94)* (1994), pp. 71–87

430. H. Hermanns, M. Siegle, Bisimulation algorithms for stochastic process algebras and their BDD-based implementation, in *ARTS'99, 5th International AMAST Workshop on Real-Time and Probabilistic Systems*. Lecture Notes in Computer Science, vol. 1601, ed. by J.-P. Katoen (Springer, Heidelberg, 1999), pp. 144–264

431. B. Herndon, P. Smith, L. Roderick, E. Zamost, J. Anderson, V. Makhija, B. Herndon, P. Smith, E. Zamost, J. Anderson, VMmark: a scalable benchmark for virtualized systems (2006)

432. I. Herrera, J. Hovden, The leading indicators applied to maintenance in the framework of resilience engineering: a conceptual approach, in *3rd Resilience Engineering Symposium* (Antibes-Juan Les Pins, France 2008)

433. Hewlett Packard, HP operations manager (2010), http://www.managementsoftware.hp.com

434. Hewlett-Packard Development Company. HP WebInspect (2011), http://www8.hp.com/us/en/software/software-solution.html?compURI=tcm:245-936139

435. Hidenets—highly dependable IP-based networks and services (project ist-fp6-strep-26979) (2006), http://www.hidenets.aau.dk/

436. J. Hillston, *A Compositional Approach to Performance Modelling*. Distinguished Dissertations in Computer Science, vol. 12 (Cambridge University Press, Cambridge, 1996)

437. J. Hillston, Exploiting structure in solution: decomposing compositional models, in *Lectures on Formal Methods and Performance Analysis*. Lecture Notes in Computer Science, vol. 2090, ed. by E. Brinksma et al., (Springer, Berlin, 2001), pp. 278–314

438. J. Hillston, Fluid flow approximation of PEPA models, in *Second International Conference on the Quantitative Evaluation of Systems*, Sept 2005, pp. 33–42

439. J. Hillston, L. Kloul, An efficient kronecker representation for PEPA models, in *Proceedings of the Joint International Workshop, PAPM-PROBMIV 2001*. Lecture Notes in Computer Science, vol. 2165 (Springer, Aachen, 2001), pp. 120–135

440. J. Hillston, L. Kloul, Formal techniques for performance analysis: blending san and pepa. Formal Aspects Comput. **19**, 3–33 (2007)

441. W. Hoarau, S. Tixeuil, A language-driven tool for fault injection in distributed systems, in *Proceedings of the 6th IEEE/ACM International Conference on Grid Computing (GRID 2005)* (IEEE, 2005), pp. 194–201

442. W. Hoarau, S. Tixeuil, F. Vauchelles, Fault injection in distributed java applications, in *Proceedings of the 20th International Parallel and Distributed Processing Symposium (IPDPS 2006)* (IEEE, 2006)

443. G. Hoffmann, K. Trivedi, M. Malek, A best practice guide to resource forecasting for computing systems. IEEE Trans. Reliab. **56**(4), 615–628 (2007)

444. G. Horton, V.G. Kulkarni, D.M. Nicol, K.S. Trivedi, Fluid stochastic petri nets: theory, applications, and solution techniques. Eur. J. Oper. Res. **105**(1), 184–201 (1998)

445. A. Horváth, M. Telek, Approximating heavy tailed behaviour with phase type distributions, in *3rd International Conference on Matrix-Analytic Methods in Stochastic Models (MAM03)* (2000)

446. A. Horváth, M. Telek, PhFit: a general phase-type fitting tool, in *TOOLS '02: Proceedings of the 12th International Conference on Computer Performance Evaluation, Modelling Techniques and Tools* (Springer, London, 2002), pp. 82–91

447. G. Horváth, M. Telek, A canonical representation of order 3 phase-type distributions, in *Formal Methods and Stochastic Models for Performance Evaluation*. Lecture Notes in Computer Science, vol. 4748, ed. by K. Wolter (Springer, Berlin, 2007), pp. 48–62

448. G. Horváth, M. Telek, A minimal representation of markov arrival processes and a moments matching method. Perform. Eval. **64**(9–12), 1153–1168 (2007)

449. Y. Huang, C. Kintala, N. Kolettis, N.D. Fulton, Software rejuvenation: analysis, module and applications, in *Twenty-Fifth International Symposium on Fault-Tolerant Computing, 1995. FTCS-25, Digest of Papers*, June 1995, pp. 381–390

450. Y.-W. Huang, S.-K. Huang, T.-P. Lin, C.-H. Tsai, Web application security assessment by fault injection and behavior monitoring, in *Proceedings of the 12th International Conference on World Wide Web, WWW '03* (2003), pp. 148–159

451. N. Huber, F. Brosig, S. Kounev, Model-based self-adaptive resource allocation in virtualized environments, in *SEAMS'11: 6th International Symposium on Software Engineering for Adaptive and Self-Managing Systems* (2011)

452. N. Huber, M. von Quast, F. Brosig, S. Kounev, Analysis of the performance-influencing factors of virtualization platforms, in *OTM 2010 Conferences—Distributed Objects, Middleware, and Applications (DOA'10)* (Springer, Heidelberg, 2010)

453. N. Huber, M. von Quast, M. Hauck, S. Kounev, Evaluating and modeling virtualization performance overhead for cloud environments, in *1st International Conference on Cloud Computing and Service Science (CLOSER 2011)* (2011)

454. M.C. Huebscher, J.A. McCann, A survey of autonomic computing—degrees, models, and applications. ACM Comput. Surv. **40**(7) (2008)

455. K. Huh, K. Han, D. Hong, J. Kim, H. Kang, P. Yoon, A model-based fault diagnosis system for electro-hydraulic brake. SAE technical paper series 2008-01-1225 (SAE International, Warrendale, 2008)

456. J.W. Hunt, M.D. McIlroy, An algorithm for differential file comparison. Online technical report, Bell Laboratories Computing Science (1976), http://www.cs.dartmouth.edu/doug/diff.ps

457. K. Huppler, The art of building a good benchmark, in *First TPC Technology Conference (TPCTC 2009)*. Lecture Notes in Computer Science, vol. 5895 (2009)

458. IBM, Tivoli Enterprise Console (2010), http://www.ibm.com/software/tivoli/products/enterprise-consol

459. IBM. IBM Rational AppScan (2011), http://www-01.ibm.com/software/awdtools/appscan/

460. S. Ibrahim, H. Jin, L. Lu, L. Qi, S. Wu, X. Shi, Evaluating mapreduce on virtual machines: the hadoop case, in *Proceedings of 1st International Conference on Cloud Computing, CloudCom '09* (Springer, Berlin, 2009), pp. 519–528

461. IEC 61508, Functional safety of electrical/electronic/programmable electronic safety related systems (International Electrotechnical Commission, 1998–2010)

462. IEEE, IEEE Std 1149.1-2001, IEEE standard test access port and boundary-scan architecture (2001)

463. IEEE, IEEE 802.3: LAN/MAN CSMA/CDE (ETHERNET) Access method (2010), http://standards.ieee.org/getieee802/802.3.html

464. IEEE, Std 1588-2008, IEEE standard for a precision clock synchronization protocol for networked measurement and control systems, http://ieeexplore.ieee.org/xpl/freeabs_all.jsp?arnumber=4579760 (2010)

465. IEEE. Systems and software engineering—vocabulary, 2010. Standard 24765 (2010)

466. IEEE Industry Standards and Technology Organization (IEEE-ISTO). IEEE-ISTO 5001 TM-2003, the nexus 5001TM forum standard for a global embedded processor debug interface, Dec 2003

467. IEEE RTS Task Force of the APM Subcommittee, IEEE reliability test system. IEEE Trans. Power App. Syst. **PAS-98**(6) 2047–2054 (1979)

468. IEEE RTS Task Force of the APM Subcommittee, The IEEE reliability test system—1996. IEEE Trans. Power Syst. **14**(3) 1010–1020 (1999)

469. M. Ihde, W.H. Sanders, Barbarians in the gate: an experimental validation of nic-based distributed firewall performance and flood tolerance, in *Proceedings of the International Conference on Dependable Systems and Networks* (2006), pp. 209–216

470. IRRIIS, European Project IRRIIS—Integrated risk reduction of information-based infrastructure systems, http://irriis.org

471. IRRIIS Consortium, Tools and techniques for interdependency analysis. Technical report, http://www.irriis.org/, July 2007

472. IRRIS Consortium, D1.3.2—list of available and suitable simulation components. Technical report, IRRIS Consortium (2006)

473. R. Isermann, Model-based fault-detection and diagnosis—status and applications. Annu. Rev. Control **29**(1), 71–85 (2005)

474. T. Israr, M. Woodside, M. Franks, Interaction tree algorithms to extract effective architecture and layered performance models from traces. J. Syst. Softw. **80**, 474–492 (2007)

475. ITU, G.8261/Y.1361 (04/2008) Timing and synchronization aspects in packet networks, http://www.itu.int/rec/T-REC-G.8261-200804-I, April 2008

476. R. Iyer, R. Illikkal, O. Tickoo, L. Zhao, P. Apparao, D. Newell, VM3: measuring, modeling and managing VM shared resources. Comput. Netw. **53**(17), 2873–2887 (2009) (Virtualized Data Centers)
477. R.K. Iyer, L.T. Young, P.V.K. Iyer, Automatic recognition of intermittent failures: an experimental study of field data. IEEE Trans. Comput. **39**, 525–537 (1990)
478. R.V.J. Delude, Analyzing quantitative data through the web, in *6th Annual Dual Use Technologies and Applications Conference*, Mohawk, Valley Section, USA (IEEE, 1995)
479. J. Kephart et al., Coordinating multiple autonomic managers to achieve specified power-performance tradeoffs, in *International Conference on Autonomic Computing* (2007)
480. J. Jackson, Networks of waiting lines. Oper. Res. **5**(4), 518–521 (1957)
481. K.R. Jackson, L. Ramakrishnan, K. Muriki, S. Canon, S. Cholia, J. Shalf, H.J. Wasserman, N.J. Wright, Performance analysis of high performance computing applications on the amazon web services cloud, in *Proceedings of 2nd IEEE International Conference on Cloud Computing Technology and Science (CloudCom)* (2010), pp. 159–168
482. T. Jansen, I. Balan, I. Moerman, T. Kürner, Handover parameter optimization in lte self-organizing networks, in *Proceedings of the 72nd Vehicular Technology Conference*, VTC 2010-Fall, Ottawa, ON, Sept 2010
483. W. Jansen, Directions in security metrics research—NISTIR 7564. Technical report, NIST (2009)
484. A. Jaquith, *Security Metrics: Replacing Fear, Uncertainty, and Doubt* (Addison-Wesley Professional, Reading, 2007)
485. M. Jayakumar, B. Das, Diagnosis of incipient sensor faults in a flight control actuation system, in *SICE-ICASE, 2006. International Joint Conference*, Busan, Korea, Oct 2006, pp. 3423–3428
486. E. Jenn, J. Arlat, M. Rimén, J. Ohlsson, J. Karlsson, Fault injection into VHDL models: the MEFISTO tool, in *Proceedings of the 24th Annual International Symposium on Fault-Tolerant Computing (FTCS 94)* (IEEE Computer Society Press, Los Alamitos, 1994), pp. 66–75
487. Y.-F. Jia, J.-Y. Su, K.-Y. Cai, A feedback control approach for software rejuvenation in a web server, in *Software Reliability Engineering Workshops, 2008. IEEE International Conference on ISSRE Wksp 2008*, Nov 2008, pp. 1–6
488. G. Jiang, H. Chen, K. Yoshihira, A. Saxena, Ranking the importance of alerts for problem determination in large computer systems, in *Proceedings of the 6th International Conference on Autonomic Computing*, Barcelona, Spain, June 2009, pp. 3–12
489. L. Jiang, X. Peng, G. Xu, Time and prediction based software rejuvenation policy, in *Second International Conference on Information Technology and Computer Science (ITCS), 2010*, July 2010, pp. 114–117
490. M. Jiang, M.A. Munawar, T. Reidemeister, P.A.S. Ward, System monitoring with metric-correlation models: problems and solutions, in *Proceedings of the 6th International Conference on Autonomic Computing*, Barcelona, Spain, June 2009, pp. 13–22
491. C.W. Johnson, Analysing the causes of the Italian and Swiss blackout, 28th Sept 2003, in *SCS '07: Proceedings of the Twelfth Australian Workshop on Safety Critical Systems and Software and Safety-Related Programmable Systems* (Australian Computer Society, Darlinghurst, 2007), pp. 21–30
492. Joint Committee for Guides in Metrology JCGM 100:2008. *Evaluation of Measurement Data—Guide to the Expression of Uncertainty in Measurement* (2008)
493. Joint Committee for Guides in Metrology JCGM 200:2008. *International Vocabulary of Metrology—Basic and General Concepts and Associated Terms*, 3rd edn. (2008)
494. N. Jones, PHP-Fusion (2009)
495. E. Jonsson, T. Olovsson (1997) A quantitative model of the security intrusion process based on attacker behavior. IEEE Trans. Softw. Eng. **23**(4), 235–245

496. K.R. Joshi, W.H. Sanders, M.A. Hiltunen, R.D. Schlichting, Automatic model-driven recovery in distributed systems, in *Proceedings of the 24th IEEE Symposium on Reliable Distributed Systems*, Orlando, Florida, Oct 2005, pp. 25–38

497. G. Jung, M. Hiltunen, K. Joshi, R. Schlichting, C. Pu, Mistral: dynamically managing power, performance, and adaptation cost in cloud infrastructures, in *Proceedings of the 2010 IEEE 30th International Conference on Distributed Computing Systems* (2010)

498. G. Jung, K. Joshi, M. Hiltunen, R. Schlichting, C. Pu, Generating adaptation policies for multi-tier applications in consolidated server environments, in *Proceedings of the 5th IEEE International Conference on Autonomic Computing*, June 2008, pp. 23–32

499. G. Jung, K. Joshi, M. Hiltunen, R. Schlichting, C. Pu, A cost-sensitive adaptation engine for server consolidation of multitier applications, in *International Conference on Middleware* (2009)

500. M. Kaâniche et al, Methodologies synthesis. EU FP6 IST project CRUTIAL, Deliverable D3, Dec 2006, http://crutial.cesiricerca.it, Public deliverables section

501. M. Kaâniche, P. Lollini, A. Bondavalli, K. Kanoun, Modeling the resilience of large and evolving systems. Int. J. Performability Eng. **4**(2), 153–168 (2008)

502. A. Kalakech, T. Jarboui, A. Arlat, Y. Crouzet, K. Kanoun, Benchmarking operating systems dependability: Windows as a case study, in *Proceedings of the 2004 Pacific Rim International Symposium on Dependable Computing (PRDC 2004)* (2004)

503. A. Kalakech, K. Kanoun, Y. Crouzet, A. Arlat, Benchmarking the dependability of windows nt, 2000 and xp, in *Proceedings of the 2004 International Conference on Dependable Systems and Networks (DSN 2004)* (2004)

504. S. Kals, E. Kirda, C. Kruegel, N. Jovanovic, SecuBat: a web vulnerability scanner, in *Proceedings of the 15th International Conference on World Wide Web, WWW '06* (2006), pp. 247–256

505. T. Kam, T. Villa, R. Brayton, A. Sangiovanni-Vincentelli, Multi-valued decision diagrams: theory and applications. Multiple Valued Logic **4**(1–2), 9–62 (1998)

506. G.A. Kanawati, N.A. Kanawati, J.A. Abraham, FERRARI: a tool for the validation of system dependability properties, in *Proceedings of the 22nd Annual International Symposium on Fault-Tolerant Computing (FTCS '92)*, ed. by D.K. Pradhan (IEEE Computer Society Press, Boston, 1992), pp. 336–344

507. G.A. Kanawati, N.A. Kanawati, J.A. Abraham, Dependability evaluation using hybrid fault/error injection, in *Proceedings of the International Computer Performance and Dependability Symposium (IPDS'95)* (1995), pp. 224–233

508. S. Kandula, D. Katabi, J.-P. Vasseur, Shrink: a tool for failure diagnosis in IP networks, in *ACM SIGCOMM Workshop on Mining Network Data (MineNet-05)*, Philadelphia, PA, Aug 2005

509. S. Kandula, R. Mahajan, P. Verkaik, S. Agarwal, J. Padhye, P. Bahl, Detailed diagnosis in enterprise networks, in *Proceedings of the ACM SIGCOMM 2009 Conference on Data Communication*, Barcelona, Spain, Aug 2009, pp. 243–254

510. H. Kang, H. Chen, G. Jiang, PeerWatch: a fault detection and diagnosis tool for virtualized consolidation systems, in *Proceedings of the 7th International Conference on Autonomic Computing*, Washington, DC, June 2010, pp. 119–128

511. K. Kanoun, Y. Crouret, Dependability benchmarking for operating systems. Int. J. Perform. Eng. **2**(3), 275–287 (2006)

512. K. Kanoun, Y. Crouzet, A. Kalakech, A.-E. Rugina, P. Rumeau, Benchmarking the dependability of windows and linux using postmark workloads, in *Proceedings of the 16th International Symposium on Software Reliability Engineering (ISSRE 2005)* (2005)

513. K. Kanoun et al., http://www.laas.fr/dbench, project reports section, project full final report (2004)

514. K. Kanoun, L. Spainhower (eds), *Dependability Benchmarking for Computer Systems* (Wiley, New York, 2008)

515. W.-L. Kao, R.K. Iyer, DEFINE: a distributed fault injection and monitoring environment, in *Fault-Tolerant Parallel and Distributed Systems*, ed. by D. Pradhan, D.R. Avresky (IEEE Computer Society Press, New York, 1995), pp. 252–259

516. W.-L. Kao, R.K. Iyer, D. Tang, FINE: a fault injection and monitoring environment for tracing the UNIX system behavior under faults. IEEE Trans. Softw. Eng. **19**(11), 1105–1118 (1993)

517. J. Kaplan, W. Forrest, N. Kindler, *Revolutionizing Data Center Energy Efficiency* (McKinsey, 2008)

518. J. Karlsson, U. Gunneflo, P. Liden, J. Torin, Two fault injection techniques for test of fault handling mechanisms, in *International Test Conference (ITC '91)* (IEEE Computer Society Press, Altoona, 1991), pp. 140–149

519. A. Karve, T. Kimbrel, G. Pacifici, M. Spreitzer, M. Steinder, M. Sviridenko, A. Tantawi, Dynamic placement for clustered web applications, in *International Conference on World Wide Web* (2006)

520. Kaspersky Lab, Kaspersky lab provides its insights on stuxnet worm, http://www.kaspersky.com/news?id=207576183. Accessed 7 Dec 2010

521. J. Katcher, Postmark: a new file system benchmark. Network appliance technical report TR3022, Oct 1997

522. Y.A. Katsigiannis, P.S. Georgilakis, G.J. Tsinarakis (2010) A novel colored fluid stochastic petri net simulation model for reliability evaluation of wind/pv/diesel small isolated power systems. IEEE Trans. Syst. Man Cybern. A: Syst. Humans **40**(6), 1296–1309

523. S. Kavulya, J. Tan, R. Gandhi, P. Narasimhan, An analysis of traces from a production MapReduce cluster, in *IEEE/ACM International Conference on Cluster, Cloud and Grid Computing*, Melbourne, Australia, May 2010, pp. 94–103

524. S.P. Kavulya, K. Joshi, M. Hiltunen, S. Daniels, R. Gandhi, P. Narasimhan, Practical experiences with chronics discovery in large telecommunications systems, in *ACM Workshop on Managing Systems via Log Analysis and Machine Learning Techniques (SLAML)*, Cascais, Portugal, Oct 2011

525. V. Kawadia, Y. Zhang, B. Gupta, System services for ad-hoc routing: architecture, implementation and experiences, in *MobiSys '03: Proceedings of the 1st International Conference on Mobile Systems, Applications and Services* (ACM, New York, 2003), pp. 99–112

526. H. Kellerer, U. Pferschy, D. Pisinger, *Knapsack Problems* (Springer, Berlin, 2004)

527. J.W. Kellington, R. McBeth, P. Sanda, R.N. Kalla, Ibm power6 processor soft error tolerance analysis using proton irradiation, in *3th IEEE Workshop on Silicon Errors in Logic—Systems Effects (SELSE-3)* (2007)

528. T. Kelly, Detecting performance anomalies in global applications, in *USENIX WORLDS*, San Francisco, CA, Dec 2005

529. J. Kemeney, J. Snell, Finite markov chains. Technical report, D. Van Nostrand Company (1960)

530. P. Kemper, Reachability analysis based on structured representations, in *Proceedings of 17th International Conference on Application and Theory of Petri Nets*. Lecture Notes in Computer Science (Springer, Heidelberg, 1996), pp. 269–288

531. P. Kemper, Numerical analysis of superposed GSPNs. IEEE Trans. Softw. Eng. **22**(9), 615–628 (1996)

532. G. Khanna, I. Laguna, F.A. Arshad, S. Bagchi, Distributed diagnosis of failures in a three tier e-commerce system, in *26th IEEE International Symposium on Reliable Distributed Systems, 2007. SRDS 2007*, Beijing, China, Oct 2007, pp. 185–198

533. E. Kiciman, Using statistical monitoring to detect failures in internet services. Ph.D. thesis, Stanford University, Sept 2005

534. D.S. Kim, F. Machida, K.S. Trivedi, Availability modeling and analysis of a virtualized system, in *IEEE Pacific Rim International Symposium on Dependable Computing* (2009)

535. R. King, How cloud computing is changing the world, in Businessweek on the World Wide Web, http://www.businessweek.com/technology/content/aug2008/tc2008082_445669.htm, Aug 2008. Accessed June 2009

536. R. Klein, Information modelling and simulation in large dependent critical infrastructures: an overview on the european integrated project irriis, in *3rd International Workshop on Critical Information Infrastructures Security (CRITIS 2008)*. Lecture Notes in Computer Science, vol. 5508, ed. by R. Setola, S. Geretshuber (Springer, Berlin, 2009), pp. 131–143

537. A.G. Kleppe, J. Warmer, W. Bast, *MDA Explained: The Model Driven Architecture: Practice and Promise* (Addison-Wesley Longman Publishing, Boston, 2003)

538. L. Kloul, Méthodes d'évaluation des performances pour les réseaux ATM. Ph.D. thesis, Université de Versailles-St-Quentin-en-Yvelines, Jan 1996

539. R.R. Kompella, J. Yates, A.G. Greenberg, A.C. Snoeren, IP fault localization via risk modeling, in *Proceedings of the 2nd Conference on Symposium on Networked Systems Design and Implementation*, vol. 2, Boston, MA, May 2005, pp. 57–70

540. P. Koopman, J. DeVale, The exception handling effectiveness of posix operating systems. IEEE Trans. Softw. Eng. **26**(9), 837–848 (2000)

541. P. Koopman, K. DeVale, J. DeVale, Interface robustness testing: experience and lessons learned from the Ballista project, in *Dependability Benchmarking for Computer Systems* (Wiley, New York, 2008), pp. 201–226

542. P. Koopman, K. Devale, J. Devale, *Interface Robustness Testing: Experience and Lessons Learned from the Ballista Project* (Wiley, 2008), pp. 201–226

543. I. Koren, C. Krishna, *Fault-Tolerant Systems* (Morgan Kaufmann Publishers, San Francisco, 2007)

544. Y. Kosuga, K. Kernel, M. Hanaoka, M. Hishiyama, Y. Takahama, Sania: syntactic and semantic analysis for automated testing against SQL injection, in *Twenty-Third Annual Computer Security Applications Conference*, ACSAC 2007, pp. 107–117

545. S. Kounev, Performance engineering of distributed component-based systems—benchmarking, modeling and performance prediction. Ph.D. thesis, Technische Universität Darmstadt (2005)

546. S. Kounev, Performance modeling and evaluation of distributed component-based systems using queueing petri nets. IEEE Trans. Softw. Eng. **32**(7), 486–502 (2006)

547. S. Kounev, Self-aware software and systems engineering: a vision and research roadmap, in *Proceedings of Software Engineering 2011 (SE2011), Nachwuchswissenschaftler-Symposium* (2011)

548. S. Kounev, F. Brosig, N. Huber, Descartes Research Project, http://www.descartes-research.net. Accessed Nov 2011

549. S. Kounev, F. Brosig, N. Huber, R. Reussner, Towards self-aware performance and resource management in modern service-oriented systems, in *Proceedings of the 7th IEEE International Conference on Services Computing (SCC 2010)*, IEEE Computer Society, Miami, Florida, USA, July 5–10 (2010)

550. S. Kounev, R. Nou, J. Torres, Autonomic QoS-aware resource management in grid computing using online performance models, in *Proceedings of VALUETOOLS-2007* (2007)

551. K. Kourai, S. Chiba, A fast rejuvenation technique for server consolidation with virtual machines, in *IEEE/IFIP International Conference on Dependable Systems and Networks* (2007), pp. 245–255

552. V. Koutras, A. Platis, Applying software rejuvenation in a two node cluster system for high availability, in *International Conference on Dependability of Computer Systems, 2006. DepCos-RELCOMEX '06*, May 2006, pp. 175–182

553. V.P. Koutras, A.N. Platis, Semi markov performance modelling of a redundant system with partial, full and failed rejuvenation. Int. J. Crit. Comput. Based Syst. **1**, 59–85 (2010)

554. V.P. Koutras, A.N. Platis, G.A. Gravvanis, On the optimization of free resources using non-homogeneous markov chain software rejuvenation model. Reliab. Eng. Syst. Safety **92**(12), 1724–1732, Special Issue on ESREL 2005 (2007)

555. M. Kovacs, P. Lollini, I. Majzik, A. Bondavalli, An integrated framework for the dependability evaluation of distributed mobile applications, in *SERENE '08: Proceedings of the 2008 RISE/EFTS Joint International Workshop on Software Engineering for Resilient Systems* (ACM, New York, 2008), pp. 29–38

556. H. Koziolek, Performance evaluation of component-based software systems: a survey. Perform. Eval. Aug 2009

557. M. Kulawiak, A. Stepnowski, Algorithms for spatial analysis and interpolation of discrete sets of critical infrastructure hazard data, in *2nd International Conference on Information Technology (ICIT), 2010*, June 2010, pp. 157–160

558. S. Kumar, V. Marbukh, On route exploration capabilities of multi-path routing in variable topology ad hoc networks, in *Proceedings of the 21st IEEE Instrumentation and Measurement Technology Conference, 2004. IMTC 04*, vol. 2, May 2004, pp. 1322–1327

559. S. Kumar, V. Talwar, V. Kumar, P. Ranganathan, K. Schwan. vManage: loosely coupled platform and virtualization management in data centers, in *International Conference on Autonomic Computing* (2009)

560. S. Kundu, R. Rangaswami, K. Dutta, M. Zhao, Application performance modeling in a virtualized environment, in *IEEE 16th International Symposium on High Performance Computer Architecture (HPCA)* (2010), pp. 1–10

561. M. Kuntz, M. Siegle, Deriving symbolic representations from stochastic process algebras, in *Process Algebra and Probabilistic Methods, Proceedings of PAPM-PROBMIV'02*. Lecture Notes in Computer Science, vol. 2399 (Springer, Heidelberg, 2002), pp. 188–206

562. M. Kuntz, M. Siegle, E. Werner, Symbolic performance and dependability evaluation with the tool caspa, in *FORTE Workshops*. Lecture Notes in Computer Science, vol. 3236 (Springer, 2004), pp. 293–307

563. J. Kurjenniemi, T. Henttonen, Effect of measurement bandwidth to the accuracy of inter-frequency rsrp measurements in lte, in *Proceedings of the 19th International Symposium Personal, Indoor and Mobile Radio Communications*, IEEE, PIMRC, Cannes, France, Sept 2008

564. M. Kwiatkowska, G. Norman, D. Parker, Prism: probabilistic model checking for performance and reliability analysis. ACM SIGMETRICS Perform. Eval. Rev. **36**(4), 40–45 (2009)

565. M. Kwiatkowska, D. Parker, Y. Zhang, R. Mehmood, Dual-processor parallelisation of symbolic probabilistic model checking, in *Proceedings 12th International Symposium on Modeling, Analysis, and Simulation of Computer and Telecommunication Systems (MASCOTS'04)*, ed. by D. DeGroot, P. Harrison (IEEE Computer Society Press, New York, 2004), pp. 123–130

566. J. Lacki, J. Niemela, J. Lempiainen, Optimization of soft handover parameters for umts network in indoor environment, in *Proceedings of the 9th International Symposium on Wireless Personal Multimedia Communications*, WPMC 2006, San Diego, USA, Sept 2006

567. J. Lala, L. Alger, Hardware and software fault tolerance: a unified architectural approach, in *IEEE FTCS International Symposium on Fault-Tolerant Computing* (1988), pp. 240–245

568. K. Lampka, A symbolic approach to the state graph based analysis of high-level markov reward models. Ph.D. thesis, University of Erlangen-Nürnberg, Technische Fakultät (2007)

569. K. Lampka, M. Siegle, Activity-local symbolic state graph generation for high-level stochastic models, in *Proceedings of 13th GI/ITG Conference on Measuring, Modelling and Evaluation of Computer and Communication Systems (MMB)*, Nürnberg, Germany (VDE Verlag, 2006), pp. 245–263

570. K. Lampka, M. Siegle, J. Ossowski, C. Baier, Partially-shared zero-suppressed multi-terminal bdds: concept, algorithms and applications. Formal Methods Syst. Des. **36**(3), 198–222 (2010)

571. C. Lamprecht, A. van Moorsel, P. Tomlinson, N. Thomas, Investigating the efficiency of cryptographic algorithms in online transactions. Int. J. Simul. Syst. Sci. Technol. **7**(2), 63–75 (2006)

572. A. Lang, J. Arthur, Parameter approximation for phase-type distributions. Matrix Anal. Methods Stoch. Models **183**, 151–206 (1996)

573. K.-D. Lange, Identifying shades of green: the specpower benchmarks. Computer **42**(3), 95–97 (2009)

574. P.E. Lanigan, S. Kavulya, T.E. Fuhrman, P. Narasimhan, M.A. Salman, Diagnosis in automotive systems: a survey. Technical report CMU-PDL-11-110, Carnegie Mellon University PDL, May 2011

575. T. Lanowitz, Now is the time for security at the application level. Online report, Gartner Group (2005), http://www.sela.co.il/_Uploads/dbsAttachedFiles/GartnerNowIsTheTimeFor Security.pdf

576. J.C. Laprie, Dependable computing: concepts, limits, challenges, in *IEEE International Symposium on Fault-Tolerant Computing: Special Issue* (1995), pp. 42–54

577. J.-C. Laprie, K. Kanoun, M. Kaaniche, Modeling interdependencies between the electricity and information infrastructures, in *SAFECOMP-2007*. Lecture Notes in Computer Science, vol. 4680 (Springer, 2007), pp. 54–67

578. N. Laranjeiro, M. Vieira, H. Madeira, Protecting database centric web services against SQL/ XPath injection attacks, in *Proceedings of the 20th International Conference on Database and Expert Systems Applications, DEXA '09* (2009), pp. 271–278

579. G. Latouche, V. Ramaswami, *Introduction to Matrix Analytic Methods in Stochastic Modeling*. Series on statistics and applied probability. ASA-SIAM (1999)

580. S. Lavenberg, M. Reiser, Stationary state space probabilities at arrival instants for closed queueing networks with multiple types of customers. J. Appl. Probab. **17**(4), 1048–1061 (1980)

581. P. L'Ecuyer, L. Meliani, J. Vaucher, SSJ: a framework for stochastic simulation in Java, in *Proceedings of the Winter Simulation Conference* (2002), pp. 234–242

582. E.E. Lee, J.E. Mitchell, W.A. Wallace, Assessing vulnerability of proposed designs for interdependent infrastructure systems, in *Proceedings of the 37th IEEE Annual Hawaii International Conference on System Sciences (HICSS'04)* (2004)

583. M. Lehn, T. Triebel, C. Gross, D. Stingl, K. Saller, W. Effelsberg, A. Kovacevic, R. Steinmetz, Designing benchmarks for P2P systems, in *From Active Data Management to Event-Based Systems and More*. LNCS, vol. 6462 (Springer, 2010), pp. 209–229

584. B. Lei, X. Li, Z. Liu, C. Morisset, V. Stolz, Robustness testing for software components. Sci. Comput. Program. **75**, 879–897 (2010)

585. L. Lei, K. Vaidyanathan, K.S. Trivedi, An approach for estimation of software aging in a web server, in *Proceedings of the 2002 International Symposium on Empirical Software Engineering, 2002* (2002), pp. 91–100

586. N. Leveson, N. Dulac, K. Marais, J. Carroll, Beyond normal accidents and high reliability organizations: the need for an alternative approach to safety in complex systems. Organ. Stud. **30**(2–3), 227–249 (2009)

587. L. Lewis, G. Dreo, Extending trouble ticket systems to fault diagnostics. IEEE Netw. **7**(6), 44–51 (1993)

588. J. Li, X. Ma, K. Singh, M. Schulz, B. de Supinski, S. McKee, Machine learning based online performance prediction for runtime parallelization and task scheduling, in *IEEE International Symposium on Performance Analysis of Systems and Software, 2009. ISPASS 2009*, April 2009, pp. 89–100

589. S. Lightstone, J. Hellerstein, W. Tetzlaff, P. Janson, E. Lassettre, C. Norton, B. Rajaraman, L. Spainhower, Towards benchmarking autonomic computing maturity, in *Proceedings of the First IEEE Conference on Industrial Automatics (INDIN 2003)* (2003)

590. T.Y. Lin, D.P. Siewiorek, Error log analysis: statistical modeling and heuristic trend analysis. IEEE Trans. Reliab. **39**, 419–432 (1990)

591. C. Lindemann, A. Reuys, A. Thümmler, The dspnexpress 2.000 performance and dependability modeling environment, in *Twenty-Ninth Annual International Symposium on Fault-Tolerant Computing, 1999. Digest of Papers*, pp. 228–231

592. J.L. Lions, Report by the Inquiry Board on the Ariane 5 Flight 501 failure. ESA/CNES, 19 July 1996

593. B. Littlewood, S. Brocklehurst, N.E. Fenton, P. Mellor, S. Page, D. Wright, J. Dobson, J. McDermid, D. Gollmann, Towards operational measures of computer security. J. Comput. Secur. **2**(2–3), 211–230 (1993)

594. B. Littlewood, L. Strigini, Redundancy and diversity in security, in *ESORICS*. Lecture Notes in Computer Science, vol. 3193, ed. by P. Samarati, P.Y.A. Ryan, D. Gollmann, R. Molva (Springer, Heidelberg, 2004), pp. 423–438

595. M. Littman, N. Ravi, E. Fenson, R. Howard, An instance-based state representation for network repair, in *19th National Conference on Artificial Intelligence (AAAI 2004)*, July 2004, pp. 287–292

596. G. Liu, A. Mok, E. Yang, Composite events for network event correlation, in *International Symposium on Integrated Network Management*, Boston, MA, May 1999, pp. 247–260

597. X. Liu, J. Heo, L. Sha, Modeling 3-tiered web applications, in *International Symposium on Modeling, Analysis, and Simulation of Computer and Telecommunication Systems (MASCOTS)*, Atlanta, GA, Sept 2005, pp. 307–310

598. Y. Liu, Y. Ma, J.J. Han, H. Levendel, K.S. Trivedi, A proactive approach towards always-on availability in broadband cable networks. Comput. Commun. **28**, 51–64 (2005)

599. Y. Liu, K.S. Trivedi, Y. Ma, J.J. Han, H. Levendel, Modeling and analysis of software rejuvenation in cable modem termination systems, in *Proceedings of the 13th International Symposium on Software Reliability Engineering, 2002. ISSRE 2002* (2002), pp. 159–170

600. Z. Liu, B. Lee, S. Kandula, R. Mahajan, NetClinic: interactive visualization to enhance automated fault diagnosis in enterprise networks, in *IEEE Conference on Visual Analytics Science and Technology*, Salt Lake City, UT, Oct 2010, pp. 131–138

601. Z. Liu, M.S. Squillante, J.L. Wolf, On maximizing service-level-agreement profits, in *Proceedings of 3rd ACM Conference on Electronic Commerce (EC'01)* (2001)

602. P. Lollini, A. Bondavalli, F. di Giandomenico, A decomposition-based modeling framework for complex systems. IEEE Trans. Reliab. **58**(1), 20–33 (2009)

603. P. Lollini, A. Bondavalli et al., Evaluation methodologies, techniques and tools (final version). EU FP6 IST project HIDENETS, Deliverable D4.1.2, Dec 2007, http://www.hidenets.aau.dk/, Public deliverables section

604. N. Looker, M. Munro, J. Xu, WS-FIT: a tool for dependability analysis of Web services, in *Proceedings of the 28th Annual International Computer Software and Applications Conference, COMPSAC '04* (2004), pp. 120–123

605. N. Looker, M. Munro, J. Xu, A comparison of network level fault injection with code insertion, in *Proceedings of the 29th International Computer Software and Applications Conference (COMPSAC 2005)* (IEEE Computer Society, New York, 2005), pp. 479–484

606. N. Looker, M. Munro, J. Xu, Simulating errors in web services. Int. J. Simul. Syst. Sci. Technol. **5**, 29–37 (2005)

607. N. Looker, J. Xu, Assessing the dependability of OGSA middleware by fault injection, in *Proceedings of 22nd Symposium on Reliable Distributed Systems (22nd SRDS'03)*, Florence, Italy (IEEE Computer Society, New York, 2003), pp. 293–302

608. S. Loveland, E.M. Dow, F. LeFevre, D. Beyer, P.F. Chan, Leveraging virtualization to optimize high-availability system configurations. IBM Syst. J. **47**(4), 591–604 (2008)

609. N. Lu, J.H. Chow, A. Desrochers, A multi-layer petri net model for deregulated electric power systems, in *Proceedings of the American Control Conference*, vol. 1 (2002), pp. 513–518

610. E. Luiijf, Scada security good practices for drinking water sector (2008)

611. E. Luiijf, M. Ali, A. Zielstra, *Assessing and Improving SCADA Security in the Dutch Drinking Water Sector* (Springer, 2009), pp. 190–199

612. E. Luiijf, A. Nieuwenhuijs, M. Klaver, M. Eeten, E. Cruz, *Empirical Findings on Critical Infrastructure Dependencies in Europe* (Springer, Berlin, 2009), pp. 302–310

613. D. Macii, D. Petri, Accurate software-related average current drain measurements in embedded systems. IEEE Trans. Instrum. Meas. **56**(3), 723–730 (2007)

614. B.B. Madan, K. Goseva-Popstojanova, K. Vaidyanathan, K.S. Trivedi, Modeling and quantification of security attributes of software systems, in *Proceedings of the International Conference on Dependable Systems and Networks (DSN'02)* (2002), pp. 505–514

615. H. Madeira, J. Costa, M. Vieira, The OLAP and data warehousing approaches for analysis and sharing of results from dependability evaluation experiments, in *International Conference on Dependable Systems and Networks* (2003), pp. 86–91

616. H. Madeira, P. Koopman, Dependability benchmarking: making choices in an n-dimensional problem space, in *1st Workshop on Evaluating and Architecting System Dependability* (Göteborg, Sweden, 2001)

617. H. Madeira, M.Z. Rela, F.M. João Gabriel Silva, RIFLE: a general purpose pin-level fault injector, in *Proceedings of the First European Dependable Computing Conference (EDCC-1)*, Berlin, Germany, Oct 4–6 1994. Lecture Notes in Computer Science, vol. 852, ed. by K. Echtle, D.K. Hammer, D. Powell (Springer, Heidelberg, 1994), pp. 199–216

618. J. Magott, M. Woda, Evaluation of soa security metrics using attack graphs, in *Proceedings of 3rd International Conference on Dependability of Computer Systems (DepCoS-RELCOMEX)* (2008), pp. 277–284

619. M. Magyar, I. Majzik, Modular construction of dependability models from system architecture models: a tool-supported approach, in *Proceedings 6th International Conference on the Quantitative Evaluation of Systems (QEST 2009)* (2009), pp. 95–96

620. A.A. Mahimkar, Z. Ge, A. Shaikh, J. Wang, J. Yates, Y. Zhang, Q. Zhao, Towards automated performance diagnosis in a large IPTV network, in *Proceedings of the ACM SIGCOMM 2009 Conference on Data Communication*, Barcelona, Spain, Aug 2009, pp. 231–242

621. A.A. Mahimkar, H.H. Song, Z. Ge, A. Shaikh, J. Wang, J. Yates, Y. Zhang, J. Emmons, Detecting the performance impact of upgrades in large operational networks, in *Detecting the Performance Impact of Upgrades in Large Operational Networks*, Aug 2010, pp. 303–314

622. R. Maia, L. Henriques, R. Barbosa, D. Costa, H. Madeira, Xception fault injection and robustness testing framework: a case-study of testing RTEMS, in *VI Test and Fault Tolerance Workshop (Jointly Organized with the 23rd Brazilian Symposium on Computer Networks (SBRC))* (2005)

623. V. Mainkar, K. Trivedi, Sufficient conditions for existence of a fixed point in stochastic reward net-based iterative models. IEEE. Trans. Softw. Eng. **22**(9), 640–653 (1996)

624. E. Marsden, J.-C. Fabre, J. Arlat, Dependability of CORBA systems: service characterization by fault injection, in *The 21th IEEE Symposium on Reliable Distributed Systems (SRDS '02)* (IEEE, Washington, 2002), pp. 276–285

625. E. Marshall, Fatal error: how patriot overlooked a scud. Science **3**, 1347 (1992)

626. E. Martin, S. Basu, T. Xie, Websob: a tool for robustness testing of web services, in *Companion to the Proceedings of the 29th International Conference on Software Engineering*, ICSE COMPANION '07 (2007), pp. 65–66

627. R. Matias, P. Barbetta, K. Trivedi, P. Filho, Accelerated degradation tests applied to software aging experiments. IEEE. Trans. Reliab. **59**(1), 102–114 (2010)

628. F. Mattiello-Francisco, E. Martins, A. Corsetti, A. Cavalli, E. Yano, Extended interoperability models for timed system robustness testing, in *IEEE Latin-American Conference on Communications, LATINCOM '09* (2009), pp. 1–6

629. J. Mauro, J. Zhu, I. Pramanick, The system recovery benchmark, in *Proceedings of the 2004 Pacific Rim International Symposium on Dependable Computing (PRDC 2004)* (2004)

630. R. Maxion, K. Tan, Benchmarking anomaly-based detection systems, in *Proceedings of the International Conference on Dependable Systems and Networks (DSN 2000)* (2000)

631. S. Mcallister, E. Kirda, C. Kruegel, Leveraging user interactions for in-depth testing of web applications, in *Proceedings of the 11th International Symposium on Recent Advances in Intrusion Detection, RAID '08* (2008), pp. 191–210

632. J.A. McCarthy, Introduction: From protection to resilience: Injecting "moxie" into the infrastructure security continuum, in *Critical Thinking: Moving from Infrastructure Protection to Infrastructure Resilience*, CIP Program Discussion Paper Series (George Mason University, 2007), pp. 1–8

633. G. McCullough, N. McDowell, G. Irwin, Fault diagnostics for internal combustion engines—current and future technologies. SAE technical paper series 2007-01-1603 (SAE International, 2007)

634. G. McGraw, B. Potter, Software security testing. IEEE. Secur. Priv. **2**, 81–85 (2004)

635. P. McLachlan, T. Munzner, E. Koutsofios, S.C. North, LiveRAC: interactive visual exploration of system management time-series data, in *Conference on Human Factors in Computing Systems, CHI*, Florence, April 2008, pp. 1483–1492

636. K. McMillan, *Symbolic Model Checking* (Kluwer, Dordruch, 1993)

637. Y. Mei, L. Liu, X. Pu, S. Sivathanu, Performance measurements and analysis of network I/O applications in virtualized cloud, in *International Conference on Cloud Computing* (2010)

638. A.M. Memon, An event-flow model of GUI-based applications for testing. Softw. Test. Verif. Reliab. **17**, 137–157 (2007)

639. D.A. Menasce, Virtualization: concepts, applications, and performance modeling, in *Proceedings of International CMG Conference* (2005)

640. D.A. Menasce, M.N. Bennani, On the use of performance models to design self-managing computer systems, in *Proceedings of the 2003 Computer Measurement Group Conference* (2003), pp. 7–12

641. D.A. Menasce, M.N. Bennani, Autonomic virtualized environments, in *ICAS '06: Proceedings of the International Conference on Autonomic and Autonomous Systems* (2006)

642. D.A. Menasce, M.N. Bennani, H. Ruan, On the use of online analytic performance models in self-managing and self-organizing computer systems, in *Self-* Properties in Complex Information Systems* (Springer, Heidelberg, 2005), pp. 128–142

643. M. Mendonca, N. Neves, Robustness testing of the Windows DDK, in *Proceedings of the 37th Annual IEEE/IFIP International Conference on Dependable Systems and Networks, DSN '07* (2007), pp. 554–564

644. J. Méreur, G. Malléus, D. Hardy, *Réseaux: Internet, téléphonie, multimédia*, Convergences et complémentarités (De Boeck, 2002)

645. J. Meserve, Sources: Staged cyber attack reveals vulnerability in power grid (2007), http://edition.cnn.com/2007/US/09/26/power.at.risk/index.html

646. J.F. Meyer, On evaluating the performability of degradable computing systems. IEEE. Trans. Comput. **C-29**(8), 720–731 (1980)

647. H. Mi, H. Wang, G. Yin, Y. Zhou, D. Shi, L. Yuan, Online self-reconfiguration with performance guarantee for energy-efficient large-scale cloud computing data centers, in *IEEE SCC* (2010)

648. Z. Micskei, I. Majzik, F. Tam, in *International Service Availability Symposium 2007*. Comparing Robustness of Ais-based Middleware Implementations. LCNS, vol. 4526, ed. by M. Malek, M. Reitenspieß, A. van Moorsel (Springer, Heidelberg, 2007), pp. 20–30

649. B.P. Miller, G. Cooksey, F. Moore, An empirical study of the robustness of macos applications using random testing. ACM SIGOPS Oper. Syst. Rev. **41**, 78–86 (2007)

650. B.P. Miller, D. Koski, C.P. Lee, V. Maganty, R. Murthy, A. Natarajan, J. Steidl, Fuzz revisited: a re-examination of the reliability of unix utilities and services. Technical report 1268, University of Wisconsin-Madison (1995)

651. S. Minato, Zero-suppressed bdds for set manipulation in combinatorial problems, in *30th ACM/IEEE Design Automation Conference* (1993), pp. 272–277

652. A. Miner, Efficient solution of GSPNs using canonical matrix diagrams, in *Petri Nets and Performance models (PNPM'01)*, ed. by R. German, B. Haverkort (IEEE Computer Society Press, New York, 2001), pp. 101–110

653. A. Miner, G. Ciardo, Efficient reachability set generation and storage using decision diagrams, in *Application and Theory of Petri Nets 1999*. Lecture Notes in Computer Science, vol.1639, ed. by H. Kleijn, S. Donatelli (Springer, Williamsburg, 1999), pp. 6–25

654. A. Miner, G. Ciardo, S. Donatelli, Using the exact state space of a markov model to compute approximate stationary measures. Perform. Eval. Rev. **28**(1), 207–216 (2000), Proceedings of ACM SIGMETRICS

655. G. Miremadi, J. Torin, Evaluating processor-behavior and three error-detection mechanisms using physical fault-injection. IEEE. Trans. Reliab. **44**(3), 441–454 (1995)

656. A.V. Mirgorodskiy, N. Maruyama, B.P. Miller, Problem diagnosis in large-scale computing environments, in *International Conference on High Performance Computing, Networking, Storage and Analysis*, Tampa, Nov 2006, p. 88

657. K. Mishra, K. Trivedi, Model based approach for autonomic availability management, in *Service Availability*. Lecture Notes in Computer Science, vol. 4328 (Springer, Heidelberg, 2006), pp. 1–16

658. Common Vulnerabilities and Exposures (2009)

659. S. Mocanu, C. Commault, Sparse representations of phase-type distributions. Commun. Stat. Stoch Models **15**(4), 759–778 (1999)

660. B. Mochizuki, I. Hadžić, Improving IEEE 1588v2 clock performance through controlled packet departures, IEEE. Commun. Lett. **14**(5), 459–461 (2010)

661. G. Mongardi, Dependable computing for railway control systems, in *3rd IFIP International Working Conference on Dependable Computing for Critical Applications* (1993), pp. 255–277

662. L. Montecchi, P. Lollini, A. Bondavalli, Dependability concerns in model-driven engineering, in *14th IEEE International Symposium on Object/Component/Service-Oriented Real-Time Distributed Computing Workshops (ISORCW 2011)* (2011), pp. 28–31 (in press)

663. L. Montecchi, P. Lollini, A. Bondavalli, Towards a mde transformation workflow for dependability analysis, in *16th IEEE International Conference on Engineering of Complex Computer Systems (ICECCS 2011)*, 27–29 April 2011 (in press)

664. R. Moraes, R. Barbosa, J. Durães, N. Mendes, E. Martins, H. Madeira, Injection of faults at component interfaces and inside the component code: are they equivalent?, in *Sixth European Dependable Computing Conference, EDCC-2006*, Coimbra, IEEE, Oct 2006, pp. 53–54

665. R. Moraes, J. Durães, R. Barbosa, E. Martins, H. Madeira, Experimental risk assessment and comparison using software fault injection, in *Proceedings of the 37th Annual IEEE/IFIP International Conference on Dependable Systems and Networks, DSN '07* (2007), pp. 512–521

666. F. Moreira, R. Maia, D. Costa, N. Duro, P. Rodriguez-Dapena, K. Hjortnaes, Static and dynamic verification of critical software for space applications, in *Proceedings of the Data Systems In Aerospace (DASIA 2003)* (2003)

667. L.M. Silva, J. Alonso, P. Silva, J. Torres, A. Andrzejak, Using virtualization to improve software rejuvenation, in *International Symposium on Network Computing and Applications (NCA 2007)* (2007)

668. J.K. Muppala, M. Malhotra, K.S. Trived, Stiffness-tolerant methods for transient analysis of stiff markov chains. Microelectron. Reliab. **34**, 1825–1841 (1994)

669. T. Murakami, Y. Horiuchi, Improvement of synchronization accuracy in IEEE 1588 using a queuing estimation method, in *International Symposium on Precision Clock Synchronization for Measurement, Control and Communication, 2009. ISPCS 2009*, Oct 2009, pp. 1–5

670. J.D. Musa, Software reliability data. Data and Analysis Center for Software, Rome Air Development Center, Rome, NY, Tech. Rep., (1979)

671. A.B. Nagarajan, F. Mueller, C. Engelmann, S.L. Scott, Proactive fault tolerance for hpc with xen virtualization, in *Proceedings of the 21st Annual International Conference on Supercomputing* (2007), pp. 23–32

672. F. Nai, M. Masera, A. De Cian, Integrating cyber attacks within fault trees. Reliab. Eng. Syst. Saf. **94**(9), 1394–1402 (2009)

673. J. Napper, P. Bientinesi, Can cloud computing reach the TOP500?, in *Proceedings of the Combined Workshops on UnConventional High Performance Computing Workshop Plus Memory Access Workshop (UCHPC-MAW'09)* (2009), pp. 17–20

674. P. Narasimhan, Vajra: benchmarking survivability in distributed systems. Technical report, CMU (2008), http://www.cylab.cmu.edu/default.aspx?id=1990

675. Metrics Data Program, http://mdp.ivv.nasa.gov (NASA/WVU IV & V Facility, 2008)

676. R. Nathuji, K. Schwan, Virtualpower: coordinated power management in virtualized enterprise systems, in *ACM SIGOPS Symposium on Operating Systems Principles* (2007)

677. National Infrastructure Advisory Council, Critical infrastructure resilience—final report and recommendations. Technical report, DHS/NIAC, 2009, http://www.dhs.gov/xlibrary/assets/niac/niac_critical_infrastructure_resilience.pdf

678. M. Nelli, A. Bondavalli, L. Simoncini, Dependability modelling and analysis of complex control systems: an application to railway interlocking, in *EDCC-2 European Dependable Computing Conference*, Taormina (1996), pp. 93–110

679. R. Nelson, Software data collection and analysis (1978)

680. C. Nemeth, R. Cook, Reliabilityversusresilience:Whatdoeshealthcareneed?, in *Symposium on High Reliability in Healthcare. Proceedings of the Human Factors and Ergonomics Society Annual Meeting*, ed. by C. Dominguez (Baltimore, 2007), pp. 621–625

681. C.P. Nemeth, Resilience engineering: The birth of a notion, in *Resilience Engineering Perspectives vol. 1: Remaining Sensitive to the Possibility of Failure*, ed. by E. Hollnagel, C.P. Nemeth, S. Dekker (Ashgate, 2008), p. 346

682. A.A. Neto, M. Vieira, Benchmarking untrustworthiness: an alternative to security measurement, Int. J. Depend. Trustworthy Inf. Syst. **1**(2), 32–54 (2010)

683. M.F. Neuts, *Matrix-Geometric Solutions in Stochastic Models. An Algorithmic Approach* (Dover, New York, 1981)

684. M.F. Neuts, M.E. Pagano, Generating random variates from a distribution of phase type, in *WSC '81: Proceedings of the 13th Winter Simulation Conference* (IEEE Press, Piscataway, 1981), pp. 381–387

685. N. Neves, J. Antunes, M. Correia, P. Veríssimo, R. Neves, Using attack injection to discover new vulnerabilities, in *IEEE/IFIP International Conference on Dependable Systems and Networks* (2006)

686. W.T. Ng, C.M. Aycock, G. Rajamani, P.M. Chen, Comparing disk and memory's resistance to operating system crashes, in *Proceedings of the 7th International Symposium on Software Reliability Engineering* (1996), pp. 185–194

687. W.T. Ng, P.M. Chen, The systematic improvement of fault tolerance in the rio file cache, in *Digest of Papers: Twenty-Ninth Annual International Symposium on Fault-Tolerant Computing (FTCS'99)*, Madison (IEEE Computer Society, New York, 1999), pp. 76–83

688. D.M. Nicol, W.H. Sanders, K.S. Trivedi, Model-based evaluation: from dependability to security, IEEE. Trans. Depend. Secure Comput. **1**(1), 48–65 (2004)

689. A. Nieuwenhuijs, E. Luiijf, M. Klaver, in *IFIP International Federation for Information Processing*. Modeling dependencies in critical infrastructures, vol. 290, ed. by M. Papa, S. Shenoi (Springer, Boston, 2009), pp. 205–213

690. Error, Fault, and Failure Data Collection and Analysis (2008)

691. R. Nou, S. Kounev, F. Julia, J. Torres, Autonomic qos control in enterprise grid environments using online simulation, J. Syst. Softw. **82**, 486–502 (2009)

692. R. Nou, S. Kounev, J. Torres, Building online performance models of grid middleware with fine-grained load-balancing: a globus toolkit case study, in *Proceedings of the 4th European Performance Engineering Conference on Formal Methods and Stochastic Models for Performance Evaluation, EPEW'07* (Springer, Heidelberg, 2007), pp. 125–140

693. NS-2. The Network Simulator ns-2, http://www.isi.edu/nsnam/ns/. Accessed Nov 2011

694. NVD, National Vulnerability Database (2010)

695. W.I. Obal, W. Sanders, State-space support for path-based reward variables, in *Computer Performance and Dependability Symposium, 1998. IPDS '98. Proceedings IEEE International*, Sept 1998, pp. 228–237

696. J. Oberheide, E. Cooke, F. Jahanian, Cloudav: N-version antivirus in the network cloud, in *Proceedings of the 17th USENIX Security Symposium*, July 2008

697. C.A. O'Cinneide, Characterization of phase-type distributions. Stoch. Models **6**, 1–57 (1990)

698. C.A. O'Cinneide, Phase-type distributions and invariant polytopes. Adv. Appl. Probab. **23**(3), 515–535 (1991)

699. A. Ogielski, W. Aiello, Sparse matrix computations on parallel processor arrays. SIAM J. Sci. Comput. **14**(3), 519–530 (1993)

700. J. Olah, I. Majzik, A model based framework for specifying and executing fault injection experiments, in *Proceedings of the 2009 Fourth International Conference on Dependability of Computer Systems, DEPCOS-RELCOMEX '09*, pp. 107–114 (2009)

701. A. Oliner, J. Stearley, Bad words: Finding faults in Spirit's syslogs. In *8th IEEE International Symposium on Cluster Computing and the Grid (CCGrid 2008)*, pp. 765–770, Lyon, France, May 2008

702. A.J. Oliner, A.V. Kulkarni, A. Aiken, Using correlated surprise to infer shared influence, in *IEEE/IFIP International Conference on Dependable Systems and Networks*, Chicago, IL, July 2010, pp. 191–200

703. Object Management Group, UML Profile for Modeling Quality of Service and Fault Tolerance Characteristics and Mechanisms Specification, v1.1, 2008, http://www.omg.org/spec/QFTP/1.1/

704. Object Management Group, UML Profile for Modeling and Analysis of Real-time and Embedded Systems (MARTE), v1.0, Nov 2009, http://www.omg.org/spec/MARTE/1.0/

705. S. Ostermann, A. Iosup, N. Yigitbasi, R. Prodan, T. Fahringer, D. Epema, A performance analysis of EC2 cloud computing services for scientific computing, in *1st International Conference on Cloud Computing* (ICST Press, 2009)

706. Open source vulnerability database (2010), http://osudb.org/

707. T.J. Overbye, X. Cheng, Y. Sun, A comparison of the AC and DC power flow models for LMP calculations, in *37th IEEE Hawaii International Conference on System Sciences (CD-ROM)*, Big Island, Hawaii, Jan 2004, p. 9

708. OWASP Top 10, 1007 (OWASP Foundation, 2007), http://www.owasp.org/index.php/Top_10_2007

709. Owasp Testing Guide v3 (OWASP Foundation, 2008), http://www.owasp.org/images/5/56/OWASP_Testing_Guide_v3.pdf

710. SQL Injection (OWASP Foundation, 2008), http://www.owasp.org/index.php/SQL_injection

711. Cross-site Scripting (XSS) (OWASP Foundation, 2009), http://www.owasp.org/index.php/Cross-site_Scripting_(XSS)

712. OWASP Foundation, OWASP WSFuzzer Project (2011), http://www.owasp.org/index.php/Category:OWASP_WSFuzzer_Project

713. S.D.P. Bucholtz, G. Ciardo, P. Kemper, Complexity of memory-efficient kronecker operations with applications to the solution of markov models. INFORMS J. Comput. **3**(12), 203–222 (2000)

714. G. Pai, J. Dugan, Automatic synthesis of dynamic fault trees from uml system models, in *Proceedings of the 13th International Symposium on Software Reliability Engineering, 2002. ISSRE 2003*, pp. 243–254 (2002)

715. X. Pan, J. Tan, S. Kavulya, R. Gandhi, P. Narasimhan, Blind men and the elephant: piecing together hadoop for diagnosis, in *International Symposium on Software Reliability Engineering (ISSRE)*, Mysuru, India, Nov 2009

716. S. Panzieri, R. Setola, G. Ulivi, An agent based simulator for critical interdependent infrastructures, in *2nd International Conference on Critical Infrastructures (CRIS 2004)*, Grenoble, France, Oct 2004

717. K. Park, S. Kim, Availability analysis and improvement of active/standby cluster systems using software rejuvenation. J. Syst. Softw. **61**, 121–128 (2002)

718. D. Parker, Implementation of symbolic model checking for probabilistic systems. Ph.D. thesis, School of Computer Science, Faculty of Science, University of Birmingham (2002)

719. R. Patton, Fault detection and diagnosis in aerospace systems using analytical redundancy. Comput. Cont. Eng. J. **2**(3), 127–136 (1991)

720. P. Pederson, D. Dudenhoeffer, S. Hartley, M. Permann, Critical infrastructure interdependency modeling: a survey of US and international research. Technical report, Idaho National lab, USA (2006), http://www.pcsforum.org/library/files/1159904563-TSWG_INL_CIP_Tool_Survey_final.pdf.

721. pentestmonkey.net (2009), http://pentestmonkey.net

722. C. Perrow, Normal Accidents—Living with High Risk Technologies (Basic Books, New York, 1984)

723. P. Peti, R. Obermaisser, H. Kopetz, Out-of-norm assertions, in *11th IEEE Real Time and Embedded Technology and Applications Symposium (RTAS '05)*, San Francisco, CA, March 2005, pp. 280–291

724. D.C. Petriu, C.M. Woodside, Approximate mva for software client/server models by markov chain task-directed aggregation, in *3rd IEEE Symposium on Parallel and Distributed Processing*, Dec 1991, pp. 322–329

725. H. Petroski, To Engineer is Human: The Role of Failure in Successful Design (St Martin's Press, New York, 1992)

726. A. Pfening, S. Garg, A. Puliafito, M. Telek, K.S. Trivedi, Optimal software rejuvenation for tolerating soft failures. Perform. Eval. **27**(28), 491–506 (1996)

727. PHP-Nuke (2010), http://phpnuke.org/

728. phpBB (2009), http://www.phpbb.com/

729. phpMyAdmin (2009), http://www.phpmyamin.net/home_page/index.php

730. G.P. Picco, Software engineering and wireless sensor networks: happy marriage or consensual divorce? in *FSE/SDP Workshop on Future of Software Engineering Research* (2010), pp. 283–286

731. M. Pizza, L. Strigini, A. Bondavalli, F. Di Giandomenico, Optimal discrimination between transient and permanent faults, in *3rd IEEE International Symposium on High-Assurance Systems Engineering (HASE '98)*, pp. 214–223 (1998)

732. B. Plateau, On the stochastic structure of parallelism and synchronization models for distributed algorithms, in *Proceedings of the 1985 ACM SIGMETRICS Conference on Measurement and Modeling of Computer Systems, SIGMETRICS '85* (ACM, New York, 1985), pp. 147–154

733. B. Plateau, *De l'Evaluation du Parrallélisme et de la Synchronisation*. PhD thesis, Nov 1984

734. B. Plateau, K. Atif, Stochastic automata network of modeling parallel systems. IEEE Trans. Softw. Eng. **17**(10), 1093–1108, Oct 1991

735. B. Plateau, J. Fourneau, A methodology for solving markov models of parallel systems. J. Parallel. Distrib. Comput. **12**(4), 370–387 (1991)

736. B. Plateau, J. Fourneau, K. Lee, Peps: a package for solving complex markov models of parallel systems, in *Proceedings of the 4th International Conference on Modelling Techniques and Tools for Computer Performance Evaluation* (1988)

737. G. Pola, M. Bujorianu, J. Lygeros, M.D. Benedetto, Stochastic hybrid models: an overview, in *International Conference on the Analysis and Design of Hybrid Systems. IPV-IFAC Proceedings* (2003), pp. 45–50

738. R.J. Pooley, The integrated modeling support environment: a new generation of performance modeling tools, in *Proceedings of the 5th International Conference in Computer Performance Evaluation: Modeling Techniques and Tools*, Torino, Italy, Feb 1991, pp. 1–15

739. P. Popov, L. Strigini, Assessing asymmetric fault-tolerant software, in *21st International Symposium on Software Reliability Engineering (ISSRE 2010)* (IEEE Computer Society Press, San Jose, 2010), pp. 41–50

740. P. Popov, L. Strigini, A. Romanovsky, Diversity for off-the-shelf components, in *DSN 2000, International Conference on Dependable Systems and Networks—Fast Abstracts Supplement* (IEEE Computer Society Press, New York, 2000), pp. B60–B61

741. M. Popovic, J. Kovacevic, A statistical approach to model-based robustness testing, in *Proceedings of the 14th Annual IEEE International Conference and Workshops on the Engineering of Computer-Based Systems* (2007), pp. 485–494

742. P. Pourbeik, P.S. Kundur, C.W. Taylor, The anatomy of a power grid blackout. *IEEE Power Energy Magazine*, Sept–Oct 2006, pp. 22–29

743. D. Powell, R. Stroud, Conceptual model and architecture of MAFTIA, in *Project MAFTIA, deliverable D21* (2003)

744. F.P. Preparata, G. Metze, R.T. Chien, On the connection assignment problem of diagnosable systems. IEEE Transact. Elect. Comput. **EC-16**(6), 848–854 (1967)

745. Project Next Generation Infrastructures, http://www.nginfra.nl/index.php?id=4. Accessed 25 Dec 2009

746. Project PSERC—Power Systems Engineering Research Center, http://www.pserc.wisc.edu. Accessed 25 Dec 2009

747. Project VITA—vital infrastructure threats and assurance, http://vita.iabg.eu/index.php. Accessed 25 Dec 2009

748. C. Pu, J. Noe, A. Proudfoot, Regeneration of replicated objects: a technique and its Eden implementation, in *Proceedings of the 2nd International Conference on Data Engineering* (1986), pp. 175–187

749. X. Pu, L. Liu, Y. Mei, S. Sivathanu, Y. Koh, C. Pu, Understanding performance interference of I/O workload in virtualized cloud environments, in *International Conference on Cloud Computing* (2010)

750. R. Pulungan, Reduction of acyclic phase-type representations. Ph.D thesis, Universität des Saarlandes (2009)

751. K. Purchala, L. Meeus, D. Van Dommelen, R. Belmans, Usefulness of DC power flow for active power flow analysis, in *IEEE Power Engineering Society General Meeting*, San Francisco, June 2005, pp. 454–459

752. R. Puttini, J.-M. Percher, L. Me, R. de Sousa, A fully distributed ids for manet, in *ISCC "04: Proceedings of the Ninth International Symposium on Computers and Communications 2004 (ISCC"04)*, Washington, DC (IEEE Computer Society, New York, 2004), pp. 331–338

753. C. Queiroz, A. Mahmood, J. Hu, Z. Tari, X. Yu, Building a scada security testbed, in *International Conference on Network and System Security* (2009), pp. 357–364

754. QCA 164-Appeal against Conviction and Sentence, Supreme Court of Queensland, May 2002

755. Acunetix. Acunetix Web Vulnerability Scanner (2011), http://www.acunetix.com/vulnerability-scanner/

756. A. Rahmati, L. Zhong, CRAWDAD data set rice/context (v. 2007-05-23), http://crawdad. cs.dartmouth.edu/rice/context. Accessed May 2007
757. A. Rajabzadeh, S.G. Miremadi, M. Mohandespour, Experimental evaluation of master/ checker architecture using power supply- and software-based fault injection, in *10th IEEE International On-Line Testing Symposium (IOLTS 2004)*, Funchal, Madeira Island, 12–14 July (IEEE Computer Society, New York, 2004), pp. 239–246
758. P. Ramachandran, P. Kudva, J.W. Kellington, J. Schumann, P. Sanda, Statistical fault injection, in *Proceedings of the 38th Annual IEEE/IFIP International Conference on Dependable Systems and Networks (38th DSN'08)*, Anchorage (IEEE Computer Society, New York, 2008), pp. 122–127
759. H.V. Ramasamy, M. Schunter, Architecting dependable systems using virtualization, in *Workshop on Architecting Dependable Systems: Supplemental Volume of the DSN'07* (2007)
760. R. Ramdhany, P. Grace, G. Coulson, D. Hutchison, Manetkit: supporting the dynamic deployment and reconfiguration of ad-hoc routing protocols, in *Middleware '09: Proceedings of the 10th ACM/IFIP/USENIX International Conference on Middleware* (Springer, New York, 2009), pp. 1–20
761. K. Rangan, The cloud wars: $100+ billion at stake. Technical report, Merrill Lynch, May 2008
762. M. Rebaudengo, M.S. Reorda, Evaluating the fault tolerance capabilities of embedded systems via BDM, in *17th IEEE VLSI Test Symposium (VTS '99)*, San Diego, 25–30 April 1999 (IEEE Computer Society, New York, 1999), pp. 452–457
763. A. Reibman, M. Veeraraghavan, Reliability modeling: an overview for system designers. Computer **24**(4), 49–57 (1991)
764. P. Reinecke, M. Telek, K. Wolter, Reducing the costs of generating APH-distributed random numbers, in *MMB & DFT 2010*. Lecture Notes in Computer Science, vol. 5987, ed. by B. Müller-Clostermann, K. Echtle, E. Rathgeb (Springer, Berlin, 2010), pp. 274–286
765. P. Reinecke, K. Wolter, Libphprng—a library to generate PH-distributed random numbers. Technical report, Freie Universität Berlin (2011) (to appear)
766. M. Reiser, S. Lavenberg, Mean value analysis of closed multichain queueing networks. JACM **22**(4), 313–322 (1980)
767. http://www.resilience-engineering.org/intro.htm. Accessed 2 Oct 2011
768. Resilience 2008. *Workshop on resilience in high-performance computing, Resilience 2008*, http://xcr.cenit.latech.edu/resilience2008/. Accessed 26 Nov 2011
769. ReSIST. From resilience-building to resilience-scaling technologies: directions. Deliverable D13, ReSIST (Resilience for Survivability in IST) European Network of Excellence 2007. http://www.resist-noe.org/. Accessed 26 Nov 2011
770. ReSIST, Selected current practices. Deliverable D39, ReSIST (Resilience for Survivability in IST) European Network of Excellence 2009
771. RESIST Consortium, D13—from resilience-building to resilience-scaling technologies: directions. Technical report, RESIST Consortium, Sept 2007
772. RESIST NoE. Resilience-building technologies: State of knowledge (2006), http://www. resist-noe.org/
773. Resist resilience for survivability in IST (project IST-0265764) (2006), http://www. resist-noe.org/index.html
774. Resist NoE, Resilience-building technologies: state of knowledge, deliverable d12, http://www.resist-noe.org/outcomes/outcomes.html
775. P. Reynolds, C.E. Killian, J.L. Wiener, J.C. Mogul, M.A. Shah, A. Vahdat, Pip: detecting the unexpected in distributed systems, in *Proceedings of the 3rd Conference on Networked Systems Design & Implementation*, San Jose, vol. 3, May 2006, pp. 115–128
776. P. Reynolds, J.L. Wiener, J.C. Mogul, M.K. Aguilera, A. Vahdat, Wap5: black-box performance debugging for wide-area systems, in *WWW '06: Proceedings of the 15th International Conference on World Wide Web* (ACM Press, New York, 2006), pp. 347–356

777. T. Rigole, G. Deconinck, A survey on modelling and simulation of interdependent critical infrastructures, in *Proceedings of the 3rd IEEE Benelux Young Researchers Symposium in Electrical Power Engineering*, Ghent, Belgium, April 2006

778. S.M. Rinaldi, J.P. Peerenboom, T.K. Kelly, Identifying, understanding, and analyzing critical infrastructure interdependencies. IEEE Control Syst Mag, Dec 2001, pp. 11–25

779. I. Rish, M. Brodie, S. Ma, N. Odintsova, A. Beygelzimer, G. Grabarnik, K. Hernandez, Adaptive diagnosis in distributed systems. IEEE Trans. Neural Netw. 16(5), 1088–1109 (2005)

780. I. Rish, M. Brodie, N. Odintsova, S. Ma, G. Grabarnik, Real-time problem determination in distributed systems using active probing, in *Network Operations and Management Symposium*, Seoul, April 2004, pp. 133–146

781. F. Ritter, N. Brausen, N. Millar, Distribution point-of-delivery interconnection process guideline—standards of service. Technical report. Alberta electric system operator (2005)

782. B. Robert, R. De Calan, L. Morabito, Modelling interdependencies among critical infrastructures. Int. J. Crit. Infrastruct. 4(4), 392–408 (2008)

783. G. Rochlin, T. LaPorte, K. Roberts, The self-designing high-reliability organization: aircraft carrier flight operations at sea. Naval War Coll. Rev. 40(4), 76–90 (1987)

784. M. Rodríguez, A. Albinet, J. Arlat, MAFALDA-RT: a tool for dependability assassment of real-time systems, in *DSN '02: Proceedings of the 2002 International Conference on Dependable Systems and Networks* (2002), pp. 267–272

785. F. Romani, S. Chiaradonna, F. Di Giandomenico, L. Simoncini, Simulation models and implementation of a simulator for the performability analysis of electric power systems considering interdependencies, in *10th IEEE High Assurance Systems Engineering Symposium (HASE'07)*, Dallas, Nov 2007, pp. 305–312

786. P. Rooney. Microsoft's ceo: 80-20 rule applies to bugs, not just features. Online report, CRN (2002), http://www.crn.com/security/18821726

787. V. Rosato, L. Issacharoff, F. Tiriticco, S. Meloni, S.D. Porcellinis, R. Setola, Modelling interdependent infrastructures using interacting dynamical models. Int. J. Crit. Infrastruct. 4(1/2), 63–79 (2008)

788. W.L. Roy Billinton, *Reliability Assessment of Electrical Power Systems Using Monte Carlo Methods*, hardbound edn. (Plenum, New York, 1994)

789. A.-E. Rugina, K. Kanoun, M. Kaâniche, Chapter a system dependability modeling framework using AADL and GSPNs, in *Architecting Dependable Systems IV*, ed by R. de Lemos, C. Gacek, A. Romanovsky (Springer, Berlin, 2007), pp. 14–38

790. J.-C. Ruiz, P. Yuste, P. Gil, L. Lemus, On benchmarking the dependability of automotive engine control applications, in *Proceedings of the IEEE/IFIP 2004 International Conference on Dependable Systems and Networks (DSN 2004)* (2004)

791. S. Ruzzante, E. Castorini, E. Marchei, V. Fioriti, A metric for measuring the strength of inter-dependencies, in *SAFECOMP 2010*. Lecture Notes in Computer Science, vol. 6351, ed. by E. Schoitsch (Springer, Berlin, 2010), pp. 291–302

792. F. Saad-Khorchef, A. Rollet, R. Castanet, A framework and a tool for robustness testing of communicating software, in *Proceedings of the 2007 ACM Symposium on Applied Computing, SAC '07* (2007), pp. 1461–1466

793. K. Sachs, Performance modeling and benchmarking of event-based systems. Ph.D. thesis, TU Darmstadt (2010)

794. K. Sachs, S. Kounev, J. Bacon, A. Buchmann, Performance evaluation of message-oriented middleware using the SPECjms2007 benchmark. Perform. Eval. 66(8), 410–434 (2009)

795. SAE-AS5506/1, Architecture Analysis and Design Language (AADL) Annex vol. 1, Annex E: Error Model Annex. Society of Automotive Engineers (2006), http://standards.sae.org/as5506/1

796. SAFEDMI—Safe Driver Machine Interface (DMI) for ERTMS trains (Project IST-FP6-STREP-031413) (2006), http://www.safedmi.org/

797. SAFEDMI—Quantitative Evaluation Methodology, Deliverable D4.1, June 2008

798. R. Sahner, K. Trivedi, A. Puliafito, *Performance and Reliability Analysis of Computer Systems: An Example-Based Approach Using the SHARPE Software Package* (Kluwe, Boston, 1996)

799. F. Salfner, K. Wolter, A queueing model for service availability of systems with rejuvenation, in *Software Reliability Engineering Workshops, 2008. IEEE International Conference on Software Reliability Engineering (ISSRE)* (2008), pp. 1–5

800. R.R. Sambasivan, A.X. Zheng, M.D. Rosa, E. Krevat, S. Whitman, M. Stroucken, W. Wang, L. Xu, G.R. Ganger, Diagnosing performance changes by comparing request flows, in *Proceedings of the 8th USENIX Conference on Networked Systems Design and Implementation*, Boston, MA, March 2011, pp. 43–56

801. W. Sanders, Progress towards a resilient power grid infrastructure, in *IEEE Power and Energy Society General Meeting (PES GM)* (2010)

802. W.H. Sanders, Integrated frameworks for multi-level and multi-formalism modeling, in *Petri Nets and Performance Models, 1999. Proceedings of the 8th International Workshop on* (1999), pp. 2–9

803. W.H. Sanders, J.F. Meyer, Reduced base model construction methods for stochastic activity networks. IEEE J. Sel. Areas Commun. **9**(1), 25–36 (1991)

804. W.H. Sanders, J.F. Meyer, A unified approach for specifying measures of performance, dependability and performability, in *Dependable Computing for Critical Applications*. Dependable Computing and Fault-Tolerant Systems, vol. 4, ed. by A. Avizienis, J. Laprie (Springer, Berlin, 1991), pp. 215–237

805. W.H. Sanders, J.F. Meyer, Stochastic activity networks: formal definitions and concepts, in *Lectures on Formal Methods and Performance Analysis*. Lecture Notes in Computer Science, vol. 2090, ed. by E. Brinksma, H. Hermanns, J.P. Katoen (Springer, Heidelberg, 2001), pp. 315–343

806. W.H. Sanders, W.D. Oball II, M.A. Qureshi, F.K. Widjanarko, The ultrasan modeling environment. Perform. Eval. **24**, 89–115 (1995)

807. Sandia National Laboratories, Information operations red team and assessments^TM, http://www.sandia.gov/iorta/

808. P. Saravakos, G. Gravvanis, V. Koutras, A. Platis, A comprehensive approach to software aging and rejuvenation on a single node software system, in *Proceedings of the 9th Hellenic European Research on Computer Mathematics and Its Applications Conference*, Sept 2009

809. P. Saripalli, B. Walters, Quirc: a quantitative impact and risk assessment framework for cloud security, in *IEEE International Conference on Cloud Computing* (2010), pp. 280–288

810. J. Scaramella, Worldwide Server Power and Cooling Expense, 2006–2010 Forecast (IDC, 2006)

811. D.C. Schmidt, Guest editor's introduction: model-driven engineering. Computer **39**(2), 25–31 (2006)

812. B. Schroeder, G. Gibson, A large-scale study of failures in high-performance computing systems, in *Proceedings of the International Conference on Dependable Systems and Networks*, Philadelphia, PA, June 2006, pp. 249–258

813. B. Schroeder, G.A. Gibson, Disk failures in the real world: what does an MTTF of 1,000,000 hours mean to you?, in *USENIX Conference on File and Storage Technologies*, San Jose, CA, Feb 2007, pp. 1–16

814. J. Schuster, M. Siegle, A symbolic multilevel method with sparse submatrix representation for memory-speed tradeoff, in *14. GI/ITG Conference Measurement, Modelling and Evaluation of Computer and Communication Systems (MMB08)*, VDE Verlag (2008), pp. 191–205

815. J. Schuster, M. Siegle, Speeding up the symbolic multilevel algorithm, in *6th International Workshop on the Numerical Solution of Markov Chains (NSMC2010)* (2010), pp. 79–82

816. N. Seixas, J. Fonseca, M. Vieira, H. Madeira, Looking at web security vulnerabilities from the programming language perspective: a field study, in *20th International Symposium on Software Reliability Engineering, ISSRE-2009*, Mysuru, India (IEEE, Nov 2009), pp. 129–135

817. R. Sekar, An efficient black-box technique for defeating web application attacks, in *Proceedings of the 16th Annual Network and Distributed System Security Symposium* (2009)

818. A. Selvam, K. Kathiravan, R. Reshmi, A cross-layer tcp protocol with adaptive modulation for manets, in *ICSCN '08: International Conference on Signal Processing, Communications and Networking* (2008), pp. 428–433

819. M. Serafini, A. Bondavalli, N. Suri, Online diagnosis and recovery: on the choice and impact of tuning parameters. IEEE Trans. Depend. Secur. Comput. **4**(4), 295–312 (2007)

820. I.G.T. Services, Ibm internet security systems x-force—2008 trend & risk report. Online report, IBM Corporation, http://www-935.ibm.com/services/us/iss/xforce/trendreports/xforce-2008-annual-report.pdf (2009)

821. K. Sevcik, I. Mitrani, The distribution of queueing network states at input and output instants. J. ACM **28**(2), 358–371 (1981)

822. K. Shen, C. Stewart, C. Li, X. Li, Reference-driven performance anomaly identification, in *Proceedings of the Eleventh International Joint Conference on Measurement and Modeling of Computer Systems*, Seattle, WA, June 2009, pp. 85–96

823. K.G. Shin, C.M. Krishna, Y.-H. Lee, Optimal dynamic control of resources in a distributed system. IEEE Trans. Softw. Eng. **15**(10), 1188–1198 (1989)

824. Y. Shin, L. Williams, T. Xie, SQLUnitGen: SQL injection testing using static and dynamic analysis, in *The 17th IEEE International Symposium on Software Reliability Engineering, ISSRE* (2006)

825. R.H. Shumway, D.S. Stoffer, *Time Series Analysis and Its Applications (Springer Texts in Statistics)* (Springer, New York, 2005)

826. L. Siegele, Let it rise: a special report on corporate IT. Economist **389**(8603), 3–16 (2008)

827. B.H. Sigelman, L.A. Barroso, M. Burrows, P. Stephenson, M. Plakal, D. Beaver, S. Jaspan, C. Shanbhagy, Dapper, a large-scale distributed systems tracing infrastructure. Technical report dapper-2010-1, Google, April 2010

828. L.M. Silva, J. Alonso, J. Torres, Using virtualization to improve software rejuvenation. IEEE Trans. Comput. **58**(11), 1525–1538 (2009)

829. N. Singh, S. Brahmjit, Effect of soft handover parameters on CDMA cellular networks. J. Theor. Appl. Inf. Technol. **B v.77**, 523 (2010)

830. S. Singh, W.H. Sanders, D.M. Nicol, M. Seri, Automatic verification of distributed and layered security policy implementations. Technical report UILU-ENG-08-2209 (CRHC-08-05) (University of Illinois, Urbana-Champaign, 2008)

831. S. Sivathanu, L. Liu, M. Yiduo, X. Pu, Storage management in virtualized cloud environment, in *Proceedings of IEEE International Conference on Cloud Computing* (2010), pp. 204–211

832. D. Skarin, R. Barbosa, J. Karlsson, Comparing and validating measurements of dependability attributes, in *Proceedings of the 8th European Dependable Computing Conference EDCC 2010*, April 2010, pp. 3–12

833. D. Skarin, R. Barbosa, J. Karlsson, GOOFI-2: a tool for experimental dependability assessment, in *Proceedings of the 40th Annual IEEE/IFIP International Conference on Dependable Systems and Networks DSN 2010*, June/July 2010, pp. 557–562

834. D. Skarin, J. Karlsson, Software implemented detection and recovery of soft errors in a brake-by-wire system, in *Proceedings of the 7th European Dependable Computing Conference EDCC 2008*, May 2008, pp. 145–154

835. P. Skomoroch, MPI cluster programming with Python and Amazon EC2, in *Proceedings of the 6th Annual Python Community Conference PyCon'08*, Chicago, March 2008, http://www.datawrangling.com/mpi-cluster-with-python-and-amazon-ec2-part-2-of-3

836. P. Skomoroch, Data wrangling image: fedora core 6 mpi compute node with python libraries (2010), http://developer.amazonwebservices.com/connect/entry.jspa?categoryID=101&externalID=705

837. C.U. Smith, L.G. Williams, *Performance Solutions: A Practical Guide to Creating Responsive, Scalable Software*, Addison-Wesley, Boston (2002)

838. W.D. Smith, TPC-W: benchmarking an ecommerce solution. TPC White paper (2001), http://www.tpc.org/tpcw/

839. W. Sobel, S. Subramanyam, A. Sucharitakul, J. Nguyen, H. Wong, A. Klepchukov, S. Patil, O. Fox, D. Patterson, Cloudstone: multi-platform, multi-language benchmark and measurement tools for web 2.0 (2008)

840. L. Song, D. Kotz, R. Jain, X. He, Evaluating location predictors with extensive wi-fi mobility data, in *Proceedings of the Twenty-Third Annual Joint Conference of the IEEE Computer and Communications Societies (INFOCOM)*, vol. 2, March 2004, pp. 1414–1424

841. L. Spainhower, J. Isenberg, R. Chillarege, J. Berding, Design for fault-tolerance in system es/9000 model 900, in *Proceedings of the 22nd IEEE FTCS International Symposium on Fault-Tolerant Computing* (1992), pp. 38–47

842. SPEC Virtualization Committee, SPECvirt_sc2010 (2010), http://www.spec.org/virt_sc2010

843. Splunk Inc., Splunk: The IT Search Company (2005), http://www.splunk.co

844. M. Sridharan, S. Ramasubramanian, A.K. Somani, Himap: architecture, features, and hierarchical model specification techniques, in *Proceedings of the Computer Performance Evaluation: Modelling Techniques and Tools*. 10th International Conference, Tools 98, Palma de Mallorca, Spain, Sept 14–18, 1998. Lecture Notes in Computer Science, vol. 1469, ed. by R. Puigjaner, N.N. Savino, B. Serra (Springer, Heidelberg, 1998), pp. 348–351

845. Standard Performance Evaluation Corporation (SPEC) Specweb99 release 1.02 documentation, Specification, Standard Performance Evaluation Corporation (2000), http://www.spec.org/web99/

846. Standard Performance Evaluation Corporation (SPEC), Website (2011)

847. A. Stefanek, Grouped PEPA analyzer, http://doc.ic.ac.uk/~as1005/GPA

848. A. Stefanek, R.Hayden, J.T. Bradley, Fluid analysis of energy consumption using rewards in massively parallel Markov models, in *International Conference on Performance Engineering (ICPE) 2011*, 14–16 March 2011 (to appear)

849. A. Stefanek, R.A. Hayden, J.T. Bradley, A new tool for the performance analysis of massively parallel computer systems, in *QAPL'10, 8th Workshop on Quantitative Aspects of Programming Languages, Electronic Proceedings of Theoretical Computer Science*, vol. 28 (2010), pp. 159–181

850. M. Steinder, A.S. Sethi, A survey of fault localization techniques in computer networks. Sci. Comput. Program. **53**(2), 165–194 (2004)

851. C. Steve, R. Martin, Vulnerability type distributions in CVE, in *Mitre report*, May 2007

852. C. Stewart, T. Kelly, A. Zhang, Exploiting nonstationarity for performance prediction, in *Proceedings of the 2nd ACM SIGOPS/EuroSys European Conference on Computer Systems 2007*, Lisbon, March 2007, pp. 31–44

853. W. Stewart, K. Atif, B. Plateau, The numerical solution of stochastic automata networks. Eur. J. Oper. Res. **86**, 503–525 (1995)

854. W.J. Stewart, *Probability, Markov Chains, Queues and Simulation. The Mathematical Basis of Performance Modeling* (Princeton University Press, Princeton, 2009)

855. A. Stock, J. Williams, D. Wichers, OWASP top 10, in *OWASP Foundation*, July 2007

856. D.T. Stott, B. Floering, D. Burke, Z. Kalbarczyk, R.K. Iyer, NFTAPE: a framework for assessing dependability in distributed systems with lightweight fault injectors, in *Proceedings of the International Computer Performance and Dependability Symposium PDS'00* 2000, pp. 91–100

857. K. Stouffer, J. Falco, K. Kent, Guide to supervisory control and data acquisition (scada) and industrial control systems security. *NIST special publication*, 800(82), Sept 2006, pp. 1–13

858. A.W. Stroupe, S. Singh, R. Simmons, T. Smith, P. Tompkins, V. Verma, R. Vitti-Lyons, M. Wagner, Technology for autonomous space systems. Technical report CMU-RI-TR-00-02, Carnegie Mellon University, Robotics Institute, Sept 2001

859. S. Strubbe, A. van der Schaft, Compositional modeling of stochastic hybrid systems, in *Stochastic Hybrid Systems: Control Engineering Series*, vol. 24 (CRC Press, Boca Raton, 2006), pp. 47–78

860. D. Stuttard, M. Pinto, *The Web Application Hacker's Handbook: Discovering and Exploiting Security Flaws* (Wiley, Chichester, 2007)

861. J.L. Sun, S. Singh, ATCP: TCP for mobile ad hoc networks. IEEE. J. Sel. Areas Commun. **19**, 1300–1315 (1999)

862. K. Sundaresan, V. Anantharaman, H.-Y. Hsieh, R. Sivakumar, ATP: a reliable transport protocol for ad-hoc networks, in *MobiHoc '03: Proceedings of the 4th ACM International Symposium on Mobile Ad Hoc Networking & Computing* (ACM, New York, 2003), pp. 64–75

863. N. Svendsen, S. Wolthusen, Analysis and statistical properties of critical infrastructure interdependency multiflow models, in *Information Assurance and Security Workshop, 2007. IAW '07. IEEE SMC*, June 2007, pp. 247–254

864. S.E.G. Systems and Operations, Telemetry and telecommand packet utilization (ecss-e-70-41a) (2003)

865. A. Takanen, J. DeMott, C. Miller, *Fuzzing for Software Security Testing and Quality Assurance*, 1st edn. (Artech House, 2008)

866. J. Tan, X. Pan, S. Kavulya, R. Gandhi, P. Narasimhan, SALSA: analyzing logs as state machines, in *USENIX Workshop on Analysis of System Logs*, San Diego, Dec 2008

867. J. Tan, X. Pan, S. Kavulya, R. Gandhi, P. Narasimhan, Mochi: visual log-analysis based tools for debugging hadoop, in *USENIX Workshop on Hot Topics in Cloud Computing HotCloud*, San Diego, June 2009

868. TCIP, NSF Project TCIP—Trustworthy cyber infrastructure for the power grid, http://www.iti.uiuc.edu/tcip

869. TCIP Team, Trustworthy cyber-infrastructure for power TCIP, in *Workshop on Research Directions for Security and Networking in Critical Real-Time and Embedded Systems*, San Jose, April 2006

870. M. Telek, A. Heindl, Matching moments for acyclic discrete and continuous phase-type distributions of second order. Int. J. Simul. Syst. Sci. Technol. **3**(3–4), 47–57 (2002)

871. M. Telek, A. Heindl, Moment bounds for acyclic discrete and continuous phase-type distributions of second order, in *Proceedings of the UK Performance Evaluation Workshop* (2002)

872. M. Telek, S. Rácz, Numerical analysis of large Markovian reward models. Perform. Eval. **36–37**, 95–114 (1999)

873. C.-W. Ten, C.-C. Liu, G. Manimaran, Vulnerability assessment of cybersecurity for scada systems. IEEE Trans. Power Syst. **23**(4), 1836–1846 (2008)

874. C.-W. Ten, G. Manimaran, C.-C. Liu, Cybersecurity for critical infrastructures: attack and defense modeling. Trans. Sys. Man Cyber. Part A **40**, 853–865 (2010)

875. G. Tesauro, N.K. Jong, R. Das, M.N. Bennani, A hybrid reinforcement learning approach to autonomic resource allocation, in *International Conference on Autonomic Computing* (2006)

876. N.X. Thang, K. Geihs, Model-driven development with optimization of non-functional constraints in sensor network, in *ICSE Workshop on Software Engineering for Sensor Network Applications* (2010), pp. 61–65

877. T. Thein, S.-D. Chi, J.S. Park, Availability modeling and analysis on virtualized clustering with rejuvenation. Int. J. Comput. Sci. Netw. Secur. **8**(9), 72–80 (2008)

878. T. Thein, J.S. Park, Availability analysis of application servers using software rejuvenation and virtualization. J. Comput. Sci. Technol. **24**(2), 339–346 (2009)

879. N. Thomas, Y. Zhao, Mean value analysis for a class of PEPA models. Comput. J. (2011) (accepted). doi:10.1093/comjnl/bxq064

880. A. Thümmler, P. Buchholz, M. Telek, A novel approach for phase-type fitting with the EM algorithm. IEEE Trans. Depend. Secur. Comput. **3**(3), 245–258 (2006)

881. O. Tickoo, R. Iyer, R. Illikkal, D. Newell, Modeling virtual machine performance: challenges and approaches. SIGMETRICS Perform. Eval. Rev. **37**, 55–60 (2010)

882. W.J. Tolone, D. Wilson, A. Raja, W.-n. Xiang, H. Hao, S. Phelps, E. W. Johnson, Critical infrastructure integration modeling and simulation, in *Intelligence and Security Informatics*. Lecture Notes in Computer Science, vol. 3073, ed. by H. Chen, R. Moore, D.D. Zeng, J. Leavitt (Springer, Berlin, 2004), pp. 214–225

883. M. Torgerson, Security metrics for communication systems, in *Proceedings of the 12TH ICCRTS* (2007)

884. Transaction Processing Performance Council, http://www.tpc.org/

885. Transaction Processing Performance Council. Tpc benchmark^{TM} app (tpc-app). Specification, Transaction Processing Performance Council, Dec 2004, http://www.tpc.org/tpcw/

886. Transaction Processing Performance Council. Tpc benchmark^{TM} w (tpc-w). Specification, Transaction Processing Performance Council (2004), http://www.tpc.org/tpcw/

887. Transaction Processing Performance Council. Tpc benchmark^{TM} c (tpc-c). Specification, Transaction Processing Performance Council (2010), http://www.tpc.org/tpcc/

888. M. Tribastone, A. Duguid, S. Gilmore, The pepa eclipse plugin. SIGMETRICS Perform. Eval. Rev. **36**, 28–33 (2009)

889. M.G. Tricker, K.-D. Lange, The design and development of spec's server efficiency rating tool (sert), in *Proceedings of the 2nd ACM/SPEC International Conference on Performance Engineering (ICPE' 11)* (2011)

890. K. Trivedi, Sharpe 2002: symbolic hierarchical automated reliability and performance evaluator, in *Proceedings of International Conference on Dependable Systems and Networks, 2002. DSN 2002* (2002), p. 544

891 K. Trivedi, M. Malhotra, in *Reliability and Performability Techniques and Tools: A Survey*, ed. by B. Walke, O. Spaniol (Springer, Aachen, 1993)

892. K. Trivedi, S. Ramani, R. Fricks, Recent advances in proceedings of the IEEE modeling response-time distributions in real-time systems, **91**(7):1023–1037

893. K.S. Trivedi, Probability and Statistics with Reliability, Queuing, and Computer Science Applications (Wiley, New York, 2001)

894. K.S. Trivedi, S. Hunter, S. Garg, R. Fricks, in *Reliability Analysis Techniques Explored Through a Communication Network Example*, Beijing, pp. 2–3

895. K.S. Trivedi, K. Vaidyanathan, K. Goseva-Popstojanova, Modeling and analysis of software aging and rejuvenation, in *Proceedings of 33rd Annual Simulation Symposium, 2000 (SS 2000)* (2000), pp. 270–279

896. T.K. Tsai, R.K. Iyer, D. Jewitt, An approach towards benchmarking of fault-tolerant commercial systems, in *Proceedings of the 26th International Symposium on Fault-Tolerant Computing, IEEE 1996*, Washington, DC (1996)

897. W. Tsai, X. Wei, Y. Chen, R. Paul, A robust testing framework for verifying web services by completeness and consistency analysis, in *IEEE International Workshop on Service-Oriented System Engineering, SOSE 2005* (2005), pp. 151–158

898. G. Tunstall, W. Clegg, D. Jenkins, C. Chilumbu, Head-media interface instability under hostile operating conditions. IEEE Trans. Instrum. Meas. **51**(2), 293–298 (2002)

899. UMTSevol, Universal Mobile Telecommunications System (UMTS); Radio Resource Control (RRC); protocol specification, release 9, v9.4.0, Oct 2010

900. P. Urban, X. Defago, A. Schiper, Neko: a single environment to simulate and prototype distributed algorithms, in *International Conference on Information Networking (ICOIN)*, 31 Jan 2001

901. B. Urgaonkar, G. Pacifici, P.J. Shenoy, M. Spreitzer, A.N. Tantawi, An analytical model for multi-tier internet services and its applications, in *Proceedings of the 2005 ACM SIGMETRICS International Conference on Measurement and Modeling of Computer Systems*, Banff, June 2005, pp. 291–302

902. M. Utting, B. Legeard, *Practical Model-Based Testing: A Tools Approach* (Morgan Kaufmann, San Francisco, 2007)

903. K. Vaidyanathan, R.E. Harper, S.W. Hunter, K.S. Trivedi, Analysis and implementation of software rejuvenation in cluster systems. SIGMETRICS Perform. Eval. Rev. **29**, 62–71 (2001)

904. K. Vaidyanathan, D. Selvamuthu, K.S. Trivedi, Analysis of inspection-based preventive maintenance in operational software systems, in *Proceedings of 21st IEEE Symposium on Reliable Distributed Systems* (2002), pp. 286–295

905. K. Vaidyanathan, K.S. Trivedi, A measurement-based model for estimation of resource exhaustion in operational software systems, in *Proceedings of 10th International Symposium on Software Reliability Engineering*, (1999), pp. 84–93

906. K. Vaidyanathan, K.S. Trivedi, A comprehensive model for software rejuvenation. IEEE Trans. Depend. Secure Comput. **2**(2), 124–137 (2005)

907. A. van de Liefvoort, The moment problem for continuous distributions. Technical report WP-CM-1990-02, University of Missouri, Kansas City (1990)

908. D. van Hertem, J. Verboomen, K. Purchala, R. Belmans, W.L. Kling, Usefulness of DC power flow for active power flow analysis with flow controlling devices, in *The 8th IEE International Conference on AC and DC Power Transmission (ACDC 2006)*, London, March 2006, pp. 58–62

909. A. van Moorsel, Y. Huang, Reusable software components for performability tools, and their utilization for web-based configuration tools, in *Proceedings of the 10th International Conference in Computer Performance Evaluation: Modeling Techniques and Tools*, Palma de Mallorca, Sept 1998, pp. 37–50

910. A. van Moorsel, W. Sanders, Adaptive uniformization. Stoch. Model. **10**(3), 619–648 (1994)

911. A. Varga, The omnet++ discrete event simulation system, in *Proceedings of the European Simulation Multiconference (ESM'2001)*, June 2001

912. F. Vargas, D.L. Cavalcante, E. Gatti, D. Prestes, D. Lupi, On the proposition of an EMI-based fault injection approach, in *11th IEEE International On-Line Testing Symposium (IOLTS 2005)*, IEEE Computer Society, Saint Raphael, 6–8 July 2005, pp. 207–208

913. V. Verendel, Quantified security is a weak hypothesis: a critical survey of results and assumptions, in *NSPW '09: Proceedings of the New Security Pradigms Workshop 2009* (ACM, New York, 2009), pp. 37–50

914. A. Verma, P. Ahuja, A. Neogi, pMapper: power and migration cost aware application placement in virtualized systems, in *International Conference on Middleware* (2008)

915. M. Vieira, N. Laranjeiro, H. Madeira, Benchmarking the robustness of web services, in *Proceedings of the 13th Pacific Rim International Symposium on Dependable Computing* (2007), pp. 322–329

916. M. Vieira, N. Laranjeiro, H. Madeira, Benchmarking the robustness of web services, in *Proceedings of the 13th Pacific Rim International Symposium on Dependable Computing*, pp. 322–329 (2007)

917. M. Vieira, H. Madeira, Benchmarking the dependability of different oltp systems, in *Proceedings of the IEEE/IFIP 2003 International Conference on Dependable Systems and Networks (DSN 2003)* (2003)

918. M. Vieira, H. Madeira, A dependability benchmark for oltp application environments, in *Proceedings of the 29th International Conference on Very Large Data Bases (VLDB 2003)* (2003)

919. M. Vieira, H. Madeira, A dependability benchmark for OLTP application environments, in *Proceedings of the 29th International Conference on Very Large Data Bases*, vol. 29, VLDB '2003 (2003), pp. 742–753

920. M. Vieira, H. Madeira, Towards a security benchmark for database management systems, in *Proceedings of the 2005 International Conference on Dependable Systems and Networks (DSN 2005)* (2005)

921. W.G. Vincenti, What Engineers Know and How They Know it: Analytical Studies from Aeronautical History, in *Johns Hopkins Studies in the History of Technology* (Johns Hopkins University Press, 1993)

922. Virtualisierung bremst Energiebedarf. Virtualisierung bremst Energiebedarf. Computer Zeitung Nr. 52, Dec 2008

923. VMWare, Vmware high availability (ha), restart your virtual machine, World Wide Web, http://www.vmware.com/products/vi/vc/ha.html. Accessed May 2009

924. E. Walker, Benchmarking Amazon EC2 for high-performance scientific computing. Login **33**(5), 18–23 (2008)

925. M. Wan, G. Ciardo, Symbolic state-space generation of asynchronous systems using extensible decision diagrams, in *Proceedings of the 35th Conference on Current Trends in Theory and Practice of Computer Science, SOFSEM '09* (Springer, 2009), pp. 582–594

926. D. Wang, W. Xie, K.S. Trivedi, Performability analysis of clustered systems with rejuvenation under varying workload. Perform. Eval. **64**, 247–265 (2007)

927. G. Wang, A.R. Butt, P. Pandey, K. Gupta, A simulation approach to evaluating design decisions in mapreduce setups, in *International Symposium on Modeling, Analysis and Simulation of Computer and Telecommunication Systems (MASCOTS)* (IEEE, 2009), pp. 1–11

928. N.J. Wang, S.J. Patel, Restore: symptom-based soft error detection in microprocessors. IEEE Trans. Depend. Sec. Comput. **3**(3), 188–201 (2006)

929. X. Wang, D. Lan, G. Wang, X. Fang, M. Ye, Y. Chen, Q. Wang, Appliance-based autonomic provisioning framework for virtualized outsourcing data centre, in *Proceedings of the Fourth International Conference on Autonomic Computing* (2007)

930. C. Warren, R. Saint, IEEE reliability indices standards. Ind. Appl. Mag. **11**(1), 16–22 (2005)

931. I. Waseem, Impacts of distributed generation on the residential distributed network operation. M.Sc. thesis, Virginia Polytechnic Institute (2008)

932. S. Weißleder, B.-H. Schlingloff, Deriving input partitions from UML models for automatic test generation, in *Models in Software Engineering*, ed. by H. Giese (Springer, Heidelberg, 2008), pp. 151–163

933. R. Westrum, A typology of resilience situations, in *Resilience Engineering. Concepts and Precepts*, ed. by E. Hollnagel, D.D. Woods, N. Leveson (Ashgate, Aldershot, 2006)

934. D. Wilson, B. Murphy, L. Spainhower, Progress on defining standardized classes for comparing the dependability of computer systems, in *Proceedings of the DSN 2002 Workshop on Dependability Benchmarking* (2002)

935. R. Wimmer, S. Derisavi, H. Hermanns, Symbolic partition refinement with automatic balancing of time and space. Perform. Eval. **67**(9), 815–835 (2010)

936. B. Winterford, Stress tests rain on amazon's cloud. IT News, http://www.itnews.com.au/News/153451,stress-tests-rain-on-amazons-cloud.aspx. Accessed Aug 2009

937. K. Wolter, G. Horton, R. German, Non-markovian fluid stochastic petri nets. Technical report 13, Technical University of Berlin (1996)

938. K. Wolter, P. Reinecke, A. Mittermaier, Evaluation and improvement of IEEE 1588 frequency synchronisation through detailed modelling and simulation of backhaul networks, in *Proceedings 8th European Performance Engineering Workshop. Lecture Notes in Computer Science*, vol. 6977, ed. by N. Thomas (2011)

939. K. Wolter, P. Reinecke, A. Mittermaier, Model-based evaluation and improvement of PTP syntonisation accuracy in packet-switched Backhaul networks for mobile applications, in *Proceedings of the 8th European Performance Evaluation Workshop (EPEW 2011)* (2011)

940. T. Wood, L. Cherkasova, K. Ozonat, P. Shenoy, Profiling and modeling resource usage of virtualized applications, in *Proceedings of 9th ACM/IFIP/USENIX International Conference on Middleware* (2008), pp. 366–387

941. D.D. Woods, Essential characteristics of resilience, in *Resilience Engineering. Concepts and Precepts*, ed. by E. Hollnagel, D.D. Woods, N. Leveson (Ashgate, Aldershot, 2006), pp. 21–34

942. D.D. Woods, J. Wreathall, Stress-strain plots as a basis for assessing system resilience, in *Resilience Engineering Perspectives vol. 1: Remaining Sensitive to the Possibility of Failure*, ed. by E. Hollnagel, C.P. Nemeth, S. Dekker (Ashgate, 2008), pp. 145–161

943. C.M. Woodside, E. Neron, E.D.S. Ho, B. Mondoux, An "active server" model for the performance of parallel programs written using rendezvouz. J. Syst. Softw. 125–131 (1986)

944. Wordpress.org (WordPress, 2010), http://wordpress.org

945. J.R. Wright, G.T. Vesonder, Expert systems in telecommunications. Expert Syst. Appl. **1**(2), 127–136 (1990)

946. F. Xin-yuan, X. Guo-zhi, Y. Ren-dong, Z. Hao, J. Le-tian, Performance analysis of software rejuvenation, in *Proceedings of the Fourth International Conference on Parallel and Distributed Computing, Applications and Technologies, 2003. PDCAT'2003*, Aug 2003, pp. 562–566

947. J. Xu, M. Zhao, J. Fortes, R. Carpenter, M. Yousif, On the use of fuzzy modeling in virtualized data center management, in *Proceedings of the 4th International Conference on Autonomic Computing* (2007)

948. W. Xu, L. Huang, A. Fox, D. Patterson, M.I. Jordan, Detecting large-scale system problems by mining console logs, in *Proceedings of the ACM SIGOPS 22nd Symposium on Operating Systems Principles*, Big Sky, MT, Oct 2009, pp. 117–132

949. Yahoo! M45 supercomputing project (2009), http://research.yahoo.com/node/1884

950. J. Yang, J. Qiu, Y. Li, A profile-based approach to just-in-time scalability for cloud applications, in *International Conference on Cloud Computing* (2009)

951. M. Yang, Z. Li, W. Yang, T. Li, Analysis of software rejuvenation in clustered computing system with dependency relation between nodes, in *Computer and Information Technology (CIT), 2010 IEEE 10th International Conference on*, July 2010, pp. 46–53

952. S.A. Yemini, S. Kliger, E. Mozes, Y. Yemini, D. Ohsie, High speed and robust event correlation. Commun. Mag. IEEE **34**(5), 82–90 (1996)

953. X. Yu, Improving tcp performance over mobile ad hoc networks by exploiting cross-layer information awareness, in *MobiCom '04: Proceedings of the 10th Annual International Conference on Mobile Computing and Networking* (ACM, New York, 2004), pp. 231–244

954. C. Yuan, N. Lao, J.-R. Wen, J. Li, Z. Zhang, Y.-M. Wang, W.-Y. Ma, Automated known problem diagnosis with event traces, in *Proceedings of the 1st ACM SIGOPS/EuroSys European Conference on Computer Systems 2006*, April 2006, pp. 375–388

955. P. Yuste, D. de Andres, L. Lemus, J.J. Serrano, P.J. Gil, INERTE: Integrated NExus-based real-time fault injection tool for embedded systems, in *Proceedings of the 2003 International Conference on Dependable Systems and Networks (DSN 2003)*, San Francisco, CA, USA, June 2003 (IEEE Computer Society, New York, 2003), p. 669

956. Q. Zhang, L. Cherkasova, E. Smirni, A regression-based analytic model for dynamic resource provisioning of multi-tier applications, in *International Conference on Autonomic Computing* (2007)

957. S. Zhang, I. Cohen, M. Goldszmidt, J. Symons, A. Fox, Ensembles of models for automated diagnosis of system performance problems, in *Proceedings of the 2005 International Conference on Dependable Systems and Networks*, Yokohoma, Japan, July 2005, pp. 644–653

958. L. Zhao, Q. Song, L. Zhu, Common software-aging-related faults in fault-tolerant systems, in *Computational Intelligence for Modelling Control Automation, 2008 International Conference on*, Dec 2008, pp. 327–331

959. J. Zhu, J. Mauro, I. Pramanick, R3—a framework for availability benchmarking, in *Proceedings of the IEEE/IFIP 2003 International Conference on Dependable Systems and Networks (DSN 2003)* (2003)

960. J. Zhu, J. Mauro, I. Pramanick, Robustness benchmarking for hardware maintenance events, in *Proceedings of the IEEE/IFIP 2003 International Conference on Dependable Systems and Networks (DSN 2003)* (2003)

961. A. Zimmermann, D. Schaffrath, M. Faber, M. Wenig, M. Güneş, Improving tcp performance through explicit corruption and route failure notification (ECRFN), in *MSWiM '07: Proceedings of the 10th ACM Symposium on Modeling, Analysis, and Simulation of Wireless and Mobile Systems*, New York, NY, USA (ACM, 2007), pp. 284–288
962. E. Zio, W.W. Krüger, Vulnerability assessment of critical infrastructures. IEEE Trans. Reliab. **59**, 449–482 (2010)